T0192590

BIOMINERALS AND FOSSILS THROUGH TIME

Fossils are essential to the reconstruction of the evolution of life and episodes in Earth history. Fossil skeletal material serves as the repository of chemical data widely used in the reconstruction of the Earth's climate–ocean system at various time scales. Knowledge of biomineralization – the processes associated with the formation of mineralized biological structures – is essential to properly evaluate data derived from fossils. Additionally, knowledge of biomineralization is critical to the understanding of major events in the evolution of faunas, such as the original appearance of skeletons and some major extinction events.

This is the first book to concentrate on aspects of biomineralization through Earth history. The book emphasizes skeletal formation and fossilization in a geologic framework in order to understand evolution, relationships between fossil groups, and the use of biomineral materials as geochemical proxies for understanding ancient oceans and climates. Approaching the subject from this viewpoint allows the authors to link the biotic, physical, and chemical realms. The focus is on shells and skeletons of calcareous organisms, although the broader impacts of these processes on other elements are also addressed, especially their roles in the global chemical cycles of carbon and silicon. The book explores the fine structures and mode of growth of the characteristic crystalline units, taking advantage of the most recent physical methodological advances. It is richly illustrated and will be of great interest to advanced students and researchers in paleontology, Earth history, evolution, sedimentary geology, geochemistry, and materials science.

JEAN-PIERRE CUIF was appointed professor of paleontology at the Université de Paris-Sud, Orsay, in 1980, after obtaining a doctoral thesis at the Museum National d'Histoire Naturelle (Paris). As a part of the Geology Department, the paleontological research team has developed a specialized approach to the mineralized units that built the calcareous skeletons of the invertebrates. In addition to basic research on distributions and compositions of the organic compounds associated with the mineral phases, this approach has led to participation of the team in multiple collaborative programs dealing with invertebrate evolutionary history and biomineralization disturbances in economically important molluscs.

YANNICKE DAUPHIN is an Assistant Professor at the Université de Paris VI (Pierre et Marie Curie). She worked in Jean-Pierre Cuif's Biomineralization research team at the Université de Paris-Sud. After focusing on cephalopod shells, she has extended her research to the structure and composition of modern and fossil molluscs, corals, and vertebrate skeletons.

JIM SORAUF taught sedimentology and paleontology at the State University of New York at Binghamton from 1962 until becoming emeritus in 2001. He has published approximately 70 papers in peer-reviewed journals on skeletal structures of modern corals, and on systematics and paleobiology of fossil corals ranging in age from Cambrian to Pleistocene. He is a former co-editor of *Fossil Cnidaria and Porifera*. He was a Trustee of the Paleontological Research Institution in Ithaca, New York, for many years, serving as President of their Board of Trustees from 1988 to 1990.

BIOMINERALS AND FOSSILS THROUGH TIME

BY

JEAN-PIERRE CUIF
Université de Paris-Sud, Orsay

YANNICKE DAUPHIN
Université de Paris VI (Pierre et Marie Curie)

AND

JAMES E. SORAUF
State University of New York, Binghamton

CAMBRIDGE
UNIVERSITY PRESS

CAMBRIDGE
UNIVERSITY PRESS

University Printing House, Cambridge CB2 8BS, United Kingdom

One Liberty Plaza, 20th Floor, New York, NY 10006, USA

477 Williamstown Road, Port Melbourne, VIC 3207, Australia

4843/24, 2nd Floor, Ansari Road, Daryaganj, Delhi - 110002, India

79 Anson Road, #06-04/06, Singapore 079906

Cambridge University Press is part of the University of Cambridge.

It furthers the University's mission by disseminating knowledge in the pursuit of education, learning and research at the highest international levels of excellence.

www.cambridge.org
Information on this title: www.cambridge.org/9781108445764

First published 2011
First paperback edition 2017

A catalogue record for this publication is available from the British Library

Library of Congress Cataloging in Publication data
Cuif, Jean-Pierre.
Biominerals and fossils through time / by Jean-Pierre Cuif, Yannicke Dauphin, and James E. Sorauf.
p. cm.
Includes bibliographical references and index.
ISBN 978-0-521-87473-1
1. Biomineralization. 2. Fossilization. I. Dauphin, Yannicke.
II. Sorauf, James E. III. Title.
QH512.C85 2011
572′.51–dc22
2010037288

ISBN 978-0-521-87473-1 Hardback
ISBN 978-1-108-44576-4 Paperback

Contents

Preface

During the last few decades, research on climate change and threats to biodiversity have drawn public attention to present-day modifications of the world environment. Thus the public is now receptive to the concept of a *changing Earth*, a vision familiar to paleontologists for roughly two centuries, as investigations into fossilized life-forms began a few decades before the nineteenth century. A method for establishing the chronological distribution of the fossil record was rapidly developed and continuously improved, thereby providing access to Earth history for the first time. From a very restricted number of research centers, convergent information resulted in a completely new view of the Earth that outlines progressive changes in geography, and reconstructs environmental and faunal modifications through time. Also, up through the first half of the twentieth century, improving our description of the fossil record was an essential activity. It was the golden age of great paleontological monographs, in which the compositions of fossil faunas were compared in minute detail.

In the middle part of the twentieth century new physical methods were developed that led to innovative uses of fossils, many of which were based on study of the chemical record preserved in biogenic calcium carbonate of shells, reflecting environments in which the fossilized organisms lived. The first application of such a new method is precisely known: the reconstruction of annual variations in water temperature in Jurassic seas (about 150 million years before present), as a result of the study of variations in oxygen isotope ratios. These were recorded from within the calcified layers of a belemnite rostrum (a fossil cephalopod). For the first time, numerical values with astonishing precision were provided for an important environmental parameter. This spectacular result was obtained by H. C. Urey and his collaborators and reported in 1951. Thus, at the start of the second half of the twentieth century, it could legitimately be expected that improvement of instrumental techniques would furnish data of increasing precision and rigor that had been lacking in geological and paleontological reconstructions up to that time. Consequently, identification of fossils which was demanding of expertise and frequently uncertain as a result, was to be replaced by interpretations based on numerical values directly obtained from the fossil carbonate. This was precisely what did occur, with the result that paleontology declined in its role as part of university training for Earth scientists.

In an interesting turn of history, this continuous improvement of measurement methods – truly amazing during the past three decades – has forced researchers to pay new attention to

ix

specific characteristics of biologically produced calcareous structures, the dominant material of fossils. Early in the development of chemical and isotopic studies, the properties of these Ca-carbonate shells were seen to be closely tied to the taxonomy of the producer organisms, that is to say, their position in the overall phylogenetic scheme. Moreover, in comparing Ca-carbonate structures from different species living in the same environment and simultaneously exposed to its variations, it was noted that environmental changes were effectively registered by each species, *but with values that are unique to each of them.* Therefore, long before understanding the biological processes responsible for shell formation, it was suspected that biocrystallization involves different mechanisms than simple chemical precipitation does as it occurs in saturated solutions. In his pioneering publication, H. C. Urey predicted that such a phenomenon may exist, and named it the "vital effect."

During the two last decades, the spatial resolution of instruments used for chemical or isotopic measurement has been increased to the micrometer range, providing possibilities for obtaining data recorded from different parts within a single shell. Resulting data clearly show that the specific controls exerted on crystallization by each of the specialized areas of the mineralizing organ lead to the recording of distinct and differing signals within each shell layer. Physical and chemical measurements now demonstrate that within a shell, several "vital effects" actively influence the formation of the distinct shell layers they contain. This provided unexpected support of early studies of Bowerbank (1844) and Carpenter (1845,1847), who first showed that calcareous shells are built by superposed layers of differing sizes, shapes, and spatial arrangements of skeletal units.

Multiple physical and chemical methods now produce convincing evidence that not only taxonomy must be considered when studying biogenic Ca-carbonate structures as potential sources of environmental data. In addition, the precision of these measurements requires that close attention is paid to distinct shell layers and their species-specific spatial arrangement. More fundamentally, evidence of these layer-specific structural and chemical properties leads us to a re-examination of the biocrystallization process itself. Improved analytical instruments now provide convergent evidence indicating that the former concept of crystallization by saturation within a fluid that more or less resembled seawater does not account for the detailed properties of the skeletal units that make up shells.

Therefore, most of this book is dedicated to an in-depth examination of the structural and chemical properties of the crystal-like units which are found forming shells and other calcareous skeletons. Morphological diversity of shells is familiar to everyone and is still emphasized by the very precise species-specific three-dimensional arrangements of their skeletal components, described at the microscopic level since the middle of the nineteenth century. Here their mode of growth is described at the micrometer and nanometer scales, providing information and examples for developing more accurate methods of microstructural analysis. Beyond the variety of sizes, shapes, spatial organizations, and growth modes of skeletal units, modern instruments also reveal striking similarities in these parameters when seen at submicrometer scales. Subunits with dimensions in the range of 10 nm can be recognized within all of these Ca-carbonate structures. Their internal structures observed at still higher resolutions provide the first direct evidence regarding the sequence of events leading to the crystallization process itself, where organic molecules are involved. When compared to equivalent data concerning phosphatic and siliceous biominerals, the specificity of the calcareous biocarbonates is still more instructive.

It is of great importance to improve our understanding of this biochemically driven crystallization process characterized by a permanent interplay between organic and mineral components. This is apparent, not only because of multiple applications, but, more essentially, to gain an increasingly profound understanding of the process of shell calcification. Recently, the genomic approach to the mineralizing process has established an amazing complexity to this aspect of biology. Most surprisingly, it is now clear that, in contrast to the rather low mineralogical diversity of shell carbonates, the calcification processes developed within major biological groups involve very different groups of genes. Therefore, the question now arises as to how genomic diversity active in the cellular mechanism of calcification is transferred to the hierarchical organization of these calcareous units in such a specific form. It is this form that allows characterization of even the most common shell. Fundamentally, deciphering the structure and mode of growth of biocrystals contributes to creating a consistent scheme that coordinates the intracellular preparative processes of calcification and the following step, the still largely unexplored organomineral interplay that drives Ca-carbonate crystallization. Remember that this is operating outside of the cell membrane. Here lies the origin of the "vital effect."

Thus, just a short time after the decline of paleontological studies due to an oversimplified application of physical methods to fossils, there is a touch of irony in the observation that, both in basic research and in its application, the understanding of shell structures and the biological mechanisms controlling shell crystallization have become a prerequisite for rational use of these newer methodologies. In addition, when applying these methods to fossil materials, relevant structural and chemical comparisons with their extant representatives now can be carried out at the micrometer scale, allowing unprecedented accuracy in research on the fossilization process, thus improving the reliability of the multiple categories of information brought to us by the fossil record.

Therefore it is noteworthy that, transferred to a new base in advanced analytical methods, the continuing dialog between the ancient and the modern worlds that has been, since its beginning, the essence of paleontology, is today no less important than it was two centuries ago.

Introduction

Milestones in the study of biominerals

A summary of the discontinuous and somewhat erratic path of research on their formation and properties

The microscopic study of living organisms did not develop until the first half of the nineteenth century, in spite of earlier observation of "cells" by Robert Hooke (1665). It was only in 1831 that Robert Brown described a new structure in plant cells: the "nucleus." In transferring microscopic observations of Mathias Schleiden on plant tissues (1838) to his own studies on animal tissues, Theodor Schwann (1839) made a major contribution by proposing the theory of universal cellular organization in living organisms. In this same year the Microscopical Society of London was founded, and among the founding members was James Scott Bowerbank.

In his paper, "On the structure of the shells of molluscan and conchyferous animals" (1844), Bowerbank showed that far from being simple aggregates of unorganized mineral particles, these shells are formed of sharply distinct units, each possessing a definite geometry whose regular arrangement demonstrates that their order is completely controlled by the producing organism. The rapid diffusion of this new type of analysis is attested to by the publication the following year of Carpenter's paper, "On the microscopic structure of shells" (1845, with a second part in 1847).

These two studies represent much more than a simple change in the scale of observation. For two or three decades the microscope had already been used to observe fossil morphology, but now, by focusing on the organization of the mineral constituents themselves, beneath the surface of the sample, study began of the functions of the cells producing structures capable of being fossilized (e.g., the mantle of molluscs, the basal epithelium of coral polyps, etc.).

The extracellular formation of crystalline structures

In confirming the high level of control exerted by living organisms on their mineralized hard parts, microscopic observation indicates equally that these structures form exceptions to the theory that regards cells as basic components of all parts of metazoans. This is particularly obvious for the category of macroscopic calcareous structures produced in various invertebrate phyla, such as shells of molluscs, carapaces of arthropods, or scleractinian corallites. It is an impressive paradox that Bowerbank, as a founding member of the Microscopical Society of London, published the first such studies, and these resulted in a conclusion that was in complete opposition to the fundamental working concept of his colleagues; namely, his conclusion that mineralized structures of living animals are not themselves cellular.

1

This surprising conclusion was progressively extended to other groups of organisms, but was not always readily accepted. Thus, when the fibrous organization of the coral skeleton (Scleractinia) was discovered by Pratz in 1882, a lively controversy developed as to whether the cells of the polyp (in its basal ectoderm, in contact with the skeleton) were themselves calcified (the position of von Heider, 1886), or if indeed they only directed external calcification (the position of von Koch, 1886). It is interesting that it was the best-informed investigator of coral structures at that time who opposed the second hypothesis most vehemently. In 1896, M. Ogilvie published her seminal analysis of the scleractinian corals, showing the possibility of systematic treatment based on precise observations of the spatial arrangements of the calcareous crystallites that form the coral skeleton. At the same time, however, she accepted the erroneous cellular hypothesis of von Heider (and one could find adherents until much more recently). It seemed impossible to her that these fiber orientations, so precise that one derived from them basic criteria for taxonomic identification, could be developed by different mechanisms from those ensuring the precise coordination of cellular tissues during the ontogenetic development of any organism. This is the key point in understanding these features. Shells and other calcified structures must be interpreted as resulting from secretion processes, and crystallization occurs external to the secreting cells.

As to the terminology used, although its use is quite widespread, the expression "mineralized tissue" should *not* be applied to calcified biogenic structures, since these structures themselves are not formed of cells. This can be argued concerning bones, as the secreting cells are included within the mineralized material (this is why fossil nucleic acids can be found in fossil bones), but the mineralized material itself is actually extracellular. More readily visible in mollusc shells, for instance, the cellular layers producing mineralized structures move freely over the mineral surface (for example, the oyster's mantle or gastropods that retract within their shells).

The heavily calcified tests of the Foraminifera may appear to be obvious exceptions to this rule, as here the calcareous structures are located within the cell membrane. However, these calcified structures are carefully isolated by continuous organic envelopes from the cell compartment where metabolic reactions occur. Perhaps more significant is the case of the unicellular algae of the prymnesiophyte Coccolithophoridae, in which the individual coccoliths are formed within vesicles issuing from the Golgi apparatus. Little is known about the molecular mechanism providing for the transit of Ca-carbonate though cell cytoplasm to the vesicles, but during the entire crystallization process organic membranes prevent contact between crystallized Ca-carbonate and living cytoplasm.

The corallinacean algae (Rhodophyceae) also appear to be exceptions to this carefully maintained exclusion between intracellular physiological processes and biocalcification. Here the cellular frame itself is incrusted by carbonate, leading to the formation of a wall completely separating adjacent cells. This process leads to the death of these cell layers at a short distance below the surface of the organism.

In addition to descriptions of calcified materials, the great importance of another mineral component was established during these early nineteenth-century decades: biomineralization based on silica. In the instructions formulated by Alexander von Humboldt for exploration of the Antarctic seas (for an expedition led by J. C. Ross in 1839), the methodical sampling of Si content of marine waters was recommended. Just previously, in 1838, C. G. Ehrenberg had shown the geological importance of siliceous unicellular algae by his microscopic observation of sands known as the "Kieselgur." J. D. Hooker participated in the Ross expedition, and his "Flora Antarctica" (1847) constitutes the first analysis of the production of biogenic silica in the ocean.

Thanks to microscopy, siliceous biomineralization, principally associated with sponges, radiolaria, and diatoms, rapidly was recognized as second in importance quantitatively, only outranked by carbonates. Although it occurs both in animals (sponges and radiolaria) and plants (diatoms), the essential character common to all siliceous biomineralization is that the silica is always present in an amorphous state, just the opposite of the two combinations of calcium (carbonate and phosphate) that are always produced in a crystalline form. The term "amorphous" applies to the molecular level of organization of the mineral material, in that the silicon–oxygen tetrahedra are associated through the intermediary of water and organic molecules to form a specific mineral named *opal*. It does not indicate that the siliceous structures are not controlled at the level of their overall morphology. Just the opposite, the extreme precision with which Radiolaria, Diatomacea, and Porifera build their mineralized structures, all with specific morphologies, has been known for a long time.

There exists a strict correlation between the nature of the material utilized and overall crystallographic characteristics of the structures produced, a situation all the more striking in that it totally ignores their taxonomic position, even at the highest level (above the phylum level). On a practical level, the result of this constant contrast between crystalline Ca-carbonate and amorphous silica biomineralization is that only the former can be the subject of detailed microstructural analysis.

The third major group of common mineralized biomaterials, far behind Ca-carbonate and silica from a quantitative standpoint, are those composed of calcium phosphate. These occupy a somewhat ambiguous position between calcareous and siliceous mineralization. Due to their resulting from extracellular crystallization in the greatest part of vertebrate bone, they are crystalline and thus parallel invertebrate carbonates somewhat. There are important differences between groups, however, and the most obvious is the dimensional scale of crystallized units. Calcareous systems can produce units with a monocrystalline aspect capable of reaching great size, commonly several hundreds of micrometers. In mineralized phosphates, crystals are always much reduced in size, so that characterization of their morphology is difficult even at present. Therefore, it is not surprising that early observations of microstructure were made on carbonate biominerals.

This situation does not rely on particular taxa. Phosphates are almost exclusively the skeletal material of the Vertebrata, but they also are present in some invertebrates. In the latter, crystalline units never attain dimensions comparable to those in carbonates. Conversely, some vertebrates also produce extremely interesting calcareous structures, e.g., otoliths and eggs. In these cases, their structures present crystallization aspects (size of microstructural units and layered growth mode) that are comparable to those character-istic of invertebrates.

Factors retarding the study of biominerals

The study of the mineralized structures formed by organisms began at practically the same time (the decades around 1840), and in the same circles as the study of purely organic tissues and cells. This simultaneity of the time of origin is noteworthy because it underlines the striking difference between the more rapid rate of progress in understanding cellular structures, as compared to that of the mineralized components of organisms. The origin of this is readily apparent. The techniques of microtomy, complemented by numerous and varied staining techniques for the characterization of tissues and cell organelles, rapidly made possible the analysis of histological and cytological phases, but have proved to be much less applicable to mineralized structures.

Preparation of rock thin sections and polarization prisms, both due to the work of William Nicol (1815 and 1828, respectively), were available already at the time of Bowerbank, and during the following century, only the increasing quality of optical microscopes provided marked improvement of the study of mineralized hard parts of organisms. In terms of methodology, prior to the 1950s no significant innovation was introduced to change the concepts of biogenetic mineralization of structures. Even X-ray diffraction (W. L. and W. H. Bragg 1912) did not have the same fundamental importance for biominerals that it did in the development of general mineralogy, because the true originality of the biominerals is not mineralogical diversity.

Optical studies carried out during the nineteenth century resulted in some high quality syntheses during the first part of the twentieth century, by Schmidt (1924), Bøggild (1930), and Taylor *et al.* (1969–1973). These firmly established that living organisms exert such precise control on the organization of their skeletal structures that families, genera, and sometimes species can be identified by reconstructing their three-dimensional arrangements of calcareous skeletal units. Bøggild, for example, succeeded in distinguishing genera of the Mollusca by their characteristic shell microstructures. However, from the standpoint of methodology, compared to Bowerbank's pioneering observations, no significant innovations were yet introduced to generate changes in existing concepts regarding the formation and, specifically for Ca-carbonate, the mode of growth of the crystal-like mineral units forming the mineralized structures of most invertebrates.

The geochemical era

It is fascinating to realize that, in this stagnant situation, the first event that contributed to a major renewal in research on biominerals resulted from the development of a specialty that until then was unknown in biology: thermodynamics, and more particularly, the thermodynamics of isotopic fractionation. In 1947, H. C. Urey, the chemist already famous for his discovery of deuterium (1934 Nobel Prize), produced a mathematical expression for the difference that would be established during chemical reactions between proportions of isotopes of a chemical element *before* it entered into the reaction, and the proportions of isotopes of this same element *after* the reaction, i.e., within new molecules resulting from the reaction. Urey showed that *this change of proportion varies as a function of the temperature* at which the reaction producing the compound has occurred, and suggested that, reciprocally, the measure of isotopic ratios in a given material could provide an indication of the temperature at which it was formed. Urey stated: "I suddenly found myself with a geological thermometer in my hands." In 1950, the technology of mass spectrometry permitted the measurement of a difference of 0.2 parts per thousand, which is the variation of the ratio between stable isotopes of oxygen per degree centigrade during precipitation of Ca-carbonate. By carrying out a series of in vitro Ca-carbonate precipitations, McCrea (1950) experimentally checked Urey's calculations and, in 1951, the result of the first practical application of the isotope-based paleotemperature reconstruction was published (Urey *et al.* 1951).

In the history of paleontological research, there are no other examples of innovation having resulted in such marked reverberations, as did this presentation in 1951 of the reconstruction of the life of a Jurassic belemnite coming from the Isle of Skye (Scotland). Belemnites are representatives of an extinct group of the Cephalopoda (Mollusca). Their most common structure is a massive conical rostrum made of radially diverging carbonate units in which concentric growth layering is easily visible. By carrying out regularly spaced sampling of carbonate material along an identifiable ray in the section, they could thus measure the ratios of $^{12}C/^{13}C$ and $^{16}O/^{18}O$ isotopes recorded in the successive layers produced by the belemnite during its lifetime. In a few lines, the announcement of a truly astonishing precision was enunciated thus: "This Jurassic belemnite has recorded three summers and four winters ... It lived during its youth in warmer water than in its adulthood, and died in the spring at an age of about four years. The mean seasonal variation of the water was approximately 6 °C, with a mean temperature of 17.6 °C." Never before had any fossil been made the object of such detailed analysis. This opened the age that has been called that of "isotopic paleontology" (Wefer and Berger 1991).

The question of a "vital effect" – establishing an overriding taxonomic influence in the mineral realm

Chemical paleontology may be a more appropriate name for studies that began in the middle of the twentieth century. With more accurate mass spectrometers providing better measurement of isotopic fractionation, chemical analysis made additional striking progress with the establishment of atomic absorption spectroscopy (Walsh 1955). As a result, it became possible to make multiple and important series of measurements of concentrations of minor chemical elements associated with biogenic minerals, such as strontium found in aragonitic material and magnesium in a calcitic one. These notable phenomena now became the focus of interest for numerous investigators, as experiments of coprecipitation showed that concentrations of minor elements incorporated in carbonates vary as a function of temperature. Thus, sampling this variation permits using concentrations of magnesium, strontium, and numerous other elements as sources of information on environmental parameters. Methods of paleoenvironmental reconstruction now appeared that were much more rigorous than traditional and older qualitative ecological observations.

Starting in the 1950s, Chave, Lowenstam, and numerous others undertook the exploration of this phenomenon in comparing results of experimental chemical precipitation with measurements on biominerals of organisms originating in natural environments. During the following decades, the growth in sensitivity of analytical techniques has responded very well to the needs of researchers seeking to accumulate data on the composition of biogenic mineralized materials. At the same time, the results obtained also bring with them a conclusion that contrasts markedly to views that had initially been proposed. Through innumerable measurements carried out on the chemical and isotopic properties of biominerals, a biological influence has been progressively established with a strength and precision comparable to those that had been shown by their optical characterization during the initial phase of biomineralization studies.

The conclusion now imposed by modern data is that the chemical compositions and the isotopic fractionations observed in biominerals cannot be explained by straightforward precipitation phenomena. Chemical partitioning and isotopic fractionation that occur during formation of biominerals are undoubtedly sensitive to temperature and other environmental parameters, but they have been shown to be primarily dependent on taxonomy.

This conclusion regarding chemical composition and isotopic fractionation in biominerals not being explicable by ordinary precipitation phenomena was foreseen from the beginning. It reflects the foresightedness of the initial authors of isotopic methods to have formulated explicitly the existence of a "vital effect" (Urey *et al.* 1951, p. 402). It is perhaps also of interest to note that, prior to becoming oriented towards physical chemistry, H. C. Urey had acquired a zoology degree at the University of Montana in 1917. Possibly his initial education determined his particular sensitivity towards the properties of living systems, a quality lacking in some of those who later applied chemical/isotopic methods to biological minerals. Forty years after Urey's remark, Morse and Mackenzie indicated that the progress of knowledge concerning the mechanisms for precipitation of carbonates had not led to development of a satisfying overall theory (1990, p. 602). It is of note that this conclusion was formulated with regard to sedimentary carbonates, regarded as reflecting purely physical and chemical mechanisms, as it is far more obvious concerning their biogenic forms.

In practice, this property has led to the necessity for "calibration" of each species utilized, and for doing this for each environmental parameter analyzed in ancient biomineralization. Inversely to the rational and rigorous views that motivated technical applications during the decades of the 1950s to 1970s, a purely empirical practice has developed to deal with this. Simply stated, the methods of formation of biominerals do not correspond to conditions observed in laboratory precipitation experiments. It is clear that the progress of instrumentation, which seemed to be the only limitation during initial phases of research on the chemical composition of biominerals, can now only be properly exploited when accompanied by better understanding of the fundamental biological processes that control the recording of environmental conditions in biominerals.

The development of biochemistry in bimineral studies: a bottom-up approach leading to an hypothesis of biochemically driven crystallization

The period during which the first applications were established for measurement of isotopic ratios and trace elements in biogenic minerals is also the time when research began that would progressively establish the *biochemical diversity as well as the taxonomic specificity* of the organic materials that are always associated with mineral structures produced by living organisms.

The presence of organic compounds in biogenic mineralized materials was recognized shortly after the observations of Bowerbank. The word "conchyolin," coined as early as 1855 by Frémy, has long been used as a general designation for these compounds as a whole. However, their concentrations are in general low enough that for a long time the analysis of these organic components was hardly possible. From a practical viewpoint, it was only a century later that electrophoresis and chromatography became the main sources of bio-chemical data, enabling consistent analyses to be made of these organic compounds, and accurate biochemical comparisons carried out on them.

At the same time, the number and diversity of known biominerals has increased rapidly. When Lowenstam began his methodical investigation of biomineral diversity (1963), the number of mineral combinations known in living organisms was roughly ten. In 1989, Lowenstam and Weiner recorded more than 60 biologically produced minerals, of which approximately half are combinations based on calcium, and a quarter are phosphates. Of these, 24 combinations are produced by procaryotes, 11 by plants, 10 by fungi, and an additional 10 more by the Protista and 37 by other animals. It is impressive that all these minerals appear as "ordinary" compounds in the sense that no unique or peculiar mineral combinations seem to have been developed by living organisms. This leads to a paradoxical conclusion, which is that, *properly speaking, there are no biominerals* per se.

Progress in our knowledge of the organic compounds associated with these various mineral features has resulted in a shift of research interest away from the minerals them-selves to the properties of their associated biochemical compounds. During decades, investigators have focused on the analysis of skeletal "matrices." Progressively, these organic compounds have been shown as diverse and specific as the shapes and compositions of the mineralized structures from which they were isolated.

This led to the suggestion made by Wada (1961), and a little later by Watabe (1965), that deposition of mineral material could only be carried out *because* of substrates formed by preexisting organic compounds. This hypothesis has been invariably confirmed by analytical advances that have been progressively extended to cover a whole array of biogenic mineralized structures, and reinforce the conclusion that biological control is always exercised by the previous emplacement of an ensemble of organic compounds.

This organic phase shows singular variety in its relative proportions of glucidic and proteic compounds, and this is clearly dependent on the taxonomic position of the producing organisms, even though environmental conditions also influence its general composition. To model the organic architectures that can furnish sites capable of attracting mineral ions is presently a fundamental approach to biocrystallization at the molecular level. Even though numerous points remain unexplained that pertain to the mechanisms by which these organic materials are able to control the growth of biogenic mineral materials, there is no longer any doubt that they do constitute an essential agent of the phenomenon called biomineralization.

It is also important to note that studies developed at the molecular scale have not yet explained the process of crystallization as it results in forming species-specific morphologies of biocrystal units. The recent development of new methods of observation and analysis, providing access to submicrometer-scale imaging, suggests that, in addition to the chemical bottom-up approach, high-resolution studies of skeletal structures can contribute to bridging the gap between traditional optical microscopy and interpretations based on the molecular approach.

Collecting information on mineral and organic components at micrometer to nanometer scales by use of the "new microscopes"

The first of these new instruments relies on production of "new light," i.e., the electromagnetic spectra that are produced when rapidly moving particles (accelerated close to the speed of light) are obliged to change their direction of propagation. This is obtained in synchrotrons, which are actually magnet-driven polygonal circuits for particles, each angle being the source of a wide electromagnetic spectrum with highly specific properties. Associated with high-resolution monochromators, selected X-ray wavelengths allow characterization not only of chemical elements but, additionally, of the chemical state of given elements. For instance, distinction can be made between the sulfur in proteins (sulfated amino acids) and the sulfur in polysaccharides, both distinct from the sulfate form in gypsum. Synchrotron mapping has established a correlation between mineral growth zonation (made visible by etching and SEM observation) and biochemical zonation. It is through this type of experiment that a mechanism has been found to produce stratified growth simultaneously involving organic and mineral phases. Taken together, this insures organic control of developing microstructural units.

Conformity between the mineral and organic phases within a given growth layer implies that the relationship between the two components occurs at a submicrometer level. The instruments grouped under the designation of "atomic force microscopes" or "near-field microscopes" now acquire images of spatial arrangements of the two components within micrometer-thick growth layers. Actually, this is presently a family of instruments that are still undergoing important development and diversification, yet they provide a method and perspective that suggests that we can look forward to acquiring data at extremely fine scales (nanometer to molecular). Different modes of observation (tapping or contact, for example) and different signals that can be acquired simultaneously provide descriptions of microstructural characteristics and distributions of associated organic phases on a submicrometer scale. The result has profoundly changed our reconstruction of biomineral structures, with important consequences to our understanding of the biomineralization process itself.

Looking for a working model to explain the variously scaled properties of calcareous biocrystals

The decade of the 1950s epitomized the disparity of scale in research dealing with calcareous biominerals. At that time, they were described as ordered crystals as seen in polarizing microscopes complemented by X-ray diffraction, both techniques emphasizing their crystalline patterns. But, simultaneously, the beginnings of studies carried out at the isotopic level revealed species-specific fractionation that suggested enigmatic growth mechanisms. Only Urey as a Nobel Prize chemist was able to hypothesize such an obscure mechanism as that referred to as the "vital effect," that is possibly involved in what was (and sometimes still is) represented as simple crystallization due to precipitation from a saturated solution. More than a half-century later, our data has been improved greatly, but in spite of the considerable influx of this new information, diversity remains the dominant impression. A consistent integrated scheme cannot yet be established, owing to the diverse interpretations concerning the respective proportions of crystallographic processes and biochemically driven mechanisms in the formation of these materials. The scale gap of the 1950s between optical morphology and isotopic and molecular levels is still yet to be totally removed. Therefore we here investigate the major types of calcareous structures in order to examine what their level of commonality may be at which similarities occur between these very different structures. Through a top-down approach and specific preparative processes, this investigation results in a different view of growth mode for most of these structures, directly linked to the cyclical secretional functioning of mineralizing organs, allowing neither free space nor time available for uncontrolled crystallization. Truly, the micrometer-thick growth layer appears to be the common characteristic of biogenic calcareous structures, whatever their morphology. As an example of multiple consequences, shaping of skeletal units by the controlled development of growing single crystals, long admitted as a key process in biocrystal formation, no longer seems an operative concept.

This growth layer is also the domain where the interplay between organic and mineral materials actually occurs. Because it is formed in direct contact to the mineralizing organ, the micrometer-thick growth layer is also a most essential location, where molecules resulting from intracellular biochemical processes are transferred to the cell exterior and become operative. Atomic force microscopy associated with transmission electron microscopy (imaging and diffraction) provides us with initial information about the sequential development of the biocrystallization process at the elementary growth level. In-depth understanding of the difference between those crystalline structures generated by living organisms, and crystalline materials that result from chemical precipitation relies on a reliable reconstruction of this stepping mode of crystallization.

From an environmental viewpoint, analyzing biochemical consequences of gene regulation that make the biomineralization process sensitive to environmental conditions will thus establish a more precise relationship between oceanic waters and various taxonomy-linked calcified structures. Understanding the relationships between environments and biomineralization offers paleontologists a new basis for re-examination of the overall evolution of fossil faunas over millions of years. The reactions of different phyla to long-term (and large-scale) environmental variation may well have influenced their capabilities during skeletogenesis, with additional consequences in the character and intensity of diagenetic processes. Not only is it the interpretation of individual fossil samples that can benefit from fine scale study of their calcareous microstructures. By applying the methods used to characterize biocrystals of related extant species, understanding of faunal development throughout the fossiliferous portion of the geological column can be improved.

1

The concept of *microstructural sequence* exemplified by mollusc shells and coral skeletons

Similarity of growth mode and skeletogenesis at the micrometer scale

At the time that Heinz Lowenstam began an investigation bearing on the different minerals that living organisms can produce, only a dozen or so biogenic minerals were known. Thirty years later, their number surpasses 60. Well before the appearance of the synthesis uniting the essential data that had been established during the 1980s (Lowenstam and Weiner 1989), Lowenstam had proposed (1981) a fundamental distinction concerning the mode of formation of biogenic minerals, and more particularly the precision of the controls on their deposition exerted by the producing organism. Lowenstam proposed then to distinguish the "matrix-mediated minerals," characterized by formation very precisely controlled by the action of an organic component specifically produced, and those that, although equally produced by a living organism, are developed in a more autonomous way; they are only "biologically induced." In these last, mineral elements can be developed according to methods and arrangements quite close to those that can be observed in purely chemical precipitated materials.

Among the calcified structures belonging to this category were placed the calcareous skeletons produced by corals. This opposition between molluscan shells, examples of "matrix-mediated minerals," and coral skeletal carbonate has remained generally accepted and very recently has still been formulated in reference journals (Veis 2005). This is however, truly surprising. The precision of the biological control on the skeleton by the coral polyps had been well established at the end of the nineteenth century by M. Ogilvie (1895, 1896), who carried out a pioneering study of the specific arrangements of the fibrous aragonite units there. But it is true that the difference between these and mollusc shells, the almost exclusive focus of research on methods of biomineralization has long remained poorly established. One should note in addition that this concept of being "biologically induced" in the formation of the coral skeleton is perfectly convenient for geochemists utilizing these materials as archives of ancient environments, because they can thus base their interpretations on a chemical theory lacking any biological influence, in spite of contradictory evidence long since published (i.e., Weber and Woodhead 1972).

During recent years, a convergent set of observations and analyses led to re-examination of this view. An additional objective of this chapter is to explain the concept of the microstructural sequence and its relationships with the mode of growth of calcified structures, beginning with two groups whose anatomical and physiological organizations are sharply distinct. Microstructural analysis shows that in spite of the geometrical differences that characterize the structures produced in these two groups, the similarities that one can determine in their methods of emplacement can be established, and allows the inappropriate differentiation to lapse.

1.1 Basic shell microstructure observed in the pearl oyster: *Pinctada margaritifera*

The great majority of molluscs produce a shell and invariably the same organ builds it: the mantle. The mantle is highly visible in an open *Pinctada* (Figs. 1.1a and 1.1b: "M"). However, it is only the exterior face of the mantle that possesses mineralizing activity; therefore, shell growth is carried out entirely *from the interior*. The internal surfaces of the two valves represent the last-mineralized surfaces produced by the animal.

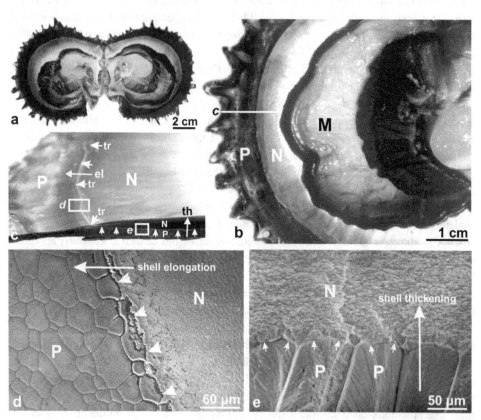

Fig. 1.1. The two-layered shell of *Pinctada margaritifera*. (a) Overall view of an open shell; (b) The mantle ("M"), and its black pigmented lip; (c) The position of the prismatic ("P") and nacreous ("N") layers viewed on the internal surface and transverse section of the shell; (d) A close-up view (scanning electron microscopy) of surfaces of the two layers and their zone of contact (arrows); (e) Superposition of the external layer (prismatic calcite) and the internal layer (nacreous aragonite). a, c: Cuif *et al.* (2008a); b, d: Dauphin *et al.* (2008).

On this internal surface, two regions are distinguishable, each readily separated by its coloration (Figs. 1.1b and 1.1c). At the periphery, a zone is present with intense black coloration, to which this species owes its common name (the black lip pearl oyster; its name actually derives from the equally black border of the mantle periphery). A section cut perpendicular to the plane of this shell (Fig. 1.1c) demonstrates that this black region is the visible part of a continuous layer formed by the mineralizing activity of the peripheral region of the mantle (Fig. 1.1c: "P"). This layer forms the external part of the shell due to its being covered during growth by material produced by the more central region of the mantle. This internal layer has lighter coloration than the peripheral zone, or may even be perfectly white, and is truly remarkable in its lustrous aspect. It is formed of nacre (Fig. 1.1c: "N"), a material with properties that are exceptional among biominerals. *Pinctada* is not the only mollusc producing nacre, but its nacreous layer is so thick and so regular that this particular characteristic has been utilized for naming the species. Additionally, it is the reason that this species is utilized for the production of pearls in several regions of the world (*margaritis* means pearl in Greek, and the common name of nacre is "mother of pearl"). The nacreous layer is separated from the peripheral layer by a distinct line (Fig. 1.1c: arrows). It is important to note here that the peripheral layer is progressively thickened from its most external part, and is thickened as far as the line separating it and the internal zone. Thereafter, it is the emplacement of the nacreous layer that accounts for shell thickening (Fig. 1.1c: arrow, "th"). Thus, the mineralizing activity of the mantle assures, at the same time, both the elongation of the shell and its thickening (Fig. 1.1c: arrow, "el").

The essential aspect of the biomineralization phenomenon is thus evident. Both of the two layers constituting the shell of *Pinctada* and characteristics of the internal and external layers are maintained throughout growth of the shell. The effectiveness and precision of the control mechanism that allows for the formation of the two layers is shown in Figs. 1.1d and 1.1e, obtained by scanning electron microscopy.

The surface of the external zone is organized into irregular polygons that are closely joined (Fig. 1.1d: "P"). In Fig. 1.1e (enlarged sector of Fig. 1.1c), it is easily understood that the polygons of this region are surfaces of units that are elongate perpendicular to this internal surface. These important elongate, polygonal units are named the prisms (Fig. 1.1d: "P"). They continue to grow longer until they are covered over by nacre at a transition level (Fig. 1.1d: arrows). Equally, in Fig. 1.1e the nacreous layer (upper part of the image) is seen to have a completely different organization than the external layer, as it is formed of very thin layers that are closely superposed. The difference between these two shell layers is even more basic, because not only do the units forming the two layers have a different form but, additionally, they do not have the same mineralogy. The mineral forming the prisms of the peripheral zone (= external layer) is calcite (calcium carbonate crystallizing in a rhombohedral form), while the elements of the central region (= nacreous internal layer) are made of aragonite (a second form of calcium carbonate, crystallizing in the orthorhombic system).

Figure 1.2 summarizes the general properties of elements forming the two superposed layers in the shell of *Pinctada margaritifera*. By preparing a polished surface in the

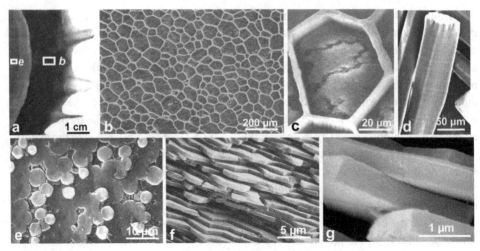

Fig. 1.2. Detailed morphology of the two microstructural types in *Pinctada*: prisms and nacre tablets. (a) Morphology of the peripheral region of the shell; (b) Surface of the prismatic region after acid etching (with 0.1% formic acid); (c) Enlarged view of one prism. A resistant organic lamella separates neighboring prisms, and this prism is here seen to be composed of several distinct mineral elements; (d) One prism freed from its organic sheath by the action of sodium hypochlorite. Growth striations are visible on lateral faces; (e) Nacre tablets viewed on the internal surface of the shell, as indicated in (a); (f) Nacre tablets seen in a section perpendicular to the shell surface; (g) Enlarged view of nacre tablets. b: Cuif and Dauphin (1996); c: Dauphin and Kervadec (1994).

peripheral zone of the shell and submitting this surface to light etching by a weak acid solution (for example, formic acid, 0.1% by weight) a polygonal network appears very clearly (Fig. 1.2b). These are organic envelopes surrounding the periphery of all prisms. In Fig. 1.1d (a natural growth surface, "P"), only their upper edges are visible. Here, it is apparent that these envelopes completely separate each prism from its neighbors (Fig. 1.2c). Equally, it can be seen that in the bottom of the cavity created by the acid etching, the structure of the prism itself is subdivided into several sectors (Dauphin 2003). In a thin section observed in a polarizing microscope, one can see that each sector has a monocrystal-line optical character, the prism being constituted by several such elements with slightly differing orientations (shown by color variations under cross-polarization). If the envelopes of the prisms are removed by an oxidizing agent (sodium hypochlorite), isolated prisms are then obtained, with their morphology as shown in Fig. 1.2d. On the faces of the prisms are observed the traces of incremental growth stages that visually reflect the progression of their construction.

Figures 1.2e to 1.2g allow precise observation of the morphology of components in the internal region: the nacre. Components of nacre are very small, thus differing from units constructed by the peripheral region of the mantle. These tablets grow progressively until joined (Fig. 1.2e). They are flattened in the plane of the growth surface to form closely stacked layers (Fig. 1.2f). The thickness of each tablet is of the order of 0.5–0.7 µm

(Fig. 1.2g), as seen on a broken surface perpendicular to the nacreous layer. This arrangement and its dimensions are important for understanding the properties of nacre, a material that has for a long time attracted attention by its iridescence (see Chapter 3).

1.2 Time-based analysis of shell construction in *Pinctada*

During ontogeny (the development of an individual), the formation of the *Pinctada* shell begins very early. Its trochophore larva swims 8 to 12 hours after fertilization of the egg, and the first shell forms after 24 hours. After two days, growth lines are easily visible, providing evidence of acquisition of a cyclical process of mineralization (Doroudi and Southgate 2003). In contrast to the adult shell, this larval shell is entirely made of aragonite-forming grains with a median diameter of 150–200 nanometers (Doroudi and Southgate 2003), but lacking any specific arrangement. It may eventually exhibit a fibrous habit (Mao-Che *et al.* 2001). This larval shell remains visible at the summit (umbo) of the juvenile shell (Fig. 1.3a).

In addition to settlement on the substrate and reorganization of the soft tissues, metamorphosis (about one month after egg fertilization) is marked by the appearance of the two typical components of the adult shell: the outer prismatic, and internal nacreous layers. On the external surface of the shell, a sharp limit separates the larval shell, simply marked by growth lines (Fig. 1.3c: black arrows) from the post-metamorphosis area of the juvenile shell (Figs. 1.3c–d: arrows). Polygons now visible on the outer surface of the young shell (Fig. 1.3d) indicate this very important change in the mechanism of biomineralization: the prism–nacre sequence is now established, and the mechanism that produces this mineralogic and microstructural differentiation continues to persist throughout the entire life of the animal, which may be 25 to 30 years, allowing *Pinctada* to attain a diameter of 25 cm in natural surroundings.

The fundamental change between the aragonitic undifferentiated larval shell and the two-layered calcite/aragonite post-metamorphosis shell exemplifies the biological control on the structural components of the shell. We can see that this biological control is exerted throughout the life of the animal by comparing the structure of the prism of an adult shell and the prism of the juvenile shell. The very early prisms exhibit a monocrystalline optical behavior (Fig. 1.3f), whereas the adult prisms are composite (Fig. 1.2c). The transition between these two distinct forms of prisms regularly occurs during the development of the prismatic layer (see Fig. 3.35). A careful analysis of the mode of growth of the shell components, from their initial stages to their finish, provides us with a first series of information regarding the method by which this control on crystallization is exerted.

The squamous external surface of the *Pinctada* shell shows prominent growth lamellae (Fig. 1.4a), overlapping growth increments that result from major contractions of the mantle towards the interior, thus separating successive growth phases. When this contraction exceeds the dimension of the peripheral zone (prismatic), it is then possible to see a doubled superposition of prisms and nacre in the cross section of one valve. This is illustrated in Figs. 1.4b and 1.4c. Here, after retraction of the mantle towards the posterior of an external

Biominerals and Fossils Through Time

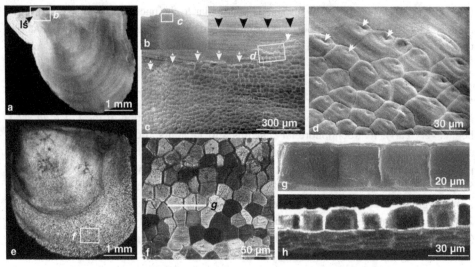

Fig. 1.3. The early prismatic layer of the post-larval shell of *Pinctada*. (a–b) Young shell of *Pinctada margaritifera* with a larval shell at its umbo ("ls"); (c) The first peri-prismatic envelopes visible on the external surface of the juvenile shell (white arrows), adjacent to the larval shell. Prismatic microstructure replaces the undifferentiated microstructure of the initial shell; (d) Enlarged view of the external surface of the prismatic layer. Note the depression close to the margin of each prism (arrows; see also Fig. 3.22); (e) Overall view of a youthful shell in polarized light; (f) Peripheral region of the juvenile shell viewed in transmitted polarized light; within the polygonal envelopes, the mineral phase forms monocrystalline prisms, each of them with a specific polarization pattern; (g) Fracture surface in the prismatic region; the prisms appear as a compact material limited by an organic envelope; (h) Fluorescence of organic laminae that surround and limit prisms. This three-dimensional reconstruction (laser confocal microscopy) allows simultaneous viewing of this section and the bases of neighboring prisms, as shown on the undersurface of this specimen (image: A. Ball, Mineralogical Dept., NHM, London).

scale or leaf (Fig. 1.4c: "ret" and "exp"), the prismatic layer of the lower lamella reappears, commencing beneath its nacreous region. When retraction is less pronounced, the prismatic layer generally shows the superposition of several sequences of joint biomineralization, becoming more and more recent *towards the interior* (Fig. 1.4e: "l1," "l2," and "l3").

At the periphery of each lamella, prismatic microstructure is directly visible (Fig. 1.4d). Organic polygons limiting prisms are only a little larger than those seen in the first stages of mineralization. At their summit, a central nodule frequently occurs, surrounded by concentric circles that develop as far as the limiting polygonal envelopes (Fig. 1.4h). This arrangement shows that, in the initial phase of prism development, the central regions play a determinant role, a role that can only be explained by more precise observation of the biomineralization mode.

In longitudinal sections of prisms (thus perpendicular to the surface of the shell), growth by superposed layers is clearly shown, in addition to perfectly synchronous growth between neighboring prisms (Fig. 1.4f). Along the entire peripheral region of the shell, the mineralizing layer that insures elongation of prismatic units is *isochronous* (= built in a single

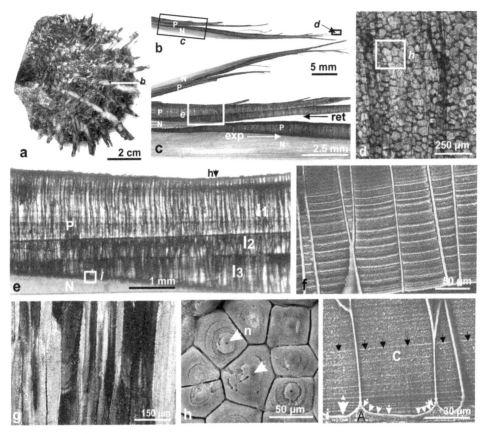

Fig. 1.4. Growth and termination of the prismatic layer of *Pinctada margaritifera*. (a) The squamous or scaly external surface of the *Pinctada* shell; (b–c) Sections perpendicular to the plane of the valves. Note the formation of the leaf-like lamella by retraction of the outer part of the mantle ("ret") followed by an expansion phase (exp) producing the new sequence (prismatic, "P" and nacreous, "N"); (d) View across one growth lamella at the shell periphery in natural light; (e) The overall structure of the prismatic layer. Layers "l1", "l2", and "l3" result from less-complete retraction of the mantle; (f) A longitudinal etched section in the prismatic layer shows the continuity of growth stages between neighboring prisms: the pallial epithelium acts in a coordinated mode on the complete growth surface; (g) A longitudinal thin section viewed in polarized light indicates that prisms have a crystallographic individuality that is maintained throughout their growth; (h) At their summit (at the beginning of mineralization) prisms show a concentric structure that bears witness to the existence of an initial phase of biomineralization around a central nucleus ("n", arrows); (i) At their base, prisms are progressively covered by an organic lamella (white arrows). The perfect isochronous nature of the mineralizing mechanism between neighboring prisms is clearly shown here at a micrometer scale (black arrows). h: Dauphin *et al.* (2003a).

mineralizing cycle). Related to the preceding figure, and demonstrating the crystalline character of each prism (Fig. 1.3f), this relationship provides the most basic proof of overall control of growth of crystals by mineralizing epithelium. The biomineralization mechanism controls the individual character of prisms in a crystallographic sense, causes

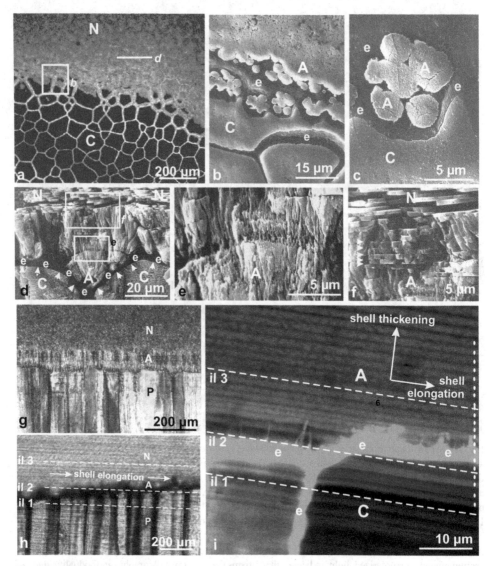

Fig. 1.5. Transition from calcitic prisms to the nacreous layer in the shell of *Pinctada margaritifera*. (a) Internal surface of a valve at the boundary between the prismatic and nacreous sectors. The fluorescence of the organic envelopes sharply separates the calcitic prisms ("C") and the nacreous aragonitic region ("N") (laser confocal microscope: blue laser); (b–c) Enlarged views of the prism surfaces being covered by aragonite. The first mineralized elements of the nacreous zone do not rest directly on calcite prisms, but instead on the organic sheet ("e") originating by expansion of the envelopes (see Fig. 1.4i); (d) Section perpendicular to the internal surface. The terminal growth stages of the prisms ("C") are seen and the start of aragonite crystallization ("A") in the form of fibrous "fan systems." Note the continuity of the organic separation ("e") between the calcitic prisms and the formation of fibrous aragonite; (e) Fibrous microstructure of aragonite prior to the formation of true nacre; (f) Appearance of the first organic lamellae (arrows), still incomplete, in the fibrous aragonite

their elongation by simultaneous growth increments and creates new prisms at the growing edge of the shell *all at the same time.*

This observation is also verified on Fig. 1.4i, showing the final stage of prism growth. The progressive covering over of the prism surface by an organic lamina is easily observable; this organic lamina originates by expansion of envelopes (Fig. 1.4i: arrows). This image also relates to Fig. 1.1d, as well as Figs. 1.5a and 1.5b. The covering of the uppermost part of the prisms corresponds to the dark layer "e" separating the peripheral region of the shell (calcitic prisms) from the central region (aragonitic nacro tablets). The microstructure of the area situated on the opposite side of the prism cover also indicates that this transition between prisms and nacre is not an instantaneous phenomenon.

Close observation of the "transition line" shows that new mineral material is appears on the organic lamella that covers the prisms (Figs. 1.5b and 1.5c–e). This material forms small granular accumulations of aragonite, but this is not yet nacre. Beginning with these initial accumulations, production of aragonite develops in the form of fibers arranged in radial bundles that are oriented perpendicular to the surface of the prisms (Figs. 1.5d and 1.5e). However, the emplacement of small, regularly spaced, parallel lamina is observed (Fig. 1.5f: arrows) that progressively develop considerable lateral extent. True nacre is formed above this fibrous aragonite, which can form a continuous layer having a thickness of several tens of micrometers above the prisms (Fig. 1.5g: "A").

It has also been observed that, just as in the initial stages of prismatic mineralization in laminae, metabolic changes in the mantle necessary to change from a system of calcite mineralization to aragonite mineralization are only carried out stepwise. Furthermore, the mechanism for production of nacre itself involves two phases: first, the formation of fibrous aragonite, and then emplacement of the cyclical secretion that forms the horizontal organic laminae and finally results in the formation of true nacre (Dauphin *et al.* 2008a). Additionally, Figs. 1.5h and 1.5i draw attention to the importance of not considering the microstructural sequence shown by Fig. 1.5g ("P–A–N") as a simple superposition of successively formed layers. In the two figures (Fig. 1.5h, thin section seen in natural light,

Caption for Fig. 1.5. (cont.)
("A"), that progressively becomes true nacre ("N"); (g) Section of the "prisms – intermediate aragonite – nacre" sequence in polarized light; (h) The same sequence in natural light. The isochronous lines ("il 1–3") make the diachronism of the microstructural sequence sharply visible, resulting from the progressive displacement of the mineralizing zones of the mantle; (i) Laser confocal fluorescence (568 nm) at the calcite/aragonite limit (image: A. Ball, Mineralogical Dept., NHM, London). The dotted line on the right side of the figure corresponds to successive positions of the isochronous growth layers. During simultaneous thickening and elongation of the shell, the limit between calcitic and aragonitic mineralization areas moves onwards. The picture shows precisely the moment (layer "il 2") at which calcite was deposited on the right side (below the organic sheet "e"), whereas aragonite started to be produced on the left side (above "e"). This is exactly what is shown on Figs. 1.1d and 1.5b when examining the transition sector at the internal surface of the shell. b, e: Cuif and Dauphin (1996); a, c, i: Dauphin *et al.* (2008).

and Fig. 1.5i, polished surface seen by laser confocal microscopy), each visible trace of an older growth surface is a true *isochronous line* ("il"), and allows following the development of microstructural units through time. Figure 1.5h indicates that, as one progresses towards the most recently formed regions of the shell (those that are now towards the interior), the isochronous lines successively traverse calcitic prisms (Fig. 1.5h: "il 1"), then the transition zone (Fig. 1.5h: "il 2"), and then only aragonite (Fig. 1.5h: "il 3"). These isochronous lines are slightly oblique to shell layer boundaries. Thus, taking into account the isochron "il 2" visible in Fig. 1.5h, it is clear that (from right to left) this growth surface starts in the upper part of the prismatic layer and successively penetrates into the non-nacreous aragonite layer "A", and only then into the nacreous layer "N".

This arrangement evidently results from the fact that the three microstructural types making up the shell are produced simultaneously by the pallial epithelium. *Thus their superposition develops through time by lateral displacement of the three specialized areas of the mantle, while each produces a different microstructure.* The resulting microstructural sequence by the mineralizing activity of the mantle during its extension is thus *diachronous*. This property must be taken into account in applications where synchroneity of measured sample areas is essential, as for example, the utilization of these materials as records of environmental parameters. This is necessary, not only to determine methods for sampling the calcareous material, but also for interpretation of the numerical data provided by its analysis.

Greater enlargement (Fig. 1.5i) allows precise observation of this diachronism. Green–yellow radiation (568 nanometers) in the confocal laser microscope produces very strong red fluorescence at the position of the organic envelope covering the surface of prisms prior to aragonite deposition. In both the prismatic and the aragonitic layer, weaker fluorescence is equally visible, corresponding to the isochronous growth layers already observed in Fig. 1.5h. At this scale of observation, one can follow the emplacement of structures forming the shell daily (white dots). For example, the duration of the process of re-covering of the surface of the calcitic prism layer by the organic layer originating from the expansion of prism envelopes can be established. At a given point, the mode of simultaneous growth of the shell elongation and thickening now can be dimensionally determined very precisely.

The microstructural organization of the nacreous layer in *Pinctada* presents a highly visible and quite surprising contrast to that of the prismatic layer. Following an initial phase of fibrous aragonite production that rests on the organic covering of the prisms (Fig. 1.6a: "A"), the first nacre tablets, 0.5 to 0.8 μm thick, are produced, but are only weakly coordinated (Cuif and Dauphin 1994). The result is a zone having a disorganized appearance (Fig. 1.6b), with a thickness that varies according to individual specimens. However, the mechanism for production of these tablets becomes rapidly more regular, ending in the formation of faces with "stair steps" (Figs. 1.6c and 1.6d). These have been described and known since the early researches of Wada (1961, 1966, 1972) and Erben (1972), Erben and Watabe (1974).

Each of these strata results from lateral coalescence of tablets whose horizontal growth begins from very small units (Fig. 1.6d: "it"). Coordination of the mechanism for generating

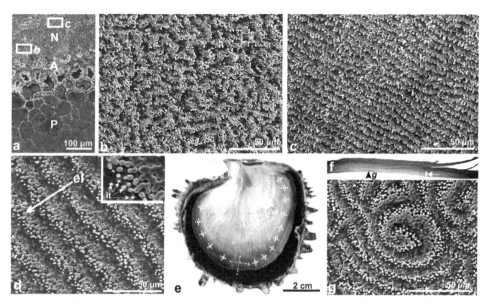

Fig.1.6. Overall growth zonation of the nacreous layer. (a) SEM image of the transitional zone (similar to that of Fig. 1.5a); (b) The peripheral area in the nacreous zone. Production of nacreous units appears to be lacking coordination, thus forming unorganized nacre; (c–d) The reverse situation in more internal areas, where generating zones succeed one another with regularity in the direction of elongation of the shell ("el"), with initial tablets ("it") shown magnified in the inset; (e) Seen overall, nacre tablets have crystallographic axes oriented in a coordinated mode throughout the nacreous layer (redrawn from Wada, 1961); (f) *Pinctada* shell in cross section perpendicular to the valve. The thickness of the nacreous layer increases only at its periphery, that is, at the right, beyond the vertical white mark; (g) In most of the internal area, the shell thickness remains constant; thus, in zones where nacre is last produced, generating zones are no longer aligned according to a growth direction, and commonly have an overall spiral arrangement. g: from Dauphin (2008).

these centers of growth is truly remarkable, because they are produced after the preceding layer (on which they are distributed) has arrived at its final stage of development to form a continuous layer of nacre. Furthermore, although they can be produced without any visible relationship between one and another, the layer of nacre thus generated is remarkably coordinated in its crystallography. By studying orientations of crystallographic axes of aragonite forming nacre tablets, it has been shown that the *c* axes are always perpendicular to the plane of the layer, while the *a*- and *b*-axes are, respectively, parallel and perpendicular to the lateral margin of the layer. The result is that the orientations of the *a*- and *b*-axes vary while corresponding to shell morphology (Fig. 1.6e). Wada (1961) has, in addition, established that this property is generally similar in most species producing nacre, regardless of their morphological diversity. It is difficult to imagine a more striking demonstration of the precision with which the nucleation and growth of crystals of biological origin can be controlled. In more central shell areas, the mineralizing activity of the pallial epithelium is reduced and the nacreous layer is not greatly thickened. However, within the large spiral figures frequently observed (Fig. 1.6f), Wada has also shown that crystallographic orientations of the nacreous tablets remain remarkably similar.

1.3 Comparison of microstructural patterns in *Pinctada* shells with those of two other pteriomorphs: *Pinna* and *Ostrea*

1.3.1 Microstructure of the shell of Pinna nobilis

Among bivalve species of the genus *Pinna*, present on all shallow shores, from temperate to tropical, the species *Pinna nobilis* of the Mediterranean is particularly remarkable. It produces a large shell several decimeters in height (Figs. 1.7a1 and 1.7a2) that is comprised of the external prism/internal nacre sequence, just as in shells of *Pinctada*. However, many differences are readily visible in the general construction of the shell, as well as in its microstructural details. The most visible characteristic is seen in the highly disproportional prismatic and nacreous layers. Practically speaking, the layers of the shell of *P. nobilis* are built almost completely by the prismatic layer (calcitic), while the nacreous layer has much reduced extent and it remains very little thickened. The prisms of *Pinna* are perfectly straight, perpendicular to the surface of the shell (Figs. 1.7b to 1.7d), and have a compact appearance (Fig. 1.7e). As in *Pinctada*, the prisms are surrounded by a resistant organic sheath (Fig. 1.7f).

At the microstructural level, their principal characteristic is a perfectly monocrystalline optical behavior, as easily seen both in longitudinal and transverse sections (Figs. 1.7g and 1.7i), and there are no exceptions to this. The crystalline individuality of each prism is maintained throughout its growth, with the c-axis in the elongation direction of the prism, through lengths that can reach four to five millimeters, the maximum shell thickness in this species. The prisms of *Pinna* species are the most perfect examples of structures among the Mollusca that have a monocrystalline appearance, as much by their dimensions as by the uniformity with which their crystallinity is maintained. Owing to their large size, simple shape and overall arrangement, as well as easily visible polarization properties, prisms of *Pinna* shells were among the first materials studied, thus playing a major role in formation of the concept of microstructure. This crystallographic individuality is all the more surprising because elongation of the prisms is carried out by superposition of growth layers that are directly shown on their lateral faces (Fig. 1.7i: arrows). These growth layers have thicknesses of two to three micrometers, and the most recent of them, the one in contact with the pallial epithelium, shows a growth surface that does not conform to any plan of crystal growth (Fig. 1.7i: "gs"). The condition of this surface must be related to another surprising property of these prisms: the nature of their fractured surfaces (Fig. 1.7e). In an inorganic crystal of calcite, cleavage planes produce the three classic rhombohedral faces, but this type of cleavage does not occur in prisms of *Pinna*. In these two fundamental aspects, the nature of the growth surface and cleavage properties, *Pinna* prisms indicate more clearly than any other material the paradoxical characteristics of crystals of biological origin.

Biological influence in forming these materials is shown equally well by the action of an enzymatic solution (pronase) on a longitudinal polished surface (Fig. 1.7j). In these prisms, having the appearance of being monocrystalline, this preparation shows that

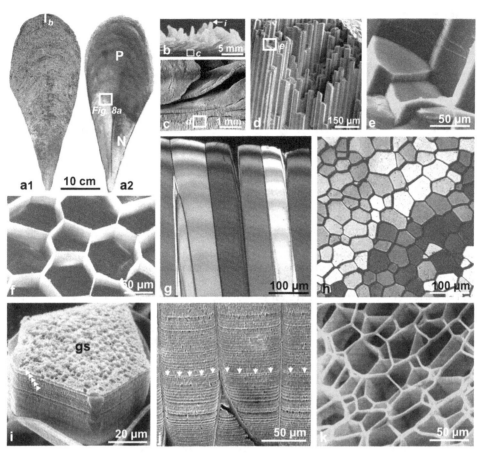

Fig. 1.7. Microstructure of prismatic layers in the shell of *Pinna nobilis*. (a) Overall view of the shell: (a1) external, (a2) internal views; (b–c) Enlarged views of the outer layer of the shell with its prismatic microstructure; (d) Regular and densely packed organization of calcite prisms; (e) Fracture surface of prismatic units. Note the absence of rhombohedral cleavage planes, contrasting with the apparent crystallinity of the prisms; (f) Prisms of *Pinna* showing organic envelopes that are visible after acid etching; (g–h) Prisms of *Pinna* with optical behavior that is apparently monocrystalline in longitudinal section (g) as well as in transverse section (h); (i) Prisms and envelopes. Superposed mineralization layers (arrows) are seen clearly, and also the granular growth surface ("gs"), very different from a purely crystalline growth face; (j) Just as in *Pinctada*, mineralization layers are deposited in a coordinated way between neighboring prisms. Each layer is thus an isochronous growth surface of the shell (arrows); (k) The prism envelopes also grow by superposition of distinct layers which can be disassociated by prolonged stay in water. Hydrolysis reveals the correlation between stepping growth of the prism envelopes and the layering of their mineralized phase (see also Fig. 5.10). a1: Farre and Dauphin (2009); d: Cuif *et al.* (1980); e: Cuif *et al.* (1981); f: Cuif *et al.* (1991); h: Dauphin (2003); i, j: Cuif *et al.* (1983).

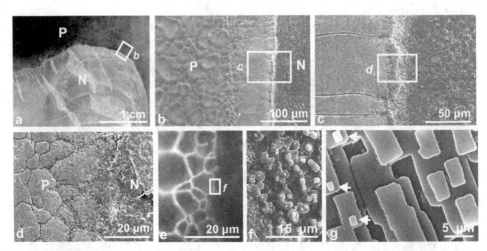

Fig.1.8. Early stages in deposition of nacre tablets in *Pinna*. (a) Covering of the internal surface of prisms by the nacreous layer; (b–e) At the growing edge of the nacreous layer, episodic small prisms may occur, exhibiting the usual fluorescent envelopes. A thin organic layer is then produced on which the first nacre tablets are formed; (f) The early aragonitic crystals are randomly oriented, and do not exhibit well-controlled morphologies; (g) In contrast, the standard aragonite tablets of nacre in *Pinna* mostly exhibit a quadrangular morphology. Note the distribution of the very latest units (arrows) appearing on the surface of the preceding mineralization cycle. e: Frérotte *et al.* (1983).

growth layers are present and that they are perfectly synchronous from one prism to another (arrows) over the complete surface of the prismatic zone. Just as in *Pinctada* species, production of strictly isochronous growth layers also indicates that here, mineralizing activities of the pallial epithelium are developed in a cyclical mode. In addition, Fig. 1.7k indicates that this cyclicity equally affects prism envelopes. After prolonged decalcification, the peri-prism sheaths disassociate, and it can be seen that they too are constructed by superposition of layers with the same thickness as mineral growth layers.

Uniformity of shell thickness in the prismatic part of the *Pinna* shell indicates that the mineralizing action of the mantle is limited to the periphery of the shell. As shown in Fig. 1.7a2, the development of the nacreous layer is produced with a long phase of mineralizing inactivity. When the internal nacreous layer commences its development, it first covers the surface of the prisms whose growth has long been halted. Frequently, it is seen that a thin prismatic zone is first formed at the front of the layer that is only transitional, but is sharply visible in UV fluorescence, with dimensions that are much smaller than those of the prisms normally formed by the mineralizing zone at the external margin of the shell (Figs. 1.8a–e). These small prisms are immediately covered by an organic layer on which the first aragonitic structures are developed (Fig. 1.8f). They are here formed of morphologically variable and irregularly distributed units, but very rapidly the mechanism of biomineralization introduces a supplementary control determining the formation of units that are regularly quadrangular (Fig. 1.8g). The orientations of these units are highly

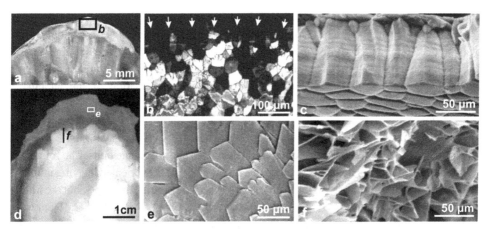

Fig. 1.9. Shell microstructure in *Crassostrea gigas*. (a–b) A prismatic growth lamella during its formation in the opercular valve: (a) shows the condition of the prismatic layer after 36 hours of growth of a new lamella, and in (b) prisms are already visible in polarized light; (c) A view of the prismatic layer after 14 days of growth (opercular valve). In spite of their crystallographic individuality and their isolation by peripheral envelopes, the overall synchroneity of their growth is readily visible; (d–e) Microstructure of the concave valve. The border is formed by foliated calcitic units (e); (f) Groups of foliated layers are imperfectly joined in oyster shells. The chambers thus created may be very well developed at times and filled in by irregularly oriented calcareous laminae.

conformable with respect to the geometry of the zone of secretion, following the rule established by Wada in 1961.

Although limited to microstructural observations only, this introductory comparison shows that even here, in two members of the pteriomorph Bivalvia, each develops an identical sequence of microstructures (calcitic prisms and aragonitic nacre), with numerous characteristics that allow the two genera to be differentiated without ambiguity. This demonstrates the very precise control exerted by the pallial epithelium on the mode of crystallization of the microstructural units, on their arrangement, and on the manner of their crystallization.

Examination of the oyster *Crassostrea gigas*, another species of pteriomorph bivalve, extends this conclusion, and increases our understanding of how, in his celebrated study, Bøggild (1930) could distinguish most genera of the Mollusca solely by examination of their microstructural sequences.

1.3.2 Microstructure of the shell of oysters

In contrast to the two preceding genera, valves of the ostreids differ morphologically and microstructurally. In *Crassostrea gigas*, for instance, only the flat valve produces a prismatic layer (Figs. 1.9a–c). Prisms exhibit their typical features – monocrystalline with polygonal sections limited by organic envelopes – but the prismatic layer remains thin and only plays a secondary role in the architecture of the shell. On the internal face, there soon appears a new type of microstructure made of juxtaposed long flat lamellae. Taylor, Kennedy and Hall (1969) named this microstructure, *foliate*. Contrary to the two preceding cases, and in spite of its appearance, which can be quite shiny, it is not nacre. Lamellae are not aragonitic, but calcitic. In addition, growth of the lamellae is quite different from growth of nacreous tablets. New mineral material is added only at the anterior margin of the lamellae. Indeed, this foliate microstructure characterizes the entire deep hollow valve of *Ostrea* (Figs. 1.9d–e).

As another difference with the two preceding genera, the oysters have a third type of mineralization that develops in spaces located between groups of foliate units. The foliation phenomenon produces calcareous laminae that are very irregularly oriented, leaving numerous empty spaces between them (Fig. 1.9f). These spaces do not remain empty, but are filled in by calcitic laminae, arranged irregularly. Spaces left between the irregularities of laminae due to this incomplete mineralization provide a locus for frequent proliferation of bacteria thus causing considerable commercial problems in oyster cultivation.

As a commercially important bivalve, the development of oysters has been well studied, and especially now, when calcification problems have been reported that are possibly related to oceanic character changes. As seen in the young *Pinctada* shell (Fig. 1.3), metamorphosis also involves mineralization mechanisms. Looking at the larval stage of *Crassostrea*, we note that a completely different shell is formed during the initial period of calcification, formed of simple superposed growth laminae, thus producing a very regular concentric striation on the external surface of the shell (Figs. 1.10a–b). These growth lamellae are carriers of mineral grains arranged in order, but not forming definite microstructural units. Remarkably, the mineral produced in this larval shell is also aragonite.

At metamorphosis, which occurs between 20 and 25 days, this very important transformation in the organism determines the zonation of the mantle, which produces the typical structures that are continued throughout the entire life of the animal: the peripheral prismatic zone (forming the external layer) and the internal lamellar zone (Figs. 1.10c–e). The mineralizing mechanisms equally change the polymorph produced, since the whole shell of ostreids is formed of calcite (except calcification developed in the muscle attachment, which is always aragonite in bivalves).

1.4 Diversity of microstructural sequences among the shells of other molluscs

In the Bivalvia (Mollusca), the microstructural sequence comprising external calcitic prisms and internal nacre, as illustrated above, is commonly considered a general model for their microstructure. Already in 1969, Taylor, Kennedy and Hall pointed out the inaccuracy of this view, certainly at least partially due to the fact that the prismatic/lamellar model is the one with the easiest representation. Actually, the microstructural type that occurs most commonly in molluscs is organized into the form called *cross-lamellar*, one that is much more difficult to represent because of the complex three-dimensional arrangement of crystallographic units. It is remarkable that this type of geometrically complex

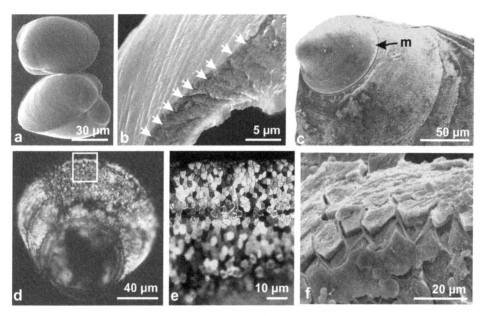

Fig. 1.10. The larval and post-metamorphosis shells of oysters. (a–b) The round-shaped larval shell is built by simple superposed layers (b), on which mineral granules (aragonite) are randomly distributed. No specific microstructure is visible (specimens from the IFREMER hatchery at Argenton, France); (c–f) In contrast, after metamorphosis (note the circular furrow: "m") the mantle produces the typical microstructural sequence of the ostreids: calcite prisms occur at the periphery, and laminar layer on the internal side. These are the early stages of the prismatic outer area (c) with the larval shell still visible at the umbo (arrow). The prismatic outer area is shown in thin section (polarized light, d–e) and scanning electron microscopy (f).

mineralization develops in comparable ways within several classes of the Mollusca: the scaphopods, bivalves, and gastropods. It is in the Gastropoda where it may be the predominant character numerically, as this class is the most numerous of all the branches of the Mollusca, and it has been for a very long time. Estimates of the number of gastropod species vary between 40 000 and 100 000, although only 8000 to 10 000 species are recognized in the Bivalvia, and scarcely more than 700 in the Cephalopoda. This preponderance is due to the great diversification undergone by the gastropods during Mesozoic and Cenozoic times, with the greatest diversity of all being developed in those families utilizing cross-lamellar microstructure in their shell.

1.4.1 The cross-lamellar microstructure: the most widely distributed biomineralization model

Although commonly occurring in the Bivalvia, Gastropoda and Scaphopoda, the cross-lamellar layer is unknown in the Cephalopoda.

Fig. 1.11. Example of cross-lamellar microstructure in *Murex* (Gastropoda). (a–b) Fractured surface of the shell, showing its organization into areas characterized by cross-stratified microstructure; (c–d) Crystal orientations within regularly alternating structural planes (arrows). The "cross-laminae" are in reality compact groups of parallel fibers with regularly alternating structural layers. Defects in crystallization patterns can occur (arrows, d); (e–f) Sections in overall (e) and closer views (f), seen in polarized light (optical microscope), illustrating the alternating orientations of microstructural layers. d: Dauphin *et al.* (2007).

Cross-lamellar microstructure in the shell of Murex (Gastropoda)

Cross-lamellar microstructure is characterized by the formation of adjoined layers whose microstructures alternate on a very regular basis. This property is easily observed in scanning electron micrographs of a simple fracture surface (Figs. 1.11a and 1.11b: arrows) (Dauphin and Denis 2000). Since these microstructural units exhibit crystallographic coherence as well, polarized light shows very strong contrasts between alternating layers, thus providing images that are the origin of the term "lamellar" (Figs. 1.11d to 1.11f). However, these lamellae are not the basic units of cross-lamellar systems. They are themselves complex structures, formed by very densely packed fibrous elements (Fig. 1.11e). More than in any other microstructural type, it is essential to understand that the coherent microstructural layers forming the cross-lamellar structure *are not growth layers*. Quite the contrary; a very strong angular discordance is always seen between the strata of mineralization that generate the isochronous lines and mineral units that result from the crystallization process. A three-dimensional reconstruction of the structure thus formed is all the more difficult to achieve because the geometry of the mineralization surface is only

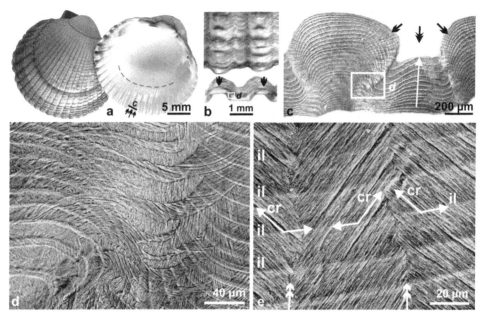

Fig. 1.12. *Cardium edule*: cross-lamellar microstructure developed from a complex surface of mineralization. (a–b) Mineralization infilling the radial grooves visible at the periphery of the shell (a: arrows), that correspond to the external ridges characteristic of the species (b); (c) Beginning with the external layer (dark arrow), deposited at the zone of the shell growth, mineralization layers are superposed *towards the interior* (white arrow), producing a practically smooth internal surface (a: dashed line); (d) Complexity of intersections between mineralization layers and the crystallization directions of crossed lamellae at a junction between a radial ridge and a valley; (e) Alternation of crystallization directions in three cross-lamellae. Note the irregularity of their limits (doubled arrows) in the absence of rigid organic envelopes, such as those that separate prisms in *Pinna* and Pinctada. e: Dauphin (2002b).

very rarely planar. In the Gastropoda, it is obviously spiral (except in a few cases, as in the Patellidae or limpets), and very often is strongly ornamented (as in the case of *Murex*).

Cross-lamellar microstructure in the Bivalvia

Cross-lamellar microstructure is just as common in the Bivalvia. In *Cardium*, for example (Fig. 1.12a), the growth surface visible on the internal face of the valve shows an alternating series of ridges and depressions oriented in the direction of elongation of the shell. This series of radial ridges and depressions constitutes the true surface of active mineralization. Practically speaking, the shell is not thickened further, since only the radial depressions are filled in (Fig. 1.12a: behind the arrows). Correlating to this, the radial depressions in the interior correspond to the ornamented ridges seen on the external surface of the shell (Fig. 1.12b). While the mineralizing epithelium is depositing successive growth layers, the anterior portion of the active zone produces the external components of the shell (ornamentation). Then, during extension of the mantle, interior radial depressions are

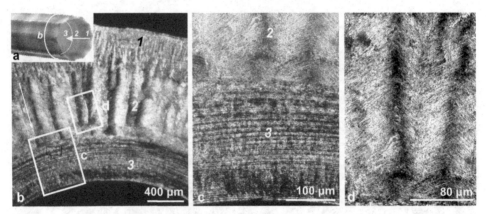

Fig. 1.13. A microstructural sequence made of three distinct cross-lamellar layers in the tube shell of a *Dentalium*.

progressively filled in by the cycles of mineralization that produce the middle zone. A section (Fig. 1.12c) clearly shows the progressive infilling of radial internal depressions by deposition of successive growth layers (in the direction shown by the arrow). A more precise examination of relationships between growth layers and the structural orientation of mineral components (Figs. 1.12d and 1.12e) clearly establishes a marked angular relationship between isochronous planes, forming the limits of growth layers (Fig. 1.12e: "il"), and the morphological orientation of mineral components (Fig. 1.12e: "cr").

The specific and truly unique character of cross-lamellar shell material lies in the perfect repetition of microstructural groups with regularly alternating crystallographic orientations. This material provides additional evidence of the precise biological control exerted on the crystallization process, and a great adaptability in the development of three-dimensional architecture in the shells and in their ornamentation.

Cross-lamellar microstructure in the Scaphopoda

A transverse section of the tube-like shell of a scaphopod shows three distinct layers (Figs. 1.13a–b), all three built by cross-lamellar structures. The medial layer preponderates, organized into radially oriented and crystallographically coherent groups of lamellae (Fig. 1.13c). Although not so consistently organized, the external (Fig. 1.13b: "1") and internal (Fig. 1.13b: "3") layers are also built by cross-lamellar units. Thus, the concept of microstructural sequence is applicable to the diachronous deposition of these three distinct cross-lamellar layers.

1.4.2 Two atypical gastropods: Haliotis and Concholepas, remarkable examples of microstructural diversity

Haliotis (the Abalone) consists of a group of gastropods with flattened shells that are easily recognized by their row of circular siphonal openings that have progressively increasing

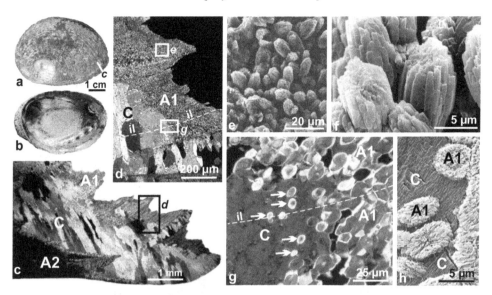

Fig. 1.14. Microstructure of the shell of *Haliotis tuberculata*. (a–b) External (a) and internal surface (b); (c) Section perpendicular to the shell at the pallial border. The sequence of three distinct microstructural layers is seen as follows: "A1" external aragonite, "C" medial calcite, "A2" internal aragonite; (d) Enlargement of the two external layers: "A1" aragonite occurs as granules and calcite as large prisms having a monocrystalline appearance ("il": isochronous incremental line); (e–f) Aragonitic granules of the external layer, with ovoid clumps of them formed by parallel prisms (f); (g–h) At the contact between the "A1" external and "C" medial layers, aragonitic grains are found included within the calcite of prisms. At the intersection between a mineralization surface ("il") and the zone of contact between the two layers, factors controlling the two types of crystallization were present at the same time. c: Dauphin *et al.* (2005); f: Cuif *et al.* (1987); g: Mutvei *et al.* (1985).

diameters (Figs. 1.14a–b). Representatives of this ancient family are distributed in all seas, and their shell is also remarkable from a microstructural point of view. Their structure generally has three well-differentiated layers: two aragonitic layers separated by a medial calcitic prismatic layer (Fig. 1.14c: "A1–C–A2"). In species from warm seas, however, the medial layer practically disappears and the shell can be entirely aragonite (Mutvei *et al.* 1985, Dauphin *et al.* 1989).

   The external layer is formed of very small ovoid nodules distributed without preferential orientation (Figs. 1.14d–f). Their arrangement in successively mineralized strata is visible in sections perpendicular to the growth direction (Fig. 1.14d: "il"). The contact between the external aragonite layer and the medial calcite layer is very irregular; one can see aragonite nodules that formed continuously, so that on this same surface, crystallization of the medial calcitic layer had already begun. Thus, some aragonite nodules can be completely isolated within the interior of calcite crystals (Figs. 1.14g–h: arrows). This arrangement indicates that a biological mechanism is capable of determining the polymorph of calcium carbonate crystallizing in a given part of the mineralization stratum, with a control effective enough to maintain a perfect separation between the two types of biominerals.

Fig. 1.15. Microstructure of the shell of *Haliotis tuberculata*; the contact between calcite and nacreous aragonite. (a) Section of the pallial margin (natural light). Note the irregularity of the surfaces of contact between the three successive layers ("A1–C–A2"); (b–c) Calcitic prisms in longitudinal and transverse section (thin sections, polarized light). In the absence of a firm organic envelope, neighboring calcite prisms have very irregular surfaces of contact; (d–e) At the base of the calcite prisms (arrows), an undifferentiated aragonitic layer "A2" precedes formation of true nacre ("N"); (f–g) Growth surface of the nacreous layer. The superposition of crystals produces irregular pyramids; (h) Fracture surface of the nacreous layer; the tablets, after joining, preserve the trace of their initial organization in regular stacks (an arrangement never occurring in the nacre of the Bivalvia). f, g: Dauphin *et al.* (2007).

Contrasting with this observed irregularity of aragonitic units in the external layer, the large calcite crystals of the middle layer form an ordered and compact bed (Figs. 1.14a–c). Here, the crystal units, elongate and slightly incurved as a function of a morphological change in the growth surface, are closely imbricate without being separated by thick organic envelopes, such as those that separate calcitic prisms of the pteriomorph bivalves. However, there is a boundary between these crystalline units that is very irregular in transverse sections (Fig. 1.14c), but each unit has a very well defined crystallographic orientation.

At the base of the calcitic crystals, shifting of the three mineralizing zones of the mantle is clearly visible during the retraction phase that precedes the formation of an accentuated growth phase, resulting in formation of a major growth "scale" on the surface. In addition to

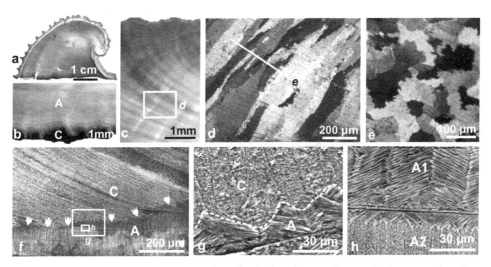

Fig. 1.16. *Concholepas* shows an association of calcitic prisms and aragonite in a cross-lamellar structure. (a–b) Overall section of a *Concholepas* shell and an internal view of the pallial border, showing both calcitic ("C") and aragonitic "A" layers; (c–e) The external layer of the shell is formed of long, incurved calcite prisms, noticeably perpendicular to the orientation of the growth layers (c–d). Note the irregular limits of prisms, in longitudinal (d) as well as in transverse section (e); (f–h) The terminal surface of the prisms is here shown covered by the cross-lamellar layer (f). Note also the irregularity of the surface of insertion (g) and the two distinct cross-lamellar microstructures "A1–A2" of the aragonitic layer (h). a, c–e: Guzman *et al.* (2007).

this, the internal aragonitic layer can also be again covered anteriorly by a calcitic micro-structural component of the middle layer (Fig. 1.14d: "il"). It is seen here that the calcite crystals that formed beneath the aragonitic layer maintain the same crystallographic orientation they had before retraction of the mantle.

Formation of the internal nacreous layer follows the deposition of a transition layer that is also aragonite, but remains irregularly granular (Fig. 1.15e). This intermediate layer conditions the surface of crystal growth of the calcite, and it is only following this that typical nacre is deposited. This sequence is easily comparable to the formation of the fibrous aragonite layer already seen in *Pinctada margaritifera*, where it also precedes progressive deposition of morphologically characteristic nacre in the same way.

The growth surface of the nacreous layer of *Haliotis* provides a good illustration of organization of nacre tablets in which successive layers have centers that are relatively well aligned, producing a structure of rectilinear piles (Fig. 1.15h). This structural organization also exists in the cephalopods, but never occurs in the Bivalvia.

The species *Concholepas concholepas* originates on the Chilean coast, where it is particularly well known because it constitutes one of the principal resources for marine aquaculture in this region (it is the "false abalone"). Additionally, the shell, which occurs abundantly as fossils in uplifted marine terraces, potentially provides a great amount of historical information concerning thermal variations affecting currents that flow past

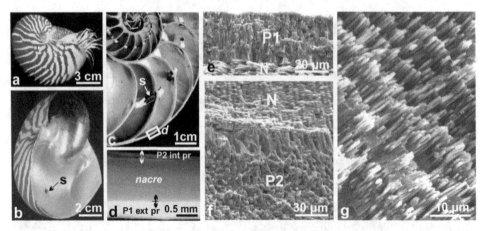

Fig. 1.17. *Nautilus*. (a–b) Overviews of *Nautilus* shell; (c) Enlargement of *Nautilus* shell illustrating siphuncle ("s"), chambers and septa; (d) Wall structure in *Nautilus*; the main component of the shell is the medial nacreous layer (*nacre*), with a thin external spherulitic prismatic layer ("P1 ext pr"), and a still thinner internal prismatic layer ("P2 int pr"); (e) Thin inner spherulitic prismatic layer "P1" of the outer wall; (f) Nacreous layer (top) and outer spherulitic prismatic layer "P2"; (g) Detail of the columnar nacreous layer.

the Chilean coast from south to north, participating in the formation of El Niño events. This species builds a shell with a thick external calcitic layer and a thin internal aragonitic layer (Figs. 1.16a–b). The calcite forms elongate crystals, incurved to conform to the curvature of growth surfaces (Figs. 1.16c–d), and irregularly imbricate in transverse section (Fig. 1.16e).

The *Concholepas* shell bears an internal aragonitic layer having a cross-lamellar microstructure. This is in contact with the internal face of the calcitic layer, with a very irregular surface at the junction (Figs. 1.16f–g). Contrasting with the homogeneity of the calcitic layer, the internal layer produces variously sized sublayers, always of this cross-lamellar type (Fig. 1.16h). As an important economic resource, this species is intensively studied in Chilean biological stations, where labelling experiments using calceine have revealed a surprisingly high periodicity in the formation of growth layers (see Fig. 3.14).

1.4.3 The formation of skeleton in Nautilus and Sepia

The shell of *Nautilus* constitutes the most popular illustration of the calcareous structures built by the Cephalopoda, but if this class of the Mollusca is still to be regarded as important in modern oceans, it would be due to a very different group, the Coleoidea, whose best known representatives are the sepioids (cuttlefish), the squid and the octopus. However, the common characteristic of the modern Coleoidea is that their shells are very much reduced in comparison with those of ancient cephalopods. For the latter, *Nautilus* still provides an exact analog, in spite of its simple coiling and its independent evolution. The coleoid shell is reduced to chitinous, nonmineralized lamellae in the squid and has completely disappeared in the octopods. Of these three main groups of modern cephalopods, only the sepioids or cuttlefish still build an important calcareous structure, but its form and architecture do not resemble the shell of the nautiloids at all. It is a cone coiled on itself in a single plane (Figs. 1.17a–b), with its internal space broken up by septa that represent the preceding back walls of successively occupied positions (Fig. 1.17c). The animal in effect only occupies the anterior part of the cone. An essential organ traverses all the septa, from the living chamber to the embryonic chamber at the summit of the cone. This is the siphon (Figs. 1.17b–c: "s"), that insures the function of the organ of flotation by determining the proportion of gas and water in the chambers. The wall of the cone and the septa possess the same microstructural sequence; essentially a layer of nacre between two layers having a prismatic microstructure (Figs. 1.17d–f) (Mutvei 1964).

With respect to consistency of microstructural nomenclature, it is important to note that, in spite of the similarity in the terms utilized, the arrangement of the microstructural components of the Cephalopoda is far different from that of the Bivalvia, for example. This observation concerning the shell structure is in accord with general taxonomic considerations that classically separate the Gastropoda and Cephalopoda from the other classes of the Mollusca.

In the Coleoidea, another evolutionary phenomenon has completely transformed the position of the shell with respect to the animal. Following modifications in development during youthful ontogenetic stages, the calcareous structures (when they exist) are placed in an *internal* position. Even the shell of *Spirula*, the only one in the coleoids to have features approaching those of ancient cephalopods, is separated from the exterior by soft tissue, so that it evidently cannot play the role of protection that it still does in *Nautilus*. This change of organization that characterizes the Coleoidea is very ancient; it goes back to the Paleozoic, and has produced important fossil groups, whose interpretation is directly related to their method of mineralization (see subsection 7.5.3). The calcareous structure of modern sepioids forms our reference for the comparison of numerous groups that have resulted from this innovation and for a discussion of their relationships.

The shell of *Sepia* (Figs. 1.18a–c) is a flattened oval structure. Its upper face (dorsal) is formed of a compact lamella bearing at the rear a point called a mucron. This lamella is slightly convex transversely, and within the curvature thus formed, a structure built of fine parallel laminae develops, equally incurved, but in the inverse sense of the dorsal shield (Fig. 1.18d). From a microstructural point of view, the dorsal shield is composed of three layers: two prismatic layers (whose thicknesses are nonsymmetrical) on both sides of a layer of nacre (Fig. 1.18e) (Barskov 1973, Dauphin 1979, 1981). This is the fundamental microstructural sequence of the cephalopod shell. The interpretation of the dorsal shield as a remnant of the ancestral cephalopod shell is reinforced by the presence of nonmineralized chitinous borders forming the periphery of the dorsal shield (Fig. 1.18c: "ch"), equivalent to the organic structure that, in squid, represents the last vestige of the formation of a calcareous cone. Interpretation of the ventral part of the cuttlefish shell (Figs. 1.18f–g) is provided by examination of the junction between a pillar and the layer on which it rests (Fig. 1.18h). The sequence of prisms/nacre/pillars typical of the cone and the septa of the cephalopods is readily seen. It is also true that although the pillars appear as truly original

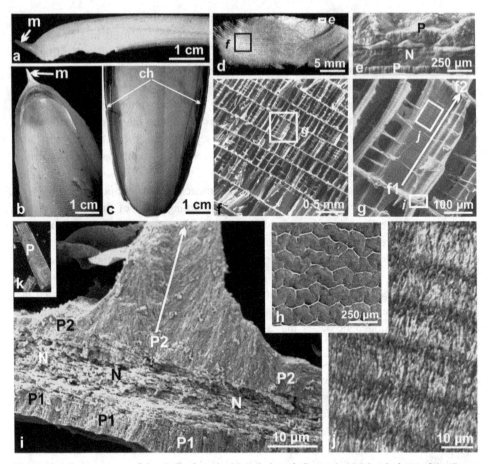

Fig. 1.18. Microstructure of the shell of a coleoid cephalopod: *Sepia*. (a–c) Morphology of the *Sepia* shell. (a) Profile of the posterior portion bearing the mucron ("m"), (b) back part of the ventral surface, and (c) anterior part of the dorsal surface. On the flanks the nonmineralized portion is a chitinous layer ("ch") similar and equivalent to that of the squid; (d) Transverse fractured surface of the shell. Under the compact wall of the upper surface, the structure in superposed lamellae forms the principal mass of the shell; (e) The compact wall of the upper surface is formed of three distinct layers, two prismatic layers ("P"), separated by a layer of nacre ("N"). This is the typical structure of the shell wall in the Cephalopoda; (f–g) The principal structure is formed of layers with regular spacing maintained by parallel pillars; (h) Lines of pillar insertions on the lamellae; (i) Microstructure of the contact between a lamella and a pillar. The ensemble forms a structure with three layers typical of structures built by the mantle of the Cephalopoda; one layer of nacre between two layers of prisms. The pair consisting of one lamella plus one pillar is then the equivalent of one septum in the *Nautilus* shell; (j) The microstructure of a pillar shows a typical "stepping growth," with synchroneity of the growth layers between neighboring pillars exactly equivalent to that seen in the preceding examples of the closely jointed prisms of *Pinna* (Fig. 1.7j); (k) Monocrystalline behavior of a pillar ("p": polarized light), equivalent to a prism. h: Dauphin (1996); i: Dauphin (1981); j: Cuif *et al.* (1983).

structures from a morphological point of view, they are constructed entirely out of the upper prismatic layer, which develops a very special form of mineralization. Instead of being a layer having an equal thickness throughout its entire surface, there are areas of preferred mineralization. On these narrow spots that form roughly parallel angular lines (Fig. 1.18i), the process of mineralizing the prismatic layer is uninterrupted (Fig. 1.18h: arrow). It continues until a new horizontal sheet can be formed (Fig. 1.18g: "f2"). Even though they are exceptionally strong, pillars formed by mineral growth in these preferred zones develop according to the usual incremental mode that is easily seen on their lateral faces (Figs. 1.17g and 1.18j). The pillars contain 45 to 50 biomineralization cycles, each one producing an increase in height of approximately 10 μm. Additionally, in each pillar the mineral phase is oriented overall so as to constitute a pseudo-monocrystal (Fig. 1.18k).

It appears that in spite of the great architectural differences that characterize them, the shells produced by the sepioids are built without profoundly modifying the basic biomineralization mechanism of the Cephalopoda. The shell itself is preserved in its fundamental structure, but mineralization there is reduced to a limited sector, the chitinous margins that represent a zone of intermediate regression. The microstructural sequence belonging to the group is also present at the base of the leaf-like structure that, in its entirety, is homologous to a series of septa.

The shell of the sepioids thus shows how a structure appearing to be completely original and innovative can in reality result from very limited modification of basic mechanisms of mineralization. These have remained essentially unchanged since their origin. Examination of the coleoid group illustrates the extent of the microstructural variations that must have developed along these lines since the late Paleozoic.

1.5 Microstructure and the mode of growth of skeleton in the scleractinian corals

As usually accepted, the term "corals" includes two modern groups that are a part of the Cnidaria Anthozoa, but the groups are sharply distinct from an anatomical point of view, as well as in their method of building calcareous skeletal structures. The Scleractinia are the most important, as much due to the number of taxa that have resulted from the complex structural evolution of their aragonitic skeleton as by their important capability as constructors.

Within the phylum, the order Alcyonaria contains several families that produce calcified structures (called *spicules* or *sclerites*) that are isolated within the tissue of the animal, but some are also capable of constructing a massive calcareous axis (Corallidae, Melitheidae). These structures invariably are formed of high magnesian calcite. Additionally, the Family Helioporidae provides a remarkable exception because they have an aragonite skeleton and sometimes also form calcitic spicules (cf. Section 4.3).

The scleractinian polyp lives at the *surface* of the carbonate structure that it produces (Figs. 1.19a1–a2). This structure (called the *corallite*) thus forms a support for the animal, which also can partially retract itself into it (Fig. 1.19b), finding some protection between the radial blades (*septa*) that form one of the basic structures of the skeleton (Fig. 1.19c: "sept"). The available space between the radial septa is generally limited, because of the vertical

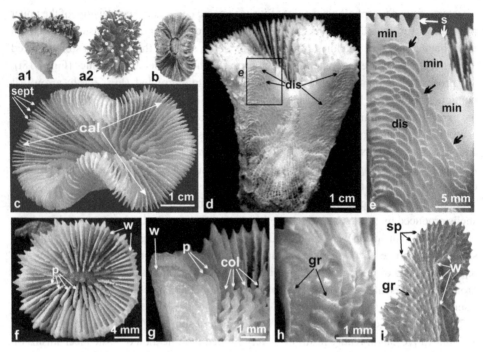

Fig. 1.19. General organization of the scleractinian skeleton. (a–b) Position of the extended polyp on the corallite (a1) and (a2), (b) polyp retracted between the septa; (c) The septa generally form the main components of the corallite. Variations in the morphology of the upper part of the septa result in a diversity of shapes for the calices ("cal"); (d–e) Between the septa, the polyp forms a structure of superposed and curved laminae, the dissepiments ("dis"), which are formed as the septa grow. The zone on which mineralization of the septa is carried out is thus restricted to the surface between the most recent dissepiments ("min", black arrows) and the summit border ("s"); (f–g) At the periphery of the corallite, the septa are united by a generally compact enclosing wall ("w"). Septa can be subdivided, bearing internal elements, the pali ("p"). In the center of the corallite there sometimes exist vertical elements forming the columella ("col"); (h–i) The summit margin of the septa is commonly broken up into spines ("sp") having varied forms, and lateral surfaces are granular ("gr"), whose forms and arrangement form part of the definition of species.

growth of the polyp, so the animal progressively abandons its first-formed areas and periodically builds temporary platforms in the form of thin, convex carbonate lamellae called *dissepiments* (Figs. 1.19d–e: "dis"). Underneath the last-formed dissepiments (Fig. 1.19e: black arrows) the calcareous skeleton is no longer in contact with the tissue of the polyp. Therefore, the biologically active surface (where the process of mineralization causes addition of new carbonate material) now is limited to the space shown between the black and white arrows in the figure (the base is formed by the latest dissepiments, and the summit is formed by the crest line of the septa).

At their periphery, the septa are connected by an enclosing structure called the *wall*, most often compact (Fig. 1.19f: "w"). To these three fundamental elements – septa, dissepiments

Fig. 1.20. Diversity of corallite morphologies. (a) Upper surface of a corallite of *Fungia* sp., an exceptional case of a solitary coral that has a very large diameter; (b) In the great majority, the scleractinians build colonies produced by division of the initial polyp. In this case (*Goniastrea*), each polyp still builds its own corallite; (c) In this colony (*Meandrina*), corallites are fused, thus it is difficult to determine the relationships between polyps and colony skeleton; (d–f) In coral colonies, the polyps can be small, and their corallites visible only with difficulty (d). Here, in *Montipora* they only measure 0.25 mm in diameter (e). Therefore, the main mass of the colony (f) is formed by undifferentiated colonial skeletal material.

and wall – are added accessory structures that are less constant. At the center of the circle of septa (sometimes called the *calice*) can exist a *columella* (Fig. 1.19g: "col"), and the septa themselves can bear lobes called *pali* (Fig. 1.19g: "p"). Morphologically, the septa are not simply undifferentiated carbonate blades. Just the opposite, their growth border is often cut into teeth or spines (Fig. 1.19i: "sp") and, additionally, their lateral faces contain nodules or granules (Fig. 1.19h: "gr") whose morphologies and arrangements are highly varied. Since the first systematic description of the scleractinian corals by Milne-Edwards and Haime (*Histoire Naturelle des Coralliaires* 1857), the taxonomic value of these features and their arrangement, sometimes considered as "ornamental" (= of secondary importance), has been progressively recognized (Alloiteau 1952, 1957). Microstructural analysis and, even more, the study of the biomineralization process in septa explains the origin of this taxonomic value of these properties, the ornamentation of the growth border and lateral faces of septa.

Beginning with this small number of structural elements, arranged according to a simple geometric scheme, the corals have developed a remarkable diversity of morphology and architecture by varying the proportions of their constituent elements, and above all, by manifesting an immense aptitude for building *colonies*. For the most part, coral polyps do not remain simple (Fig. 1.20a). Very early they commence the process of asexual multiplication leading to the formation of groups that may number thousands of polyps, all issuing from the same initial parent individual and attaining dimensions in meters (sometimes several meters, as in colonies of *Porites*). This process affects the formation of innumerable architectural types because, in these colonial forms, the individuality of the polyps is not necessarily transferred to their support structures. Certain colonies clearly show a correspondence between polyps and calices (Fig. 1.20b), but frequently one observes the formation of continuous rows of calices in which it can be difficult to recognize the position of individual polyps (Fig. 1.20c). Another widely developed evolutionary tendency in colonial Scleractinia is the formation of colonies with small and numerous calices. Additionally, this architecture seems to be favorable on the biological level, because colonies of two taxonomical groups with this architectural style play a major role in modern coral faunas: the Acroporidae and the Poritidae. An extreme example is the colony form of the genus *Montipora* (an acroporid). Here, the total colony presents a massive aspect (Fig. 1.20d), while at its surface, calices each have submillimeter dimensions (Fig. 1.20e). Thus, the main mass of the colony is formed as an undifferentiated mineral network (Fig. 1.20f).

As a result, in the Scleractinia, one can find polyps whose diameters vary between decimeters (the polyp of a solitary *Fungia*, for example) and a fourth of a millimeter (a polyp of *Montipora*). These differences associated with the architectural variability resulting from varying modes of colony construction are at the origin of difficulties, still unresolved, in the classification of scleractinian corals. Classification of the Scleractinia rests on criteria originating in the characteristics of their mineralized structure, quite different from the Mollusca, where classification is based largely on the anatomy of the animals. During the last decades, several research projects have used methods of molecular phylogeny that have yielded results differing greatly from the classification of the Scleractinia still in use, that produced by Vaughan and Wells (1943) and then by Wells (1956). In a parallel way, microstructural research has resulted in significant modification of the relative importance of different components in the corallites. Now it will be necessary to develop a method of classification that utilizes criteria originating from the two approaches in order to integrate this biological data and equally important data furnished by fossil material.

1.5.1 Fibrous structure of the coral skeleton

In spite of their great architectural diversity, all Scleractinia produce their skeleton according to identical methods, with no exceptions. The polyp is embryologically simple, and only contains two cellular layers, *ectoderm* and *endoderm*, separated by an intermediate gel, the *mesoglea*. This triple layer remains very thin but is capable of completely coating the septa, even the most serrate septa. In a simple view of the surface (Figs. 1.21a–c), the three layers cannot be separated, and only the overall fluorescence in ultraviolet light indicates the presence of the tissues in contact with the mineralized structure. The ectodermal layer is composed of very flat cells with a thickness of only a few micrometers (TEM images in Tambutté 1996).This cell layer is in direct contact with the upper surface of the calcareous skeleton (the calice or calyx), thus it has been named the calicoblastic ectoderm. It is equally important to note that carbonate mineralization develops *on the external side* of the ectoderm. This obviously is of the greatest importance because this controls the constructional properties of scleractinian corals. The calcareous structure produced by the lower surface of the ectoderm, under the polyp, can be added on top of the previously produced exoskeleton, without any limitations on size. Each polyp participating in the formation of a colonial skeleton (corallum) thus can build a tubular corallite with a length of several decimeters, or even meters, without changing its own dimensions.

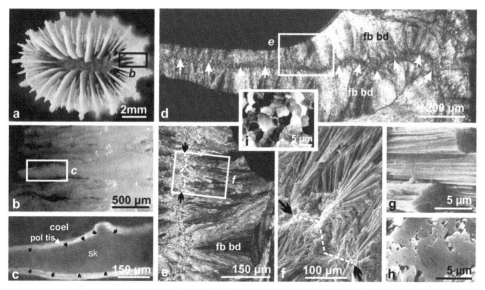

Fig. 1.21. Fibrous structure of the corallite. (a–c) The septa are permanently covered by the three basal layers: ectoderm–mesoglea–endoderm, made visible here (c) by fluorescence of the tissues observed in ultraviolet light; (d) A septum with a simple microstructure is built by bundles of fibers ("fb", "bd") arranged on both sides of a medial line, which is here continuous (arrows) and almost straight or rectilinear (thin section perpendicular to the growth direction of the corallite, seen in polarized light); (e) Enlarged part of (d); (f) Bundles of diverging fibers (scanning electron microscope); (g–h) Fibers appear as massive elements with grooved lateral faces (g), irregularly imbricate laterally with a compact internal structure (h); (i) Fibers in a section perpendicular to their elongation, with a monocrystalline appearance (polarized light). f: Cuif and Dauphin (2005b).

Therefore, in spite of the geometric complexity of scleractinian corallites, the basal ectoderm of a polyp produces their calcareous structure in an analogous way to that of the external epithelium of the molluscan mantle. In both cases, the epithelium produces the carbonate structure towards its exterior (Fig. 1.21c: arrows). The complexity of the coral skeleton is derived from the fact that the undersurface of the polyp is geometrically complex as it develops in three dimensions. The surface of the external epithelium of the molluscan mantle, on the contrary, generally remains simple, although it shows undulations, as at the pallial border of the *Cardium* shell (Fig. 1.12). Therefore, the three-dimensional orientation of the microstructural components in coral skeletons is much more difficult to reconstruct than are the layers of the mollusc shell. At the time the mineralizing role of basal epithelium was discovered, the relationship between ectodermal cells and the mineralized skeleton was the object of a dispute between von Heider (1886), who considered that cells of the ectoderm were themselves calcified to form the skeleton, and von Koch (1886), who proposed that crystallization occurred externally to these cells. The views of von Koch prevailed. The importance of this controversy also derives from its date, because 1882 was the year when Pratz carried out the first microscopic observations on coral skeleton. This established the fibrous structure of coral aragonite and furnished some preliminary indications of the three-dimensional arrangement of aragonite fibers in septa, thus initiating the microstructural analysis of corals.

A septum, observed in a plane perpendicular to its direction of growth, reveals fibers that are grouped into clusters on both sides of a medial line (Figs. 1.21d–f). In each half of the septum, fibers are grouped in divergent bundles whose origin is at the margin of the medial line. The fibers themselves appear as elongated units with grooved or fluted morphology of their external surface (Fig. 1.21g). The fibers, whose transverse section can attain 4 to 5 μm (Fig. 1.21h), are formed of groups of fibrous subunits whose crystallographic properties are likewise homogeneous, so that they present an optically coherent image in the polarizing microscope (Fig. 1.21i).

1.5.2 Fibers and zones of distal mineralization: the existence of a microstructural sequence in coral skeletogenesis

Based on the observations of Pratz, the first truly complete microstructural study of the modern scleractinian coral skeleton was published in 1896 by M. Ogilvie. She established the diversity of the three-dimensional arrangement of bundles of fibers and the possibility of utilizing these as taxonomic criteria. For the description of coral microstructure, Ogilvie took the starting points of the bundles of fibers as reference points ("the points from which fibers diverge"), and she named these "centers of calcification." Starting with this purely geometric definition, she thus introduced a concept whose biological value remained in doubt for a long time. As a result, in the description of the coral skeleton that introduces the classification of the Scleractinia for the *Treatise on Invertebrate Paleontology* (the one still mainly used), Wells (1956) always used quotation marks when he spoke of the "centers of calcification." In effect, optical microscopy does not truly allow the imaging resolution necessary to establish their structural peculiarities, just as their specific chemical nature could only be established commencing with electron microanalysis of X-ray spectra or X-ray fluorescence. These recently acquired structural and chemical data have removed most uncertainty about "centers of calcification." Accordingly, in establishing the presence of a system of specific mineralization with multiple aspects at the summits of septa and walls, it was found that the first step in modifying the interpretation of their role must be to name them using more objective terminology.

1.5.3 The growing edge of the septa: specific mineralization and diversity

The genus *Lophelia* (Fig. 1.22a) contains scleractinian corals whose polyps lack algal symbionts (zooxanthellae); thus they can live at great depths. The majority of them have retained simple skeletons, with their septa forming compact blades. These have a summit line (also called the distal border) that is regularly incurved axially. It is at the position of this summit line that vertical growth occurs in septa (Fig. 1.22b: "gr edge"). On the lateral faces, growth striations are visible that reflect previous positions of the summit line and bear

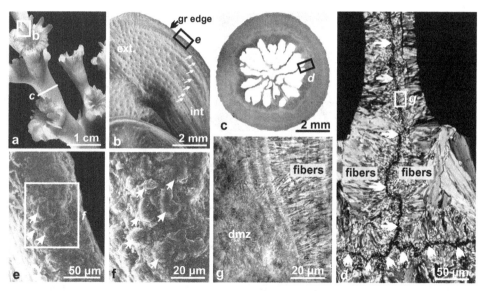

Fig. 1.22. The zone of distal mineralization and the surrounding fibers. (a–b) Individuals of *Lophelia* build branching colonies where the corallites are sharply distinct (a). There septa (b) are simple with a growing edge ("gr edge") that curves downward axially. On the lateral surfaces of the septa, the development of fibers covers and masks traces of preexisting growth edges (arrows); (c–d) In thin sections, traces of the superposed growing edges are easily identified (arrows) between the fibrous layers (c: natural light, d: polarized light); (e–f) In *Lophelia*, the specific mode of mineralization at the septal growth edge is well shown. The zone of distal mineralization of the septa ("dmz") is entirely formed of densely packed nodules (arrows) that are clearly different from the fibers forming the lateral flanks of the septa; (g) From a microstructural viewpoint, fibers are sharply distinct compared to the zone of distal mineralization ("dmz") with its granular mineralization in this section (SEM, etching; formic acid, 0.1% by weight, 40 seconds). a, b, e, g: Cuif *et al.* (2003a); c–d: Cuif and Dauphin (2005b).

witness to the regularity of growth (Fig. 1.22b: white arrows). These striations are easily visible in internal regions ("int") of the septa only, because in external zones ("ext") they are masked by the fibers that develop on both sides of the medial region. Thus, a section perpendicular to growth direction (Figs. 1.22c–d) shows the material formed at the growing edge symmetrically covered by the two lateral fiber layers (Cuif and Dauphin 1998, 2005).

Closer observation of this growing edge material reveals that its microstructure is sharply differentiated (Figs. 1.22e–f). Here nodules are observed with diameters of from 10 to 15 μm, and these are densely accumulated. A scanning electron micrograph of a section through this zone (Fig. 1.22g) establishes the microstructural peculiarity of this zone of distal mineralization even better. The closely associated nodules are formed of granular microcrystals, an observation indicating that septa of *Lophelia* result from the superposition of two distinct systems of crystallization. The first one, as a zone of distal mineralization (Figs. 1.22d–g: "dmz"), builds the septal growing edge, and the second forms the fibrous layers on both sides of the material produced by the zone of distal mineralization. During

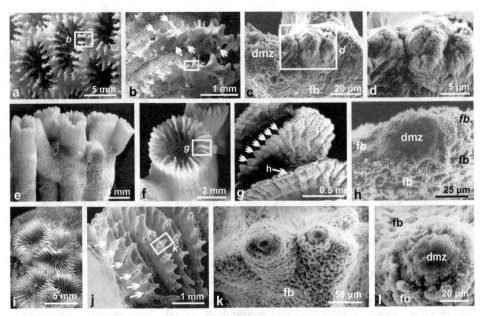

Fig. 1.23. Distribution of zones of distal mineralization on the growing edge of septa. (a–d) Corallites of *Favia stelligera* (a) have septa exhibiting a regular and continuous border (b), forming a continuous zone of distal mineralization (c), over which granular mineralization is easily visible (d). This zone of distal mineralization also gives rise to lateral axes, which are developed perpendicular to the plane of the septum (b: arrows); (e–h) In the septa of *Cladocora caespitosa* (e–f), the zone of distal mineralization is no longer continuous, but rather is formed of isolated regions at the summit of the spines on the growing edge (g: arrows). But each of these zones preserves its characteristic mineralization (h: "dmz"), sharply distinct from the fibers that developed around it (h: "fb"); (i–l) In *Favia fragum* (i) the zone of distal mineralization is also formed of disjointed centers, grouped in short series *perpendicular* to the growing edge of the septa (j–k). However, the specific structures of the distal mineralization surfaces are preserved (l). b–d: Cuif *et al.* (1997b); g, h: Cuif and Dauphin (1998); k: Cuif *et al.* (2003b); l: Cuif *et al.* (2005b).

vertical growth, the covering by fibrous layers of the structures produced by the distal mineralization zone thus forms a true diachronous microstructural sequence. This presence of a mode of crystallization unique to zones of distal mineralization can be extended to all of the Scleractinia, a generalization that can be verified by taking into account morphological observations on the distal border of the septa (the summit). A key point lies in the fact that the zone of distal mineralization only rarely shows the simple linear and continuous arrangement such as that conveniently shown by *Lophelia*. For example, in *Favia stelligera* (Figs. 1.23a–d), the zone of distal mineralization is equally continuous, but is only repre-sented by a single linear series of joined centers (Fig. 1.23c), readily characterized by their microgranular crystallization (Fig. 1.23d) (Cuif and Dauphin 1998). In this species there still is a correlation between the summit line of the septa and the zone of distal mineralization. Additionally, the zone of distal mineralization produces numerous offshoots, lateral axes that extend horizontally into the interseptal space (Fig. 1.23b: arrows).

In corallites of *Cladocora caespitosa* (Figs. 1.23e–h) the distal borders of septa form regularly spaced teeth (arrows). At the summit of each tooth, the zone of distal mineralization is reduced to a circular spot with a diameter of 30 to 40 μm ("dmz"). Here, this situation no longer corresponds perfectly to a zone of distal mineralization. Distal mineralization has now become discontinuous; it has been reduced to a series of small and isolated circles at the points of teeth on the summit line (Fig. 1.23g: arrows). The fibers developing around them are thus organized into radial systems, centered on the vertical axes resulting from the superposition of these circular zones of distal mineralization. This is the origin of the *trabecular* arrangement of coral fibers.

In *Favia fragum* a comparable arrangement exists (Figs. 1.23i–l); the zone of distal mineralization is equally discontinuous, formed of centers that are separated from one another. However, it can be seen in addition that these centers are no longer arranged in a simple linear series within the axial plane of septa, as they were in *C. caespitosa*. In *F. fragum*, zones of distal mineralization are grouped in short parallel series, with each of them oriented perpendicular to the septal plane (Fig. 1.23f). Here, the structural peculiarities of these surfaces of initial mineralization are readily visible (Figs. 1.23g–h). Frequently, crystallization here is not strongly developed (the opposite of that observed in *Favia stelligera*, for example). In certain cases, it can even appear as though all secondary crystal formation can be absent (see also Fig. 1.28l: *Porites*).

1.5.4 The stepping growth mode of fibers

Since the observations of Pratz (1882), aragonite fibers have been considered as the elementary crystalline components of coral structures. In a parallel way, taxonomic research has confirmed the views of M. Ogilvie (1896) and W. Volz (1896), the two pioneers of research utilizing three-dimensional arrangements of these fibers as tools for the classification of the Scleractinia. Thus, it is obviously necessary to have a biological mechanism that controls development of these crystalline structures.

After a comparison with radial crystal clusters formed by minerals precipitated from saturated solutions, Bryan and Hill (1941) suggested that the coral skeleton resulted from spherulitic crystallization carried out around the centers that had previously been recognized by Ogilvie. In this hypothesis, the centers of calcification take a direct role in organizing septal structure; it is their position that determines the orientations of fiber crystallization. Following Bryan and Hill, the growth of fibers remains the result of purely chemical precipitation. In the same way, Barnes (1970) stated that relationships between neighboring bundles of fibers are established by "crystal growth competition." It is true that radial crystallization is commonly seen in the coral skeleton, and that may fit with this interpretation (Fig. 1.24a). However, a simple experiment suggests a different view of the growth process in the coral skeleton. Beginning with two surfaces resulting from the fracture of a piece of coral skeleton, make an ultrathin section from one side of the break, and from the other side, prepare a polished surface and apply an enzymatic solution to this surface. In these two preparations, the arrangements of the fibers are practically identical due to their

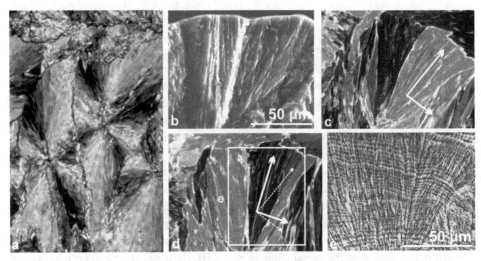

Fig. 1.24. Incremental crystallization of coral fibers. (a) Example of the spatial arrangement of fibers that served as a model for Bryan and Hill when they advanced the hypothesis of "spherulitic crystallization" of coral fibers; (b–d) Fracture surface of a corallite of *Montastraea* (b), here showing fibers in scanning electron microscopy. An ultrathin section made starting on the side of the previously illustrated fracture is here shown in polarized light (c–d). The optical coherence of the bundles of fibers is verified; (e) Traces of the incremental growth of fibers are here exposed by enzymatic etching. Note that the successive growth layers are in perfect continuity between neighboring fibers. The basal ectoderm of the polyp here controls the growth of fibers very precisely. b–d: Cuif and Dauphin (2005a); e: Cuif *et al.* (1997b).

close proximity where they were joined before fracturing. On the thin section, one can easily verify that the fibers are very clearly occurring as polycrystalline structures (Figs. 1.24b–d). But these same fibers, submitted to the action of an enzyme solution (Fig. 1.24e) reveal, within the interior of the crystalline structure, a fine growth zonation (Cuif *et al.* 1997a). Also, it is important to note that these growth lines, obviously marking successive stages of fiber development, are *in perfect geometric continuity between adjacent fibers*. This clearly indicates that fiber development cannot be due to a process of purely chemical growth where each fiber grows according to the space available, but on the contrary, it is a process strictly controlled by the functioning of the basal ectoderm of the polyp.

1.5.5 Architectural diversity versus similarity of growth mode: the overall layered structure of scleractinian skeleton

This method of observation, extended to the whole of the coral skeleton, establishes the separation of the two domains of mineralization (zones of distal mineralization and fibers) controlling formation of the septa and walls. This dual skeletogenetic process can be observed in all corallites by careful preparation of the studied surfaces; for example, the incremental growth of fibers, as noted by Sorauf and Jell (1977) in *Desmophyllum*, or granular micro-crystals in the distal mineralization zones, as noted by Gladfelter in *Acropora* (1982). This

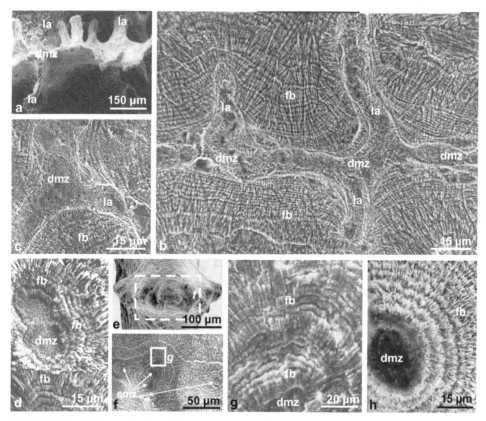

Fig.1.25 Controlled growth of coral fibers. (a–b) Morphology (a) and section (b) of the upper part of a septum in *Favia stelligera*. The control over growth of fibers by a cyclic mineralizing activity of the basal ectoderm of the polyp is made obvious by etching of the polished surface; (c) Microstructural continuity between the zone of distal mineralization ("dmz") and the lateral axes ("la"). The structure formed by the zone of distal mineralization and lateral axes is progressively covered by the incremental growth of fibers ("fb"); (d) Distinct zones of distal mineralization ("dmz") surrounded by incremental growth of fibers in *Cladocora caespitosa*; (e–f) Incremental growth of fibers in *Favia fragum*; (g–h) Growth stratification of fibers can be established by the action of enzymatic solutions (g: pronase), and also by weak acid solutions including a fixing additive (h: formic acid, 0.1% plus 3–4% glutaraldehyde). a–c: Cuif and Dauphin (1998); d, h: Cuif *et al.* (2005b).

approach facilitates precise reconstruction of the method of growth for any corallite, because it reveals the development of the skeleton *through time* at micrometer length scale. It also helps evaluate the microstructural diversity that is the apparent result.

Some examples will suffice to establish the precision with which the formation of the septa can be established by analysis of the functioning of the calicoblastic ectoderm.

In the septa of *Favia stelligera* (Figs. 1.25a–c), the zone of distal mineralization is practically continuous, formed by joined centers of mineralization, and also giving rise to lateral axes that extend horizontally into the interseptal spaces. On a polished surface prepared to show the incremental growth of fibers, one sees this zone of distal mineralization very clearly

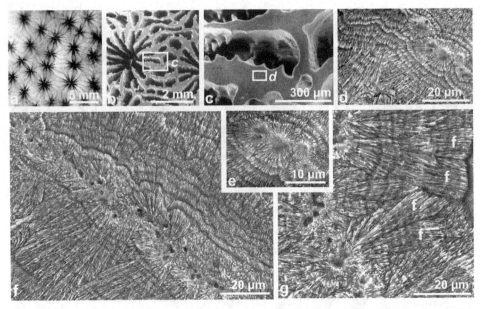

Fig. 1.26. Incremental growth of the skeleton of *Pavona* sp. (a–c) Morphology and arrangement of the septa; (d–f) In *Pavona* the zone of distal mineralization is formed by small, distinct centers (diameters less than 10 μm) arranged in a single linear series in the medial plane of the septum. They only form the crest of the septa, and the development of the fibers begins very early, with incremental growth readily visible. The growth of the complete septum is strictly coordinated on both sides of the medial line, demonstrating the overall control by the polyp; (g) Fibers ("f") are in reality bundles of fibrils with a diameter less than one micrometer.

(Fig. 1.25b: "dmz"), as well as the lateral axes that branch from it ("la"). The growth stratification allows observation of the developmental process – the covering over of the initial structure by fibers that form on both sides. One here verifies how the layered growth process controls the formation of the fibers at an overall scale. The latter do not develop simultaneously at all points of the zone of distal mineralization, as suggested by the concept of direct inorganic crystallization of the fibers. The first mineralized layers appear very selectively, in order to fill in the more deeply indented portions of the space between the median line of centers and the lateral axes. As for any mollusc shell, septa and the whole corallite are diachronous structures that can be understood by careful time-based examination at the micrometer scale.

This method of growth in the fibrous phase of the coral skeleton by superposed biomineralization of growth laminae also occurs in *Cladocora caespitosa* (Fig. 1.25d), and in *Favia fragum* (Figs. 1.25e–f), in spite of very different arrangements of their zones of distal mineralization. This apparently is a general characteristic of the second phase of skeletogenesis. Distinguishing zones of distal mineralization and fibers is the key to understanding septal development in the Scleractinia.

Pavona (Figs. 1.26a–c) has septa with centers dispersed in a linear zone of distal mineralization, and forms compact septa that are little differentiated on the microstructural

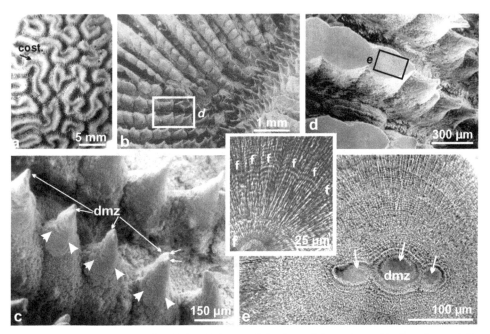

Fig. 1.27. Incremental growth of the skeleton of *Diploria labyrinthica*. (a) Overview of *D. labyrinthica*, illustrating the labyrinthine costae ("cost"); (b–c) Regular and conical teeth positioned at the summits of septa (b), with the zone of distal mineralization ("dmz") restricted to the tip of each tooth (arrows show the beginning of fiber production); (d–f) Sections illustrating the stepping-growth mode of fibers ("f"), here radially arranged (f) around complex "dmz" (e: arrows). c: Cuif and Dauphin (2005a); d: Cuif and Dauphin (2005b).

level. The zone of distal mineralization is generally linear, but differs from the preceding examples. The centers in *Pavona* are very small (less than ten micrometers in diameter) and sharply distinct from each other (Figs. 1.26d–f). Coordination of growth layers in the bundles of fibers is clear. Above all, one sees here that, even when fibers have growth directions opposite to one another, their development is insured by the crystallization layers that pass with perfect continuity from one fiber to the next (Fig. 1.26g). In spite of the appearance of "crystal growth competition," it is actually the basal ectoderm that determines their growth.

Diploria labyrinthica (Fig. 1.27a) possesses septa that, in their costal zone ("cost"), have a summit line that bears very regular conical teeth (Figs. 1.27b–d). The summit of each tooth is a center of initial mineralization that commonly shows two or three distinct elements. Around these centers, fibers are developed in a radial arrangement (Figs. 1.27e–f), clearly illustrating the concept of "trabeculae," axes surrounded by radial fibers. Preparation by etching to reveal growth layering shows the regularity of layer development around the central axes. Here again, the same preparation reveals the coordination of growth layers between neighboring fibers and the polycrystalline character of each of them (Cuif and Dauphin 2005).

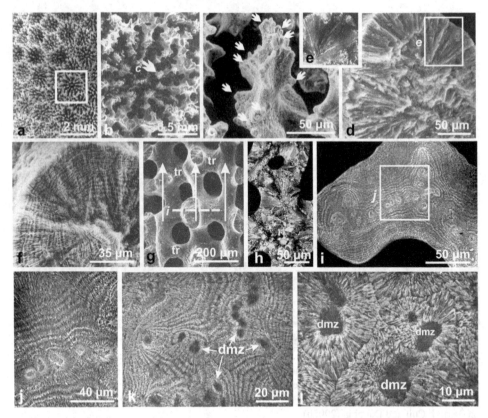

Fig. 1.28. *Porites* and the Poritidae. (a–e) The colony in its entirety is formed of distinct axes. At the summit of these axes, the "centers of calcification" are irregularly distributed (c: arrows). These axes retain a fibrous structure in radial bundles (d) of fibers, each with a monocrystalline appearance (e); (f–j) Stepping growth of the rods forming the skeleton. Sections transverse to the axis of growth clearly show the growth of fibers by continuously superposed strata (f). This mode of growth is also observable in the transverse skeletal elements that join the vertical rods and insure the solidity of the colony (g–j); (k–l) In the zones of distal mineralization, here either isolated or in small irregular groups, microgranular crystallization is commonly less well developed. The microstructural difference with the fibrous layers is then very much accentuated. a–f: Cuif and Dauphin (2005a); k: Meibom *et al.* (2007).

Species of *Porites* and related genera form one of the most important families among modern Scleractinia. This family builds colonies in which corallites are always small (Figs. 1.28a–b), but their essential characteristic is microstructural. In each corallite, walls and septa are formed of roughly parallel vertical axes. These axes have dispersed centers of distal mineralization at their upper extremity (Fig. 1.28c: arrows). Fibers develop to surround these zones of distal mineralization and assume a generally radial organization (Figs. 1.28d–f). The complete colonial skeleton is constituted of axes that can be compared to trabeculae. Their layered structure is due to zones of concentric growth, readily observable within the axes themselves, as well as in the transverse elements that join them to give the colony its rigidity (Fig. 1.28g: "tr"). In spite of the varied orientations of the fibers (Fig. 1.28h), the control exerted by the polyp on all of its structural elements is attested to by their structure of perfectly concentric growth layers (Figs. 1.28f and 1.28i–j). It should be understood by now that distinct centers of mineralization are seen in most axial regions (Figs. 1.28j–k). Commonly, precise observation indicates that they are themselves only weakly calcified (Fig. 1.28l).

1.5.6 A microstructural innovation: the fibers of the Acroporidae

The Acroporidae constitute the most abundant and diverse family of modern Scleractinia. This remarkable preponderance is partially due to several biological reasons (Veron 1986), but equally, this family is innovative in its method of skeletal construction. Like the Poritidae, acroporid colonies are formed of small corallites with millimeter diameters (the one at the growth axis being notably larger) (Fig. 1.29a). Instead of being laterally conjoined

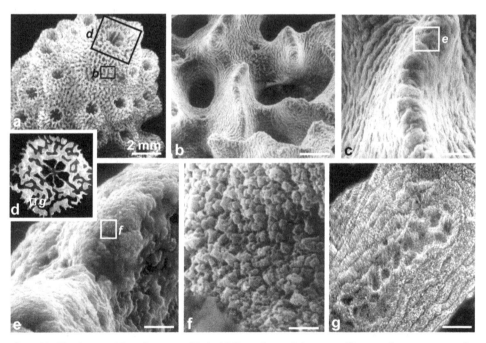

Fig. 1.29. The Acroporidae, *Acropora* (1). (a–b) Overviews of *Acropora* illustrate the occurrence of a larger axial corallite in the axis of growth of each branch; (c–f) The main part of the skeleton is constituted of anastomosing rods (c) in which can be seen the zones of distal mineralization (d–e), with a microgranular crystallization (f); (g) In the Acroporidae, in contrast to other scleractinian families, the second phase of mineralization does not produce fibers with a radial arrangement, but instead they form laminar structures that envelop the zones of distal mineralization (see Fig. 1.30f). a, c: Juillet-Leclerc *et al.* (2009).

Biominerals and Fossils Through Time

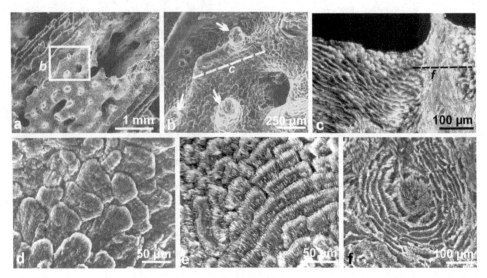

Fig. 1.30. Acroporidae (2): the laminar microstructure in the second phase of skeletogenesis in the Acroporidae. (a–c) Views at increasing magnifications to illustrate the arrangement of fibers in "scales" surrounding growth axes formed by centers of distal mineralization; (d–f) SEM views of fibrous laminae (d–e), and microscopically in cross-polarized light (f). In spite of the microstructural difference introduced by the formation of lamellae, the construction of the corallite by two superposed areas of crystallization is readily seen in the Acroporidae, and this is common to all of the Scleractinia. a, b, d: Cuif *et al.* (1996a); f: Cuif and Dauphin (2005b).

as in the poritids, the corallites are here dispersed in a skeletal mass that is commonly very extensive. All structural elements are constructed following the typical sequence of scleractinians. The zones of distal mineralization are developed as centers that are generally small (Figs. 1.29b–g) and grouped in ways that vary according to the species (Cuif and Dauphin 2005).

During the second phase of skeletogenesis, acroporids show their characteristic peculiarity. In place of bundles of fibers that develop symmetrically on both sides of the zone of distal mineralization, the second phase of skeletogenesis in acroporids produces fibrous lamellae that progressively envelop the distal structures by surrounding them by forming spiral systems (Figs. 1.30a–f; see also Nothdurft and Webb 2007).

The Acroporidae are known for their remarkable constructional activity, as their rates of linear extension are the highest among the Scleractinia. On this point, they surpass the Poritidae, who also build a porous skeleton. It is probable that this innovation in the method of mineralization, truly specific to them, is to be included in those properties by which the family has attained its marked supremacy within modern reef environments. Thus, an understanding of the evolutionary differentiation of genera and species benefits from this type of analysis based on knowledge of skeletal development. The first studies that are now applying these methods indicate the extent of these possibilities, not only for modern faunas but also for the study of the fossil record.

1.5.7 The dual microstructure of scleractinian corallites compared to the microstructural sequence of the mollusc shells

Comparison of the three-dimensional modes of growth for mollusc shells and coral skeletons reveals a surprising similarity between these two groups that are distant from a taxonomic viewpoint.

Similarity in overall mode of construction

As mollusc shells, the corallites of the Scleractinia are built by superposition of two distinct mineralizing domains, and the growth mode of the corallites always leads to covering of more distal structures by those resulting from the second step in the skeletal construction.

By studying the first stage of the polyp formation in *Pocillopora damicornis*, Vandermeulen and Watabe (1973) have made clear that the two-step mineralization process that can be shown in any corallite is directly related to the time sequence of its early mineralization, a few hours after metamorphosis of the swimming larvae. In the very young polyp, mineralization occurs over the whole outside surface of the basal ectoderm. The resulting subcircular mineral plate is then made of aragonite micrograins. Thirty-six hours after larval settlement, the mineralizing surface is reduced to radiating lines that build the embryonic septa. This truly initial mineralization of the septa is also made only of "small crystals or granules" (Vandermeulen and Watabe 1973, Fig. 10). Then, a second mineralizing process producing fibers is developed on both sides of the vertically growing granular septa, reinforcing the initial septal framework. According to Vandermeulen and Watabe, the first indication of this additional fibrous skeleton occurs 72 hours after larval settlement.

This description closely corresponds to the process of skeletogenesis as it can be reconstructed from adult corallites: the variously arranged distal mineralization zones ("dmz") at the growth edge of the septa are in direct continuity with the initial microgranular vertical layers forming the first septa in the first post-larval mineralization phase. The evolutionary processes that may have variously modified the respective positions of (i) the patches of microcrystals observed in the distal regions of septa, and (ii) the spatial arrangement of the fibrous structures developed on both sides, have maintained the sequential mineralizing mechanism.

In a different way than the simple superposition that occurs in mollusc shells (outer = external and inner layers), the geometrically complex superposition of the second domain (fibrous) to the initial one (granular) in the formation of coral septa is similar to what occurs in post-larval stages of the molluscs. It remains this way in coral families that have only undergone limited microstructural evolution (i.e., Fig. 1.22d *Lophelia* or Fig. 1.26f *Pavona*). In most coral families, modifications in the geometrical arrangement of the distal mineralization zones have led to their long-recognized microstructural diversity (see sub-section 7.4.2). Whatever its complexity, the basic structure of any corallite (solitary or part of a colony) results from the superposition of two distinct domains *simultaneously moving upwards* through a stepping growth process.

Similarity in the cyclical nature of mineralization

The first images drawing attention to a layered growth mode in the coral skeleton at the micrometer scale were presented by Ogilvie (1896: e.g., Fig. 19, p. 134), but no indications were given regarding continuity between the layers of adjacent fibers. In the context of her cellular interpretation of the formation of the coral skeleton, such notation might have been irrelevant. J. W. Wells also noted that thickening of dissepiments resulted from superposition of micrometer-thick layers (1956, Fig. 230, p. F 316), but no similar process was described concerning the growth of septal components. It is only by using scanning electron microscopy and etching polished surfaces that the presence of two superposed domains with distinct crystallization patterns has been readily shown.

As in molluscs, the mineralizing organ that produces coral skeleton – here the basal ectoderm of the polyp – functions as a whole, following a cyclic process during which the two septal components are simultaneously and repeatedly produced. As in mollusc shells, the two domains are geometrically superimposed in the skeletal structure – creating the microstructural sequence – because of the spatial separation of the mineralizing surfaces, after having been truly distinct in time during the post-larval formation of the initial skeleton. Some of the images above provide instructive examples of this synchroneity of the biomineralization process in the basal epithelium of a given septum (and possibly of the whole polyp). Here, each skeletal growth layer continuously surrounds a number of the previously formed layers (e.g., Figs. 1.25 to 1.27), creating a sort of very complex "cone-in-cone" structure (see also Fig. 3.18).

But etching studies of the layered growth process in coral septa also demonstrate that the basal epithelium controls the localization of mineralization with great precision. This approach to skeletal growth explains how modulation of the thickness of the mineralizing layers enables a given species to build the specific morphology of its corallite. In *Favia stelligera*, for instance, variation in the thickness of the growth layer may attain the almost complete reduction of mineral deposition within the internal part of the septa, producing the typical morphology: a thick and massive outer part of the septum, and a thin, branched internal part (Fig. 1.31). The ability to develop such local modulation of the mineralization

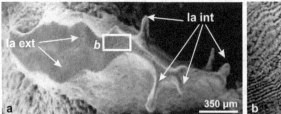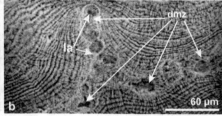

Fig. 1.31. Modulation of the layered mineralization of fibers. (a) Septum of *Favia stelligera*, showing the external region thickened by the deposition of thick growth layers, whereas they are very reduced in the internal part of the septum; (b) Although lateral axes ("la") have equal dimensions on the lateral faces, they remain visible in the internal area only (a: "la int"), whereas they are masked (but not absent) in the external part of the septa (a: "la ext"). b: Cuif and Dauphin (2005b).

process that results in the species-specific morphology of septa is, without doubt, convincing evidence that the growth of coral fibers is not controlled by simple "crystal growth competition" (Barnes 1970), but a genetically regulated mechanism.

1.6 Conclusion

The results of microstructural analysis establish without equivocation an unintended result concerning mineralized parts of the Mollusca and the Cnidaria Scleractinia. In the two cases, and in apparent contradiction to the crystalline aspect of their microstructural units, both shells and corallites exhibit a fundamentally stratified microstructure. Control of skeletal growth by the animal is carried out in the same way, by the superposition of isochronous crystallization growth layers, having thicknesses on the order of micrometers.

It is essential to realize that each micrometer-thick growth layer simultaneously contributes to the growth of different microstructural units that build the skeleton of a given species, whatever the diversity of these skeleton components. This implies that the mineralizing organs (the mantle of molluscs and the basal epithelium of corals) are composed of distinct areas. How can the complex and highly diversified microstructure-specific shells and corals be produced through this common growth process? Simply by a stepping displacement of the whole mineralizing organ. In this process, the structural units produced by distinct mineralizing areas located at the distal part of the mineralizing organ are progressively covered by structures elaborated by the more internal (in the mollusc mantle) or lateral (in the coral epithelium) zone of mineralization.

Therefore, time is a key parameter in the interpretation of the crystal-like microstructural units, and also the microstructural sequence itself. The basic time unit in the construction of any calcareous structure is the duration of the process of formation of common growth layers. In corals, for instance, the vertical growth of the septa leads the zone of distal mineralization located at the top of the septa to be covered by the fibrous domain. The covering of the internal surface of the calcite prisms in *Pinctada* or *Pinna* shells by the aragonite nacreous domains is strictly equivalent. In the gastropod *Haliotis*, three distinct areas can be identified by considering the morphology, the mineralogy, and the three-dimensional arrangement of skeletal units. In spite of such diversity, observation of common growth layers allows the progressive covering of preexisting microstructural areas to be readily recognized.

Some implications of such a cyclic mineralization process are already apparent.

(i) Contrasting with the static description of crystal-like units forming shells or corallites (the prism, the fiber, the crystal of nacre), it is now seen that these microstructural components (in the usual sense) can no longer be regarded as a single crystal that developed independently. There have been numerous attempts to influence the growth of calcite crystals by using various molecules so as to form "biomimetic" materials. Such experiments can provide useful information concerning the interaction of organic and mineral materials, but do not seem to be relevant to forming microstructural units in shells or skeleton.

(ii) Production of such distinct, micrometer-thick areas, with specific structural and mineralogic characteristics, appears to be difficult to reconcile with any concept of an "extrapallial" or "subepithelial" fluid (for molluscs and corals, respectively), the medium in which crystallization would supposedly occur. The existence of sharp limits between different minerals simultaneously precipitated is not compatible with the common liquid compartment hypothesis. In contrast, what we report here concerning the layered growth mode of calcareous materials both in corals and in molluscs fully supports Crenshaw's statement that a "direct transfer" is necessary, moving material from mineralizing epithelium to the growing mineral surface. Questions regarding the true nature of this transfer, the materials involved, and the mechanisms of the process now need to be investigated through more precise identification of the salient character of these crystal-like units and the mineralized growth layers by which they are formed.

2

Compositional data on mollusc shells
and coral skeletons

Mineral and organic components viewed from overall
characteristics to localized measurements

When Urey *et al.* (1951) raised the eventuality of biological influence on isotopic fractionation within a belemnite rostrum from the Isle of Skye, this hypothesis must have seemed
surprising because it concerned *crystallization* of Ca-carbonate, which was then regarded as
a physical process governed solely by thermodynamics. Nothing in the earlier crystallization
experiments carried out by McCrea (1950) allowed for consideration of possible biological
influences. Paradoxically though, if biological influence would then have seemed improbable at the level of isotopic fractionation, more than a century of mineralogical observations
and chemical measurements had already definitely established that crystallization of
the carbonate of shells and coral skeletons was precisely linked to taxonomy, and therefore
biologically controlled. It had been shown long before (Necker de Saussure 1839; Rose
1858) that molluscs produce both calcite and aragonite in a mode of deposition that is
specific to each family, and that in certain species these two polymorphs of calcium
carbonate are present within the same shell. Just the opposite, while corallites of the
Scleractinia are exclusively aragonitic (Sorby 1879), the axes and spicules of corals in the
Octocorallia are calcitic. It was understood, though, that the genus *Heliopora* builds its
skeleton of fibrous aragonite.

It was also known that variations in environmental conditions tolerated by each species,
and differences in physicochemical conditions which they can support, do not seem to
determine whether changes occur in the polymorph of calcium carbonate deposited.
Prisms of *Pinctada margaritifera* are always calcitic and its nacre aragonitic, whatever
latitude at which it lives. In the same way, when one looks at a group of higher taxonomic
rank, the Order Scleractinia, for example, it is seen that its mineralogy remains homogeneous even though it contains species distributed in very different environments (although
each species itself has only a limited "domain of tolerance"). Thus, scleractinian corals
that live at depths of several hundreds or thousands of meters construct their skeletons in
aragonite regardless, just as do the species of tropical reefs. This is true, even though at
these depths all aragonite particles originating at the surface have long since been
dissolved while descending by settling in the oceanic water column. Clearly, the mineralogy of these deepwater Scleractinia must be very strongly determined biologically in
order to persist in conditions so unfavorable to the formation of aragonite. From a
purely chemical viewpoint, production of less soluble calcite would appear to be an

advantageous adaptation. At a still higher taxonomic level, the tests formed by the Echinodermata have invariably been calcitic since the early Paleozoic, in spite of differences in the physical and chemical properties of the highly diverse environments to which these organisms adapted during their long history. In the middle of the twentieth century, nobody had doubts about the reality of the *biological control* over the mineralogy of shells and corallites. The unsolved questions at that time were about the way in which this control is exerted, and limits to the extent these controls can be effective versus the physical controls that normally drive the crystallization process. After half a century of investigation and production of a huge amount of literature on this question, it must be said that the controversy has not ended: no complete explanation can yet be provided concerning the link between taxonomy and crystallization.

While studying the distribution of mineralogical determinism in invertebrates with calcareous skeletons, Lowenstam (1954) drew attention to certain molluscan genera in which variation in temperature can induce variation in the *proportions* of the calcitic and aragonitic layers of their shell. *Mytilus*, for example, does this. Equally, tubes built by certain marine worms react in the same fashion. Even if there is no true mineralogical change in a given layer, but only variation in the calcite/aragonite ratio at a general scale, this modification indicates macroscopically that the mechanism of biomineralization is sensitive to thermal parameters. Beginning with the 1950s, research utilizing a great diversity of technologies focused on a more restricted topic: the incorporation of trace elements into calcareous skeletons as compared to concentrations resulting from coprecipitation experiments. Results rapidly converged to indicate a truly surprising behavior of biogenic carbonate crystals. Not only are compositional differences always observed between biological crystals and equivalent crystals chemically precipitated under similar conditions, but also, more paradoxically, it is seen that temperature changes can cause a contradictory change in the parameters measured. During the last half century, investigations were technically diverse and widely distributed from a taxonomic point of view. Information acquired to date has necessarily been fragmentary. However, owing to parallel progress in microstructural analysis and in the sensitivity and resolution of chemical measurements, at present no doubt remains regarding the reality of biological control on the incorporation of trace elements within biogenic Ca-carbonates. Now that instruments allow measurements to be made at a resolution of a few micrometers or less, the correlation between the stepping growth of calcareous structures and variations in chemical composition has become firmly established.

Of course, continuous analytical efforts of this type had an obvious objective: reconstructing paleoenvironments and climatic oscillations during the last thousands or millions of years, throughout the entire geological column. These investigations were primarily carried out with a chemical/mineralogical approach to biological specimens. What is remarkable (and to some extent rather disappointing) is that the investigations that focused on mineral chemistry had practically no relationship to research during the same period that was focusing on the organic components of biologically mineralized structures. Summarizing the results provided by half a century of investigation in these two practically separate, research areas, although dealing with the same objects, results in a complex and confusing panorama. What seems encouraging, however, is the fascinating progress of analytical techniques during recent years. With the instrumentation available at present, analytical resolution reaches the spatial dimensions at which the biomineralization process is operating, that is to say, at the micrometer and submicrometer level. For the first time, there is a real possibility that structural, chemical, and biochemical data can be truly correlated because they can be acquired from the same micrometer-sized sample of biomineral. Therefore, in this chapter, after briefly summarizing the results gained by pioneering investigators who discovered the multiple and intricate factors contributing to the complexity of biomineralization studies, emphasis here is placed on the most significant of these chemical and biochemical results. This will allow a clear correlation to be made between the structure, composition, and growth mode of skeleton building units.

2.1 Mineral compositions of Ca-carbonate from shells and corals

The minor elements normally associated with sedimentary calcite and aragonite, such as magnesium in calcite, or strontium in aragonite (Noll 1934), very often assume abnormal concentrations in carbonates of biological origin. Until the 1950s, techniques permitting the establishment of chemical compositions of biogenic carbonates remained qualitative or else relatively imprecise (such as evaluation of peaks on X-ray diffraction spectra). Staining methods, such as those proposed by Lemberg (1892) or Feigl (1937), were broadly applied, mostly by field geologists and paleontologists. Results are sometimes remarkable: Fig. 2.1 shows how precise the characterization of aragonite can be in the skeleton of a Triassic coral, a result that is comparable to mapping made by modern instruments (see Chapter 7, Fig. 7.24).

To obtain precise chemical compositions at that time necessitated the use of large quantities of material, and as a result, mineralogical and structural details could not be taken into account. However, Clarke and Wheeler (1922), Cayeux (1916), and Bøggild (1930) carried out extensive research on this subject. In addition to optically characterizing the microstructures of the different groups of the Mollusca, Bøggild identified considerable differences in magnesium concentration in various biogenic calcites and, to a lesser degree, in strontium concentration in aragonites. In 1957, Odum established experimentally that the strontium content of aragonitic structures of different species is influenced by its concentrations in the surrounding aqueous milieu. He also showed that this response varies by a factor of ten, depending on the taxonomic position of the species studied. Obviously, determination of the influence of temperature variation on incorporation of minor elements provides us with an extremely attractive prospect for paleoenvironmental applications. Followed by research on the effects of salinity by Lowenstam (1954), it was rapidly confirmed that organisms belonging to different taxa not only have different concentrations of minor elements present within their carbonate structures, but different sensitivities to identical environmental variations.

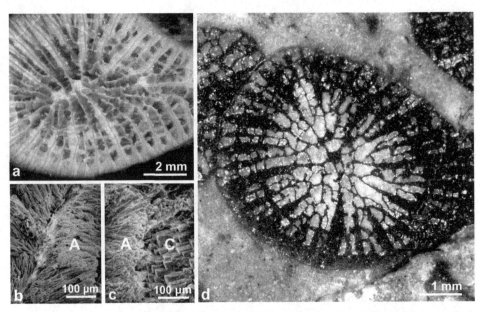

Fig. 2.1. Staining of a section of a Triassic corallite using Feigl's solution. (a) Polished section of the corallite; (b–c) Fibrous aragonite of a septum (b) and contact between fibrous aragonite and blocky calcite from diagenetic infilling; (d) Aragonite structures are black, a typical result of Feigl staining. Note the precision of this staining of the thinnest of skeletal structures (sample from the Lycian Taurus, Turkey).

2.1.1 The first syntheses dealing with trace element concentrations

One of the earliest of the post-1950 investigations, published by Chave in 1954, covered nearly all groups of geological interest in marine faunas. It only concerned magnesium concentration, but the multiple diagrams published here by Chave to indicate relationships between water temperature and magnesium concentration constitute the first fully documented study in an area that became a major theme of sedimentologic research. Historically, Chave's diagram (Fig. 2.2) clearly established that each major taxon has a unique slope illustrating the incorporation of magnesium into its calcite as a function of temperature variation.

Chave also evaluated the aragonite/calcite proportions in mixed shells by establishing ratios between X-ray diffraction peaks, and in shells with a mixed composition (calcite and aragonite), a single value for their concentrations was furnished. The report of a predominance of calcite in *Pinctada* (the "pearl oyster" well characterized by its heavy layer of nacre) shows how meaningless the representation of one simple parameter can be when dealing with generalized methods without microstructural control of the sample analyzed. As methods that are more precise have made it possible to measure chemical properties on much reduced quantities of material, the diversity of biomineral compositions has clearly been recognized, along with their observed differences from chemical equivalents within sediments. Turekian and Armstrong (1960) limited their investigation to molluscs, but

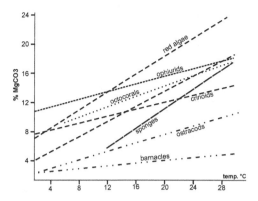

Fig. 2.2. Variation of concentration in Mg incorporated into carbonates in tests of different invertebrates (redrawn from Chave 1954).

expanded their investigation to strontium and barium. They carried out measurements on numerous specimens (several species of each genus), but did not separate the different microstructural components of the shells involved. Waskowiak (1962) included measurements of copper, boron, lead, manganese, cobalt and nickel, and the precise identity of the species analyzed, their mineralogy, and in shells with mixed mineralogy, he separated calcite from aragonite. During this initial phase of geochemical research, the magnesium content of mollusc shells was the most studied, from Vinogradov (1953) to Gunatilaka (1975). Additionally, variability in the incorporation of strontium was also the subject of analogous studies, producing results that definitely establish the taxonomic specificity of variations.

Ten years after the research of Chave, abundant documentation was available, as summarized by Harris (1965) and later by Dodd (1967). Utilization of Sr/Ca and Mg/Ca ratios in relation to temperature became a standard methodology, and practically became universal with the advent of automated instruments for measurement. In the long list of chemical elements whose concentrations were measured by Segar, Collins and Riley (1971) are found cadmium, nickel, and cobalt, etc., that later were established as being of use as pollution indicators. This work is remarkable for the precision of the identification of the species studied, although they did not take into account shell mineralogy. They here established differences between specimens with distinctly different geographic origins and even between the two valves of the same individual (for example, the nickel and manganese in *Pecten*). Segar, Collins and Riley (1971) also contributed to better appreciation of the importance of biological processes in this phenomenon that numerous researchers would much longer consider as simple coprecipitation. The classic synthesis presented by Milliman (1974) summarized the results of this early research at a time when progressive improvements in analytical techniques allowed for a major improvement in our understanding of the mineralogical diversity of biomineralized carbonates (Fig. 2.3).

In 1980, while reviewing measurements carried out during the first three decades of this research, Rosenberg raised some cogent questions regarding the possible origin of multiple difficulties that were becoming increasingly apparent at that time. His emphasis was on

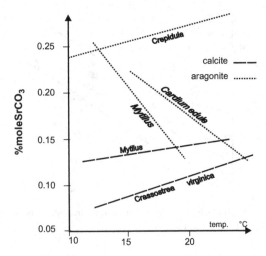

Fig. 2.3. Variation in Sr incorporated as a function of temperature in some bivalves and gastropods (Mollusca), redrawn from Milliman (1974).

insufficiently precise methods available for collecting samples of carbonate material for measurement. Very often, the main cause of the disparity between results – in fact, most often – is due to sampling layers that have different microstructures, but are not separated as different samples (even when they also have a different mineralogy!).

Also in 1980, the first publication by Masuda and Hirano (see Fig. 2.4) appeared and adequately confirmed that this methodology is a necessary one, pertaining to many different types of microstructures.

Equally important, Rosenberg stressed that, for a long time, statistics that had been thought to provide additional validity to numerical results were based on measurements obtained in a disorderly fashion, without taking into account possible variation during growth. In these conditions, results obtained were only rarely acceptable for comparison with data on modern organisms. It is no surprise that there was some divergence in the appreciation of this with respect to the determination of environmental temperatures during the life of a mollusc based on measuring concentrations of trace elements in its shell, an application already foreseen by Clarke and Wheeler in 1922.

A shortcoming that severely restricted carrying out the initial objectives of this research was quickly recognized. Far from ending with development of rules having general applicability over vast geographic regions and recognizable through geological time, investigations instead had to be limited to strictly empirical approaches while attempting to establish, *for a given species*, the correlations existing between known temperatures and measured concentration values in structures whose individual age had to be estimated. This is the process of "calibration" of species. In this calibration procedure, the composition of the seawater that forms the living environment for the animal evidently needs to be taken into account, and results obtained are only valid for the species studied.

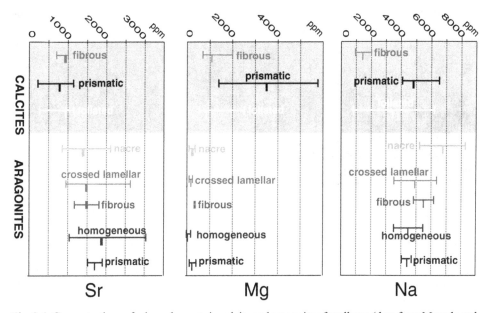

Fig. 2.4. Concentrations of minor elements in calcite and aragonite of molluscs (data from Masuda and Hirano 1980).

2.1.2 Evidence for biologically driven complexity: from calibrating a species to calibrating a shell layer

Electron microprobe analysis provided us with an initial response to the problem of localized chemical characterization. The spectrum of X-rays resulting from the interaction between the electron beam and a substrate is analyzed here in terms of wavelength (systems focus radiation by using the surface of a curved crystal, thus are called wavelength dispersive), or in terms of energy. In the latter case, radiated photons are caught by a diode, a small and lightweight device that has contributed greatly to the wide distribution of this method of analysis, as it can be easily mounted on a scanning electron microscope. Interestingly, through regular displacement of the sample, point chemical analyses can be transformed into maps allowing visualization of the distribution of chemical elements. Comparing these maps with views of the same areas obtained from secondary electrons provides access to a wealth of chemical information directly tied to the shell structure and growth process.

Compared to staining methods, not only are the characterizations much more diverse, but through a calibration process based on measurement of standards – well-known material generally similar in composition – this method is able to provide truly significant data from a quantitative viewpoint.

In the genus *Haliotis* (Fig. 2.5, also Figs. 1.14 and 1.15), images show that the two aragonitic layers (external and internal), with proportions varying according to species, occur together with a medial calcitic layer that separates them. Each shows chemical differences that correlate to specific microstructures (Cuif *et al.* 1989; Dauphin *et al.* 2007a).

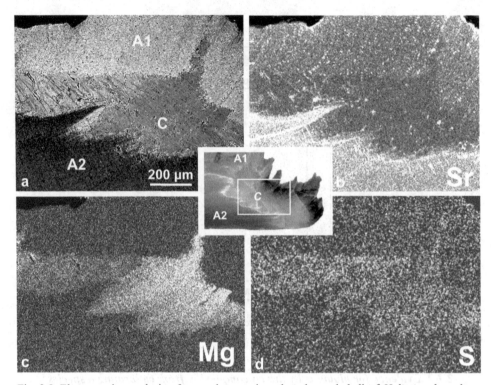

Fig. 2.5. Electron microanalysis of trace elements in a three-layered shell of *Haliotis tuberculata*. (a) Position of the sectors studied; (b) Distribution of strontium; (c) Distribution of magnesium; (d) Distribution of sulfur. Mapping by T. Williams, Mineralogical Dept., NHM, London; CAMECA SX100. a: Dauphin *et al.* (2007).

Measurements carried out on these three components confirm the specificity of crystallization in each layer very clearly. In electron microanalysis, the mapping approach remains qualitative, as residence time of the beam at each point is generally too brief to yield data with statistical precision sufficient for quantitative use. In order to obtain data of quantitative value, particular attention must be paid to the interaction between the beam and the material analyzed. During longer beam-residence times that permit quantitative measurements, penetration of the beam into substrate forms a "diffusion sphere" that reduces resolution by increasing the area analyzed. This point is particularly pertinent with respect to biogenic carbonates, with their highly variable structural properties. When this is taken into account and beam-residence times controlled, it is possible to determine the percentages of minor elements in mapped areas precisely if contents are sufficiently large, and when these can be compared to standard samples (Fig. 2.6).

In Chapter 1 (Fig. 1.14), we indicated that aragonite in *Haliotis* is crystallized in two distinct modes (granular and nacreous). Measuring their respective concentrations of strontium indicates that they are just as different on a mineralogic level. In this way, such measurements establish differences occurring on the same growth surface, meaning at the

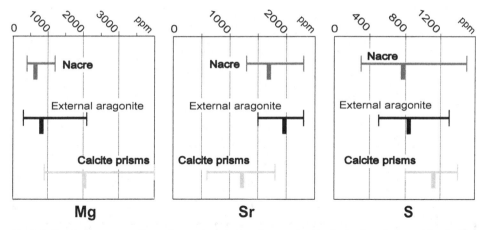

Fig. 2.6. Graph of concentrations of strontium, magnesium, and sulfur in the three layers of the shell of *Haliotis tuberculata*. Dauphin *et al.* (1989a, 2005).

same moment (in isochronous lines, illustrated on Figs. 1.14d and 1.14g), and in conditions that are identical from all points of view, the two aragonites crystallized differently. *It is not only the mineralogical nature of a microstructural component that determines the minor element concentrations found incorporated there.* Additionally, one should recall that, on a given growth surface, the two areas of aragonitic crystallization are separated by an area where calcite is being produced. The driving mechanism of crystallization thus operates with very great precision on the mode of crystallogenesis. Figure 1.14g shows aragonite granules several micrometers in diameter that are completely isolated and included within calcite crystals. This occurrence demonstrates that in each layer of biomineralization, the biologic control over crystallization exerts its influence at the submicrometer scale.

Chemical variation during shell growth: assessment of independent variation in concentrations of minor elements in the growth layer

Figure 2.7 illustrates a remarkable example of chemical stratification of prismatic calcite of *Pinna nobilis* that exactly reflects the layering of mineralization (see Fig 1.8). The reality of the biomineralization laminae as units of shell growth, already indicated by the micro-structural approach, is now confirmed by their chemical individuality.

This longitudinal section of prisms in the external layer of a *Pinna nobilis* shell illustrates equally well the *independent variation* in concentrations of sulfur and of magnesium (Dauphin *et al.* 2003a). Each layer is homogeneous for both elements, indicating that it has been formed as an isochronous unit, independent from its neighbors. Furthermore, proportions of magnesium and sulfur rapidly vary from one layer to the next. The homo-geneity *of a given layer*, combined with the rapid variation in concentrations of the chemical elements incorporated in successive layers, indicates that each of them is *one unit of crystallization* that varies from following units in recording different environmental con-ditions. The conjunction of these two properties clearly opposes growth of the prisms by

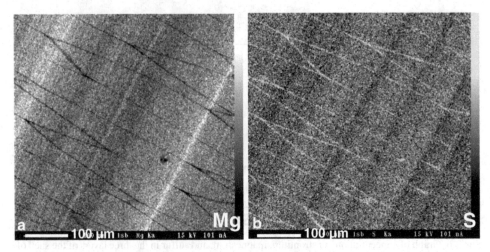

Figure 2.7. Variations in concentration of magnesium and sulfur during growth of shell prisms in *Pinna nobilis*. The prisms are cut somewhat obliquely, resulting in their elongated polygonal sections. Note that the noncorrelated chemical variations suggest that they are related to different aspects of the crystallization process. (a) Magnesium; (b) Sulfur. Maps produced by C. T. Williams, Analytical Dept., NHM, London; CAMECA SX50. a–b: Dauphin *et al.* (2003b).

crystallization in a space-filling fluid, the composition of which could not be so abruptly modified.

This example indicates the importance of detailed analysis of the distribution of chemical elements in order to establish relationships between environmental parameters. It also illustrates that overall measurements cannot help us reach this goal, except in rare and exceptional cases. This example also helps explain the variability in measurement that early researchers encountered, and provides us with a new analytical scale to establish true relationships between the chemical composition of tests and the natural environment. Taking into account the process of growth at a microstructural scale constitutes the appropriate level to establish relationships between the composition of the surrounding milieu and incorporation of chemical elements into carbonate structures. Maps of chemical distribution correlated with corresponding microstructural information thus provide records of the biological processes that successively formed multiple biomineralization laminae.

Specific composition of the zone of distal mineralization in scleractinian corals

Electron microanalysis of the skeleton of the scleractinians establishes that the zone of distal mineralization and succeeding fibrous areas show minor element concentrations that differ distinctly within the same specimen and also vary from one taxon to another (Fig. 2.8). Taking into account the widths of zones of distal mineralization and their placement (unique to each species, Fig. 1.23), microanalysis should be preceded by microstructural study to allow for precise placement of measurement locations.

Utilizing this methodology, significant compositional differences can be distinguished between the zones of distal mineralization and surrounding fibers (Cuif and Dauphin

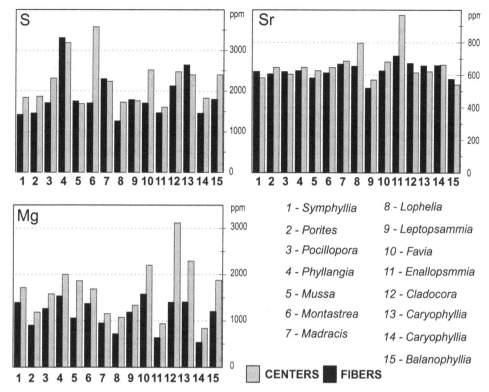

Figure 2.8. Differences in content of minor elements in zones of distal mineralization and surrounding fibers of coral septa in fifteen species of scleractinian corals.

1998). Chemical characterization of the zones of distal mineralization in corals confirms microstructural analysis. Chemical differences exist between the two types of crystals that build the septa: those in the zone of distal mineralization and those in the fibrous layers. These chemical differences confirm the morphological differences of the crystals that Wainwright had previously recognized in 1964 based on optical observations. On the other hand, when Gladfelter (1982) described groups of "tiny crystals" at the growth tips of septa in *Acropora cervicornis*, she did not use the expression "centers of calcification" and did not carry out chemical measurements on these crystals. However, there certainly is no doubt that they also correspond to the zones of distal mineralization (see Fig. 1.29f).

A chemical property is thus associated with the microstructural peculiarity of zones of distal mineralization. This association suggests once again that the surface of the basal ectoderm of corals bears two equal but distinct mineralizing domains, whose processes of biomineralization are sharply different. From a more general point of view, the microstructural and chemical heterogeneity of the aragonite produced in distinct areas of the mineralizing epithelium of coral polyps thus reinforces its functional similarity to the pallial epithelium in the Mollusca. On this level, both possess distinct zones of mineralization that operate simultaneously in a cyclical mode and produce mineral elements with specific forms and compositions.

For decades, analysts have been actively working to reduce the volume of sample size required to obtain reliable data from materials. Computer-assisted sampling (micromilling) represents the present state of the usual techniques involving the manual collection of powders produced by the sampling instrument, a small diameter drill. With this approach, a sample area of 0.1 mm diameter represents the dimensional scale accessed in the best case. In parallel, technological advances now allow acquisition of chemical data instrumentally that does not require the physical removal of samples (electron microanalysis), or in which they themselves do the sampling by erosion of the substrate through focusing a laser or ion beam of very small diameter. Material thus sampled is directly analyzed, most commonly using mass spectroscopy (secondary ion mass spectrometry, SIMS). These instruments have several advantages that are very important for accurate representation of minor and trace element distribution in biogenic mineralized structures. In the most part, these are "multi-collection" methods, and instruments can simultaneously carry out measurements of several elements; a major advantage for analysis of complex specimens in which composition can vary from point to point. Automation techniques also permit obtaining data while organizing points of measurement to produce a representative distribution map of chemical elements within microstructural units. Spectacular progress in resolutions possible has also resulted. Although methods of sampling by laser ablation still necessitate surfaces of several tens of micrometers in diameter, sampling by a focused ion beam permits measurement of chemical element content on surfaces having a diameter of less than 20 to 30 micrometers, and this can even be reduced to submicrometer dimensions in some instruments (CAMECA NanoSIMS).

In the skeleton of a reef-dwelling Caribbean scleractinian (*Colpophyllia* sp., Guadeloupe; Fig. 2.9a), the zones of distal mineralization form highly visible axes along the septal growth margin (Fig. 2.9b: arrows) and, additionally, show short lateral axes on septal flanks. The extremity of one of these lateral axes allows us to see the microstructural difference between the two structures; the granular domain forms the summit of the cone ("dmz"), surrounded by fibers ("f") (Figs. 2.9c and 2.9d). Measurement of the minor element concentrations in this coral (Meibom *et al.* 2006) utilized a SHRIMP ion probe (Stanford University). The instrument has a beam resolution of 15–20 μm, and data obtained here both confirm and extend the results of electron microprobe measurements.

Zones of distal mineralization and fibers show significant differences in their concentrations of various minor or trace elements (Sr, B, Ba, S). Additionally, as it is now possible to obtain separate measurements showing isotopic fractionation *in homogeneous microstructural areas*, the specificity of aragonite crystallization in the two different microstructural components of these corals is confirmed once again.

Figure 2.10 (magnesium and sulfur distribution maps) illustrates the presence of high-frequency compositional variation in successive mineralized growth layers that form the coral fiber, each of which has thicknesses of 2 to 3 μm. Until recently, no instrument could

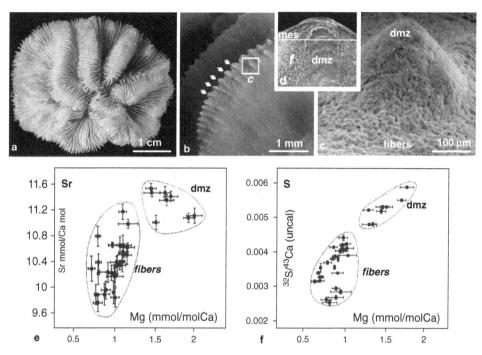

Fig. 2.9. Distinct chemical compositions of the distal mineralization zone ("dmz") and fibers in the scleractinian coral *Colpophyllia* sp. a–f: Meibom *et al.* (2006).

produce measurements having this resolution. In the evolution of measurement methods using secondary ion emission, the latest development in the series of instruments (conceived by G. Slodzian) responds to the resolution question. Thanks to sampling by an ion beam that can be focused down to a diameter of only 0.4 µm, the instrument CAMECA NS 50 now makes maps and measurements such that we can, for example, separate successive layers of crystallization that form during the growth of the *coral fibers*.

The curve shown in Fig. 2.10i illustrates the oscillations in magnesium concentration within successive laminae and suggests a specific role in the biomineralization process for this element (see also Fig. 3.13 and comment).

2.1.3 Patterns of isotopic fractionation

In 1951, when H. C. Urey and his collaborators carried out the first application of the paleothermometry method using a Jurassic belemnite, they took advantage of the particularly simple microstructure of this fossil: the massive conical rostrum (Fig. 2.11a) in which prisms of calcite are radially arranged (Fig. 2.11b). Thin sectioning or etching shows that the concentric growth layering is also preserved (Figs. 2.11c and 2.11d). Through regularly spaced sampling of carbonate material along a radius of the section (Fig. 2.11b), they were able to measure the $^{16}O/^{18}O$ and $^{12}C/^{13}C$ isotope ratios that had been recorded in the

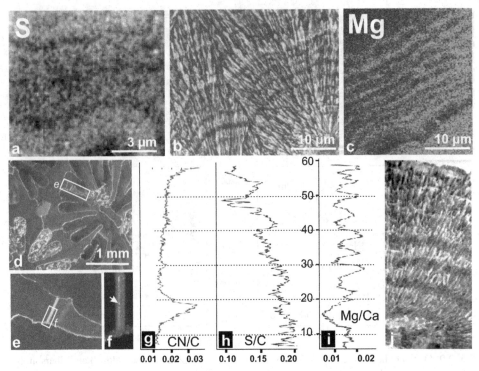

Fig. 2.10. Measurement and mapping of chemical elements at a micrometer resolution. (a) The first map of sulfur distribution within growth layers of a coral skeleton (made at the CAMECA factory (Asnières) by Hillion and Meibom using NanoSIMS); (b–c) Growth layers within fibers of *Porites* (b) and layered distribution of Mg; (d–f) Localization of a measurement transect within the septum of the coral *Pavona* sp. Section of a corallite (d); position of the transect (e) and trace of the measurement (f: arrow) in the middle of the "cleaning lane"; (g–h) Chemical signals collected during the transect: (g) CN/C, qualitative distribution of nitrogen; (h) S/C, overall change in sulfur concentration; (i–j) Regular pulses of Mg during layered growth (i). After measurement, etching was carried out and SEM allowed visualizing the layered growth mode (j) exactly at the place where the measurement was made. c–d: Meibom *et al.* (2004); g–j: Meibom *et al.* (2007).

superposed growth layers produced by this belemnite during its lifetime. In a few lines, they described precisely the living conditions for the animal that built this rostrum about 150 million years ago (see Introduction). Since this time, virtually every paleoclimate reconstruction includes measurement of isotopic fractionation.*

---

* In practice, one compares the value of the fractionation of the specimen to a reference standard. The properties of each specimen analyzed are expressed with respect to the standard under the form of a δ ratio (Ex: $\delta^{18}O = [(^{18}O/^{16}O$ specimen) $- (^{18}O/^{16}O$ reference)/$(^{18}O/^{16}O$ reference)]*1000. A specimen with a $\delta^{18}O$ positive carries at present more of this heavy isotope than the reference standard, and it is "deprived" if it carries less. In the laboratory of Urey, at the University of Chicago, the material utilized to establish this reference was a belemnite (*Belemnitella americana*) coming from the Pee Dee Formation of Upper Cretaceous age from South Carolina. This is the origin of the famous "PDB standard." The carbonate powder coming from this belemnite was completely used up a long time ago, but in relation to other standards, it is still the one that is referred to.

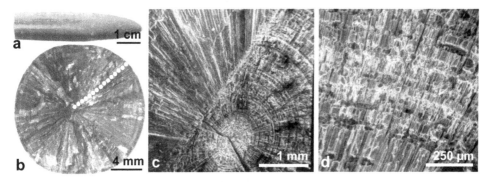

Fig. 2.11 Coexistence of radial prismatic and concentric layering in the rostrum of a belemnite. (a–b) Morphology of the rostrum (a) and a broken surface (perpendicular to the axis); (c) Half of this surface has been submitted to acidic etching; on the upper-left side the prisms are intact, whereas in the lower-right side, the concentric growth pattern is easily visible; (d) Closer view of the prisms; interpretation of their present status is a crucial step in assessing the meaning of measured chemical or isotopic values.

However, in reaching this completely new level of precision in environmental analysis, the authors of the paper truly focused on the central question of preservation of the information in the fossil records. They were obviously sensitive to this problem; the paper comprises a careful description of the rostrum prisms, allowing them to conclude that the chemical signal was a reliable witness of the temperatures in the Jurassic sea (see discussion in Chapter 7). Thus, at the same time, attention was drawn to a two-fold source of complexity: the hypothesized *vital effect*, leading to focusing on the mechanisms by which taxonomy-linked properties can be imposed on calcareous precipitation, overcoming thermodynamic rules; and the fossilization state of the specimen measured, the assessment of which depends in corals on the identification of its distal mineralization.

The "vital effect" question at the species level: empirical response

The influence of a possible "vital effect" was rapidly shown as being of fundamental importance. It was seen, actually, that biogenic precipitation of carbonates does not strictly conform to thermodynamic laws controlling inorganic precipitation. From shells produced by organisms that were kept at controlled temperatures, the first equations to permit the interpretation of measurements carried out on ancient shells were produced in 1951 (Epstein, Buchsbaum, Lowenstam and Urey), and later revised by the same authors in 1953. The impossibility of directly applying values of fractionation that were calculated and verified experimentally for synthesized nonbiogenic carbonates made an empirical method necessary, consisting of very exactly "calibrating" the species used, just as for the study of incorporated minor elements. This has resulted in producing innumerable "equations" that present exactly the same difficulties earlier pointed out by Rosenberg (1980) with regard to measurement of trace element concentrations. An example of the extreme care brought to establishing these calibration equations is furnished by a study utilizing planktonic foraminifera living in the North Atlantic Ocean today. This had as an objective the use of information acquired up to then to trace the evolution of seawater temperatures by interpreting sequences of fossils *of the same species* contained in oceanic sediments *from the same sector* of the North Atlantic (Elderfield and Ganssen 2000).

Their calibration began first with eight species of planktonic foraminifera, collected on a transect made between 30° N and 60° N latitude, with water temperatures ranging from 8 °C to 22 °C. For each data point, 20 specimens of each species were evaluated. From the point of view of modern analytical capacity, it is still considered that these 20 specimens suffice to establish both the relations of the content of $\delta^{18}O$ of the carbonates and the Mg/Ca ratio simultaneously. The authors compared these with the $\delta^{18}O$ of water collected at the same localities. This permitted them to express the difference between the water temperature where they had been collected and the temperature indicated by their isotopic composition. Even though these eight species of foraminifera produce calcitic tests, they are not all to be considered equivalent a priori. Just the opposite, the characters that reflect seasonal variation in temperature need to be verified for each species, even details such as thickening of the test at the moment of gametogenesis need to be determined. This last is a factor that over a very short time period changes the amount of carbonate secreted by the organism and, additionally, affects its recording of the temperature present during this time.

Starting with this calibration of high quality, due to the extreme care taken to account for the biological peculiarities of each species, values obtained for each scaled point of temperature varying from 8 °C to 22 °C were applied to specimens collected in sequence within cores. In this way, reconstruction of temperature changes over 28 000 years was possible (ages given by $\delta^{14}C$), thus permitting discussion of the modern oceanic configuration and changes in water mass movements in the North Atlantic, and in particular, displacement of the zone of intertropical convergence.

With the development of automatic methods for sampling and measuring, the construction of calibration equations such as these has been facilitated to a large degree, but without changing their completely empirical character at all. An example of the level of disparities observable is given by summarizing the traces obtained for individuals of the genus *Porites*, the scleractinian coral most commonly used for reconstruction of surface water temperatures in tropical regions (Fig. 2.12).

This graph illustrates the critical necessity of improving our understanding of biomineralization mechanisms and the origin of the vital effect. Discrepancies such as this, resulting from research of careful investigators, cast doubt on the possibility of obtaining reliable broadly applicable syntheses.

Variation of isotopic fractionation at higher taxonomic levels

Research carried out during the first decades after publication of the method showed that levels of the vital effect vary greatly between the principal groups of organisms that produce calcareous skeletons. Envelopes of fractionation possible for each group were assembled into one well-known diagram published by Swart (1983) that summarizes data collected between 1965 (Keith and Weber) and 1983 (Swart). Comparison of these two publications

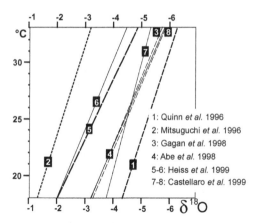

Fig. 2.12. Examples of calibrations obtained for $\delta^{18}O$ in corals of the genus *Porites* (after Watanabe *et al.* 2007).

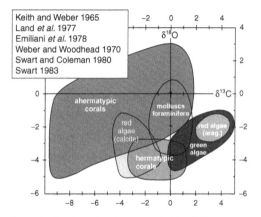

Fig. 2.13. Envelopes for isotopic fractionation data recorded in different groups of invertebrates with calcareous skeletons (redrawn from Swart 1983).

suffices to illustrate the disparity between results recorded during recent decades: values obtained during the use of different calibrations of coral skeletons belonging to the single genus *Porites* (Fig. 2.12), and those actually occurring in the domain of variability recognized for zooxanthellate scleractinians (Fig. 2.13).

By contrast, molluscs and foraminifera remain in a central position with respect to temperatures ($\delta^{18}O$), which has led to accepting that these organisms precipitate "at equilibrium" or nearly at equilibrium (Aharon and Chappell 1986; Elliot *et al.* 2003). However, research based on shells of molluscs living in controlled conditions shows results that are diverse and often contradictory. Thus, in the common bivalve *Pecten maximus*, construction of a shell in isotopic equilibrium with surrounding water is limited to the slow growth phase, whereas fractionation occurs during phases of rapid shell extension (Owen *et al.* 2002). In contrast to that, *Pinna nobilis* shows crystallization congruent with temperature only during youthful phases of development. Variation may also occur between the high and low temperatures measured and calculations from measured isotopic ratios. In *Chione correzzi*, for instance, a bivalve from the Gulf of California, Goodwin *et al.* (2001) reported that calculated and measured temperatures agreed during the summer, but diverged from recorded winter temperature.

In this group of investigations, it is apparent that, although the mineralogical nature of shells is precisely stated overall, their microstructural characteristics are only rarely pointed out. Questions of growth are defined only by external study of shell morphology, at best describing their ornamentation. In addition, analysts are evidently little inclined to draw attention to different properties seen in different microstructural shell components. However, this also seems to be the case in the study of minor elements; here the process of biomineralization itself affects fractionation of the stable isotopes within each microstructural type. As a result, determining the microstructure greatly improves the precision of the data that results.

From this point of view, it is clearly indicated that restudy of data from corals needs to include measurement points that are well located with respect to areas of differing microstructure in corallites (within both septa and walls), thus bringing a definite clarification of experimental results.

2.1.4 Isotopic fractionation as it relates to specific crystallization within microstructural sequences: examples from corals

Utilizing computer-assisted sampling techniques (the micromill), Adkins *et al.* (2003) studied fractionation of $^{13}C/^{12}C$ and $^{16}O/^{18}O$ in septa of a specimen of *Desmophyllum*, a deepwater scleractinian (Figs. 2.14a–b) that has retained a very simple microstructural plan in its septal structure. Septa are formed with a zone of distal mineralization that is very straight and continuous, and generally have a symmetrical fibrous layer on each side of it. Adkins used this microstructural detail to sample perpendicular to the central symmetry plane of septa. A very clear correlation was thus established between the different "reflectance" of the medial and lateral zones (Figs. 2.14c–d) and major differences that he observed in oxygen fractionation in these two areas. "Over a distance of 50 μm, there is a > 5 p. mil drop in ^{18}O" (Adkins *et al.* 2003, p. 1133).

It is noteworthy that, in this deep-sea coral, living in stable conditions, a 5 per mil difference between two points located at a distance of 50 μm is not at all related to temperature variation. A difference of a similar order of amplitude (even higher) was measured in the septa of a corallite of the genus *Lophelia* (Figs. 2.15a–b), also a deepwater scleractinian that lives on the continental slope off Norway. Measurements were carried out by secondary ion emission spectrometry on a CAMECA instrument IMS 1270 in Nancy (France).

The section of the measuring beam was somewhat elliptical, with a long axis of 30 μm (Fig. 2.15d). The remarkable difference in isotopic fractionation of oxygen between two very closely spaced areas within the same septum is not unique to deepwater

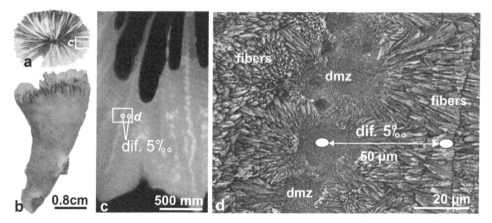

Fig. 2.14. Isotopic fractionation and microstructure in a septum of the coral *Desmophyllum*. (a–b) Morphology of the corallite; (c) Transverse section (perpendicular to growth axis) observed in reflected light. Measurements were made based on this difference in reflectivity; (d) Image in part of the previous one, here with electron microscopy. Differences in isotope fractionation are correlated to the distinct microstructures of the septum.

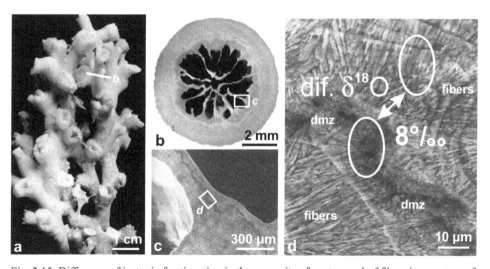

Fig. 2.15. Difference of isotopic fractionation in the aragonite of centers and of fibers in a septum of *Lophelia*. (a–c) Morphology of the colony (a): transverse section (b) and position of the median plane (= zone of distal mineralization at the top of the septum) (c); (d) Difference in isotopic fractionation of oxygen shown in respect to the difference in crystallization of aragonite. b: Cuif and Dauphin (2005b); d: Rollion-Bard *et al.* (2010).

corals (Blamart *et al.* 2002; 2005). This is present to the same degree in shallow-water, hermatypic (zooxanthellate) corals, so that, in a single septum of a corallite of the scleractinian *Montastraea* (Faviidae), $\delta^{18}O$ measurements show variation of 5–6 per mil (Fig. 2.16).

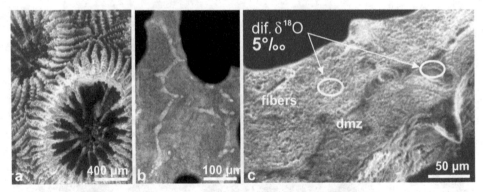

Fig. 2.16. Corallite of *Montastraea*. (a) Corallite morphology; (b–c) Trace of the zone of distal mineralization (b) and the isotopic difference between the two regions (c).

2.1.5 How to reduce scatter of measured values of isotopic fractionation: the example of Porites *(scleractinian coral)*

It was in the reef-dwelling scleractinian *Porites* that this degree of heterogeneity in fractionation was first measured, and it greatly exceeded what was explicable by the mechanisms habitually invoked for the explanation of this phenomenon. Figure 2.17 illustrates the importance of microstructure-based sampling, although it obviously can be very difficult or even impossible to do by using the usual method of drilling. This has been proved feasible in deepwater corals (Adkins *et al.* 2003) because they have only weak microstructural differentiation and a broad distal zone of mineralization, while septa of shallow-water corals have evolved to develop a great diversity of microstructural configurations (see Fig. 1.23 and Section 7.4). From a practical point of view, only the precise understanding of septal microstructure and positioning of the electron beam of the probe, based on this septal structure, can provide representative measurements. This likewise provides the fundamental objective of furnishing a realistic basis for the interpretation of scatter commonly observed in measurements resulting from the usual techniques.

Sampling by ion beam allows the successful study of differences in fractionation within scleractinians with microstructural characteristics that do not readily lend themselves to mechanical sampling, even when computer assisted (the micromill). This is particularly obvious for corallites in the genus *Porites*, a favorite for geochemical sampling, but also corals that have zones of distal mineralization ("centers of calcification") that are seen as irregularly dispersed, rod-like, small-diameter nodules that only rarely exceed 10–15 μm. Using a CAMECA IMS 1270 instrument, Rollion-Bard (2001) carried out a series of measurements on a longitudinal section of *Porites* sp., where she found that, in a field corresponding to vertical growth of 5 mm (Fig. 2.17), there was demonstrated a highly variable $\delta^{18}O$.

Porites skeletons characteristically have a microstructure of small vertical rods (Fig. 2.17b) formed by bundles of irregularly oriented fibers. Measurements carried out without preliminary microstructural study, with a beam having a long axis length of 25–30 μm, therefore result in an extremely large amount of scatter of isotopic

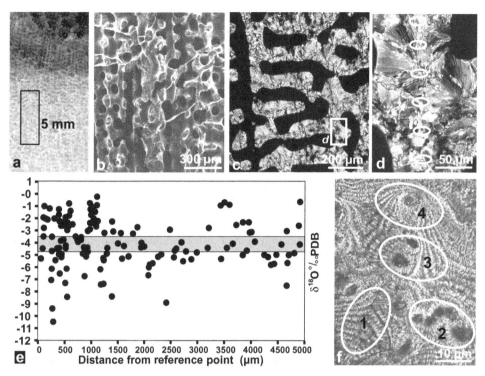

Fig. 2.17. Origin of scatter in results of isotopic fractionation measurements within skeletal carbonate of *Porites* sp. (a) Length of the sector on which measurements were made; (b–d) Microstructure of the *Porites* sp. skeleton. SEM view (b) showing the rods and connecting segments typical of this genus, and (c) crossed-nicols view of a thin section. Measurements were made, as far as possible, on the same longitudinal rod; (e) Black dots are the actual results; note the wide scattering of the value. Gray area is the value obtained by the usual drilling method (holes about 1 mm in diameter) (redrawn from measurements by Rollion-Bard 2001 and Rollion-Bard *et al.* 2003); (f) Among the successive measurements (spot 30 μm wide), variable proportions of "dmz" vs. fibrous tissue explain the broad scatter of measured $\delta^{18}O$. Conventional drilling method results in a reduced amplitude of the isotopic signals due to mixing of the distinct values. Detailed and reliable information can be collected by using fibrous structures only. f: Meibom *et al.* (2007).

measurements: about 10 per mil (Fig. 2.17e). Such a large amount of variation is quite surprising, compared to levels of fractionation ($\delta^{18}O$) usually measured when sampling is carried out by traditional drilling methods. In this case, measured values ranged between −3.5 and −4.5 ppm (the shaded field on Fig. 2.17e). Note that, under similar experimental conditions, measurements carried out on *Lophelia pertusa* (Blamart *et al.* 2002; 2005) clearly established that the $\delta^{18}O$ variance between fibrous regions and zones of distal mineralization ("dmz") also reached equivalent amplitudes (8–10 ppm).

From a practical point of view, it is noteworthy that these results from *Porites* sp. and *Lophelia* sp. were acquired under quite different conditions. Measurements carried out on *Lophelia* were done following study of the microstructure, and loci for measurements were located as a function of the microstructure. In addition, in *Lophelia*, the zones of distal mineralization are linear, continuous, and broad enough that a beam of 30 μm furnishes measurements *sharply separated* from those located in the fibrous regions. By contrast, the first sequence of measurements made by Rollion-Bard (2001) was lacking any previous microstructural analysis, but rather, measurements were made with a uniform spacing of points (Fig. 2.17d: spot positions). As a result of this method of positioning measurements without relating them to microstructure (Fig. 2.17f), the 30 μm electron beam was in part positioned on regions composed exclusively of fibers (Fig. 2.17f: "1") and some almost exclusively on centers (Fig. 2.17f: "2") or alternatively, on both in variable proportions (Fig. 2.17f: "3," "4"). Points of mixed measurement are evidently most numerous, and the fractionation values they furnish are necessarily combinations of values of fractionation characteristic of centers and exclusively fibers. This is exactly what is shown in Fig. 2.17e.

As a result, the dispersal of points as shown by the diagram in Fig. 2.17e *does not express* the actual variability of $\delta^{18}O$ fractionation in the *Porites* corallite. Most of them have been obtained on mixed beam-spots, including various proportions of the two distinct mineralizing domains, due to an inadequate sampling method.

2.1.6 Chemical forcing due to "crystallization efficiency" in aragonite (Cohen et al. 2006) as seen in the deep-sea coral Lophelia pertusa

The difficulty of accounting for the chemical properties of crystals resulting from biological processes is very well illustrated by two complementary studies concerning the incorporation of Sr and Mg in aragonite produced by the nonzooxanthellate scleractinian *Lophelia pertusa*. In order to integrate these signals into a purely mineralogical frame of reference, Gaetani and Cohen (2006) carried out a study in vitro of the coprecipitation of Sr and Mg in aragonites precipitated directly from seawater. In this case, variation in temperature determined a parallel modification of Sr and of Mg, *both decreasing in concentration as temperature increased*. Later, Cohen *et al.* (2006) compared the amount of these two elements incorporated in skeletal carbonate of an individual of *Lophelia pertusa*. The coral was collected at a site where the temperature had been measured for more than a three-year period (Tsiler Reef, Skagerrak, 129 m deep, average temperature variation: minimum 4.8 °C, maximum 8.7 °C). Microstructural study of the wall of this *Lophelia* specimen allowed reconstruction of its mode of growth, and a double series of measurements (transverse and longitudinal) were made, following the growth of wall carbonate over time. Several remarkable results were obtained.

(1) Experiments with the crystallization of aragonite in vitro (Kinsman and Holland 1969; Gaetani and Cohen 2006) have shown that variation in strontium incorporation was low in artificially precipitated aragonite (approximately −0.039 mmol/mol/°C) when resulting from thermal change. However, incorporation of Sr in the corallite wall of *Lophelia pertusa* is more than four times as great (−0.18 mmol/mol/°C). The thermal dependence thus appears stronger for this nonzooxanthellate scleractinian than for the zooxanthellate corals themselves. Crystallization of aragonite by this deepwater coral evidently is not influenced by the photosynthetic activity that causes accentuation of thermal dependence.

(2) Although in the experiments of in vitro precipitation, Sr and Mg show similar occurrences, in the aragonite of this *Lophelia* individual the amount of Mg and Sr incorporated has an inverse relationship to variations in temperature. In order to allow integration of these results into the geochemical model proposed by Gaetani and Cohen (2006) based on Rayleigh diffusion, Cohen *et al.* relied on the artificial attenuation of measured variations by utilizing a "filter" intended "to eliminate high-frequency variations." However, in spite of the elimination of "excessively" high and low values – a debatable methodology, as these values represent *real* records and obviously should be explained – Cohen *et al.* could not fit the partitioning of strontium in *Lophelia* aragonite into the nonbiological model of crystallization. This was resolved by admitting in their interpretation of the results that there was intervention of a factor named "precipitation efficiency," although their purely chemical model for crystallization could not furnish any indication of its nature.

The observations and measurements carried out in this analysis of the skeleton of *Lophelia* are interesting, not only because the deepwater corals are very important sources of information on oceanic circulation, but by the exceptional quality of the sampling and measurements. The method of sampling is very precise and trustworthy (ion probe CAMECA 3f), allowing a sampling area with only a 20 μm diameter. The measurements are equally well located according to microstructure. The section studied took as its reference line the zone of distal mineralization (primary nucleation sites) and the series of measurements were carried out towards the exterior. Thus they take into account exclusively the fibrous thickening around the calice properly stated (Fig. 2.18: "ect"). It is known that in *Lophelia*, polyps are positioned only at the end of each branch, and do not cover the external surfaces of colony branches (Fig. 2.18b). These surfaces are covered by a layer of mucus that flows out of the peripheral portion of the polyp that can also envelop organisms (worms, foraminifera) or grains of sediment that are attached to the colony (Figs. 2.18h–i). Freiwald and Wilson (1998) have demonstrated that the thick fibrous layer built around the calices of *Lophelia* is built by crystals developed within this layer of mucus. The rhythmicity of mucus secretion is related in part to seasonal nutrient variation, but equally, there is a very fine concentric stratification related to slower and faster flow of this organic phase.

The data reported by Cohen *et al.* directly measured features not primarily built by the polyp (walls and septa) in most part, but rather, on fibers that resulted from chemical properties of the mucus. This is truly a singular case of crystallization induced biochemically. Regardless, this series of analyses is interesting in a methodological sense because in spite of the objective stated by the authors, to integrate chemical variations into a purely thermodynamic model (Rayleigh diffusion), they were obliged to recognize the necessity of an additional factor that they named "crystallization efficiency." There could be no better way to illustrate the insufficiency of chemical models to interpret properties of these weakly controlled calcareous structures. Their fractionation level was far below the values noted by Blamart *et al.* 2005 on epithelium-controlled carbonate of the same species. This was also after the elimination of the extreme values in the analyzed series, and having exhausted all the resources of the purely chemical approach.

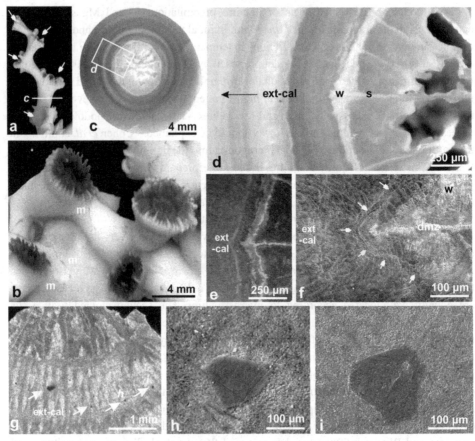

Fig. 2.18. Structure of the extra-calicinal wall of the corallite of *Lophelia pertusa*. (a–b) The polyps of *Lophelia* are completely separated one from the other and rest only on the border of the calices at the extremity of the branches. Between the calices, the branches of the colony are covered in mucus that forms a continuous envelope ("m"); (c–d) The mucus determines the formation of bundles of fibers oriented radially, *towards the exterior* ("ext-cal"); (e–f) Laser confocal fluorescence in the calicinal region (at right) and the fibrous thickening (at left). Note the decreasing fluorescence from the bright distal mineralization zones of septa and wall, the lowered response from skeletal fibers and the very low signal in the mucus crystallization; (g–i) The extra-calicinal origin of the crystallization medium (mucus) is attested to by the presence of many sedimentary grains (g: arrows) exclusively located outside the contact line between the corallite wall and the mucus.

2.1.7 *Contribution of trace element and isotopic data to understanding biogenic crystallization in corals*

It is a singularly complex series of concepts that have been progressively established by the analytical approach to biologically produced carbonates. Far from having developed and grown in precision to keep pace with the remarkable progress in developing analytical techniques, application of experimental results to natural biological materials has only

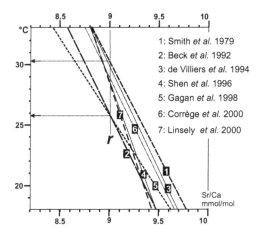

Fig. 2.19. Variation in the relationship Sr/Ca vs. temperature in the skeleton of the genus *Porites* (redrawn from Watanabe *et al*. 2007).

produced a bewildering biological diversity. Even at the macroscopic scale of sampling (such as was practical in the first decades of this research), it was rapidly apparent that simple multiplication of numbers of measurements carried out would confer no additional credibility to results obtained. In the single genus *Porites* (as an example), recent calibration of data gathered regarding the incorporation of strontium has resulted in a large number of different slope values (as shown in Fig. 2.19).

It is noteworthy that all of the authors here referred to have operated using the traditional empirical method, with the properties of biological carbonates interpreted as purely chemical. Establishing these multiple "equations" thus represents the ultimate result of the empirical methodology that began more than 50 years ago. No insight that can lead to the resolution of long-recognized difficulties can be developed simply from this approach without methodological improvements, even by carrying out a "multiproxies" approach. Lough (2004), after detailed comparison of a series of measurements obtained on scleractinian corals, suggested the development of a strategy to improve the dependability of data collected from these environmental records in biogenic, calcified skeletal material. As previously stated by Rosenberg (1980), it is critical to take the biological properties of these "archives" into account prior to using them as a source of numerical data. Surprisingly, it was beginning at this time that technological progress finally began to allow a response to Rosenberg's conclusion. While taking into account the methods of biomineralization unique to each group, one can develop the new strategy recognized as necessary by Lough.

From a methodological viewpoint, comparable results provided by conventional methods (Adkins *et al*. 2003) and measurements carried out on the SIMS instruments (Rollion-Bard 2001; Blamart *et al*. 2002) clearly indicate that each time these methods are applied to well characterized microstructures, *the amplitude and consistency of signals collected are greatly increased*.

The 1980 Crenshaw statement is also valid for coral biomineralization

This result is very significant for the determination of methods of biomineralization in corals. In traditional geochemical representations, from McConnaughey (1989a, b) to Sinclair and McCulloch (2004), the extracellular "subectodermal space" in which mineralization apparently occurs was conceived as a fluid zone whose composition is "close to seawater."

Depending on the author, variable attention is paid to the organic phase and its possible role, but crystallization itself always occurs within this intermediate space, sometimes thought of as being in a direct and continuous relationship with seawater (Fig. 2.20b). Noticeably, even the true physiologists among these authors in the end assume that crystallization itself occurs as a simple chemical process. Sometimes they suggest a role for organic matrix (Fig. 2.20a), without asking the question of how it is that such sharp boundaries between distinct mineralizing areas can persist within the common fluid layer.

In contrast to these diagrams that propose a single region for skeleton formation, the two-step mode of growth displayed in the post-larval stage of formation of the coral skeleton by Vandermeulen and Watabe (1973) has been shown to be a general characteristic of all coral skeletons. In these two domains that later become superposed during the growth process, differences in trace element partitioning and isotopic fractionation between fibrous and distal growth zones provide chemical evidence that these two distinct regions that are visible in microstructural analysis actually correspond to two distinct crystallization processes. Owing to the sharp limits observed between distal mineralization zones and surrounding fibrous tissues, whatever the diversity of microstructural patterns resulting from evolution, the common liquid infraepithelial (or subectodermal) compartment cannot any longer be accepted as a working hypothesis to explain these two very distinct and contiguous crystallization areas.

To account for both microstructural and chemical patterns, models of coral biomineralization have to account for Crenshaw's (1980) observation, formulated for crystallization in the mollusc shell. Between the mineralizing epithelium and the growing mineral surface of the skeleton, "the transfer of material is essentially direct."

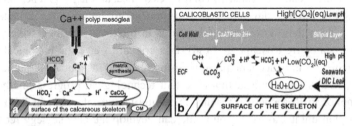

Fig. 2.20. Ca-carbonate crystallization in subectodermal spaces hypothesized below the coral polyp. (a) This model (redrawn from Furla *et al.* 2000), pays attention to organic matrix (its synthesis and transfer to the skeleton), but has crystallization occurring in the fluid layer, supposedly homogeneous and thus not explaining the simultaneous occurrence of distinct and contiguous crystallization of "dmz" and fibers; (b) A purely geochemical model (redrawn from Adkins *et al.* 2003) in which the liquid layer is provided with a direct link to seawater (DIC leak). No explanation is suggested for the double crystallization of "dmz" and fibers.

2.1.8 Mineral ions: partitioning and isotopic ratios in the Mollusca

What has been established in the corals is equally true for the Mollusca. Shell micro-structures are much more diverse by far among molluscs than in corals, although one sees a much narrower range of variation concerning chemical partitioning and isotopic fractionation. Reexamination of methods of coprecipitation of Sr and Mg ions in the chemical crystallization of carbonates characterized the decades of the 1980s and 1990s (from Mucci and Morse 1983, and Ohde and Kitano 1984, to Carpenter and Lohmann 1992). Using methods comparable to those initiated by preceding generations of researchers, the chemical standards thus established were compared to new measurements – and these are presently carried out with very great care in their application – taking into account details of growth mode and using instrumentation focused nearer to the life size of the organisms. This now allows establishing precise correlations between environmental conditions and the geo-chemical records within shell materials. Frequently, research has been simultaneously carried out on chemical signals (Sr/Mg partitioning) and isotopic fractionation ($\delta^{18}O$ and $\delta^{13}C$). Since this research is carried out very precisely, such analyses commonly establish properties of distribution or fractionation intensity that then oblige researchers necessarily to permit one or more additional mechanisms to intervene in the crystallization of biogenic carbonates. The manner of their action is presented, as in the case of the study by Klein, Lohmann and Thayer (1996a) that focused on variation in Mg/Sr and Sr/Ca concentrations within the calcitic layer of *Mytilus trossulus (= M. edulis)* on the British Columbia coast (Cortes Island, Georgia Strait).

This work of Klein *et al.* illustrates simultaneously the experimental rigor of their study and their extremely exact control of the growth of the organisms studied. In the course of a year, mussel shells (*Mytilus*) were artificially kept at a depth at which they were not subjected to temporary tidal emersion, and they produced 160 to 180 growth lines over the year; thus these can be considered as "daily growth bands". For this species, in effect, growth is interrupted when temperature drops below approximately 6 °C. In order to date the stage of formation of shell in their specimens, the authors established a correlation between measurements of $\delta^{18}O$ and a "predicted isotopic composition" for the calcite. Measurements were carried out on 80/100 mg of calcite. Obtaining this necessary measured quantity evidently represents different lengths of time, according to whether they came from the axial zone of the shell (with maximal growth speed) or from their lateral borders. Each sample thus represents "one day to two weeks of shell growth."

After having established a mathematical expression to summarize the amount of magne-sium incorporated as a function of temperature, the authors compared it with indicated temperatures derived from $\delta^{18}O$ variation. They summarized perfectly the amount of offset between these experimental results and those of theoretical approaches based on the experi-ments of coprecipitation, noting that, "If our mussels had been collected without prior knowledge of temperature and seawater chemistry at Squirrel Dowe, and if we had assumed that they grew in normal seawater (i.e. constant salinity at 35 p.p.mil.) isotopic temperature estimates would be 10 to 18° higher than observed." By contrast, the incorporation of magnesium, known to be much less affected by the flux of fresh water than isotopic fractionation of oxygen (Livingstone 1963; Drever 1988), furnishes positive results with respect to these objectives of the authors. However, differences remain between temperature observed in the field and interpretation of Mg/Ca ratios recorded in the shells of *M. trossulus*. Surprisingly, no correlation was established between the Mg/Ca and Sr/Ca ratios (Klein *et al.* 1996b). The highest values of Sr/Ca (and concurrently those of $\delta^{13}C$) were found in the area of maximal extension of the shells (= ventral in Fig. 2 of the Klein *et al.* paper), so that reciprocally, border areas to one side or the other of the growth axis show the lowest values in Sr/Ca.

In order to explain this distribution of the Sr/Ca ratio, which was expected to remain identical at all points within the shell (since the authors noted on page 4215 that the extrapallial fluid is "chemically similar to seawater"), they therefore considered that these localized differences tied to shell geometry are due to "the metabolic activity" of the mantle. This concept of metabolic activity came from the research of Rosenberg, Hughes and Tkachuk (1989) that established variations (in *Mytilus edulis*) of glucose consumption in different regions of the mantle, where a gradient can be established between the area of maximal shell extension and lateral areas in which there is less growth. Contrary to what one might expect, it is precisely in this axis of maximal growth that mantle metabolism is weakest. They proposed a model in which the mantle is considered as the regulator of Sr/Ca concentration, while forming a hypothesis that, by two different pathways (intracellular and extracellular), the mantle is capable of locally modifying the extrapallial fluid composition and, consequently, the composition of calcite that precipitates within it.

It is remarkable that this interpretation was developed in the traditional framework of purely chemical crystallization. Their conclusion concerning the possible influence of metabolic activity offers a striking parallel to Cohen's hypothesis of "biological forcing" to explain trace element concentrations in *Lophelia*. What is still more surprising about this interpretation of the *Mytilus* shell is that their hypothesis of metabolic activity was presented about 30 years after the remarkable publications of Travis on the presence of a fine organic network within the fibers themselves (see Section 2.2 below). Even though the model proposed by these authors is based on simple chemical influence by the mantle and not by its mineralizing function through an intermediary of an organic matrix, the extremely detailed analysis carried out by Klein *et al.* does however constitute a convincing demonstration of the existence of a true "vital effect." Through numerous examples, we have seen in Chapter 1 that between the different crystallizing areas that build a mollusc shell, boundaries are as sharp as they are in coral skeletons between distal mineralization zones and fibrous lateral layers. However, schemes based on the existence of common crystallization domains are still being used, even in biologically oriented investigations (Fig. 2.21).

Comparison of Figs. 2.20 and 2.21 that represent the commonly accepted models of "crystallization spaces" for coral and mollusc shells, respectively, reveals that these figures both lack the ability to account for the microstructural or chemical/isotopic properties of the microstructural sequences specific to these skeletons. Some parts of this scheme are disproved

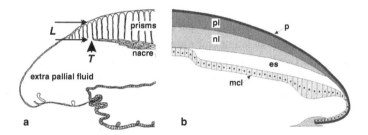

Fig. 2.21. Examples of models of common fluid chambers for the simultaneous crystallization of structurally and chemically distinct domains. (a) On this figure (redrawn from Petit 1978), the length of prisms ("*L*") is continuing to increase long after covering of their internal surface by the nacreous layer ("*T*"). No indication is provided to explain the way by which material is added to the internal surfaces of prisms, allowing for their continued growth; (b) On this sketch (redrawn from Volkmer 2007), although the dimensions and morphology of the common crystallization space are attenuated, the scheme is essentially similar.

by basic microstructural observations, such as the growth of the prisms forming the outer layer of the shell after their internal surface has been covered by nacreous layers.

High-resolution imaging techniques now allow *in situ* characterization of the biochemical compounds associated with each of these microstructural domains. The resulting biochemical mapping, when correlated to microstructural data, provides additional evidence that makes these schemes even more inadequate to account for the biomineralization process as it operates on macroscopic scales.

2.2 Characteristics of the organic components extracted from mollusc shells and coral skeletons

The presence of an organic phase in mollusc shells has long been recognized, beginning in 1855, when the term *conchyolin* (Frémy 1855) was first used to designate this nonmineral component. In early research, the components were called "scleroproteins" because analyses were limited for evaluating the insoluble compounds resulting from complete decalcification of shells. Then, in 1856, Schlossenberger established the glycol-proteic composition of these organic compounds by comparing their nitrogen content with that in reference compounds, such as chitin. The observation of these organic structures in optical microscopy furnished almost no data on their structure, although their study in polarized light allowed the establishment of birefringence in some of them (Schmidt 1924). The usual histological stains (carmine, picro-carmine, hematoxyline), while coloring them strongly (Moynier de Villepoix 1892; Schmidt 1924), did not bring further structural information to their simple identification by optical microscopy.

In contrast, beginning in 1955, Grégoire, Duchateau and Florkin carried out a long series of observations by electron microscopy on the "proteic textures" originating within crystals of nacre and prisms of some bivalves, gastropods and cephalopods, thus revealing specific morphological states (fibrous structures, different types of porosity) that made their recognition possible in fossils. The resistance of these insoluble organic structures to the process of fossilization allowed them to persist in sedimentary rocks for a long time. Grégoire even succeeded in isolating and illustrating in the transmission electron microscope, membranes derived from nacre of Paleozoic molluscs (Pennsylvanian, Upper Carboniferous, 290–325 Ma).

It should be noted that at this time, these materials were considered as being "extracrystalline" components of calcified structures, practically speaking, envelopes surrounding mineral components, but an active role in calcification had also sometimes been shown. Watabe and Wilbur (1960) thus utilized these insoluble materials isolated by decalcification of nacre both in fresh water (*Elliptio complanatus*) and marine bivalves (*Pinctada martensi, Atrina rigida*). They succeeded in inducing formation of aragonite by inserting them beneath the mantle of shells that only produced calcite (*Crassostrea virginica*). They applied the term "*matrix*" to them; this quickly came to be the generally used term.

Contrasting this rapid recognition of the presence of organic compounds in mollusc shells, the presence of a nonmineral component in coral skeletal material has remained a matter of discussion for a long time. For numerous analysts, the existence of an organic phase associated with "centers of calcification" has been considered a possibility, because the zones of distal mineralization seen in thin sections of septa have an opaque aspect commonly attributed to the accumulation of organic material. However, the fibrous areas (of far greater quantitative importance) are very commonly, even today, considered as exclusively mineral, resulting from simple chemical precipitation from a saturated solution.

In addition, regarding simple quantitative aspects, estimates concerning percentages of nonmineral components of coral skeletal material have long been confined to remarkably small values. One estimate commonly used was that organic phases represent 0.1% of total weight. This has been the value utilized starting with Wainwright (1964) and continuing until very recently. Concerning this very basic, but long-controversial point of view, thermogravimetry, a method based on a very simple principle, furnishes incontestable data about the respective weights of mineral and nonmineral components of shells and coral skeletons. Infrared absorption spectrometry additionally provides the essential advantage of being applicable without any specific preparation, except for careful cleaning of specimens and checking their status with respect to microstructure. Results of research using these techniques significantly modify traditional views on the proportions of nonmineral materials within biogenic carbonates, and in particular, in coral skeletons.

2.2.1 Measurement of the organomineral ratio by thermogravimetry

The sample, under the form of a calibrated powder (having particle diameters of 40 µm), is placed in a suspended basket and heated progressively in an oven. Loss of weight is continuously recorded. An inert gas (argon or nitrogen) is generally used to replace air in order to limit the influence of oxygen on reactions during heating. When pure aragonite is studied in this fashion, no changes are recorded below 650 °C, the temperature at which weight loss begins. Then decomposition is very rapid, beginning at 800 °C, and being complete at 900 °C.

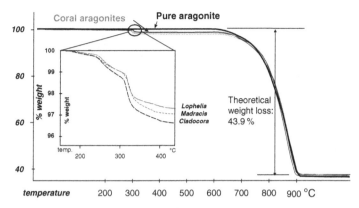

Fig. 2.22. Traces of weight loss during heating for pure aragonite and for three coral aragonites.

Three coral aragonites were sampled from the upper part of specimens collected alive and carefully cleaned of organic tissue by sodium hypochlorite; specimens in which the absence of perforating endobionts was verified by observations in the SEM. These have furnished decomposition curves which are generally comparable (Fig. 2.22). At about 300 °C there occurs an inflection of the trace, indicating the decomposition of a thermally sensitive material (Cuif *et al.* 2004).

The sensitivity of modern-day instruments allows these inflections to be followed with great precision, and the three coral species (*Lophelia pertusa*, *Madracis pharensis*, and *Cladocora caespitosa*) do display some diversity in their nonmineral components at this initial level of observation. For the three species, the loss of weight is much greater than the 0.1% generally cited. By approximately 400 °C the greatest part of the thermally sensitive material has disappeared, with a resulting reduction in sample weight of approximately 3%.

It is also now possible to acquire data on the nature of products emitted during heating. In effect, the gases resulting from the decomposition of coral material can be characterized by their infrared absorption spectrum. In experimental thermal decomposition of coral material, two significant wavelength absorptions are used: $1508\,cm^{-1}$ allows the study of water molecule emission, and $2363\,cm^{-1}$ allows recognition of CO_2. Tracing the emission of water vapor (Fig. 2.23a) shows a very characteristic peak, centered on 300 °C. This massive and sudden outgassing follows two weak inflections of the curve; the first, at close to 100 °C corresponds to the loss of water adsorbed on surfaces, and the second, at around 200 °C, indicates the existence of weakly bound water.

The principal peak for water emission (at $1508\,cm^{-1}$) corresponds to the steepest downward slope of the weight-loss curve. However, this correspondence cannot be interpreted as loss of water exclusively. The diagram of infrared absorption at $2363\,cm^{-1}$ (Fig. 2.23b) shows a first peak of CO_2 emission (shadow area). The main water-vapor peak also corresponds to the first decomposition of the organic phase. This suggests that the water and the organic phase were strongly associated in the coral structure. The existence of a hydrated organic phase thus is probable.

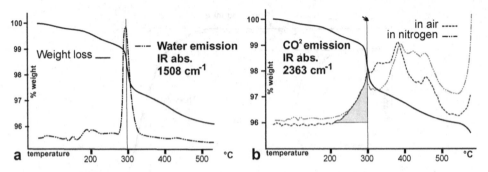

Fig. 2.23. Infrared absorption identification of gaseous products emitted by coral skeletal material during heating. (a) Evolution of infrared absorption of water during heating shows that the rapid weight loss at about 300 °C is due to emission of water. This emission is produced during a very short time. Further weight loss is certainly due to decomposition of another component; (b) Profiles of the CO_2 emission by the same material in air or nitrogen during the same experiment. In both cases, the first emission of CO_2 occurs prior to the water loss and continues, with decreasing intensity, up to about 500 °C. This suggests that the water was linked to organic components in the skeleton, and was liberated in their early decomposition. Cuif *et al.* (2004).

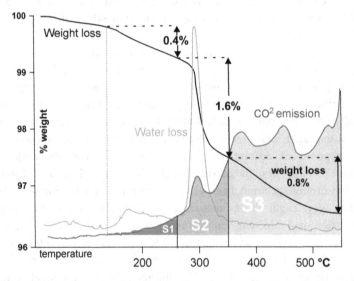

Fig. 2.24. Thermogravimetric and infrared absorption diagrams for *Favia* skeleton. This diagram allows an excellent estimate of the respective weights of water and organic components of the skeleton. Cuif *et al.* (2004).

The thermogravimetric profile obtained for aragonite in the *Favia* skeleton presents the advantage of a terminal phase in which the loss of weight is due to the organic phase only, water being virtually removed in total at the end of the major peak (Fig. 2.24). This permits establishing a correlation between the S3 surface, limited in its upper part by the CO_2 trace and the corresponding loss of weight (0.8%). Additionally, this allows estimation of the weight loss corresponding to the S1 and S2 regions of the spectrum, areas in which the weight loss is due to the simultaneous departure of CO_2 and water.

Thermogravimetric measurements (Cuif *et al.* 1997b, 2004) thus establish that the coral skeleton contains hydrated organic compounds representing 2.5–3% of its total weight. By infrared absorption, Gaffey (1988) had estimated that they should contain up to 3% water (in conformity to the widely shared crystalline concept of coral fibers at that time, Gaffey wrote that, "scleractinian corals contain only aqueous fluid inclusions," p. 412). The coincidence of the two values is noteworthy, but in associating thermogravimetric analysis with infrared absorption, it is confirmed that *the nonmineral component is not constituted solely of water.* It is a mixed organic/hydrated phase, in which the proportion of water can reach nearly half of the organic phase itself, strictly speaking. Thus, the estimate proposed by Cohen and McConnaughey (2003) which suggested that the proportion of the organic component of coral skeletal material can attain the "range of 1% weight" is supported by ATG measurement if considering only the organic molecules themselves. But, actually, water that is structurally associated with organic compounds must also be taken into account.

It is important to consider that, in an aragonitic structure, the presence of 2.5–3% of a hydrated organic component is an important element in two respects.

– On a structural level, even if one accepts only a density of 1.5 for this hydro-organic gel, we are led to calculate that, from the point of view of its volume in the coral structure, this nonmineral component of a coral skeleton must represent approximately 5% to 6% of the space in the carbonate structure (as the density of aragonite is 2.93). Thus, it is essential to determine its nature and its distribution within the carbonate structure.

– The presence of this hydrated organic compound has a considerable impact from the point of view of *stability through time* of the organomineral system. This role appears even more important when taking into account the biochemical characteristics of this organic phase and the structural level at which it is associated with the mineral phase.

Additionally, it is important to note that measurements of isotopic fractionation in these occluded waters from shells have been made on a series of marine and freshwater molluscs (Lecuyer and O'Neil 1994). Results have shown that isotope ratios of these waters *are quite different from those of the ambient water* for each species, suggesting that the water associated with biominerals of the shell is a remnant of body fluid.

In spite of its very great simplicity, the ATG method shows equally well the diversity of the organic phase present in coral skeletal material and in molluscan shells. Not only do different genera show different profiles of thermal sensitivity, but some differentiation is also possible at more-elevated taxonomic levels. It is remarkable, for example, that only the coral profiles show the very strong slope at around 300 °C, which we have seen is caused primarily by a massive loss of water. The molluscs, and even the red coral (*Corallium rubrum*, an octocoral), do not show equivalent patterns (Fig. 2.25).

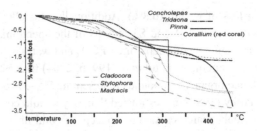

Fig. 2.25. Weight losses during heating in some corals and mollusc skeletons.

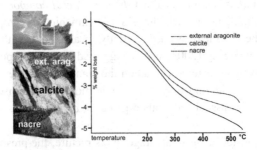

Fig. 2.26. Differences in thermal degradation profiles for materials from each of the three layers of the *Haliotis* shell. Redrawn from Dauphin *et al.* (2007).

The precision of the method should not be underestimated, taking into account the sensitivity of modern instruments. Beginning with samples that do not exceed 50 mg, profiles of thermal degradation of each of the three layers comprising the shell of *Haliotis tuberculata* (Gastropoda, see microstructure in Figs. 1.14 and 1.15) can be sharply differentiated (Fig. 2.26.).

These methods of physical characterization now allow it to be established that, contrary to a widely held concept, organic components present in the same order of magnitude quantitatively do exist both in coral skeleton and in the shells of molluscs.

2.2.2 Infrared absorption spectroscopy: direct characterization of the organic phase in calcareous hard parts

Absorption of infrared radiation provides the most practical method of detection of organic material in calcareous structures. For a long time, IR studies were constrained by inconvenient preparation protocols. After extraction, the organic phase of biominerals had to be dispersed into KBr pellets, thus causing a risk of the modification of their initial state due to the high pressure and temperature involved. Because of this, infrared absorption was limited for a long time to the characterization of the mineral phase, as analysts limited their observations to a very narrow domain of wavelengths (400–2000 cm^{-1}). Now, though, the DRIFT method (Diffuse Reflectance of Infrared by Fourier Transform) allows operation

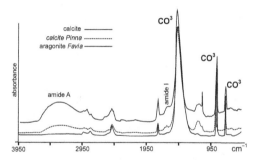

Fig. 2.27. Chart presenting infrared absorption curves for inorganic calcite, calcite from *Pinna* sp. and coral aragonite from *Favia* sp.

directly on small quantities of mineral powder (5–10 mg), dispersed in KBr, i.e., without any decalcifying or heating process. Thus, on specimens where different microstructural types are separated, the presence of organic compounds is easily detectable, particularly in the domain of amide A ($\sim$3300 cm^{-1}) (Fig. 2.27).

Infrared spectrometry, in addition to determining mineralogy, is a technique that allows an estimate of the chemical composition of organic compounds to be rapidly acquired. It is now possible to acquire spectra on complex mixtures, whether they are soluble or insoluble phases, although the attribution of bands cannot be precisely known. However, the majority of the bands, with respect to organic compounds, can be correlated to proteins or to glucides, as there is a zone between 950 and 1150 cm^{-1} that is considered as characteristic of glucides alone.

Worms and Weiner (1986) have shown that beginning with three molluscs with differing microstructures, diagrams of infrared absorption of the soluble organic phases were not identical. Carried out in a saline-liquid milieu, spectra show that these compounds are rich in aspartic acid, thus conforming to amino acids. Using the same method, Constantz and Weiner (1988) were able to distinguish the spectra of soluble compounds of two corals (*Acropora* and *Platygyra*) by infrared absorption.

The comparison of spectra of soluble organic compounds extracted from three genera of corals (Fig. 2.28) thus shows sharp differences, not only in the amino acid bands (A, I and II), attributable to proteins and/or glucides, but equally well in the region corresponding to glucides alone. Finally, certain bands indicate the presence of sulfated sugars (Cuif *et al.* 2003).

Although these types of compounds extracted from biominerals are always mixtures, the detailed analysis of absorption bands permits estimation of the dominant types of proteins. In organic matter extracted from prisms of *Pinna* and *Pinctada* (pteriomorph Bivalvia), the dominant conformation is similar to an α-helix (band at 1653 cm^{-1}). Four bands attributable to aspartic acid (1717 and 1575 cm^{-1}) and to glutamic acid (1712 and 1558 cm^{-1}) are present in soluble matrix material extracted from the prisms of *Pinctada* (Fig. 2.29), while in *Pinna*, multiple bands of aspartic acid are present, but only one of glutamic acid is present (Dauphin 2003).

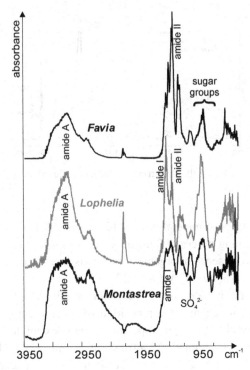

Fig. 2.28. Infrared spectra of soluble organic matrices from three corals, showing major differences in amide bands and in the zone of glucides and sulfated sugars. Cuif *et al.* (2003a).

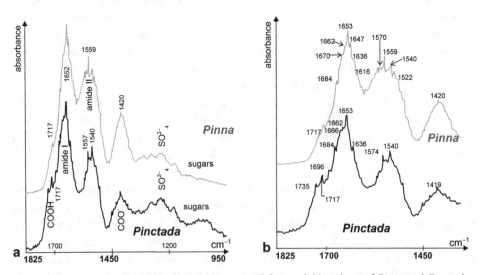

Fig. 2.29. Comparison of soluble compounds extracted from calcitic prisms of *Pinna* and *Pinctada*. (a) Overall spectra; (b) Detail of amides I and II, showing differences in the composition of the compounds. a, b: Dauphin (2003).

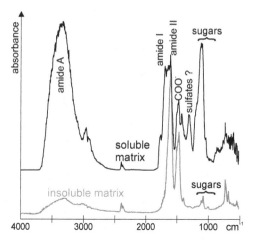

Fig. 2.30. Infrared spectra of soluble and insoluble organic phases in nacre of septa of a nautiloid. Dauphin (2006b).

There are fewer analyses on insoluble compounds than on soluble compounds (see paragraphs directly below). Infrared spectrometry allows the differences between these two types of compounds extracted from the same sample to be unraveled. Thus, the region characteristic of glucides is more intense in the soluble matrix of nacre from the septa of a nautiloid (Fig. 2.30). Always, taking account of the lack of specificity of amide bands, one can deduce that the content in glucides is more elevated in the soluble fraction (Dauphin 2006b).

2.2.3 Progress of biochemical methods in characterizing the organic compounds associated with mollusc shells and coral skeletons

By fractional extraction, Grégoire *et al.* (1955) showed that these organic compounds were not composed of a single type of molecule, as had been accepted until that time. Tanaka *et al.* (1960a,b) and then Voss-Foucart (1968) confirmed this, but more precise biochemical characterization of these organic materials was first carried out in 1951 (Roche *et al.*) by establishing their amino acid compositions.

Discovery of "soluble matrices"

Different observations carried out during preparation of samples for microstructural observation led to identifying the existence of a "soluble" portion of the organic component in biogenic carbonates (Watabe 1965; Mutvei 1970; Meenakshi *et al.* 1971). In 1972, Crenshaw established the importance and the potential role of these organic components that had remained very poorly characterized until then, attention being above all focused on the insoluble compounds isolated by centrifuging (Crenshaw 1972b). In some cases, the preparation methods even brought with them a precipitation phase that ended in reuniting in

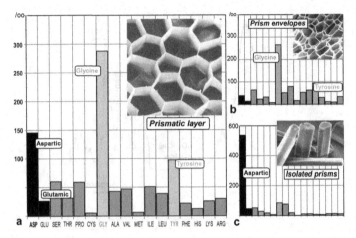

a ASP GLU SER THR PRO CYS GLY ALA VAL MET ILE LEU TYR PHE HIS LYS ARG   **c**

Fig. 2.31. Difference in amino acid composition between the organic envelopes and the mineralized phase within the prisms of *Pinna nobilis*. (a) Amino acids in the total organic content of the prismatic layer; (b–c) Amino acids of prism envelopes (b) and from isolated prisms (c).

a single insoluble ensemble, materials that were already naturally occurring, and some that could have remained insoluble.

Crenshaw pointed out that in *Mercenaria mercenaria* (Mollusca, Bivalvia), this organic phase that could not be sedimented in spite of centrifugation (at 37 500 g) represented as much as 15% of the total organic phase. The importance of this "nonsedimentable" organic phase also draws on the experiments of Crenshaw on preparation methods of these components, which led him to conclude that this set of molecules is closely associated with the mineral material itself. In effect, they are only progressively liberated, and the quantity one obtains is proportional to the degree of dissolution of the mineral phase. Reciprocally, this very close association suggests that this organic phase has played an important role in genesis of the mineral itself.

Following the research of Degens (1979), Krampitz and Witt (1979), and Samata *et al.* (1980), the dual nature of soluble and insoluble organic phases is now accepted as a general law. Models of mineralization implicating these organic phases now seek to represent the possible roles of these two organic ensembles. Insoluble matrices as observed in electron microscopes led to considering them as peripheral features in microstructural units, while the intimate relationship of soluble matrices with the mineral phase directly suggests involvement in crystallization itself.

In the prismatic structures forming the external layer of certain of the Bivalvia (Pteriomorphs, Unionids, Trigonids), the distinction between soluble and insoluble organic phases is more easily observable (Fig. 2.31). Complete decalcification allows effective isolation and separate analysis of the solid network that forms the periphery of prisms (Fig. 2.31b). Inversely, one can destroy these same peripheral networks with an oxidant (sodium hypochlorite, for example), and separately decalcify the disjointed mineral prisms (Fig. 2.31c). The organic material contained by them can thus be compared to that enveloping the prisms.

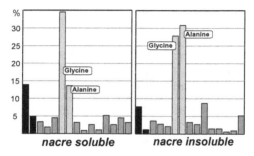

Fig. 2.32. Amino acids of soluble and insoluble components of nacre in *Pinctada margaritifera*.

The envelopes of the prisms constitute exceptional features by their compactness and stability that allow them to be easily isolated. However, in the nacreous layer of *Pinna* or *Pinctada*, insoluble organic phases can also be identified by their distinct biochemical compositions (Fig. 2.32), even if they are not as obvious morphologically as the envelopes of the prisms.

The high proportion of the "acid" amino acids in soluble organic phases of carbonate biominerals has been verified by extensive investigations, and does clearly seem to form a universal characteristic developed by all of the groups producing Ca-carbonate. From the soluble organic phase extracted from prisms and from nacre of five species having different microstructures that were isolated from shells of Bivalvia, Samata (1990) showed that each of them possesses a relatively specific spectrum of amino acids (Table 2.1, Fig. 2.33).

It rapidly became clear that it is these soluble organic materials that, when they are analyzed independently from the insoluble compounds, show the most elevated concentrations of aspartic and glutamic acids, justifying the role that was attributed to them in Crenshaw's interpretation (1972b). Among the 20 amino acids of the eucaryotes, aspartic and glutamic acids have the lowest isoelectric points – aspartic: 2.98; glutamic: 3.08. In normal physiological pH conditions, these molecules thus are most susceptible to assuming a state of elevated charge, producing sites susceptible for fixing the cations utilized in mineralization. In corals, the amino acid composition of soluble organic phases shows a marked similarity to those in molluscs, primarily by the preponderance of these acidic amino acids (Table 2.2). Young (1971), after having analyzed the amino acid contents of 14 species of scleractinian corals, pointed out this close approach to the results already obtained for mollusc shells by Hare (1963), Degens *et al.* (1967) and Travis *et al.* (1967). The extension of analyses and improvement of their precision by the separation of microstructural categories have largely confirmed this conclusion.

It was now clear that in the molluscs and corals, regardless of what the mineralogy or the microstructure of components may be, the soluble organic phases always show concentrations in aspartic and glutamic amino acids that are much higher than their median abundance in cellular proteins. This elevated proportion of the amino acids with very low isoelectric points leads one to assume that they possess a similar capacity to fix Ca ions in these two major groups of calcifying organisms (Mitterer 1978).

Table 2.1 *Amino acids in the soluble organic phase of gastropods, bivalves, and cephalopods.*

	H.d. nacre	*T.c.* nacre	*P.m.* nacre	*A.v.* nacre	*N.p.* nacre	*H.d.* prisms	*T.c.* prisms	*P.m.* prisms	*A.v.* prisms	*N.p.* prisms
ASP	2655	2746	2638	2679	2923	2953	2640	2895	2855	2585
THR	344	570	312	512	331	295	372	303	481	333
SER	622	730	648	815	498	800	915	1075	909	396
GLU	887	972	1073	956	944	841	1127	942	1059	955
PRO	681	662	624	424	617	482	540	510	514	501
GLY	2248	1541	2315	1955	1775	2546	1763	2154	1799	1868
ALA	647	758	658	595	740	410	396	528	581	256
VAL	359	354	252	293	441	207	262	280	278	277
MET	146	126	20	85	33	–	–	13	–	42
ILE	166	200	127	168	266	81	162	156	191	199
LEU	329	395	284	284	489	229	298	355	22334	260
TYR	101	155	235	157	116	216	165	80	172	918
PHE	198	252	301	277	255	244	383	146	224	300
LYS	376	219	180	262	181	3154	262	133	207	549
HIS	40	100	43	121	102	57	184	9996	147	221
ARG	282	178	184	270	239	275	333	205	186	367

H.d.: *Haliotis discus*; *T.c.*: *Turbo cornutus*; *P.m.*: *Pinctada martensii*; *A.v.*: *Atrina vexillum*; *N.p.*: *Nautilus pompilius*; aragonitic and calcitic prisms (after Samata 1990).

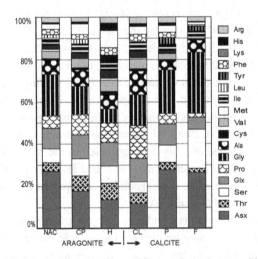

Fig. 2.33. Proportions of amino acids in molluscan biocrystals, both aragonitic and calcitic.

Table 2.2 *Amino acid composition of the soluble organic phase extracted from aragonite of 12 species of corals.*

	P.d.	*P.m.*	*S.h.*	*A.l.*	*M.v.*	*P.b.*	*F.s.*	*F.e.*	*H.l.*	*P.c.*	*C.o.*	*L.c.*
ASP	1390	2310	1730	1570	1640	2300	2010	1150	2020	2230	1660	1770
THR	220	480	360	tr	340	340	770	420	630	420	250	400
SER	440	1160	880	370	200	840	1000	880	780	540	480	550
GLU	1030	1250	1260	1180	1110	1310	1140	1180	950	1120	1020	1060
PRO	640	0	590	550	570	tr	690	710	700	500	610	560
GLY	970	1450	1110	930	960	1200	1260	1340	1140	1000	1090	1310
ALA	770	640	700	840	1060	920	730	940	790	750	760	780
VAL	760	700	570	740	930	630	630	690	540	860	610	630
MET	250	tr	tr	240	100	tr	60	120	50	80	25	100
ILE	520	410	510	530	6210	500	440	490	40	600	410	390
LEU	1120	970	820	1270	1060	1330	730	1150	680	190	990	840
TYR	500	0	220	300	110	tr	120	300	100	200	200	330
PHE	490	620	48	660	570	630	410	640	400	600	610	470
LYS	700	0	380	560	480	0	tr	tr	250	610	840	430
ARG	200	0	380	260	200	0	tr	tr	440	270	200	390

P.d.: *Pocillopora damicornis*; *P.m.*: *Pocillopora meandrina*; *S.h.*: *Seriatopora hystrix*; *A.l.*: *Acropora loripes*; *M.v.*: *Montipora verilli*; *P.b.*: *Psammocora brighami*; *F.s.*: *Fungia scutaria*; *F.e.*: *Fungia echinata*; *H.l.*: *Herpolitha limax*; *P.c.*: *Porites compressa*; *C.o.*: *Cyphastrea ocellina*; *L.c.*: *Lobophyllia corymbosa* (Young 1971).

2.2.4 Molecular weights and isoelectric specificity of soluble skeletal compounds

From the viewpoint of reconstructing the role of these organic phases during the crystallization process, their molecular mass characteristics, as well as the properties that result from their very specific compositions, have been explored by chromatography and electrophoresis. These two methods result in the separation of the organic phase components by utilizing as the principle for fractionation the dimensional properties of the molecules concerned, transposed into molecular mass, or else the electrochemical properties of the same compounds. These two methods differ essentially in the physical method by which they produce the movement of molecules, and the differences in speed by which one can obtain their separation. In the chromatographic method, it is the liquid flux (called the eluant) which displaces the molecules, thus one carries out a separation in time. Electrophoretic methods, on the contrary, utilize an electrical field applied to the margins of a gel containing all the components; one here carries out a separation in space, sometimes even in two dimensions. Based on these two principles, numerous techniques for identification have been developed, in the end arriving at very informative results.

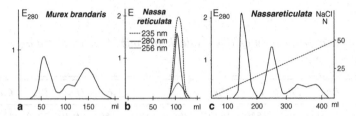

Fig. 2.34. (a) Elution profile in gel-filtration showing the separation of three molecules with distinct weights (detection by absorption UV, 280 nm); (b–c) A compound that shows an elution of the monomolecular type (b) is subdivided into three distinct molecules (c) by an increase in the ionic force of the mobile phase (by progressive addition of sodium chloride). a–c: redrawn from Krampitz *et al.* (1976).

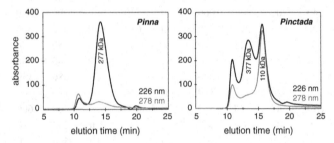

Fig. 2.35. Elution curve of organic components extracted from prisms of *Pinna* and *Pinctada* – two pteriomorph bivalves. Dauphin (2003).

One of the earliest chromatographic studies was carried out on gastropod shells, in which Krampitz *et al.* (1976) showed the presence of several proteins by using a method combining fractionation as a function of molecular mass (gel-filtration) and also as a function of their properties of electrical charge (ion exchange) (Fig. 2.34).

Information provided by the chromatographic method not only depends on operating conditions that determine the method of fractionation, but also on the methods of detection of the eluted compounds. Thus a 280 nm wavelength essentially detects the aromatic amino acids that are only present in small proportions and are practically absent in matrices extracted from molluscs and corals. Quantitative estimates of organic compounds based on these methods are necessarily much underevaluated. Utilization of UV detectors having several wavelengths (Krampitz *et al.* 1976), associated with refractometers (sensitive to other compounds), has confirmed the diversity of matrices present in skeleton, as well as their complexity. Thus, in spite of morphological and mineralogical similarities, matrices extracted from prisms of *Pinna* and of *Pinctada* clearly differ (Fig. 2.35).

A very similar situation exists when examining matrices of the layers of molluscs with cross-lamellar shells, whose mineralogical monotony is opposed to the diversity of their organic components, so that differences seem to depend on molecular weight or acidity. Samata *et al.* (1980) expanded the range of molluscan species studied, while characterizing their polyphenol oxidases, and Weiner (1983) showed that in the same shell, soluble matrices of different layers do not have the same composition. The complexity of the

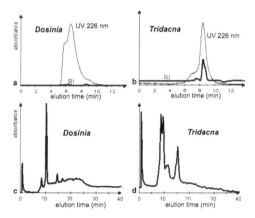

Fig. 2.36. Elution curves of the crossed-lamellar layers of two Bivalvia. (a–b) Molecular weights; (c–d) Ion-exchange liquid chromatography. a–c: Dauphin and Denis (2000).

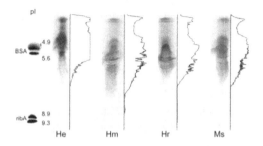

Fig. 2.37. Isoelectrofocusing and intensity profiles of the soluble matrices of *Hydnophora exesa* (He), *Hydnophora microconos* (Hm), *Hydnophora rigida* (Hr), and *Merulina scabricula* (Ms). Dauphin *et al.* (2008).

soluble phase has also been shown in scleractinian corals by a combination of amino acid analyses by liquid chromatography and infrared absorption (Constantz and Weiner 1988). The nature of soluble organic components is also demonstrated using 1D and 2D electrophoresis. In 2D gels, relationships between acidity and molecular weights are shown, as well as the presence of sugars (neutral or acidic), proteins, and glycoproteins by the use of specific stains (Dauphin 2003; Dauphin *et al.* 2008b). Isoelectrofocusing gels of soluble matrices extracted from different species of corals of the Merulinidae show they are all mainly composed of sulfated acidic sugars (stained with Acridine Orange). Nevertheless, each species has a specific profile (Fig. 2.37).

The difference in composition of the soluble matrices extracted from the large calcite prisms of *Pinna* and *Pinctada* are seen in 2D gels (Fig. 2.38). In both matrices, high molecular weights are acidic sulfated sugars, as shown by Alcian Blue staining in the first-dimension (acidity) strip on the upper part. However, the second-dimension gels

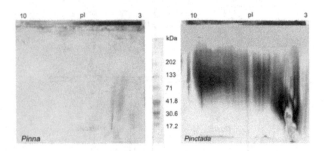

Fig. 2.38. 2D electrophoresis gels, where *Pinna* and *Pinctada* soluble matrices are stained with Alcian Blue (for sulfated acidic sugars) and then silver (for proteins and glycoproteins). Dauphin (2003).

(molecular weights) clearly reveal marked differences. The soluble matrices of *Pinna* stain with Alcian Blue, but only faint smears are stained by silver staining for glycoproteins. In contrast, the main part of the second-dimension gel of the matrices from *Pinctada* is silver stained, the Alcian Blue stain only being visible in the low molecular weight acidic portion (right low corner) of the gel.

2.2.5 Insoluble organic components: envelopes, lipids, chitin

Amino acid compositions of the insoluble organic matrices of coral skeletons and mollusc shells have frequently been studied since the first papers of Voss-Foucart and Grégoire (1971), Young (1971), and Grégoire (1972a). Results have shown that amino acid contents differ in both soluble and insoluble matrices of a single layer (Figs. 2.31 and 2.32), but compositions also differ between comparable structural layers from different taxa.

Research on lipids in skeletons has only rarely been undertaken, even though the lipids were identified in the coral skeleton by Young *et al.* (1971). Since they only concern corals or molluscs, analyses have only confirmed the diversity and the complexity of the compounds present in soluble organic matrices (Fu *et al.* 2005). To date, there is a lack of information concerning the presence, abundance and composition of these components in coral skeletons and mollusc shells. Goulletquer and Wolowicz (1989) estimated that lipids represent 0.8% to 2.9% of the organic matrix of mollusc shells. Fatty acids, cholesterol, phytadienes and ketones have also been described in modern and fossil shells (Cobabe and Pratt 1995). Using a procedure to extract intra- and intercrystalline organic matrices, Collins *et al.* (1995b) detected n-alkanes, n-alcohols, fatty acids and sterols in modern shells. It is suggested that the contents and ratios of these components are dependent on environment and phylogeny. Lipids of the nacreous layer of *Pinctada* are diverse, with cholesterol, fatty acids, triglycerides and an unknown component (Rousseau *et al.* 2006). Farre and Dauphin (2009) have shown using FTIR and thin-layer chromatography that lipids are present in the aragonite nacreous layer of *Pinctada* and in the calcitic prismatic layer of *Pinna* (Fig. 2.39). Lipids are also present in the aragonitic cross-lamellar layer of

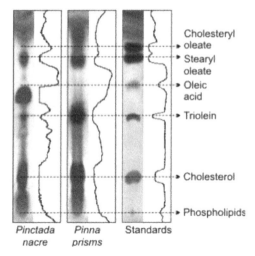

Fig. 2.39. Thin-layer chromatography of the lipids extracted from mollusc shells. Farre and Dauphin (2009).

various molluscan species, and in coral skeletons where they differ both in quantity and quality between species.

Chitin, a polymer of N-acteylglucosamine, has been described in some molluscan shells (Jeuniaux 1965; Peters 1972) using decalcification. Since this date, chitin has been identified and quantified in various molluscan species (Weiss *et al.* 2002; Nudleman *et al.* 2007; Bezares *et al.* 2008; Furuhashi *et al.* 2009). Beta-chitin has been identified in *Nautilus* (Weiner and Traub 1980). The monosaccharide composition of the insoluble matrices of mineralized shell of cephalopods shows high glucosamine contents (Dauphin and Marin 1995). Not only does the insoluble organic matrix of mollusc shells contain proteins, lipids and chitin, but Furuhashi *et al.* (2009) have shown that insoluble polysaccharides (e.g., glucan) are also present (Fig. 2.40). From infrared spectrometry of enzymatically etched shells (using pepsin and chitinase) insoluble extracts have shown that the quantity of sugars differs in different taxa (Furuhashi *et al.* 2009).

This short survey of numerous investigations dealing with organic components of calcareous skeletons reflects the somewhat erratic progress in collecting this information. Until the 1980s and 1990s, chromatography was the primary source of this data, leading to a serious lack of quantification. With respect to this important parameter, thermogravimetric methods allowing precise measurement of the weight of small quantities were only available much later, as well as Diffuse Reflectance Infrared absorption Fourier Transform (DRIFT). In addition to the rather scattered quantitative results, no information was available regarding the localization of these organic compounds with respect to mineral materials. The identical timing of these two factors helps to explain the long period of uncertainty about the mode of growth of skeletal units and the diversity of concepts concerning the organomineral relationships within calcareous skeletons.

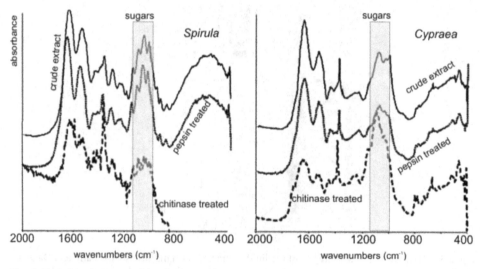

Fig. 2.40. Infrared spectra of insoluble matrices extracted from the aragonite shell of a cephalopod (*Spirula*) and of the cross-lamellar layers of a gastropod (*Cypraea*). Redrawn and modified from Furuhashi *et al.* (2009).

2.3 Access to the distribution of organic components within skeleton units

Following several decades during which the organic materials extracted from biogenic carbonates were biochemically studied after having been isolated through various preparative processes, Johnston (1977, 1980) noted that the models proposed to reconstruct the role of these organic compounds in the mineralization process had been formulated "in complete ignorance" of the physical position that these organic compounds occupy within the carbonate structures. The possible interactions between these molecules and mineral ions have been hypothetically proposed at a very fine scale; that of amino acids or peptides sometimes with an attempt at reconstruction at the supramolecular scale, including the polysaccharides. Nevertheless, in no case have chemical models resulted in forming a biomineral comparable to the microstructural units such as produced by invertebrates.

If the three-dimensional organization of biogenic crystals has been poorly known for such a long time, thus slowing progress in their interpretation compared to cellular structures, it is for the obvious reason that the ordinary techniques of microscopy (microtome sections and staining) cannot easily be applied to them. The presence of the mineral phase works against successful microtome sectioning. Additionally, the application of histochemical staining, which demands very precise pH conditions, also encounters a major obstacle in the presence of the calcareous material. As a result, in parallel to staining mineral components (as shown in Fig. 2.1), numerous attempts have been made to test

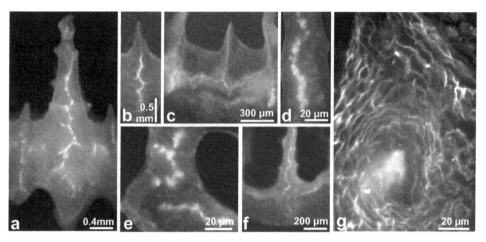

Fig. 2.41. Staining by a fluorochrome (Acridine Orange) in the coral septum. The arrangements of the zones of distal mineralization are clearly reactive. (a) *Montastrea* sp.; (b) *Favia stelligera*; (c) Wall and septa of a specimen of *Lophelia* sp.; (d) The line of centers within a septum of *Lophelia* sp.; (e) Irregularly grouped centers of *Porites* sp.; (f) *Desmophyllum* sp.; (g) One center and fibrous envelopes in a specimen of *Acropora* sp. a: Cuif *et al.* (1999); (c– f): Cuif and Dauphin (2005b); g: Gautret *et al.* (2000).

staining dyes able to characterize the organic components on surfaces that are simply polished (Gautret *et al.* 2000). This method suffers from the usual inconvenience of any staining process when dealing with complex and poorly known materials. Qualitative differences in light emission cannot be quantitatively correlated to the concentration of the product sensitive to the dye used, even if its interaction with a given receptor is well established at the molecular level. In natural materials, the reasons that lead these receptor sites to be active or not on composite reactive surfaces are unknown. However, fluorochromes (Acridine Orange, for example) can indicate the distribution of skeletal organic compounds and have provided noteworthy results (Fig. 2.41), as shown by the characterization of distal mineralization zones in the coral skeleton (Figs. 2.41a–f) or the presence of organic envelopes surrounding the fibers of coral septa (Fig. 2.41g).

The presence of organic envelopes surrounding coral fibers is worth emphasizing because these skeletal components are, by far, not as easily visible here as in the mollusc shells, with calcite prisms and envelopes commonly used to illustrate the concept. Therefore, comparable results obtained by immunological characterization (although not a staining method properly said) deserve to be emphasized as a very positive indication of the presence of organic envelopes at the periphery of coral fibers.

In situ *characterization by use of the immunological method*

Figure 2.42 shows the immunological localization of antibodies primarily developed to act against soluble matrix extracted from the skeleton of the scleractinian coral *Stylophora*

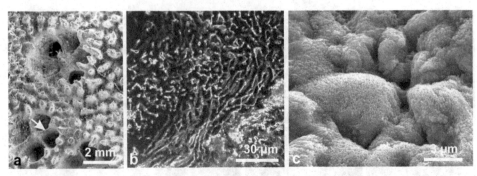

Fig. 2.42. Immunological response on a polished surface in the skeleton of the coral *Stylophora pistillata*. (a) Calicinal surface of *Stylophora pistillata*; (b) Microscopic observation (laser confocal) of a polished surface of the septal structure (a: arrow) submitted to a solution containing antibodies developed to act against soluble skeletal matrix. In the lower-right part of the image, the median line of the septum is strongly reactive, whereas the upper-right half of the image shows a reticulate pattern. In a comparison of objects with similar dimensions, this reticulate pattern corresponds to fiber envelopes; (c) SEM view of the growing surface of fibers in *Stylophora pistillata*. Fibers show their strongly convex growth surfaces. Organic envelopes are located in the depressed areas between each fiber, forming the reticulate structure illustrated in (b). b: Puverel (2004 with permission).

pistillata (Puverel 2004). Application of the antibody-bearing solution onto a polished surface of the coral skeleton reveals a reticulate pattern, the dimensions and overall organization of which exactly correspond to the sizes and arrangement of aragonite fibers as seen on the natural growing surface of the corallite.

This significant result also illustrates the complexity (and resulting uncertainties) of immunological methods. The reticulate pattern corresponding to fiber envelopes was obtained by using antibodies developed to counter soluble matrix extracted from the fibers, i.e., the organic compounds extracted from the mineral units themselves. Although it can be anticipated that these resulting antibodies would be preferentially fixed on the mineral surface of the fibers, they show higher concentrations at the periphery of the fibers. This unexpected result is however extremely useful, confirming the presence of fiber envelopes (compare Figs. 2.41g and 2.42b), and providing additional evidence of the overall similarity between skeletal units in molluscs and scleractinian corals. Additionally, some diagenetic patterns (see Chapter 7) can also be explained by the presence of this thin (generally unnoticed) organic envelope surrounding each coral fiber.

2.3.1 Distribution of the organic matrix observed in transmission electron microscopy

The previous *in situ* characterizations are apparently somewhat approximate when compared to the results obtained by those very rare investigators, including Johnston himself, who have succeeded in obtaining transmission electron microscope (TEM) micrographs of the intraskeletal organic phase. The first attempts to use the TEM were carried out using shadow casting and carbon replication methods that allowed observation of an artificial reproduction of the object. During the 1950s and 1960s, Grégoire, Duchateau and Florkin (1955), as well as Towe and Lowenstam (1967), obtained images using this method, and established the structural complexity of mineralized materials, largely at the submicrometer scale.

By carrying out a complex technique of preparation, starting with a first phase of prolonged infiltration and impregnation (3 months), then a first thin-sectioning (100–300 μm), which was followed by a second phase of imbedding, Travis (1968) succeeded in direct observation of the ultrastructural organization within the external calcitic layer of *Mytilus edulis*, and produced a series of reference images of very great value. The shell of *Mytilus* consists of two mineralized layers; an external fibrous layer of calcite comprises the major part of the shell. It is backed on the interior by a thin nacreous layer. The three-dimensional arrangement of these microstructural units is simple. Fine calcitic fibers (2 to 4 μm in section) are closely associated and remain parallel to one another during their complete growth (Fig. 2.43). They develop with an orientation that is oblique to the external surface of the shell. The active surface of mineralization is restricted to a narrow area, with the result that the shell remains thin (Figs. 2.43a–c). At the extreme margin of the shell, the extremities of the fibers are sharp (Fig. 2.43e) and are surrounded by an organic gel.

The images of Travis show that not only are fibers surrounded by an organic layer of quite regular thickness (approximately 0.1–0.2 micrometers), but that within these envelopes, organic and mineral phases are organized into two, closely complementary, networks that result in a finely granular structure. According to Travis, the median dimensions of the elementary crystalline components would be from 25 to 30 nanometers long and have a "roughly rectangular" morphology. To Travis, this suggested that the organic component within the calcitic fibers is a perfect "carbon copy" of the elementary crystalline lattice.

An equivalent arrangement was revealed by sections that were made within decalcified prisms of the external layer of oysters (Fig. 2.44; Travis and Gonsalves 1969). They show that, in the internal space circumscribed by the thick organic envelope (i.e., the place occupied by the crystal-like mineral material), a very similar arrangement of organic compounds can be observed. Decalcified sections (Figs. 2.44a–b) show the dense network of parallel organic layers even more clearly, with an orientation that is parallel to the growth axis of the prisms. It is surprising that in summarizing these images, which show with great precision the very close relationships between organic and mineral phases, little commentary was made regarding the role of the organic compounds in the biomineralization process. A few sentences suggested that their role consisted above all to provide limits for the development of crystallization. The formation of calcite was still conceptualized as an ordinary chemical process, for which the preformed organic network would constitute a framework that guided crystallization.

These truly exceptional images constituted the first direct evidence of the distribution of an organic phase entering into the mineral constituents of microstructural units themselves,

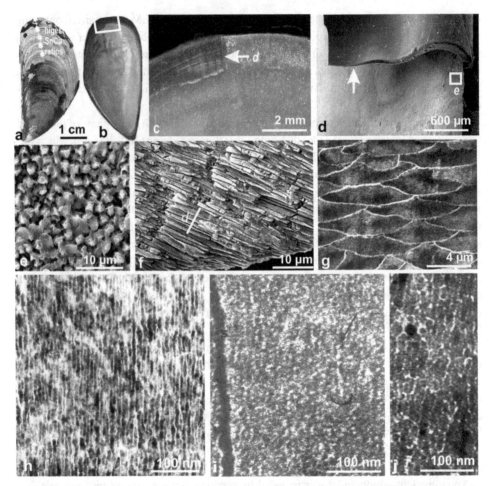

Fig. 2.43. Organic network within the calcitic fibers of the shell of *Mytilus*. (a) Shell growth bands (visible after stripping of the periostracum) and "axis of metabolic activity" as seen by Klein, Lohmann and Thayler (1996b) in their investigation on relationships of growth rhythms and trace element incorporation (see subsection 2.1.6); (b–d) The pallial border of *Mytilus* (internal side); reflected part of the periostracum. Arrows indicate the part of the periostracum that was in contact with the internal fold of the mantle. Growth of fibers is towards below; (e–f) Views of the growing surface of fibrous calcite (e), and (f) the same layer in longitudinal view; (g) Transverse sections of fibers; peripheral envelopes are translucent to electrons; (h–j) Direct illustration of the intracrystalline organic network by transmission electron microscopy. This network is formed by the internal organic phase *within* one fiber after decalcification; longitudinal section (e), transverse section (i); arrangement of microcrystals (j). g–j: Travis (1968) with permission.

at a largely submicrometer scale. Because of the difficult techniques of preparation that caused some uncertainty in their results, this method of observation by transmission electron microscopy was not extended to the whole spectrum of the calcifying organisms. These observations did, however, play a major role in the evolution of thinking regarding the

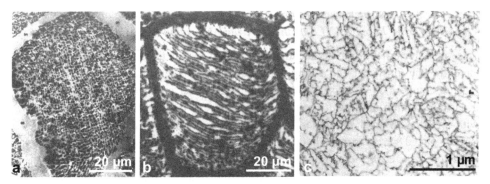

Fig. 2.44. Submicrometer-size organic networks in biogenic calcite and aragonite. (a–b) Lamellar organization of the intraprismatic organic network in prisms of *Ostrea*; (c) Reticulate pattern in the organic network observed by Johnston (1980) in the upper part of the aragonite fibers of the coral *Pocillopora damicornis* (transmission electron microscopy). Johnston's original figure (Fig. 18, p. 201) also shows an interval between this organic mesh and the basal cell layer of the polyp. Johnston carefully indicated that this interval is an artifact due to the preparative process. a–b: Travis and Gonsalves (1969) with permission.

mechanisms of biomineralization. Almost immediately, an interpretation was proposed that made the organic phase not only a framework for crystallization but, additionally, a determinant factor in the process of crystallization itself. This is the *organic matrix concept* proposed by Towe (1972), who developed a model in which the organic phase plays an active role in genesis of the mineral phase, the concept that Lowenstam would formalize even more clearly (see Chapter 3).

Next, Johnston (1977) presented an equivalent and seminal publication, apparently the only such study carried out on the coral skeleton (Fig. 2.44c). It is with this point of view that evidently it will be interpreted. Thanks to sample preparation quite comparable to that which had enabled the success of Travis and her collaborators, Johnston was able to establish a very fine organic network in the growth zone of a corallite with a mesh of submicrometer dimensions and a thickness of three micrometers. The very slight penetration of fixative agents into the mineralized structure only permitted observation of this network in the superficial three micrometers. However, it was clear to Johnston that this network had been secreted by the metabolic activity of basal epithelium in the polyp, which thus played an essential role in the crystallization process. This network was integrated into aragonite fibers during crystallization. Interestingly, Johnston also made special note of the space between the surface of this mineralizing network and the surface of the epithelial cell layer. Clearly, to Johnston, this space (which is in the exact position of the frequently hypothesized "subectodermal compartment") was the result of the preparative process that included microtome slicing. These essential results acquired during the years of the 1970s were not used generally until much more recently, when the development of newer technologies now provide easier and more exact methods of observation and analysis.

2.3.2 **In situ** *characterization and mapping of organic compounds by*
synchrotron-based XANES methods (X-ray Absorption Near Edge Structures)

It is only during the last decade that data could be obtained via synchrotron-based methods that provided, in a single measurement, the chemical characterization of organic compounds in calcareous structures, and better still, obtained a map of their occurrence within microstructural units. Synchrotrons produce electromagnetic spectra with specific physical properties that enable broad and diverse investigations to be carried out. The experiment here utilizes the domain of its lower-energy X-rays in order to characterize the coordination state of atoms. The success of the determination depends on the precision with which wavelengths can be selected and the degree of focusing of X-rays by specially designed lenses. Taking advantage of high technical standards at the ID 21 beam-line of the European Synchrotron Radiation Facility (ESRF) at Grenoble, France, it was possible to associate a focused beam in the submicrometer range (0.3 μm) and very precise selection of energy, allowing distinct characterization of the different oxidation states for given elements. Then, scanning the carefully polished sample surface, which can be as large as several hundreds of square micrometers, one can record the distribution of the various chemical combinations in which a particular element is bound.

The case of sulfur is a good example of this role of synchrotron instruments in the analysis of calcareous biominerals. The presence of sulfur has been recognized in carbonates built by molluscs and corals, as well as in numerous sedimentary carbonates. Geochemical interpretations had concluded that its position was within the mineral lattice (review in Pingitore *et al.* 1995). However, the results of electrophoresis have established (Dauphin 2001a) that there also exists in molluscs and in corals (particularly abundant in the latter) a sulfated organic compound with a compositional form that makes it sensitive to staining by Alcian Blue. This is characteristic of acidic sulfated polysaccharides. Measurement of amino acids has additionally shown that sulfated amino acids are present here as well, even though they may not be abundant.

The extremely precise energy selection permitted by the synchrotron monochromator allows differentiation of various combinations of sulfur by XANES characterization (Fig. 2.45). The peak wavelength for the sulfur of amino acids is obtained at 2.473 keV (Fig. 2.45a) and for the double S–C bond of cystine, but with a different form (Fig. 2.45b), whereas it occurs at 2.4825 keV for sulfur in the sulfate state (Fig. 2.45c).

Additionally, for a given combination, the principal responses at its absorption peak is followed by a series of oscillations. The conformation of these extended oscillations results from the atomic environment of the element. Interestingly, if the sulfate peak of the mineral sulfate (gypsum) approximates the peak of organic sulfates, such as sulfated polysaccharides (although some minor differences can be found on the ascending and descending slopes), the extended oscillations provide quite distinct spectra (Fig. 2.45c). This leads us to the important conclusion, *that the sulfur found*

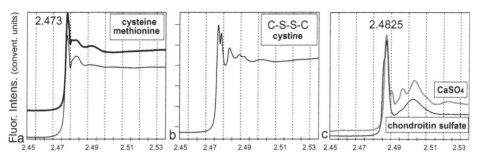

Fig. 2.45 XANES profiles obtained for different sulfur as departures from reference molecules. (a) Sulfur of amino acids (note the difference of the XANES domains for the two amino acids); (b) Di-sulfur bonds; (c) Sulfates of sulfur in a polysaccharide and in a sulfate mineral (gypsum). a–c: Cuif *et al.* (2003a).

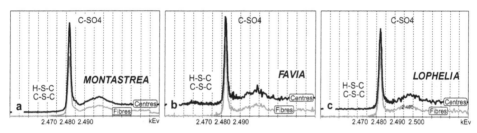

Fig. 2.46. XANES spectra obtained on fibers (black) and in zones of distal mineralization (gray profiles) of three species of corals. a–c: Cuif *et al.* (2003a).

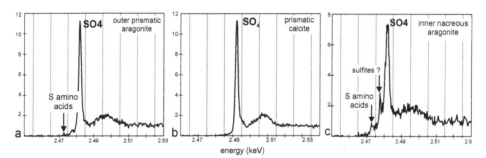

Fig. 2.47. XANES spectra obtained in the three distinct microstructural regions of the shell of *Haliotis tuberculata*. a–c: Dauphin *et al.* (2005).

in calcareous biominerals is not associated with the mineral part but with the organic components.

The first result of the application of synchrotron radiation to carbonate material produced by corals (Fig. 2.46) and molluscs (Fig. 2.47) was to determine with certitude that the sulfur is associated exclusively with polysaccharides, confirming in the first place the results of electrophoresis. One sees also that there is a great difference in the amount of sulfur originating in the polysaccharides and that from the amino acids.

The sensitivity of the method is clearly shown by results obtained in the three distinct regions forming the shell of *Haliotis tuberculata* (see Figs. 1.14 and 1.15 for the micro-structural patterns and Fig. 2.5 for mineral characterization). Although the characteristic peak of organic sulfates is always very easily visible, the sulfated amino acids are only apparent in nacre (Fig. 2.47c).

During these measurements, one should also pay close attention to the risk of chemical modification resulting from the action of the beam. Here for example, the "sulfite" peak of Fig. 2.47c is certainly a derived product resulting from decomposition of a part of the nacreous organic phase. Additionally, this form of sulfur does not appear in the two other microstructural types, thus indicating that it may be tied to the protein phase, which is more abundant in nacre than in the external aragonite or calcite of *Haliotis*.

In addition to this physical identification and description of the polysaccharide sulfur, obtained by measurement of points, the mapping function of the X-ray microscope at the ID 21 beam-line allows construction of a detailed map of a surface of 100×100 micrometers, with a resolution of 1 μm. A map such as this has been made on a portion of a septum from a corallite of *Montastrea* (Fig. 2.48a), where the polished surface was prepared with its orientation perpendicular to the direction of growth (Fig. 2.48b). To define the position of the mapped area with certainty, observation in UV epifluorescence has also been carried out, and this permits the determination of distal zones of mineralization (Fig. 2.48c). The mapping area has been positioned on one of these zones, allowing us to observe the beginnings of zonation of fibrous growth on both sides of it (Cuif *et al.* 2003a).

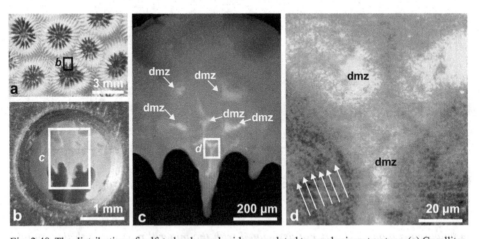

Fig. 2.48. The distribution of sulfated polysaccharides correlated to coral microstructure. (a) Corallites at the surface of a *Montastraea* colony; (b) Polished surface on a sample holder of the X-ray microscope at ESRF, ID 21 beam-line, at Grenoble, France; (c) UV epifluorescence of zones of distal mineralization (arrows); (d) XANES mapping of a zone of distal mineralization and the adjacent first cycles of fiber growth (arrows). a: Dauphin *et al.* (2006); b–d: Cuif *et al.* (2003a).

Thus, it is possible to identify directly the type and presence of sulfated polysaccharides throughout the complete section of the corallite and to make the variations in concentration of these macromolecules visible, together with their relation to microstructure. Owing to the complexity of the structures in which the signal originates, quantitative measurements are not yet possible. However, the distribution of organic compounds alone is extremely informative, as the polysaccharide concentrations are clearly more abundant in zones of distal mineralization (Fig. 2.48d). Additionally, it is important to observe that the banding pattern of polysaccharides around the zones of distal mineralization (Fig. 2.48d: arrows) corresponds exactly with the arrangement of the fiber crystallization layers.

The major interest of this biochemical mapping at one-micrometer resolution is that it permits a direct relationship to be established between the geometric characteristics of the mineralization layers and the distribution of organic materials at the same scale and within the same field of view. This possibility has proven to be of particular interest with respect to the models for formation of coral aragonite that are still currently accepted. Correspondence between the layered distribution of the biochemical compounds, such as sulfated polysaccharides, and the mineral banding patterns shown by microstructural analysis leads to accepting that, in the coral fibers, as in molluscs, the mineral phase is permanently associated with the organic matrix.

With an identical objective, very high resolution mapping was performed on the fibrous skeleton of another coral, *Goniastrea* sp. (Fig. 2.49, a specimen from New

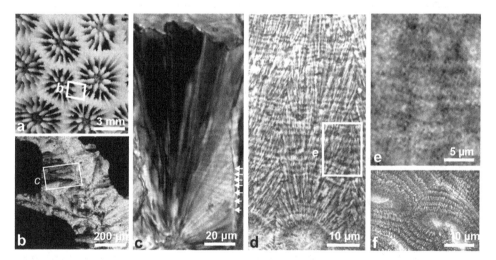

Fig. 2.49. Correlation between mineral and organic layering in the skeleton of the scleractinian coral *Goniastrea*. (a) Morphology of individual corallites of the colonial skeleton; (b–c) Fibrous bundles. Note the growth layers visible in polarized light (c: arrows); (d) Stratified structure of these fibrous bundles (post-measurement etching of the polished surface on which mapping was done); (e) XANES mapping of sulfated sulfur obtained on a polished surface of the corallite. The distribution of sulfated polysaccharides corresponds to the mineral zonation visible in (d) (image from the ID 21 beam-line, ESRF, Grenoble); (f) A closer view of growth layering in fibers of this genus, around zones of distal mineralization (etched polished surface). c, e: Cuif *et al.* (2008b).

Caledonia). An ultrathin section was prepared in order to illustrate the fibrous aragonite bundles as clearly as possible using optical methods (Figs. 2.49a–c, crossed nicols). Growth layering is clearly visible on this section. An equivalent sector from the same specimen also shows identical structural and growth patterns (fibrous bundles and transverse growth layering). The next image is the high-resolution map obtained using energy of 2.4825 keV, characteristic for sulfates. Associated with the XANES profile specific for organic sulfate (Fig. 2.49c), this result clearly indicates the correlation between mineral layering and the banding pattern and distribution of organic sulfate.

This correlation between distribution of the polysaccharide and crystal zonation implies that the association between mineral and organic phases occurs within the micrometer-thick growth lamellae, i.e., *at a submicrometer scale*. This result cannot be overemphasized in the discussion of any mineralization model. At this analytical scale of mapping, if the coral fibers were formed of an alternation of organic layers and purely mineral layers (the "template" model) the high resolution of this instrument would provide us with maps that differ greatly from the two shown above. Note for instance that in Fig. 2.49e, visible zonation is due to a higher concentration of sulfated polysaccharides that corresponds to mineral banding. An identical conclusion can be drawn from the calcite prisms of *Pinna*, long the reference model for biologically controlled crystallization (Fig. 2.50).

To appreciate Fig. 2.50f we must go back to the remarkable sulfur maps obtained with the electron microprobe (Fig. 2.7), and here emphasize the interesting properties of synchrotron beam light. The extreme precision in tuning this X-ray frequency allows us to characterize different chemical states of sulfur. Mapping of sulfur as a chemical element did not allow us to resolve formally the question of its coordination, nor its eventual association with the mineral phase itself. Thanks to XANES characterization of sulfur as a component of polysaccharide molecules, the association of both organic and mineral

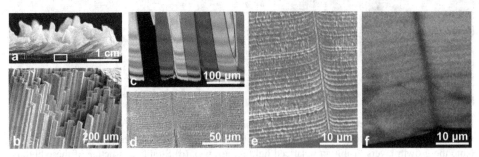

Fig. 2.50. Layered distribution of polysaccharides in prisms of *Pinna nobilis*. (a–c) Repeat of the morphological and microstructural features of this standard of reference for a monocrystalline biocrystal (details, Fig. 1.7); (d–e) Growth stratification made visible by enzymatic etching of a polished surface; (f) Map of sulfated sulfur on the same surface. f: Dauphin *et al.* (2003b).

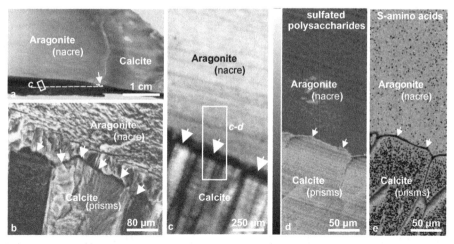

Fig. 2.51. The transition from the calcitic prismatic layer to the aragonitic nacreous layer in *Pinctada margaritifera*: its morphology, microstructure, and biochemical characterization. (a) Morphological view of a sectioned shell; (b) SEM view of the transition: the arrows mark the covering of internal surfaces of the prisms by the organic layer, which is then followed by deposition of fibrous aragonite and finally nacre (for detailed views, see Chapter 1); (c–e) Optical view (c) and synchrotron-based characterization (d–e) of the same region. The layered pattern of the organic compounds within prisms is particularly visible through the distribution of polysaccharides. Additionally, the characteristics of the organic layer covering the prism surface (arrows) are shown here. a, d–e: Dauphin *et al.* (2008).

components at the submicrometer level can now be firmly established, both in corals and in molluscs.

More generalized mapping is of equal interest, even though such maps do not provide strictly quantitative measurements. Images of broad surfaces at the prism/nacre transition in the shell of *Pinctada* (Fig. 2.51) illustrate clearly the metabolic change related to the long-recognized mineralogical change typical in this shell, a standard for biomineralization studies.

2.3.3 Infrared absorption mapping utilizing synchrotron beam light

The great brilliance of the synchrotron beam offers another mapping possibility usually not accessible in stand-alone instruments. Due to the low intensity of classical infrared sources on Ca-carbonate biominerals, some of the significant wavelength domains are very close to calcite and/or aragonite wavelengths, so that in these areas of the infrared spectrum, mapping is not feasible. Infrared wavelengths also represent a serious limitation to mapping resolution. On the other hand, good maps can be obtained for some specialized domains, such as the lipids, for example.

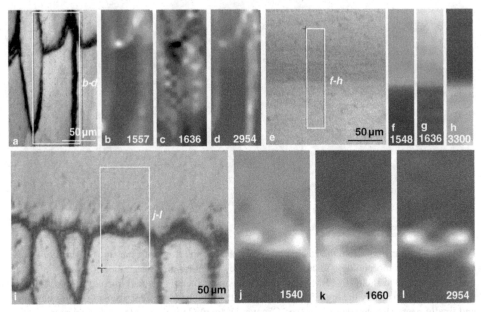

Fig. 2.52. Examples of mapping by synchrotron IR beam. (a–d) *Pinna* prisms, longitudinal section: (b): 1557 cm^{-1} (amide II), (c): 1636 cm^{-1} (amide I), (d): 2954 cm^{-1} (lipids); (e–h) *Mytilus* prism–nacre transition, glycoprotein bands; (f): 1548 cm^{-1}, (g): 1636 cm^{-1}, (h): 3300 cm^{-1}; (i) *Pinctada* prism–nacre transition; (j) 1540 cm^{-1} (amide II); (k) 1660 cm^{-1} (amide I); (l) 2954 cm^{-1} (lipids). i: Dauphin *et al.* (2008).

2.3.4 Mapping of organic molecules using TOF-SIMS: the Time of Flight Secondary Ion Mass Spectrometer

Secondary ion mass spectrometry (SIMS) is basically different from microprobe and synchrotron methods in that, instead of using electromagnetic radiation emitted by the studied material through the action of electron or X-ray energy, it is the matter itself constituting the sample that is eroded by a specialized ion beam; as most of these particles are detached from the sample as ionized particles (secondary ions), they can be attracted to a detector or screen. The first instruments based on this method were developed in the early 1960s (Castaing and Slodzian 1962). They were first conceived as true optical instruments, forming their images through electrostatic lenses, magnetic prisms, and slits that select ions of a given mass. We can thus obtain direct chemical pictures of sample surfaces (Fig. 2.53a). These instruments have proved to be extremely effective in study of the diagenesis of fossil structures. In the first generation of these instruments, electric charges resulting from beam ions were grounded and removed by a metal grid deposited on the surface to be analyzed, prior to observation (Fig. 2.53b). Now, accumulated electric charge is compensated for by depositing a flux of oppositely charged ions on the sample surface, thus enabling observation of a broader field of view.

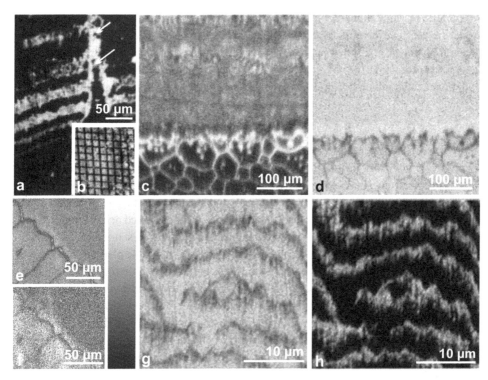

Fig. 2.53. SIMS pictures: from early instruments to TOF-SIMS. (a) Distribution of Sr in the growth layer of a belemnite (fossil Cephalopod from the Triassic, see additional examples in Section 7). In contrast to the following images, where distribution is obtained by surface scanning, this was obtained through optical imaging, showing the layered distribution of Sr (imaging by R. Dennebouy and G. Slodzian, LPS Orsay); (b) The metal grid at the sample surface. It was eroded during observation (see a: arrows) reducing the available time for observation; (c–d) Distribution of molecular masses, (c) glycine, (d) Ca, at the growth front of the nacreous layer. Comparing this picture to SEM views of the spreading of aragonite onto the surface of the prismatic layer provides precise chemical information on metabolic changes concerning this major transition in shell formation; (e–f) Molecular mapping of the calcite/aragonite transition in a *Pinctada* shell: (e) glycine, (f) alanine; (g–h) High-resolution mapping of molecular fragments in the nacreous layer: (g) Ca, (h) glycine (imaging by A. Brunelle, ICSM-CNRS Gif/Yvette). a–b: Cuif *et al.* (1977).

More recent instruments are designed for increased resolution (see Fig. 2.10), and for reduction of the quantity of material that is detached from the sample surface. Distribution of selected elements or molecules is obtained then through a scanning system with stepping displacement of the sample surface and computer reconstitution of positions and their distribution.

In the time-of-flight (TOF) instruments, detached ions are separated according to the duration of their transit from the sample surface to an ion collector. One major advantage of this in the study of biominerals is that much larger molecules can be managed in the system and their distributions mapped. No longer is the distribution of amino acids the only one that can be imaged (Figs. 2.53c–d).

Ongoing research is increasing the database bearing on molecular fragments, many of which are yet to be identified. Beyond important analytical work, yet to be done in order to take full advantage of this new approach to mineralizing matrix composition, the TOF-SIMS method also commonly provides additional evidence that, from a dimensional point of view, both organic and mineral components are associated at the submicrometer level.

2.4 Conclusion

Applied to corals and molluscs, chemical and biochemical characterizations complemented by investigations using localized measurements and mapping lead us to the recognition of an overall similarity between the layered growth mode in the skeletons of the two phyla.

It is conclusive that the contribution of SIMS and synchrotron imaging methods to our understanding of the structure of microstructural units is their repeated demonstration of the continuity of the distribution of organic components within the crystal-like mineral units. Since the general application of Crenshaw's discovery of the soluble phase in *Mercenaria* (1972b), no doubts remained with regard to the presence and role of this organic component in the biomineralization process. However, determining the *localization* of the soluble matrix is essential to the interpretation of its mode of action. It is now clear that the soluble matrix neither has a location surrounding the microstructural units (hypothesis of a shaping role by modulating the relative speed of growth of crystal faces), nor does it occur as an alternation with pure mineral layers (acting as a laminar template for growth of a purely mineral component).

The soluble organic component exists throughout the entire volume of any microstructural mineral unit in both coral skeletons and mollusc shells. Results suggesting such an interpretation were formerly obtained by using etching techniques; the fine organic network visible within the growth layers of a *Pinna* prism (Cuif *et al.* 1980) can now be better understood (Fig. 2.54). This can be compared to Johnston's imaging of the organic network that he illustrated as occurring in the distal growth layer of a coral skeleton (see Fig. 2.44c). Both images certainly suffer from rather comparable artifacts: the organic matrix components have been separated from the mineral phase and the present network is only an approximate image. However, further observations (Chapter 3) provide additional data supporting this.

Owing to widely shared opposing views noted previously, it is worth mentioning again that all of the results reported in this chapter have been found to occur in identical ways in both molluscs and corals, thus demonstrating that the two groups have well-controlled structures with organic matrix associated with mineral components at a submicrometer level. The long-standing opposition between hypotheses of "biologically induced" and "matrix-mediated" Ca-carbonate deposition, one that hypothesized weaker control on the formation of coral skeleton, simply cannot be maintained any longer.

Another conclusive point is the following: mapping using SIMS and synchrotron imaging, with micrometer resolution, has repeatedly shown an apparent superposition in the distribution of organic and mineral components of microstructural units. This implies that

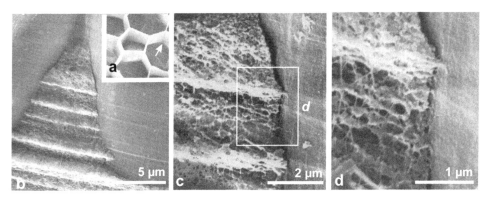

Fig. 2.54. Examples of imaging a network of soluble matrix within a prism of *Pinna*. (a–b) Internal side of the prism envelope after fixative decalcification: the overall distribution of the intraprism organic component corresponds to growth layering of the shell; (c–d) Within each growth layer, remains of organic components appear as a fine submicrometer-scale network. c–d: Cuif *et al.* (1980).

the spatial interplay between the mineral material and its associated organic component is established at a submicrometer scale. Owing to its new method of imaging, atomic force microscopy allows reconstruction of sample surfaces with resolution easily reaching the 10 nm range, thus providing new and stimulating information about the fine structure of elementary growth layers and the mode of association presented by organic and mineral components. Not only the continuous interplay of these, described in the two first chapters of this book, is here illustrated, but furthermore, we can show the remarkable and obvious similarity between crystallization patterns in both molluscs and corals. These remarkable images thus provide us with decisive data necessary to advance the addressing of questions regarding the crystallization process itself. Only observation carried out at magnifications in the submicrometer range can provide us with the necessary data to determine more precisely the exact mode of organomineral relationships from both descriptive and functional points of view.

3

Origin of microstructural diversity

Facts and conjectures regarding the control of crystallization during skeletogenesis

Chapters 1 and 2 have established that the microstructural units forming mollusc and coral skeletons exhibit very comparable growth patterns. Coral fibers and mollusc prisms grow by the repeated production of mineralized layers, the thickness of which is in the micrometer range. Clearly, the unit of biomineralization is not the often-described microstructural component that forms each shell layer (prismatic, foliaceous, fibrous, granular, etc.), but rather it is the mineralization cycle that produces the elementary growth layer for each of the different components of the shells or corallites.

From a functional point of view, numerous examples show that distinct mineralized domains are simultaneously produced (e.g., Fig. 1.5i). The growth layer is isochronous, thus, the cyclical mineralizing process is developed on the whole mineralizing surface during the same time period, whatever the differences (microstructural and/or mineralogical) between contiguous mineralizing areas. In the first part of this chapter (Sections 3.1 and 3.2), the structure of the mineralizing growth layer here was investigated in both corals and molluscs, taking advantage of instruments that provide resolutions in the nanometer range. The spatial distribution of organic and mineral components is thus described within the elemental growth layer itself, allowing a functioning model to be inferred. The great similarity revealed by this high-resolution imaging fully confirms the results of the two previous chapters.

However, in contrast to the uniformity which can result from this common growth mode, attention has long been drawn to the potential of skeletal structures for providing information regarding the evolutionary process both in corals and molluscs. Therefore, explanation of shell and corallite formation must include studies dealing with the origin of taxonomy-linked microstructural features. Many examples indicate that although growth layer crystallization does play the major role in producing the main part of the skeleton from a quantitative point of view, it is essentially a repetitive process lacking much innovative potential. In contrast, the mechanism driving initial crystallization appears considerably more important (Sections 3.3 and 3.4). In addition to an improved understanding of skeletogenesis, investigations concerning the early crystallization stages of microstructural units can help us create consistent nomenclature, and result in more reliable phylogenetic hypotheses. At a higher level, exploration of the microstructural variations during individual skeletogenesis demonstrates that crystallization of the skeletal units is a fully integrated

119

process, with multiple hierarchical mechanisms ensuring the permanent control of the mineralizing process by the living system (Section 3.5) in both molluscs and corals.

3.1 Crystallization within the elemental growth layer in both mollusc and coral skeletons

A completely new method of observation has been developed during the last decade, field force microscopy, commonly referred to as atomic force microscopy (AFM). The principle of this method can be simply stated. A very fine point (ideally mono-atomic at its tip) is brought into proximity with the specimen under study. The distance between the sample surface and the tip of the microscope point must be zero for operation in the "contact mode," or else reduced to a very small interval for operation in a vibrating mode, called "tapping." In the latter, the part that bears the point (cantilever) vibrates at a precisely controlled frequency. The instrument vertically moves the tip in the direction of the substrate and measures the van der Waals force established between the substrate and the extremity of the point when it is in the close position. The proximity between specimen and point thus is defined by the value assigned to this parameter, a value that remains constant during description of minute surface features of the specimen.

In the two instances (contact or tapping mode), description of the specimen surface to be studied is accomplished by moving the specimen along equidistant parallel lines; thus it is actually a form of two-dimensional (x–y) scanning. Because the level of its surface (in the z direction) varies during lateral displacement of the specimen (at the molecular scale, of course), the instrument adjusts by making the distance of the cantilever permanent, thus keeping the point in contact with the surface (in the fixed point mode) or maintaining its value as identical to that assigned to the van der Waals force. This constant adjustment of the z distance during scanning in the x–y direction provides us directly with a description of the relief on the surface.

An essential property of this method of observation is that imaging of the object surface does not require any specific preparation, such as a conductive coating, as is the case with the scanning electron microscope, since in the way that it is utilized for morphological observations the process does not involve any electron exchange. All of the images shown in this chapter and the next have been obtained by use of the atomic force microscope in the "tapping" mode to allow simultaneous collection of three image types. The first is the simple description of surface relief, expressed by variations of vertical position during scanning, thus producing an image of real surface topography (height image); generally it is not diagnostic. A second type of image is obtained by deriving the height signal; the derived image, conventionally known as an amplitude image, emphasizes the variation of the surface topography. A third type of image can also be obtained, based on analysis of the actual vibration of the tip and, more precisely, the influence of the substrate on this vibration during scanning. Since the extreme low position of the point occurs at the base of the oscillation (where its proximity to the sample is defined by the assigned value of the van der Waals force), interactions develop with the surface of the specimen being studied that modify the oscillation. This introduces a lag in the oscillation phase with respect to the nominal frequency. This phase lag is measured and used to create a specific phase image that is of very great interest.

Actually, the phase lag varies according to the nature of the material interacting with the point. On a given surface, certain materials produce strong interactions, depending on their properties of viscosity, elasticity, and cohesiveness, all related to their chemical composition and structure, while other components of the surface are more neutral substrates and only cause a weaker phase lag. The images produced by phase lag are therefore an expression of the heterogeneity of the material encountered by the point during scanning. This ability to establish heterogeneous substrates is shown very well in the results of studies on carbonate skeletons of corals and mollusc shells.

3.1.1 Submicrometer organization of the coral skeleton

In Chapter 2, a corallite from a *Goniastrea* colony illustrated the correlation that can be established between successive growth laminae and the distribution of organic matrices mapped by X-ray microscopy and by fluorescence of sulfated polysaccharides (Figs. 3.1a–e,

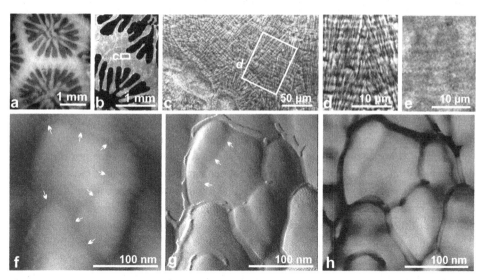

Fig. 3.1. Growth layers and nanostructures in skeletal aragonite of the genus *Goniastrea*. (a–c) Section of corallites and layered microstructure of the fibrous material; (d–e) Correlation between growth layering (visible after etching) and distribution of sulfated polysaccharides mapped on the same polished surface; (f) Height image: topography of the sample surface at the nanometer scale; (g) Amplitude image; (h) Phase image: interpreting the strong contrast between the central grain and the dark cortex is essential to understanding the crystallization process.

NB: In Fig. 2.49e (broader view of Fig. 3.1e), a granular heterogeneity is visible in each growth lamina. These points are the pixels resulting from the micrometer stepping displacement of the sample during the mapping process. The granular pattern thus has nothing to do with the reticulate structure visible by using AFM. b, d, e: Cuif *et al.* (2008b).

with more details in Fig. 2.45). Application of atomic force microscopy to a surface cutting through the fibrous part of the *Goniastrea* wall or septum provides a representation of the structure of the fiber in the nanometer range.

On the height image (Fig. 3.1f), coral skeletal aragonite appears formed of nodular grains with their long axis approximately 100 nm, and between them there is a continuous layer with a fairly regular thickness (4 to 6 nm), forming an envelope around groups of grains (arrows). This organization is clearly verified by the "amplitude" image (Fig. 3.1g), that accentuates the composite character of the grains. The linear zones with indistinct boundaries (arrows) suggest that major grains may result from the assembly of smaller elements, a possibility that the phase image shows is even more probable.

Taking advantage of the phase imaging technique (Fig. 3.1h), envelopes are very clearly distinguished from central grains. The very strong interactivity of the intermediary layer (which results in a maximal "phase lag") signifies that it has levels of elasticity and adhesiveness that are much greater than that of the grains themselves. The great contrast between these interaction patterns originating in the grain and the cortical materials suggests that, for the first time, we are able to observe the spatial arrangement of the two main components of the coral fiber; both organic and mineral. The highly interactive material can be interpreted as a dominantly organic material, whereas the less interactive part should then represent the mineral component. Taking into account the dimensions of the grains (with sizes in the order of 50–120 nm) it is understandable that fluorescence mapping, where resolutions are in micrometers, as a result only produces an appearance of having a continuous distribution of their glycoprotein components within a single growth lamina (Fig. 2.50e).

A highly versatile instrument, the atomic force microscope also allows precise measurement of forces to which the tip is subjected all along its interplay with the sample surface. In this particular functioning mode, the mobile part of the microscope (i.e., the cantilever and its tip) executes a series of back and forth movements with respect to the sample surface (Fig. 3.2a). During each of them, deflection of a laser beam indicates the position of the flexible part of the cantilever. During forward movement, the mobile part is moved towards the sample surface, up to the point of contact. Continuing this "descending" movement causes a bending of the flexible tip-bearing blade; this curvature is made visible by deflection of the laser spot (the ascending part of the trace after "contact" in Fig. 3.2b). In the reversing part of the measurement movement, the laser trace is at first exactly inverse to the initial deflection but, in the case of an adhesive substrate, the tip remains in contact with the substrate for a duration that depends on the physical property of the substrate. The backing movement continues and the tip is finally detached (see Fig. 3.2b), providing us with the possibility for an exact measurement of the adhesive forces.

Interpretation of these phase images is thus positively established through physical characterization of the material surrounding the nanograins, providing us with experimental evidence that the grain cortex can be considered as organic in the most part.

Phase images also indicate that grain surfaces are not actually homogeneous with respect to their own interactivity. Here, the zones already differentiated by "amplitude" images appear effectively endowed with properties by their composition, again suggesting that they

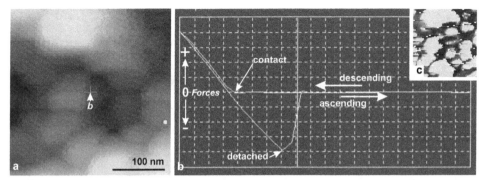

Fig. 3.2. Measurement of forces caused by adherence of the cantilever tip to the sample surface. (a) Sample surface (height image). Arrows indicate the point where the measurement figured in (b) was made; (b) Trace of the movement of the cantilever tip and indication of the related forces during a back and forth movement. The descending part of the movement places the tip in contact with the sample surface (arrow), then the descending movement continues, causing the cantilever to be curved (to the point of breaking if no stop order is given). During the reverse movement (ascending), the tip of the cantilever exits the contact point, remaining fixed to the sample surface (adhesiveness) up to a certain position (arrow) where the tip is forcibly detached from the surface; (c) By fixing a given value to adhesive forces, it is possible to obtain a scanned map for distribution of equivalent properties. Comparing (a) (height image), (c), and the previous phase image in Fig. 3.1h clearly establishes the grain cortex as being a region of strong interaction between the tip and the sample surface. a–b: Julius Nouet, IDES Orsay.

cannot be completely mineral. In this way, the possibility of "fusion" of more elemental grains receives additional reinforcement. This new information regarding the dual composition of the coral skeleton at the nanometer scale clearly explains our inability to distinguish between organic and mineral phases within a single growth layer when commonly used analytical instruments have a resolution in the range of a micrometer at best.

The organization of granules within coral aragonite at the nanometer scale can be observed in all Scleractinia, not only in the zooxanthellate (hermatypic) forms (Figs. 3.3a–f), but also in deeper water forms lacking symbionts (Figs. 3.3g–l) (Cuif and Dauphin 2005; Dauphin *et al.* 2006). In all of these cases, and whatever the location at which the observation is carried out, the fibrous carbonate of the coral is seen to be organized in the same way: composed of rounded granules, densely compressed and held together by a strongly interactive material. Thus, these granules can be considered as the *building blocks* of growth layers and, naturally also, of the fibers at the same time.

This type of organization is a true paradox; it contrasts completely with the conventional concept of coral fibers being considered as monocrystalline units. There are several important points to consider:

(1) It is noteworthy that these grains do not show any traces of a growth plan that would reflect their having a crystalline nature. As the building block of a crystalline unit, some visible crystallographic properties such as faceted growth surfaces could reasonably be expected. No such feature has ever been observed, although it is obviously necessary that these units comprise a crystalline phase.

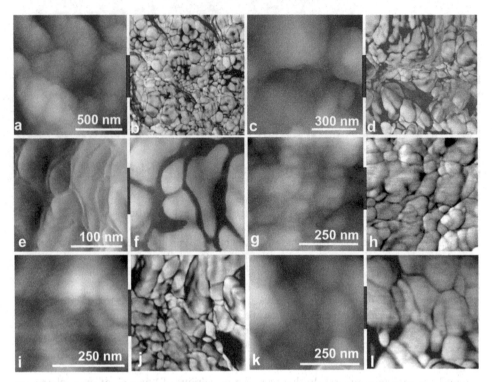

Fig. 3.3. Corallite nanostructures for six species of corals: three zooxanthellate (a–f) and three nonzooxanthellate (g–l). (a–b) Height and phase image of *Pavona* sp.; (c–d) *Cladocora caespitosa*; (e–f) Amplitude and phase image of *Merulina scabricula*; (g–h) Height and phase image of *Desmophyllum* sp.; (i–j) Height and phase image of *Caryophyllia* sp.; (k–l) Height and phase image of *Lophelia* sp. c–f: Cuif and Dauphin (2005a).

(2) Also, it should be kept in mind that these nanostructures are produced within isochronous growth layers. The nanograins of one growth layer crystallize simultaneously but, although crystallographic orientations change between neighboring fibers (at least in their *a*- and *b*-axes), no change in grain arrangement is visible.

(3) In each fiber thus formed by superposition of elemental growth layers, the nanogranular structures must crystallize in such a way that crystallographic uniformity is maintained between successive growth strata of a given fiber (Fig. 2.51).

To depict more precisely the sizes of structures to which AFM gives us access (compared to the microscopic structures we are more familiar with), a thin section was made in a sample, then the same sample was submitted to the AFM in tapping mode, to provide images of the granular structures shown in Figs. 3.1f–h. The densely packed nodules are precisely the material that constitutes the layers that are clearly visible as superposed strata forming fibers in Fig. 3.4, and thus strictly equivalent (within the same sample).

Other examples seen in Fig. 3.3 indicate that this nodular organomineral constitution of the growth layers in corals is a general feature. To date, no exception has been found, even

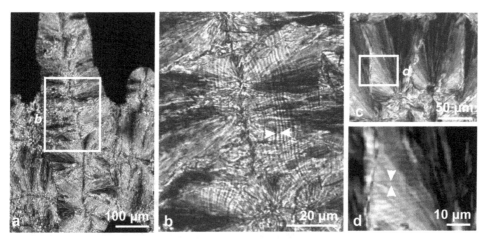

Fig. 3.4. Growth layers and fiber crystallinity in the skeleton of *Goniastrea*. (a) Fiber bundles in the septum of *Goniastrea*; (b) Superposed sinuous growth layers that have formed the fiber bundles. Each of these layers (2 to 3 μm thick, arrows) is made of the densely packed nodular grains shown in Fig. 3.1 (same sample); (c–d) Greater enlargement of a group of fibers showing strongly oblique growth layering (note the distinct gray-colored layering in each of them = a polarization color).

though some diversity in the size of the nodules has been observed. At present, this is not yet interpretable in terms of taxonomy or environmental influence, etc. Once again, this spatial distribution of organic and mineral components explains very well why it was impossible to obtain distinct chemical images by making growth layer maps utilizing instruments having a micrometer-sized step in map construction.

3.1.2 Nanostructure of the molluscan skeleton

As the most studied example of prismatic structures, the internal organization of the calcitic prisms forming the external layer in *Pinna* was first explored by using etching methods. The main features of these prisms were presented in Fig. 1.7. Figure 3.5 recalls their initial massive aspect (Figs. 3.5a–b), directly visible from the exterior (Fig. 3.5c), and within the interior after action of an enzymatic solution (Fig. 1.7g), and additionally, their remarkable monocrystalline appearance under cross-polarized light (Figs. 3.5d–e).

Submitting the prisms of *Pinna* to the action of a dilute sodium hypochlorite solution to remove the intraprismatic organic compounds makes a granular structure visible (Fig. 3.5f). Superposed growth layers appear separated from one another due to removal of organic compounds by chemical action, and possibly also (owing to the roughness of the method) removal of the more weakly associated grains in the prism structure. In this respect, surfaces marking the limits between successive growth layers were submitted to an accentuated dissolution (Fig. 3.5f: arrows). Interestingly, removal of the organic components distinctly separated the grains within each of the growth layers (Fig. 3.5g). In some respects, this image is complementary to those in Fig. 2.50 that illustrated the continuity of

Fig. 3.5. Structure of the growth layers in the calcite prisms of *Pinna nobilis*. (a–b) Prisms in the shell lamellae (a) and (b) enlarged view of the same; (c) The layer surface, visible when its organic envelope has been removed; (d–e) Reminders of the remarkable monocrystalline optical property of the prisms in this species; (f) Dilute sodium hypochlorite allows the degradation and removal of organic compounds from within prisms, revealing the discontinuity between layers of successive growth (arrows); (g) Strata themselves, formed of grains arranged in layers oriented obliquely to the growth axis of the prisms; (h) AFM view of an equivalent zone (tapping mode). Here the black network surrounding the white to gray grains is the result of the high interactivity of the material (and not a void space due to action of the sodium hypochlorite). b: Cuif *et al.* (1981); c: Cuif *et al.* (1983); e: Dauphin (2003); g: Cuif *et al.* (1980).

distribution of organic components at an equivalent enlargement. Due to this accentuated dissolution, the spatial arrangement of carbonate grains within a growth layer is readily identified (Cuif *et al.* 1980).

Applying atomic force microscopy in the tapping mode to a polished longitudinal section of a *Pinna* prism has provided practically equivalent data (Fig. 3.5h), however, with a considerable difference. Shown in Fig. 3.5h, the black, reticulate network seen encircling the grains is due to the physical/chemical properties of the organic component, which is still present, and not to a void space (as in Fig. 3.5g) that might have resulted from the dissolving action of the sodium hypochlorite. This illustrates the essential advantage of the AFM imaging as compared to etching methods.

However, although both Figs. 3.5g and 3.5h (separated in time by more than 25 years) represent validation of the hypothesis for an organic nature of the material surrounding the grains, we shall see below that additional observations (Fig. 3.10) lead to modification of this first interpretation.

A similar granular structure in shell material viewed at the nanometer scale has been observed in every species of mollusc examined thus far, regardless of microstructural

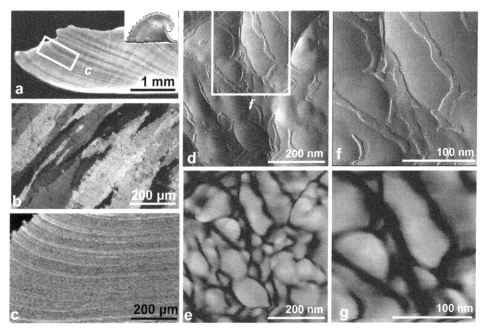

Fig. 3.6. Nanostructure of the calcitic prisms in *Concholepas concholepas* (Gastropoda). (a) Section of the shell border, where a daily pigmented striation is visible; (b) Longitudinal section of the calcite prisms; (c) Map of X-ray fluorescence at the energy level that is characteristic of sulfur in contained polysaccharides; (d–f) Amplitude images; (e–g) Phase images of the same area, showing the typical arrangement of nanograins surrounded by their cortex. b–e: Guzman *et al.* (2007).

organization. The following published figures provide examples of the principal micro-structural components of shells: prisms, nacre, and cross-lamellar structures (Dauphin 2001c, 2006, 2008; Dauphin *et al.* 2003c; Grefsrud *et al.* 2008). In all of these, a relatively comparable nanogranular organization was established. In spite of the diverse types studied, and very marked differences in their character at the microstructural level of analysis, it is even more noteworthy that their resemblance is so similar at the nanometer scale.

As an example of the calcite prism, Fig. 3.6 shows the outer layer of the gastropod *Concholepas concholepas*. In this species, the calcite prisms forming the outer layer are also perfect crystal-like units (Figs. 3.6a–b), built by the superposition of micrometer-thick layers in which interlayered organic material has here been mapped by the synchrotron fluorescence method (Fig. 3.6c). Although their morphology is somewhat different from *Pinna* prisms (being slightly curved from their origin towards the internal surface), the nanogranular organization of the calcite in this gastropod shell (a rather uncommon miner-alogy among the gastropods) is closely comparable to the prismatic calcite of *Pinna*. Amplitude images (Figs. 3.6d and 3.6f) and phase images (Figs. 3.6e and 3.6g) provide direct evidence of their similarity in the relationship between organic and mineral compo-nents when observation is made at the 10 nm scale.

3.1.3 *The general nature and diversity of submicrometer-sized units forming molluscan microstructures*

Applying AFM methods of study to microstructural units over a broad spectrum of shell types leads directly to the hypothesis that, in the 10 nm range, whatever their mineralogy, size, and shape, they are all formed of a rather comparable type of basic unit. Figures 3.7 and 3.8 are illustrations of this.

These images provide the first examples illustrating the contrast between the morphological diversity of microstructural units and the similarity of their internal organization at

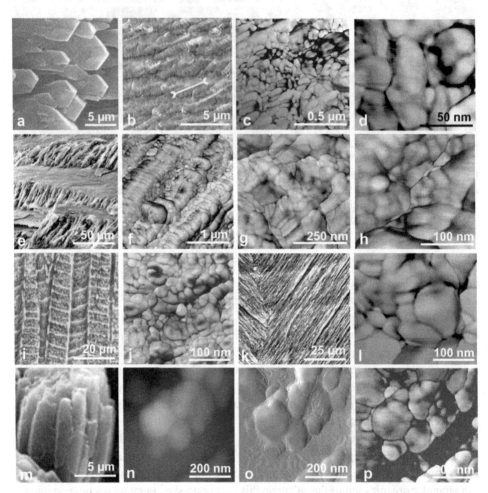

Fig. 3.7. Similarity of submicrometer features in various shell microstructures. (a–d) Fibrous calcite of the outer layer of the *Mytilus* shell; (e–h) Cross-lamellar aragonite in *Tridacna*; (i–j) Cross-lamellar aragonite in the internal layer of *Concholepas*; (k–l) Cross-lamellar aragonite in the *Cardium* shell; (m–p) Outer layer of *Haliotis tuberculata* (aragonite). i, j: Dauphin *et al.* (2003); m: Cuif *et al.* (1987); p: Dauphin *et al.* (2005).

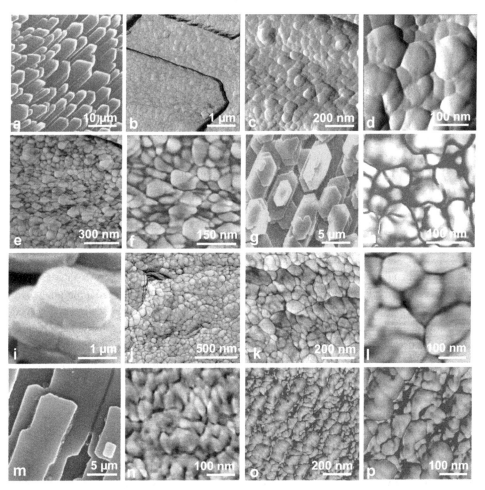

Fig. 3.8. Nanograins in some foliate and nacreous units. (a) Foliate calcite in the main shell layer of *Pecten*. A conventional crystallization process is strongly suggested by the angular limits; (b–f) Some units present aspects that are truly crystalline when looking at their horizontal boundaries, such as the ends of the calcite fibers in the outer layer of *Mytilus* (Figs. 3.7a–d), or the foliated layer of *Pecten* (a–f). At the submicrometer level, the included organically coated grains indicate that a different process has been involved (e–f: phase images); (g–h) A similar observation can be made here regarding growing nacreous tablets of *Nautilus*; (i–l) Nacreous tablets of *Haliotis* are less regularly shaped, but their internal submicrometer structure is closely comparable; (m–n) Nacreous tablets of *Pinna*: with their regular quandrangular shape, (m) might be suggestive of chemical crystallization of this aragonitic unit, but (n) reveals a reticulate structure; (o–p) Nacreous subunits of *Trocha* also show the typical assemblage of coated aragonite granules. b: Grefsrud *et al.* (2008); g: Dauphin (2006b).

the submicrometer scale. Some units present aspects that are truly crystalline when looking at their horizontal boundaries, such as the ends of the calcite fibers in the outer layer of *Mytilus* (Figs. 3.7a–d), the foliated layer of *Pecten* (Figs. 3.8a–b), or the nacreous tablets of *Nautilus* (Fig. 3.8g). It is noteworthy that, in contrast, these skeletal units are without a doubt formed of the universal, composite, organically coated granules. Additionally, when looking at transverse sections of them, the morphology of lateral faces is not at all crystal-like (as a good example of this, see Figs. 3.7a–d for *Mytilus* fibers).

Innumerable images can be added without modifying the overall result: *even those units that are the most crystal-shaped are built by assemblages of composite, organically coated granules.* This is obviously consistent with the multiplicity of identifications of the specific organic components within all these materials. What is being added here is data concerning the dimensional scale at which this interplay occurs.

On the other hand, these densely packed assemblages of organically coated granules are exceedingly intriguing with respect to the crystallization process itself. Clearly none of these 10-nanometer-size grains show faceted surfaces, which leads to suggesting that formation of microstructural units does not result from the crystallographic arrangement of preformatted, nanometer-sized units, as proposed by models of "nonclassical crystallization." An additional factor is thus added to the developmental sequence in the crystallization process.

At this observational level, evidence also suggests that, provided that the presence of irregular rounded nanograins is a feature common to all microstructural units known thus far, this uniformity of morphology of the 10 nm "building block" raises questions about morphological differences of the microstructural units themselves. How can such diversely sized and shaped microstructural units result from the assemblage of such similar and irregularly shaped subunits?

The molluscs that simultaneously produce both calcite and aragonite are particularly demonstrative. The best known of these is *Pinctada margaritifera* (the pearl oyster). In the calcite of its prisms (Fig. 3.9a) as well as in the aragonite of its nacre (Fig. 3.9c), the nanograins forming the basic units of both of these two truly distinct microstructural units have sizes and shapes that are practically indistinguishable in their morphology (Figs. 3.9d–i). This phenomenon occurs within the same shell, and therefore occurs in two mineralizing areas situated on both sides of a line of transition (Fig. 3.9b).

Thus, in areas of both calcite and aragonite deposition positioned within a single isochronous growth layer, similar coated grains are produced that are indistinguishable from a morphological viewpoint, but with well-established differences in the biochemical compositions of their organic constituents and, of course, the long-recognized mineralogical contrast in Ca-carbonate polymorph between prisms and nacre.

It is hard to find a clearer indication of the directive role of organic matrix in determining the mineralogy of basic constituents of biocrystals. Additionally, this result suggests a clearly distinct origin for the crystallization process itself, and the mechanism that determines the sizes and shapes of microstructural units.

3.2 Formation of the elementary growth layer: a two-step process that reconciles the crystalline behavior of microstructural units with their compositional properties

Crystalline behavior, the first observed and most obvious character of the microstructural unit, now appears to be an intriguing feature, because our observations at the highest

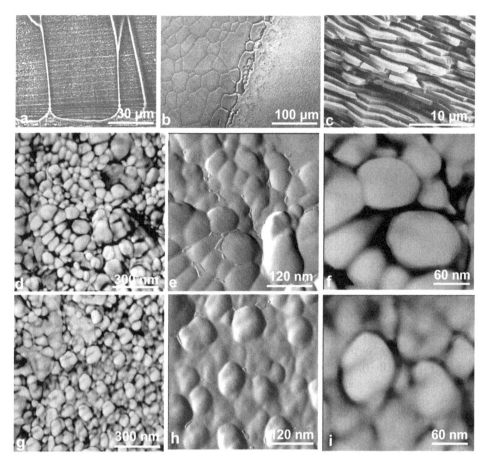

Fig. 3.9. Nanostructure of calcite and aragonite as they occur in *Pinctada margaritifera*. (a) Longitudinal section in calcite prisms; (b) Line of separation between prismatic and nacreous regions; (c) Tablets of nacre (aragonite); (d–f) Nanogranular structure of the calcite of prisms; (g–i) Identical organization of nacre: (d, f, g, i) phase images, (e, h) amplitude images. d: Baronnet *et al.* (2008); f, i: Dauphin (2008).

possible resolution do not allow us to see what could reasonably have been expected, that is, the presence in a given skeleton unit of nanometer-scale building blocks exhibiting morphologic patterns corresponding to the mineralogic polymorph of the sample studied. Such an observation has never been made during the examination of tens or hundreds of different shell materials. A question thus arises with regard to the process by which such undoubtedly crystalline material can be produced. The explanation of this is based on closer observation of the relationship within the growth layers of what seem to be the common patterns of any mollusc shell or corallite component: the coated grain, with dimensions in the 10 nm range.

Accessing the scale relevant to such observations is possible by transmission electron microscopy (TEM). In this method, an electron beam passes through the material to be studied, which must be prepared using ion thinning, consisting of controlled abrasion by projecting an ion beam at the sample. The energy of these particles must be very carefully controlled during the process in order to avoid structural modification of the sample. The very first prisms formed at the growing edge of juvenile *Pinctada margaritifera* provide a particularly useful case study, owing to their micrometer thickness.

3.2.1 TEM imaging and diffraction: evidence concerning the crystallization process at the nanometer scale

The larval *Pinctada* animal produces an aragonite shell that lacks a definite microstructure (Fig. 1.10) whereas the post-larval animal (Fig. 3.10a) begins to produce the classical sequence of calcite prisms followed by aragonite nacre. Within these youthful forms, size increase is very rapid and the distal border of the shell here is formed of a very fine prismatic layer (Figs. 3.10b–c). At this very young stage, such prisms already behave optically as perfect monocrystals. Under cross-polarized light, those at the margin of the prismatic layer exhibit first-order coloration up to gray at the growing edge (Fig. 3.10e). Observations in AFM show the normal structure of composite microgranular areas, in each of which highly interactive peripheral cortex covers a central rounded granule (Figs. 3.10e–h).

Transmission electron microscope observation of grains at the crystallization growth front (Fig. 3.10i) is of particular interest. TEM imaging in a series of increasing enlargements (Figs. 3.10j–l), combined with electron diffraction in the same area, permits demonstration of the sharp contrast that exists between the peripheral and central parts of these grains (Figs. 3.10m–n). Calcium carbonate in the central region of granules is well crystallized (Fig. 3.10m), whereas it appears to be amorphous in peripheral regions (Fig. 3.10n). Two different diffraction patterns, in the center and at the periphery of the same grain respectively, clearly illustrate the developing crystallization which eventually results in the highly coherent and crystalline network of the central region (Baronnet *et al.* 2008).

These results suggest that we must reconsider the interpretation of the cortical region of the nodular microdomains. This was first considered to be purely organic material because of the intensity of its interactivity with, and resulting contrast to, the central portion of the grain. TEM also indicates that Ca-carbonate is present as amorphous mineral material within the organic coating of grains. Amorphous Ca-carbonate was detected early on, and its possible role in the biomineralization process emphasized (see Chapter 5). Here, direct imaging provided by TEM is likewise strengthened by direct comparison to crystalline Ca-carbonate in the well-crystallized areas located more centrally.

These observations are of great importance in evaluating the concept of biochemically controlled crystallization within microstructural units. These indicate that crystallization develops from mineral components that were previously included within the organic phase. They also suggest that formation of the crystallized structure results in a relatively complete segregation of organic and crystallized materials; no organic molecule can occupy a place in the growing mineral lattice itself. As a result, organic material is pushed to the exterior of

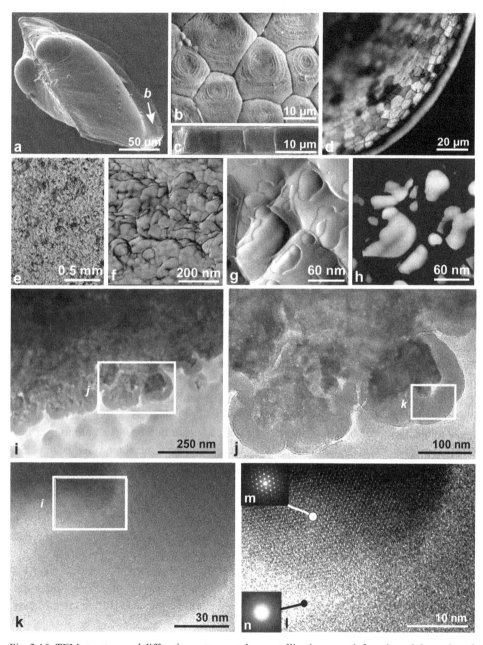

Fig. 3.10. TEM structure and diffraction patterns at the crystallization growth front in nodular grains of the calcitic prisms of *Pinctada margaritifera*. (a–c) Morphology of an early juvenile (i.e., post-metamorphosis specimen of *Pinctada margaritifera*); (d) Shell growth margin in transmitted light (crossed nicols); (e–h) Internal part of prisms observed by AFM (tapping mode). Coated granules are individually visible in (g) amplitude image, and (h) phase image. Image (g) illustrates the sharp contrast between grains and coating, leading to the interpretation of this coating as organic; (i–l) A

crystallized zones, but having no way to escape, it forms the irregular "coating" of crystal-lized regions, with its typical high phase lag observed in the AFM tapping mode.

In this interpretation, the role of organic components appears to be of particular impor-tance. Previous observations showed that every coral or mollusc skeleton contains a portion of hydrated organic compounds, the proportion and composition of which varies from species to species. Among crystallization models based on the existence of *common extrapallial space* (in molluscs, or *subectodermal space* in corals), no precise role is attributed to the organic component, although its direct relationship to the mineral phase is sometimes accepted (Furla *et al.* 2000). TEM imaging and associated diffraction figures support the hypothesis that a major role for this organic phase may be that some of these organic compounds act as Ca-carbonate carriers during the process of growth layer for-mation, so that crystallization itself occurs as the final step in the process of producing a newly mineralized growth unit.

3.2.2 A model for sequential, two-stage formation of the mineralized growth layers

In order to produce this somewhat paradoxical structure of polycyclic biocrystals, their stratified growth by isochronous layers must be reconciled with the crystalline continuity that is successfully maintained during production of superposed layers. Establishing that Ca-carbonate changes from an amorphous state to a crystallized one as the final step in the mineralization cycle of a growth layer (Fig. 3.10l) suggests a two-stage model for the growth of these crystal-like units.

Initially, in the first phase of emplacement of a new growth layer (Fig. 3.11b), mineraliz-ing epithelium produces nodular structures composed of nanometer-sized spherules of an organic hydrogel, which are carriers of calcium carbonate that is still amorphous at this stage. These nanograins accumulate between the secreting cell membrane and the mineral-ized surface of the previous crystallization cycle. In addition, it is emphasized that in each of the distinct mineralizing areas of active epithelium, the composition of the organic hydrogel varies.

In agreement with Crenshaw's 1980 concept (and in contrast to models based on a common extra-epithelial liquid compartment), no space exists here that could be filled with a fluid in which uncontrolled crystallization can develop. The organic phase, with a specific composition for each mineralizing area of the secreting epithelium, thus plays the primary role of transporting agent for amorphous calcium carbonate; nodules are produced external to the cell layer and accumulate until reaching the thickness of the growth layer

---

Caption for Fig. 3.10 (cont.)

sequence of progressive enlargements in TEM. Images (i) and (j) allow the recognition of individual granules, whereas (k) and (l) reach sufficiently high magnifications that the internal organization of grains can be seen; (m–n) Diffraction figures in the core (m) and peripheral parts of grains (n), illustrate the crystallographic contrast between the well-crystallized center and amorphous diffuse mineral material at the periphery. g–n: Baronnet *et al.* (2008).

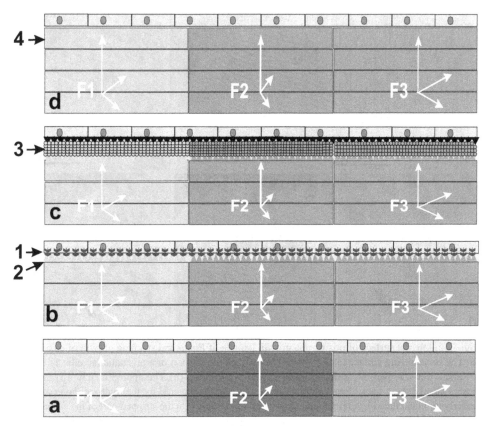

Fig. 3.11. The sequence of steps illustrating a growth cycle for three polycyclic units. (a) The mineralizing epithelium is in contact with three microstructural units: F1, F2, F3 (fiber, prism, etc.); (b) Cells of the mineralizing epithelium begin secreting mineralizing materials (1), which come into contact with the upper surfaces of the three fibrous units F1–F2–F3 (2); (c) The next, preparative step (accumulation of mineralizing material) having reached the necessary thickness (3), the crystallization process then occurs, and is controlled by some overall signal (4) (see the Mg map in *Porites* (Fig. 3.13) for discussion); (d) In each compartment that faces F1, F2, F3, the crystallization process adds one growth layer to the underlying microstructural unit, with more or less precise lateral continuity depending on the bounding properties of surrounding organic envelopes. Modified from Cuif and Dauphin (2005a).

(Fig. 3.11c). Obviously, the organic phase involved in this process is associated with water. It can thus be assumed that it is this water that is emitted at a temperature of about 300 °C during ATG experiments (see Fig. 2.23). With respect to the origin of this water, an observation of major interest was made by Lécuyer and O'Neil (1994). Studying the oxygen isotopic ratios of these "skeletal waters," compared to those found in normal seawater, they showed that the occluded waters are by no means seawater: instead they originate from body fluids as a part of the secreted mineralizing compounds. It is unnecessary to emphasize again how important this conclusion is in regard to biomineralization models. It is fully

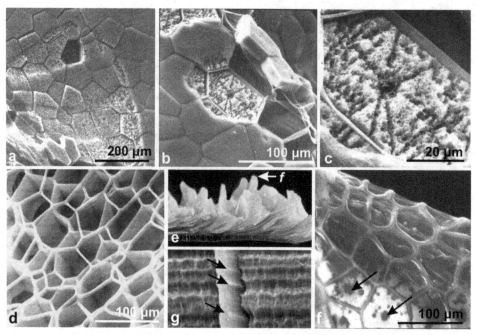

Fig. 3.12. Internal surface of prisms in a growth lamella of *Pinna* and stepping growth of envelopes. (a) The mineralizing surface (about 1 mm²) in the growing portion of a prismatic lamella. The thick organic sheet was partially stripped away during preparation; (b) The approximately 15 µm thick prisms clearly show the difference between the soft surface covered by the hydrogel layer and the underlying mineralized layer (in an initial stage of a prism); (c) On the stripped part of the surface one sees that their envelopes surround the growth layer that is as yet noncrystallized; (d) On a completely decalcified shell fragment, the last growth layer of insoluble envelopes has been detached, confirming the stepping growth mode of the insoluble component; (e–f) At the extreme distal margin of a growth lamella (e: arrows), the polygon of organic material is ready to accommodate the mineralizing gel (f), but the process has been interrupted by a premature retraction of the animal's mantle. Only some polygons from a more internal position show development of the mineralizing phase (arrows); (g) The stepping growth process is visible on this view of the internal side of an organic envelope (arrows), complementing (d). b, c, f: Cuif *et al.* (1983).

compatible with the hypothesis of an internal origin of all of the components of elementary growth layers (including water) and it reinforces the conclusion of the ATG experiments that have suggested a physical link between organic components and these body-fluid waters. Figures 3.12a–b best illustrate this state of continual accumulation of noncrystallized material superposed on the previously mineralized layer.

The second stage in the formation of a growth layer must then involve a process that produces the crystallized structure, as it will finally exist. To reproduce successively the specific crystal form and orientation characteristic of each adjacent microstructural unit (F1–F2–F3), the process of crystallization must be based on there being a close proximity between the organic nanogranules (carrying amorphous Ca-carbonate) and the previously crystallized surface of the underlying microstructural unit. Some crystallization signal must exist (see Fig. 3.13) that leads to the breaking of chemical linkages between the organic component and Ca-carbonate within accumulated nanograins. Crystallization can then progress through the complete growth layer. During the crystallization process, the organic phase, which was until then the carrier for isolated molecules of Ca-carbonate, apparently is pushed to the periphery of the crystallized zones. Continuity of the mineral network with the underlying mineralized compartment is thus established.

Only then does the crystallographic coherence develop that is appropriate for each microstructural unit, with the organic compounds found distributed at the periphery of crystallized elements. Thus, in spite of its overall monocrystalline behavior, it is not a true compact crystal that is formed, but instead, *a reticulated crystal* that includes spaces occupied by remaining hydro-organic gel.

From a functional point of view, the organic phase of the nodular structures now appears to be an essential element in the mineralization process, assuring the stabilization of amorphous Ca-carbonate prior to crystallization.

3.2.3 Crystallization within the gel layer: the first indication of a crystallization signal

The sequential model of growth layer crystallization, that was initially proposed to explain crystallization in layers of coral skeletons (Cuif *et al.* 2005), is clearly also valid as a model for the formation of microstructural units in mollusc shells. The observations on which it

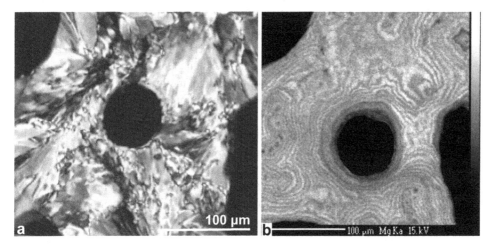

Fig. 3.13. Evidence of the presence of a common signal related to the layered crystallization process in a *Porites* skeleton. (a) Thin section of the skeleton: fibrous bundles growing in various directions, apparently justifying the concept of *concurrent growth*; (b) Mapping of Mg within the same skeleton provides evidence of layering on an overall scale. In this field of view (about 250×250 μm), the pattern of Mg banding indicates that control of the stepping growth of the skeleton is exerted over broad surfaces. (Image: C. T. Williams, Mineralogical Dept., NHM, London; CAMECA SX100.)

was based are in complete agreement with the conclusion of Crenshaw, formulated in the 1980s. His model concerned the process of transferring organic compounds and minerals from epithelial cells to the internal surface of the shell during growth. It is worth remembering that, after having carried out numerous measurements of physicochemical conditions in the extra-epithelial space of molluscs and analyses of any fluids that could be extracted from it, Crenshaw recognized that the numerous such attempts of this type "have contributed little to the understanding of shell formation." Thus, to Crenshaw, and contrary to the hypothesis of precipitation from a liquid phase, basic to all prior investigations of this type, "the transfer of materials to the inner shell is essentially direct" (1980, p. 120). It is remarkable that Crenshaw's observation is found applicable to corals as well as molluscs, two groups that for a long time were considered as very different from a biomineralization point of view, grouped differently, one as "biologically induced" and the other as "matrix mediated."

Figure 3.12 shows the growth surface of some prisms of *Pinna* that have been carefully prepared to avoid dessication features (critical point drying). These provide very direct evidence of the secretion cycle that functions here. Beneath the partially stripped sheet of hydrogel, the mineralized surface of the previous cycle can be seen. Since this picture was taken at the distal edge of a growth lamella, the particular images that correlate to early mineralization steps are still visible (centers of crystallization, see below).

Synchronous action of the mineralization process within a given growth layer is a practically unexplored concept because dominant opinion until recently has agreed on the completely opposite view: crystal growth competition. Barnes (1970) was perhaps the first to develop this idea concerning the coral skeleton. According to this model, directly inspired by the influential paper of Bryan and Hill (1941), the principal factor determining the specific three-dimensional organization of microstructural units is the presence of crystallization occurring freely, with competition between crystals for orientation and rate of growth occurring within a Ca and carbonate ion-saturated, liquid-filled space. More recently, interpretations concerning microstructures in mollusc shells have commonly been developed in the same way (see model for a "common mineralizing chamber" in Fig. 2.21). However, all of the data and images accumulated at the micrometer scale that provide evidence of coordinated growth argue against such a view of an "unregulated" growth process (or more exactly stated, regulated by physical forces only).

However, considering biologically regulated processes, we must find evidence from some chemical or structural pattern that a regulating signal has been active on wide areas of growing skeletons, ensuring the synchronous nature of growth layer crystallization. Figure 3.13 provides significant data on this point.

On this section of a *Porites* skeleton (Fig. 3.13a), fiber bundles are growing in various directions, supporting an interpretation of it being the result of freely occurring crystallization processes. However, microprobe mapping of cyclical change in magnesium concentration provides evidence of synchroneity on a broader scale (compare this to Fig. 2.10, a higher-resolution picture made by NanoSIMS). Although the role of magnesium in the crystallization process is not clear in this instance, the broad mineralizing area where the Mg signal occurs definitely indicates that a mechanism ensuring coordinated crystallization is functioning throughout broad areas.

3.2.4 Some indications of the frequency of cyclic growth layer formation

Cyclicity in the growth of molluscan shells

We now turn to the kinetics of the biocrystallization process, an important parameter for the interpretation of this phenomenon. We focus here on examples where the periodicity of deposition of these basic growth layers has been suitably established, or at least estimated. Marking experiments conducted along with microstructural analysis of formation of calcitic prisms in *Concholepas concholepas* (gastropod of the Chilean coast) now provide us with data that are truly reliable (Fig. 3.14).

These gastropods are raised in experimental aquaria in the biological station at Antofagasta (Chile). The species has an alimentary regime that is very specialized; in the adult stage it only consumes small intertidal mytilid bivalves of the species *Perumytilus purpuratus*. The animals were maintained in optimal nutritional conditions by furnishing them with natural nutrients, and then marked with calceine (Guzmann *et al.* 2007). By temporarily placing the animals in seawater containing 50 mg/l of calceine, the fluorescent molecules are then incorporated in mineralization layers formed during 3 to 6 hours (the duration of each experimental bath) and a very precise growth chronology established. As a result, examination of a polished transverse surface under light that is a mixture of fluorescent UV (400 nm) and tungsten light allows simultaneous observation of both the

Fig. 3.14. Timing of crystallization of growth layers in *Concholepas concholepas*. (a–b) Section of the shell and a polished surface; (c) Position of calceine markers after an interval of one week. Note the irregularity of growth layers and also the distinction between nocturnal layers (dark) and daytime layers (light); (d–e) Growth layers corresponding to one day's growth. a–e: Guzman *et al.* (2007).

fluorescent calceine tags and differences in pigmentation corresponding to nyctithermal alternations (Figs. 1.15f–h). One day's average growth provides an elongation of calcite prisms of approximately 100 µm. SEM observations after etching of the same surface establish that this daily growth is accomplished by superposition of growth layers, each having an approximate mean thickness of 2.5 µm (Fig. 1.15h). This result, by accurate and direct measurement, bears witness to the high periodicity that cyclical biomineralization can reach in optimal conditions.

Observation in UV fluorescence of sections prepared in these shells (Fig. 3.14) provides very precise location of the markers, and confirms the day/night alternation that allows clear reconstruction of the mode of shell growth between each successive marker. During optimal living conditions, a daily growth layer attains a thickness of approximately 100 µm (Figs. 3.14c–d). Microstructural study (on polished and etched surfaces) reveals that this thickness contains several tens of elementary strata (in practice, approaching 50), reflecting the extremely rapid cycles of diurnal crystallization.

Pinctada margaritifera also has been very thoroughly studied because of its utilization in production of commercial pearls. It was early established that the growth doublet present in these pearls has a quotidial periodicity, with a dark/light doublet having a thickness of 5 to 6 µm (Figs. 3.15a–c), and each corresponds to the emplacement of 9 to 12 layers of nacre per day. This number is markedly influenced by environmental conditions, among which environmental constancy is a permanent preoccupation in the rearing of *Pinctada* because irregularities within nacreous layers are a major factor in pearl quality.

In oysters belonging to *Crassostrea* species, an estimation of their elementary cyclicity can be obtained from the growth lamella produced in one day. Taking into account the value for calcite birefringence, the thickness of prisms can be measured by their polarization color; that attained by the thickest being approximately 15 µm, which provides an estimate of the number of growth cycles of at least 7 to 10 each day.

An equivalent number is provided by the study of growth striae and lamellae formed during larval stages of *Ostrea edulis* (Fig. 3.16). In larval stages, these oyster larvae produce an *aragonitic* shell with a microstructure that is reduced to a series of organic lamellae in which mineral granules are present, but without any particular arrangement (Figs. 3.16c–d). These growth lines are visible on the larval shell exterior (arrows) and are the result of

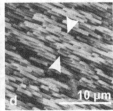

Fig. 3.15. Frequency of the mineralization cycle in *Pinctada margaritifera*. (a–b) Growth stratification common to nacre and to prisms (see also Fig. 1.1); (c) Thickness of a growth doublet (dark/light) in the nacreous layer; (d) Strata of nacre corresponding to the thickness of a daily doublet. a, b: Dauphin *et al.* (2008).

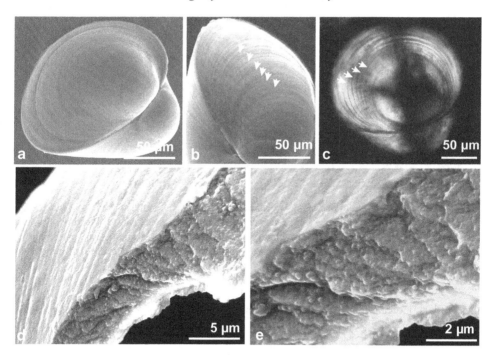

Fig. 3.16. Larval shell of *Ostrea edulis*, 13 days after fertilization (IFREMER, Argenton hatchery). (a–b) Growth striations are readily visible from the exterior. Major concentric lines correspond to daily growth; (c) Image of the complete larval shell under polarized light (crossed nicols). The black cross indicates that a unifying structure exists to form the uniaxial cross, perhaps due to uniformly arranged calcite biocrystals and organic macromolecules (such as chitin); (d–e) Broken shell, illustrating the relationship between external striations and the layered organization of the shell.

deposition of the lamellae. As the ontogenesis of the larva is well known, rhythmicity in the formation of lamellae can be precisely measured: 20 to 25 growth striations daily. To appreciate the importance of larval metamorphosis with respect to shell formation, it should be noted that after metamorphosis the complete shell is constructed of calcite, and is then comprised of three mineralized forms: calcite prisms, lamellar calcite, and "chamber lamellae" (the last, structures specific to oysters).

In *Sepia* (Cephalopoda), pillars of its calcareous structure (Fig. 3.17) also furnish very precise indications of the rapid periodicity with which cycles of mineralization are produced. In spite of the transformation required for the formation of this very original microstructural type (see Fig. 1.18), the fundamental mechanism of calcification has preserved its rhythmicity; approximately equivalent to that observed in shells of other molluscs. Evaluating the influence of environmental parameters on growth and the number of calcareous layers produced each month, Le Goff *et al.* (1998) determined that during optimal growth periods, 18 lamellae are produced in 30 days. The pillars that join these lamellae are thus constructed in two days, and possibly even more quickly, taking into account the time necessary for formation of the lamellae themselves. Growth striations, highly visible on

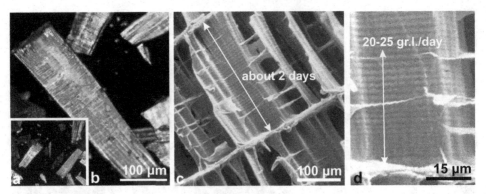

Fig. 3.17. Frequency of the mineralizing cycles in the pillars of the cuttlefish bone of *Sepia*. (a–b) Interlamellar pillars of the cuttlefish bone, each crystallized as a monocrystal; (c) Position of pillars between two successive lamellae; (d) Growth layers of the pillars. The same number of growth layers can be readily counted on multiple pillars from a given interval between two lamellae, reflecting their synchronous formation. c, d: Dauphin (1981).

these pillars, indicate that the pillars are constructed by at least 25 growth cycles per day (Fig. 3.17c).

Cyclicity in the growth of scleractinian corals

The growth of corals is generally measured at the millimeter scale, most of the time by using X-ray radiography that establishes density differences in skeletal aragonite during an annual cycle (Fig. 3.18). These radiographs also indicate that cycles of an infra-annual nature also exist. This chronological method is adapted to research that utilizes corals as paleoenvironmental records. Such measurement of linear growth does not directly reflect rates of mineralizing activity because the growth layers forming skeleton are virtually always oblique with respect to the direction of skeletal extension. This obliquity of the mineralizing area is easily seen on a vertical axis of *Porites*, the summit of which generally has a conical morphology (Fig. 3.18d), and this growth surface morphology is again seen in longitudinal sections of these axes (Figs. 3.18e–f). Longitudinal growth corresponding to the number of cycles of mineralization can be estimated, but in the absence of a time base adapted to this scale of observation, transposing such an estimate of annual growth is only an approximation that is not exact enough to establish the periodicity of cycles correctly. In particular, variations in the rhythm and phases of slowing or stoppage will be integrated into averaged estimates, making them nonrepresentative.

Measurement of mineralization cycle frequency in scleractinian corals thus makes marking experiments obligatory. These are made during brief intervals of time, and must be carried out on specimens cultivated in an aquarium under well-regulated conditions. To provide significant information, growth estimation at the micrometer scale requires that the bundles of fibers utilized are oriented in the plane of the observation (cut perpendicular to growth strata).

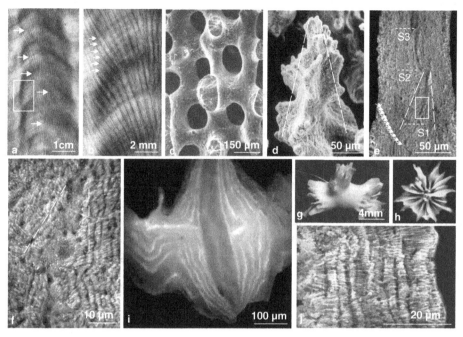

Fig. 3.18. Periodicity of the crystallization cycle in scleractinian corals. (a–b) Radiograph of a longitudinal section through a colony of *Porites* sp. The longitudinal dark lines visible in (b) correspond to sections through walls between corallites (X-rays are absorbed more by walls than the axial cavities); (c–d) Longitudinal section of a corallite. The vertical rods have been formed by the mineralizing epithelium; image (d) shows the last surface of mineralization produced; (e–f) On these longitudinal sections are seen oblique traces of successive positions of the mineralizing epithelium; (g–j) Calceine-c marking in a specimen of *Galaxea fascicularis*: (g, h) morphology of the specimen, (i) traces of successive calceine markings; fluorescence under 435 nm lighting (calceine marking by E. Tambutté, CSM Monaco), (j) growth strata observed on the calceine-marked specimen.

At the Science Center of Monaco, marking with calceine has been carried out on *Galaxea fascicularis* (Figs. 3.18g–i). In sectors where the growth direction of fibers corresponds to the orientation of the observation surface, fiber elongation of 50 μm is the result of 12 to 14 mineralization cycles. In addition, traces of calceine marking carried out at weekly intervals are seen to be separated by 35 to 50 μm of skeleton (within zones of maximum thickening). This suggests that, in a week, *Galaxea fascicularis* has deposited a maximum of 10 to 15 cycles of mineralization in these areas; a much lower number than was determined in molluscs.

Very little experimental data exists concerning the true rhythmicity of coral growth at the scale of biomineralization cycles. It is probable that the frequency of mineralization cycles is dependent on nutritional conditions (as it is in molluscs), and influenced by light for the zooxanthellae in most scleractinians. However, it is also certain that microscale growth cycles also exist in nonzooxanthellate, deep-water corals (*Lophelia, Caryophyllia,*

Desmophyllum, etc.), each of which live in very stable conditions. This suggests that a micrometer-scale cyclic crystallization process is basic for most (if not all) of these organisms, allowing them to maintain permanent control on the formation of their skeleton.

3.3 Formation and control of growth of microstructural units: examples from prismatic structures

The schematic model for the layered mode of growth (shown in Fig. 3.11) allows us to account for the structural, biochemical and crystallographic properties of coral and mollusc skeletons. It should be observed that this is applicable only to the part of the growth process in which geometrical features and the three-dimensional arrangement of skeletal units are already well defined. In other words, it is essentially a repetitive process, which does not provide an explanation of the origin of taxonomy-linked peculiarities.

Actually, the formation of the principal characteristics of skeletal organization occurs at the extreme anterior margin of the mineralizing region. In this generative zone, newly formed structures insure the lateral extension of mollusc shells or elongation of coral skeleton. In both cases, evolutionary mechanisms continuously produce innovative patterns of three-dimensional features deposited in the distal mineralization zones, resulting in their being of major taxonomic significance. Thus, a contrast is apparent between those structures created in the distal mineralization zone and those deposited as a second phase through a simple repetitive process, as previously described.

In this part of the discussion, a more precise examination focuses on the early stages of microstructural units in some typical examples of bivalve shells, usually simply described as the superposition of two distinct layers, prismatic in the outer, and nacreous in the internal layer.

3.3.1 *The intial stages of formation of calcitic prisms: identification of a "center of crystallization"*

The distal border of a growth lamella in a *Pinna* shell (Figs. 3.19a–c) exhibits a polygonal network that forms the limits for prisms, but also reveals that the early development of each prism starts with a series of concentric circles surrounding well-differentiated nodules at the center of each polygon. There is a remarkably regular geometry formed by the positions of these centers and the positions of the segments that make up the polygons. Each segment is perpendicular to a line that joins two neighboring centers so that, in drawing the network thus formed, a classic figure is obtained, called a "Voronoi Tesselation System" (Figs. 3.19d–e). This arrangement is very significant in determining the manner in which the polygonal network forms. A Voronoi arrangement is observed each time a diffusion process is produced that begins from distinct centers; the median segment perpendicular to the line joining each pair of centers represents the relative influence of each diffusion center. In actuality, each of these segments provides a *line of equal influence* of the two centers (Fig. 3.19e).

Figure 3.20 presents the opposite side of the growth lamella, the internal face showing the temporary condition of the latest mineralizing activity of the mantle, just before its

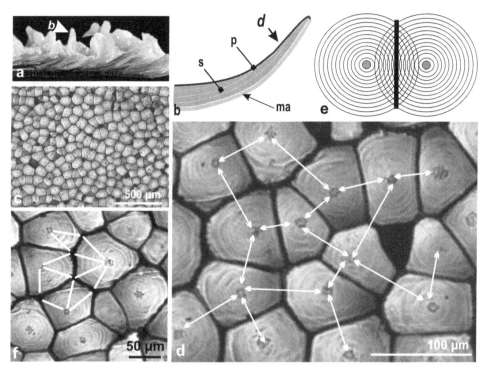

Fig. 3.19. Initial stages of prism formation in *Pinna* sp., here seen from the external (periostracal) side. (a) Growth lamellae at the margin of the *Pinna* shell; (b) Schematic drawing of a lamella ("p": periostracum, "s": growth sheet, "ma": mantle); (c–d) Prism envelopes (c) and geometric relationship between the centers of the polygons and their segments. The polygons are organized around "centers" allowing a Voronoi tesselation system to be drawn; (e) Construction of a line of equal influence radiating from two diffusion centers; (f) Inverse pattern obtained by viewing segments from the centers.

retreat, prior to formation of a new lamella deposited during a more extended position. Figures 3.20a–e show that in each polygon, mineralization develops in a concentric fashion around a center, *which is not itself mineralized* (Fig. 3.20: arrows). During formation of the initial structure of a prismatic layer, these centers play a double role. They direct the formation of polygons that define the prismatic structure, and equally, the crystallization that develops around these centers. At this stage, one would imagine that the polygonal envelopes could just as well be interpreted as the result of crystal growth, ending with the contact between polygons during crystallization of the mineral phase. Figure 3.20f shows that this is not the case. At the lateral extremity of a growth lamella (Fig. 3.20b), one can see that the polygonal network already exists prior to beginning crystallization of the mineral material. When the crystallization phase is complete, it fills spaces that were previously defined by the organic network. Thus, it is certain that even if they are not visible on Fig. 3.20f, the centers already exist and at this stage have fulfilled

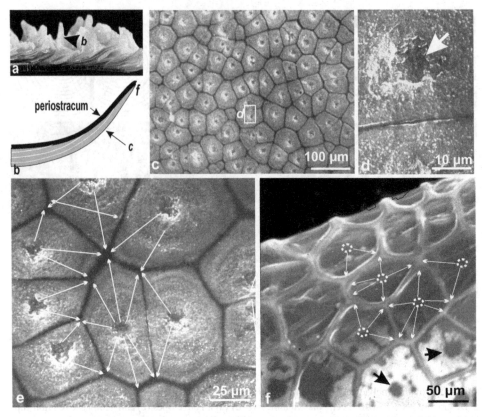

Fig. 3.20. Views of the internal face of a growth flake of a *Pinna* shell. (a–b) Position of the surface observed; (c–d) After having been an undetermined time in contact with seawater, the prism centers appear as hemispherical empty cavities; (e) Concentric development of the mineral material is still visible; (f) At a similar observation site, but on a freshly collected and carefully prepared sample (critical point drying), polygonal envelopes are visible and early mineralization can be interpreted as the result of the action of "centers" (dots and arrows). On this image, it is seen that the formation of typical polygonal organic envelopes clearly precedes mineralization. c, e: Cuif *et al.* (1983).

the first part of their role: the creation of polygons to define the form and arrangement of the prism limits.

These two figures have been obtained by study of the growth extremity of a lamella of a *Pinna* shell. They illustrate the structure established at the moment when the process of biomineralization stops immediately preceeding retraction of the mantle, the step preliminary to deposition of a new more extended growth lamella.

It is concluded that the question of prism formation has to shift to the examination of the origin of these central and highly influential structures that determine the morphology of yet to be formed mineral units, and that also control the very early stages of crystallization. Examination of the prismatic structures produced in the most youthful portion of the *Pinctada* shell leads us to focus on the role of the periostracum, the organic covering of the shell, commonly described as being a simple protective layer that isolates a mineralization compartment from seawater.

3.3.2 Prism formation in **Pinctada**: *crystallization of initial stages of the prism and determination of its crystallographic orientation*

On the upper surfaces of prisms in juvenile specimens of *Pinctada* (i.e., post-metamorphosis), a system of concentric grooves organized around a "center" is visible (Figs. 3.21a–b: arrows). During this phase of rapid shell length development, the groove

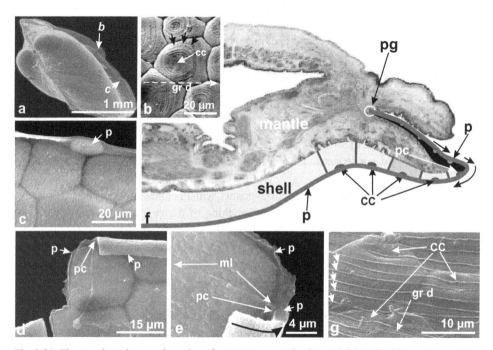

Fig. 3.21. The mantle at the growing edge of a post-metamorphosis youthful shell of *Pinctada* sp. (a) In the juvenile shell it is commonly possible to observe both the external (b) and internal (c) side of the growth border; (b) Concentric grooves surround the "center" on the external ("cc") side of the prisms; (c) The internal sides of these same prisms have undifferentiated growth surfaces ("p": a very young prism, yet to be completed); (d–e) The periostracum at the extreme growing edge of the shell valve. Note the organic layer covering the internal surface of the prism until the start of periostracum curvature ("pc"); (f) Schematic presentation of the formation of initial stages of the internal face of the periostracum. Produced by a group of cells localized in the periostracal groove ("pg"), the periostracum transports organic materials progressively deposited on its internal surface by the internal lobe of the mantle. Observation of the shell growth edge indicates that material forming the "centers" of the prisms arrives just before the distal incurving of the periostracum ("pc": periostracal curvature); (g) Parallel striations of the periostracum, here visible (arrows) on the external surface, indicate that its growth is a stepping process (histological section of the mantle: courtesy N. Schmitt, University of Polynesia, Papeete).

system is strongly eccentric. In this genus, in contrast to that observed in the lamellae of *Pinna* (Fig. 3.20), mineralization begins around centers even before the limits of prisms are complete. As a result, the most recently formed prism now generates the segments of its envelope, while maintaining its geometry with respect to the centers of the two neighbors (Fig. 3.21c).

In Figs. 3.21e–f, the two layers forming the periostracum are seen to be formed prior to the first prisms, and incurving of the periostracal lamellae can be observed. These cover the upper face of the prisms. The organic layer originates at the bottom of the periostracal groove (Fig. 3.21: periostracal groove, "pg"). Simultaneous growth of the animal mantle and shell is allowed by production of periostracum. In addition, the features that constitute centers are deposited by the cells of the mantle lobe onto the internal face of the growing periostracum and thus transported towards the shell border. Practically speaking, it is at this level that the structures are generated and ordered; subsequently, they determine the microstructural organization.

Immediately after incurving of the periostracum occurs (Figs. 3.21d–f: "pc"), the mineralizing activity of these structures begins, and forms the concentric mineralized circles that are readily visible on the external faces of prisms once the periostracum is removed (Fig. 3.21b). On intact shells, the rhythmic mode of growth of the periostracum is very clearly visible (Fig. 3.21g: arrows), although the centers cannot be discerned except by the presence of a very weak depression in a nearly central position, typical of rapidly expanding shells.

It is remarkable that in these very youthful prisms, with thicknesses of only approximately 3 to 4 µm, the internal face does not show anything of the concentric system that is so obvious on the upper surface. This indicates that the stage of concentric crystallization is limited to the very external surface of the prism only. Very quickly, it is replaced by stratified growth, illustrated here by the thin organic layer visible on the internal surface of these prisms (Fig. 3.21e: "ml").

At the growth margin of juvenile *Pinctada* shells, prisms are monocrystalline, just as are the prisms of *Pinna* (Figs. 3.22a–b). They then become polycrystalline during a later stage of their development (Figs. 3.35e–f). Observation of the external surface of the prisms indicates conclusively that regardless of having a granular appearance (probably due to early alteration of the surface), the initial layer of mineralization exhibits typical crystallographic patterns (Figs. 3.22c–f). Traces of the ever-expanding crystallization are now distributed into four regular sectors (or at times, three unequal ones) that are always centered on the developmental center of the prism. This concentric organization of granules within these sectors makes it obvious that an expanding crystallization process is present and active here.

In order to identify more precisely the relationship between the superficial traces of early crystallization and the progress of crystallization beneath the surface, an ultrathin section passing through the center of a prism and perpendicular to its external surface (Fig. 3.22f: line) was thinned by the use of the focused ion beam (FIB) method.

In the transmission electron microscope (Figs. 3.22g–h), the top of the section (that is, positioned at the external surface of the prism) shows a small nodule (Fig. 3.22g) that

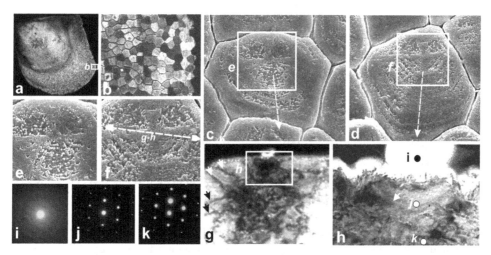

Fig. 3.22. Development of initial crystallization in a single prism of *Pinctada margaritifera*. (a–b) Juvenile shell and (b) true monocrystalline prisms (typical for youthful stages); (c–d) and (e–f) Morphological figures showing development of crystallization from the offset "center" on the tops of prisms; (g–h) and (i–k) Images and location of diffraction patterns in a focused ion beam through a thin section in the prism (axial plane comprising the nonmineralized central nodule). Arrows in (g) and (h) show the Bragg's fringes. (Data: A. Baronnet, Marseille.)

provides us with a completely amorphous diffraction diagram (Fig. 3.22i). An obvious parallel can be made with the cavities visible at the top of the prism when periostracum has been eroded away and the underlying organic material decayed. When observed by X-ray diffraction in TEM, the FIB thinned section shows Bragg fringes organized around this central nodule, suggesting that, although nonmineralized itself, it acts as the *center of crystallization*.

Beneath this organic nodule, the first mineralized layer has been developed (Fig. 3.22h: arrows). In contrast to the organic nodule, this layer furnishes a diffraction diagram reflecting its crystalline state (Fig. 3.22j). The structures beneath this correspond to successive phases that progressively occupy the complete volume of the underlying space. This observation, along with diffraction images, supports the interpretation that an initial organic element, transported by the internal periostracal layer acting as a "conveyor belt," defines the crystallinity of each prism as it passes the periostracal curvature. Depending on the orientation of this crystallization-inducing element, prisms develop variable orientations (however, always with the c-axis more or less perpendicular to the shell surface).

This sequence of images that establishes the occurrence of crystallinity within the first mineralized layer of prisms makes clear the difference between the initial stage of development, when crystallization patterns are established, and the elongation stage which simply repeats the crystal characteristics established in the initial process. However, in cases where very large prisms are formed, as in the outer layers of *Pinna* or in oysters, we can also see an additional internal structure that is most probably related to the perfect monocrystalline nature of these prisms (Figs. 2.44b and 3.24a–d).

3.3.3 Axial structure within prisms: an organic framework that maintains the crystalline character through superposed growth layers

It has long been observed (Tsuji *et al.* 1958) that among those bivalves that build large calcite prisms exhibiting strictly monocrystalline behavior, there also exists an axial organization in addition to the banding that results from patterns of the thickening process. These axial structures are certainly linked to the growth direction of the shell, and can easily be seen by looking at the growing surface of the *Pinna* shell. The ridge and groove series, visible in both transmitted light and SEM views (Figs. 3.23a–b), are always parallel to the shell border. After natural or experimental decay of their organic envelopes, *Pinna* shell prisms show that this growth pattern is not just limited to the growth surface, but rather, is seen to continue throughout prism growth, as the same pattern is readily visible on their entire lateral surfaces (Figs. 3.23c–d).

Preparing polished longitudinal sections in prisms of *Pinna* and submitting them to surficial decalcification with a decalcifying solution containing a fixative dye (for instance, Mutvei's mixture of acetic acid, glutaraldehyde, and Alcian Blue), allows us to see that these superficial patterns are actually related to the internal three-dimensional structure of the prisms (Fig. 3.24). First, the envelopes of the prisms cut longitudinally are exposed (Fig. 3.24a: "e") and growth lamellae are seen to correlate perfectly between adjacent prisms. This image corresponds structurally to Fig. 1.7j, obtained by enzymatic etching of prisms, but of course with the reverse image. Figures 3.24a and 3.24d show that growth lamellae are separated by *ridges* (white horizontal arrows), rather than grooves that result

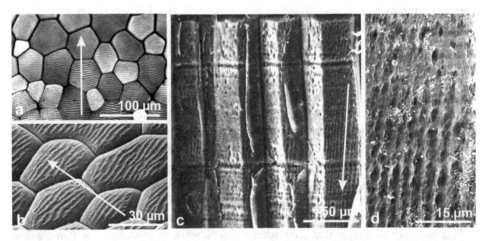

Fig. 3.23. Ridge and groove systems in prisms of *Pinna*. (a–b) View of growing surfaces. Orientation of crests and grooves parallel to the margin of the shell: in the optical microscope in transmitted light (a) and in an SEM image (b). Arrows indicate the growth direction of the shell; (c–d) After decay of prism envelopes, the crest and groove system is also visible on lateral faces of the prisms (probably here accompanied by partial alteration of the external surfaces of the prism itself). Note the general obliquity of the crest and groove system with respect to the limits of the prisms (arrows); also visible internally in Fig. 3.24a. c–d: Cuif and Raguideau (1982).

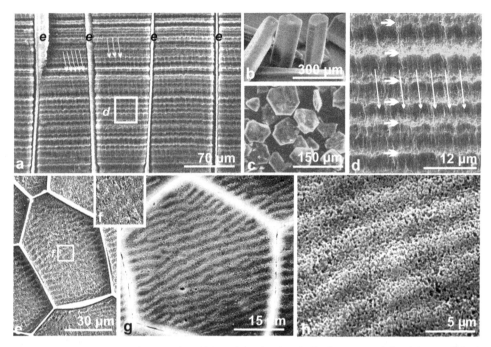

Fig. 3.24. Axial organic framework within prisms in the shell of *Pinna nobilis*. (a) A polished longitudinal surface of prisms that has been submitted to a decalcifying and fixative solution. The internal organic structure is made visible by controlled decalcification. Note its oblique orientation with respect to prism, axes with slightly varying angles from prism to prism (arrows); (b–c) Organic-rich limits between the superposed growth layers of prisms are easily destroyed by heating, allowing the separation of growth units (c); (d) Enlarged view of the longitudinal (axial) framework within the prisms, made visible by the decalcifying and fixative mixture. Arrows focus on the perfect continuity of orientation through successive growth layers; (e–h) The geometric pattern of the organic framework of the organic laminae in (a) and (d) is correlated dimensionally to variations in the mineral phase; observed in a section oriented 90° to the prism axis obtained after the slow destruction of intraprismatic matrix (diluted hypochlorite) a: Cuif *et al.* (1983); d: Cuif *et al.* (1987).

from the use of enzyme solutions (Fig. 1.7j). Ridges are due to preservation of the organic component by the fixative dye in the etching solution. These organic-rich boundaries between growth layers are also notably preferred dissociation zones during pyrolysis of prisms previously stripped of their organic sheaths by an oxidant such as sodium hypochlorite (Fig. 3.24b). When submitted to progressive heating (10 °C/min up to 450 °C) in a nitrogen atmosphere, polygonal elements corresponding to the superposed growth layers are separated (Fig. 3.24c). This separation occurs far below the temperature of calcite decomposition.

The decalcifying plus fixative method also reveals a second organic network (Figs. 3.24a and 3.24d) transversely oriented to growth layering. Orientation of this network is perfectly constant within a given prism and differs only slightly from prism to prism (Fig. 3.24a: arrows). Higher magnification of images (Fig. 3.24d) allows the measurement of minimal distances between successive crests on this longitudinal network. These can be estimated at 3–4 μm, depending on orientation of the network with respect to the plane of section. This length also corresponds to the distance between two ridges on the prism flanks (Fig. 3.23).

It also is established that the continuity of orientation in a given prism and its diversity from one prism to the next are exactly the properties that polarized light has revealed in thin sections of calcite prisms in the shell of *Pinna* (see Fig. 1.7). Simply from correlations such as this, it would be expected that this series of parallel internal organic planes have some relationship to the perfect continuity of crystallographic orientations of the prisms in *Pinna* shells. There appears to be an organic framework for the prisms, with properties that are exactly equivalent to those of the overall mineral (i.e., the polarizing properties of each).

A series of specific decalcifications using Mutvei's solution (acetic acid, glutaraldehyde, and Alcian Blue) or a solution of chrome sulfate provides complementary images that support this hypothesis. Sections perpendicular to the growth axis of the prism indicate that the organic component within a given growth layer is a very dense feature (Fig. 3.24e), much more compact than the distinct lamellae illustrated by Travis and Gonsalvez (1969; Fig. 2.44). In actuality, these complementary preparations remind us that the mineral components of prisms are the same submicrometer-sized grains easily visible with AFM, but also in some cases visible with SEM (Figs. 3.24g–h). The axial organic framework observed by removal of the mineral phase not only has broad distribution, but also a fine-scale structure corresponding to that of the basic mineral component. Thus, one can hypothesize that the axial organic framework within prisms plays an important role in maintaining a constant crystal lattice when crystallization of the growth layer develops on the broad surfaces shown in transverse sections of *Pinna* prisms.

A very similar equivalent organization is also seen within ostreid prisms, likewise with mean dimensions reaching approximately 10 μm.

Figure 3.25 illustrates the organic network previously identified in oyster shells by Travis *et al.* (see Fig. 2.40), showing that it forms a very dense longitudinal framework. In addition to the layered mode of growth, this longitudinal framework provides an additional structural method for controlling skeletogenesis at a submicrometer scale. Some authors have interpreted these layered longitudinal structures of oyster prisms as a purely crystalline "lath" (Checa *et al.* 2009b). Atomic force microscopy (Figs. 3.25d–g) as well as SEM examination of growth surfaces of freshly collected oysters thus provides data that fully negate any interpretation of these prismatic subunits as crystals undergoing "crystal growth competition."

3.3.4 Improvement of microstructural terminology: "centered prisms" compared to calcite prisms in the outer shell layer of Mytilus edulis (the blue mussel)

Prisms with an initial stage that formed according to the methods outlined above are present in various families of the Bivalvia (Mollusca): the calcitic prisms of pteriomorph bivalves, as well as the aragonitic prisms of the Trigoniidae and Unionidae (Figs. 3.26a–c). Features

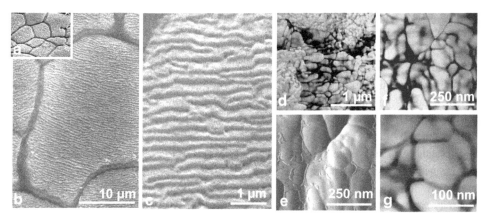

Fig. 3.25. Micrometer and submicrometer structure of calcite prisms in the oyster shell. (a–b) SEM view of the growth surface of prisms; note the remarkable density of parallel grooves (to be compared to Fig. 2.44b); (c) Greater enlargement allows us to estimate their frequency of about three grooves per micrometer; (d–g) AFM pictures of the surface of the same sample: (d) phase image at low magnification (note heterogeneity of organic component), (e–f) amplitude and phase images of the same area, (g) enlargement of (e) showing that organically coated grains appear very similar to those of other bivalve shells.

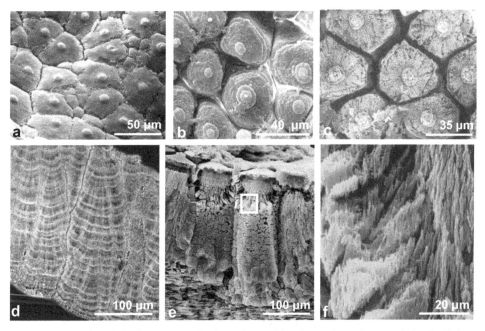

Fig. 3.26. Examples of bivalve molluscs having prisms with polygonal envelopes and clearly distinct central structures in their initial concentric stage. (a–c) *Pinctada* (a), *Unio* (b), and *Trigonia* (c). NB: In Unionids and Trigoniids, prisms are aragonitic; (d–f) Longitudinal sections of prisms in *Unio* (d) and *Trigonia*, revealing well-differentiated organization, a result of morphological modification of growth layers. b: Cuif *et al.* (1983); e, f: Ben Mlih (1983).

seen on their external surface confirm that their summit region is deposited in an analogous way to that already described in *Pinctada*.

This method of formation strongly suggests that utilization of the term *prism* should be restricted to describe only those microstructural components formed in this fashion. It is also noteworthy that this similarity in initial stages is not necessarily followed by longitudinal growth features that are similar. By way of illustration, Figs. 3.26d–f clearly show that prisms of the Unionidae and those of the Trigoniidae, although both aragonitic, are sharply differentiated by their internal structure.

These three prisms have similar polygonal cross sections, with thick lateral organic membranes and clearly differentiated centers during their initial stages. However, their internal organization differs considerably, thus emphasizing the importance of changes that may have occurred in their mineralizing matrices. To establish a coherent terminology for microstructural descriptions and, still more importantly, for reliable use of skeletal microstructures in research on the phylogeny of molluscs (extant or fossil), considerable attention must be paid both to their initial stages and growth mode. At present, micro-structural designations obviously lack such consistency, as shown by the calcitic outer layer of the blue mussel (*Mytilus edulis*), where calcitic units of the shell are usually called *prisms*, in spite of their completely different method of crystallization. These "prisms" (Figs. 2.39 and 3.27) have long drawn the attention of researchers. On the exterior of the valve, periostracum is present in the form of an organic lamella issuing from the bottom of the mantle groove (Figs. 3.27a–b). It contains an external lamina with well-marked stepping growth (Fig. 3.27b: black arrows, and Fig. 3.27c: "p1"), and additionally, a layer of organic gel that thickens progressively during the transit of the periostracum past the active secretory cells of the mantle lobe (Figs. 3.27b–c). This organic layer is highly reactive to UV light (Fig. 3.27d), and when examined on its internal face, is characterized by a large number of grains whose size increases in the direction of growth (Figs. 3.27e–f), that is, the growth margin of the shell. When the mantle is carefully removed (Fig. 3.27g), this double structure of the periostracum is clearly visible.

3.4 From control of crystallization of individual units to the coordination of crystallography at the shell level: evidence from the internal shell layer

Mollusc shells, generally speaking, are constructed of two superposed layers, with the internal one most commonly composed of lamellar units flattened parallel to the shell surface. Nacre is the best known of the internal lamellar microstructures, although not the most frequently occurring, by far. In contrast to the classical scheme (outer layer prismatic, inner layer nacreous), true nacre (i.e., made of aragonite tablets) is only produced in a restricted number of molluscan taxa. In numerous molluscs, the internal layer is formed of lamellar calcitic layers, as exemplified by Ostreidae and Pectinidae. Additionally, an equivalent role in shell construction can be played by the occurrence of an aragonitic cross-lamellar layer, as exemplified by shells of *Concholepas* (see Fig. 3.14).

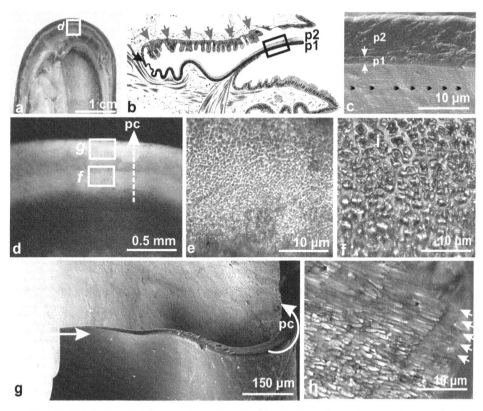

Fig. 3.27. Calcitic "prisms" of the outer layer of *Mytilus edulis*. (a) Position of the mantle on the interior face of the valve; (b) Origin of the two layers of periostracum. The external layer is produced by a group of specialized cells (black arrows); the internal side of the periostracum is covered by progressively thickening organic material, produced by specialized cells of the internal lobe (gray arrows); (c) SEM view of the periostracum. "p1": external layer; note the stepping growth (small black arrows). "p2": the thick organic layer being transported to the shell growth edge; (d) Optical view of the internal side of the periostracum (UV fluorescence). Two distinct areas are visible between the producing zone of the mantle (below) and the upper part closer to the shell; (e–f) Optical view in transmitted light. In the thickening organic layer, small differentiated units are growing during the transit of the periostracum from the mantle to the shell; (g) This material is transported to the growing edge of the shell, at the periostracal curvature ("pc"); (h) At the growing edge, fibers crystallize in close contact with the internal layer of granular material transported in the internal portion of the periostracum (arrows).

3.4.1 Nacreous units: diversity of structure and mode of formation

In research on calcareous biominerals, nacre occupies a paradoxical position. Habitually, descriptions treat molluscan microstructure as reduced to that of the Bivalvia, and at times, this is reduced to being just the superposition of an external prismatic layer above an internal layer formed of nacre. In reality, true nacre is a microstructure that only rarely occurs. Additionally, characteristics of development and microstructure are distinctly different in each of the three main classes, Bivalvia, Gastropoda, and Cephalopoda. In contrast to common opinion, there is no "the" nacre, but different types that have differing, taxonomy-linked microstructures.

The exceptional interest that nacre arouses evidently is tied to its remarkable appearance, the result of the structure of the units that compose it. Tablets of nacre have a thickness that is of the same order of size as the length of light waves in the visible spectrum. It is this property that causes the phenomenon of irridescence (the "luster"), particularly remarkable in *Pinctada* and *Haliotis*. From this point of view, not all nacres are equal. That of *Haliotis* has iridescence that is superior to that of *Pinctada* because of greater regularity of the nacre layer thickness, leading to a cumulative effect; as a result, the coloring of its nacre is much more vivid. In the Cephalopoda, the nacre of *Nautilus* presents a remarkably lustrous appearance not seen in the Coleoidea. The success of *Pinctada* in pearl production is largely because, using a part of the mantle, it is possible to carry out a grafting operation on them (see Chapter 5). However, this cannot be done either in gastropods or in cephalopods. For this reason, in spite of the small size of the units that constitute nacre, no other micro-structural type has inspired so much study as nacre tablets.

Nacre is often described as being similar to a brick wall: the mineral tablets (the bricks) are stacked, and are separated by organic layers (the mortar). In a vertical section, the resemblance is clear (Figs. 3.28a–b). The median thickness of these nacre tablets ranges from 600–700 nm, and their diameter is generally less than 10 μm.

Even in a simple vertical broken face, major differences in types of nacre can be seen, thus indicating taxonomic control by organisms on their shell. In gastropods and cephalopods, nacre tablets are stacked in such a way that their centers are almost aligned. At low magnifications, one thus sees a regular and columnal aspect to them (Figs. 3.28a–d). In bivalves, the centers shift regularly in successive layers of tablets, so that at low magnification (Figs. 3.28e–f), they have the appearance of steps. The manner of growth is obviously different between these two groups (Wise 1970; Erben 1972). In the growth zone of the Cephalopoda and Gastropoda, there appear columns (Figs. 3.28c, h, i), while in the Bivalvia, stair steps appear (Figs. 3.28j–k). It is perhaps pertinent that gastropods and cephalopods all have a single-valved shell, with a margin that grows both in thickness and in length, but the shell does not thicken after reaching its mature size, except that some gastropods form an external callus near their aperture. Contrary to this, bivalves have a shell with two valves that, after having grown lengthwise, continue to thicken.

In cephalopods, nacre can be even more differentiated. Although in modern nautiloids the nacre of the wall and septa is as described above, in *Spirula* and *Sepia* (the cuttlefish), the only other two representatives possessing a calcified shell, their nacreous structure is markedly different, and therefore it has been called Type 2 nacre. Although the layered nature of this type of nacre persists, the hexagonal shape of the tablets has disappeared, as well as boundaries between tablets (Figs. 3.28g, l). The layers are composed of acicular crystals whose orientation shifts in successive layers (Grégoire 1961; Erben 1972; Mutvei 1979; Dauphin and Keller 1982). The internal structure of the crystallites differs in such a way that they develop an acicular shape, and thus vary in this respect from nacre of Type 1

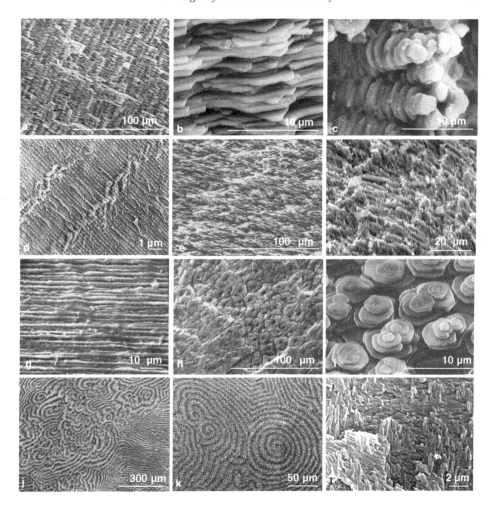

Fig. 3.28. Structure of nacreous layers in Cephalopoda, Gastropoda, and Bivalvia. (a) Vertical section of nacre showing the columnar pattern in *Nautilus* (Cephalopoda); (b–c) Columnar nacre of *Haliotis* shell (Gastropoda); (d) Vertical etched and fixed section of *Haliotis*: interlamellar and intercrystalline organic membranes are clearly visible; (e–f) Vertical sections in *Pinctada* (e) and *Modiolus* (f) (Bivalvia); (g) Vertical section in Type 2 nacre of *Spirula* (Cephalopoda); (h) Tangential section showing contiguous tablets in *Nautilus*; (i) Growing tablets in *Haliotis*; (j–k) The growth motif featuring spirals and steps typical of the bivalves (here, the internal surface of *Pinctada*); (l) Tangential section illustrating the absence of nacre tablets in *Sepia*. a: Dauphin (2005); b, d: Dauphin *et al.* (2005); g, l: Dauphin and Keller (1982); h: Dauphin (2002c); k: Farre and Dauphin (2009).

and Type 2 (Mutvei 1970) (see also Figs. 3.29b, d and h, respectively). At our present state of knowledge, it is unknown whether this ultrastructure results from an adaptation of the shell as it became internal, or from phyletic differentiation between the tetrabranchiate and dibranchiate cephalopods.

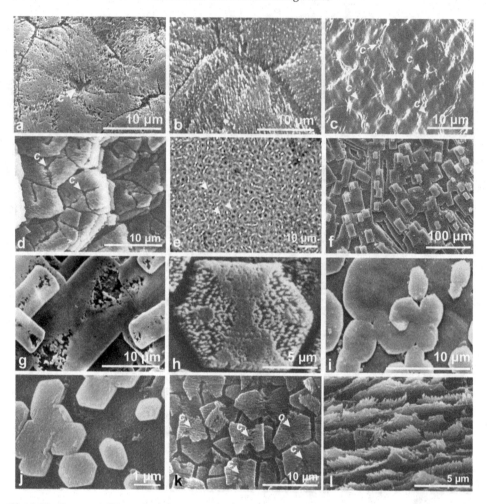

Fig. 3.29. Centers and the polycrystalline organization of nacreous crystals. (a) Nacre tablets of the septum showing the dissolved "center of calcification" (c) and radial sectors. *Nautilus*, treated with buffered HEPES; (b) Septal tablets of *Nautilus* showing the acicular tablets composing each sector, treated with a protease; (c) Multiple centers (c) in acicular tablets; (d) Tablets composed of several sectors in *Haliotis*, treated with pronase; (e) Dyed organic membrane isolated from nacre of *Haliotis*. At the center of each polygonal envelope of one tablet, centers (dyed black) have become readily visible (arrows); (f–g) Nacre from the bivalve *Pinna*: an untreated sample (f), and (g) after enzymatic hydrolysis; (h) Nacre from the bivalve *Unio*, showing two easily soluble sectors and growth in concentric layers (from Mutvei 1977). This image is similar to those in Figs. 3.22c–f; (i) Nacre from *Pinctada*, observed at the beginning of the screw growth pattern described by Wada (1966); (j) Hexagonal aspect of the tablets in the growth zone of an untreated specimen; (k) Lengthy etching with trypsine makes two soluble sectors visible, as well as acicular crystals in each tablet. The composite structure of the tablets appears equally well in section; (l) The interlamellar organic membranes and acicular structure are visible after fixation and partial decalcification. a, b: Dauphin (2001b); g: Dauphin (2005).

The differences between nacre in the Gastropoda, Cephalopoda, and Bivalvia are not limited to the manner of tablet arrangement. In each type, the form of the tablets is different, hexagonal, rectangular, etc. (Figs. 3.29a, d, e, i), with equally diverse internal structures. In cephalopods and gastropods, each tablet is divided into a variable number of sectors, with each sector itself being composed of acicular crystals (Figs. 3.29a–d). These features are clearly visible when samples of them are partially decalcified. In bivalves, acid and/or enzymatic etching establishes that each tablet is composed of four sectors of unequal size, symmetrically placed and with unequal resistance to dissolution (Figs. 3.29f–g). Each sector is composed of acicular crystallites whose orientations change in adjacent sectors. At the center of each tablet there is a zone enriched in organic compounds (Mutvei 1978).

The major part of ultrastructural information derives from observations carried out with the transmission electron microscope, either on decalcified or thinned samples. Although the relationships between organic and mineral components are better established by study with the atomic force microscope, only a small number of species have been examined with this microscope to date. The complex internal structure of tablets is shown very clearly with the atomic force microscope. The tablet surface of *Pinctada* is irregular (Figs. 3.30a–b) and composed of granules, the largest of which have a size of approximately 100 nm, surrounded by cortex that is clearly visible in phase images. This cortex has a variable composition, and can be simply organic or be organic with the addition of amorphous Ca-carbonate. Sections show that each tablet is itself composed of several layers (Figs. 3.30c–d), thus confirming Mutvei's observations. Phase images of the thickness of tablets show in addition the presence of interlamellar organic membranes (Fig. 3.30d: "ilm"). These membranes apparently are composed of several layers, as has been described in the Gastropoda by Nakahara (1983). The structure of tablets in *Haliotis* does not show this organization (Fig. 3.30e: "t"), even though, at higher magnifications, granules surrounded by cortex can also be visible (Fig. 3.30f). The nacre of *Trochus* is the same (Fig. 3.30g). In cephalopods, the arrangement of granules appears to differ somewhat according to taxon, but statistical studies are lacking regarding the form, size, and arrangement of these granules. Additionally, they are themselves composites, that is, mixtures of mineral and organic materials, as indicated by phase images showing internal irregularities within them. The granules on the surface of nacre tablets in *Nautilus* do not appear to be as regularly oriented as those of *Pinctada*, even when the limits between tablets are clearly visible (Fig. 3.30h). On interlamellar membranes, the granular structure is sharply visible (Fig. 3.30i). The acicular structure observed in the scanning electron microscope is revealed because the interlamellar membrane has been removed by use of an aqueous solution of sodium hypochlorite (Fig. 3.30j). These same acicular crystals are present in the nacreous layer of the septa in *Spirula* (Fig. 3.30k), and their interior is equally granular (Fig. 3.30l). As has been the norm in all material examined to date, none of these granules forming various types of nacreous tablets have a geometric form corresponding to the basic lattice of nonbiogenic aragonite crystals.

The composition of nacre reflects both its "mineralogy" and its biogenic origin. Like other aragonitic layers in molluscs, it is low in strontium (Masuda and Hirano 1980), from 1000 to 2000 ppm Sr (in contrast to the coral skeleton, as an example, that commonly contains about

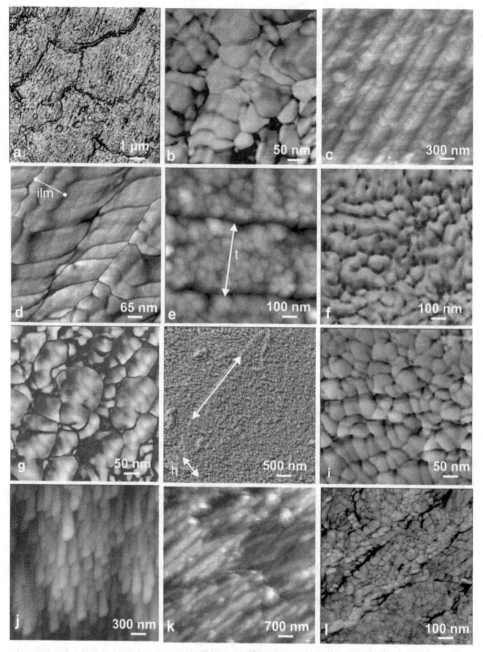

Fig. 3.30. Nacreous tablets as seen in the atomic force microscope. (a) Nacre tablets from the internal surface of *Pinctada*, fixed and partially decalcified, here showing a granular structure; (b) Enlargement of a portion of the preceding image; (c) Section showing several layers of tablets in *Pinctada*; (d) Enlarged detail of the preceding (phase image); "ilm": interlamellar membrane; (e) Tablet of nacre with a granular structure in *Haliotis*; "t": thickness of one tablet; (f) Detail of the granular structure on the

7000 ppm Sr). It is also magnesium-poor (200 to 1000 ppm Mg), but enriched in sodium (6000 to 8000 ppm Na), and may additionally contain sulfur.

Diverse types of organic matrices have been described as being associated with nacre, depending on the criteria chosen. Based on topography, there are interlamellar and inter-crystalline organic matrices. Based on solubility during decalcification, there are soluble and insoluble matrices. Membranes and insoluble matrix are commonly considered as being equivalent, but this is an oversimplification and erroneous.

Based on observations made on gastropod nacre, Nakahara demonstrated (1979, 1983) that both interlamellar and intercrystalline membranes are structured. The decalcification necessary for observation in the TEM establishes that this structure is three-layered: two electron-dense external layers enclosing a middle layer that is electron-transparent (Fig. 5.1e). Watabe (1965) also observed a diffuse intracrystalline organic matrix.

Interlamellar membranes are typically rich in aspartic acid (Weiner 1979), and the median layer in the Gastropoda is chitinous (Nakahara *et al.* 1982). Crenshaw and Ristedt (1976) pointed out that sulfate enrichment of the central zone occurs in the interlamellar membrane of *Nautilus*. Since then, four zones have been distinguished by using a series of specific stains, most notably a central zone rich in carboxylates and a central annular zone rich in sulfates (Nudelman *et al.* 2006, 2007; Fig. 5.1f). Furthermore, the insoluble component has a structure similar to that of the silk fiber, and appears to be a hydrated gel.

The amino acid composition of "soluble" and "insoluble" matrices depends in part on the method of their extraction, but the soluble matrix is always rich in acidic amino acids (both aspartic and glutamic) and glycine (Samata 1990; Gotliv *et al.* 2003). The presence of polysaccharides in soluble matrices is indicated by the presence of peaks detected in liquid chromatographs at 226 nm (Fig. 3.3a) and through analysis by infrared spectrometry (Dauphin 2001c; Fig. 5.3b). The insoluble fraction is rich in glycine and alanine (Marin and Dauphin 1992). The average calculated isoelectric point (pI) for the amino acids here is 4.5 (Cuif and Dauphin 1996; Dauphin 2002), with this level of acidity confirmed by electrophoresis. This is in addition to the dominance of proteins in soluble matrices (Marin *et al.* 1994). The molecular weights of these proteins are generally heavy, some being greater than 1000 kDa (Dauphin 2001b). The majority of studies have focused uniquely on proteins, and most of all, on their sequences. Weiner *et al.* (2003) thus identified diverse groups of proteins, but their role in the emplacement of the structural elements of nacre is still unknown.

In liquid chromatographs and mass electrophoresis, peaks or bands are often poorly separated when "adult" nacre has been used for the analysis. However, extraction of soluble matrices from nacre during its formation in fresh shells allows discrete bands to be obtained

---

Caption for Fig. 3.30. (cont.)

surface of *Haliotis*; (g) The granular structure of *Trochus*; (h) Surface of one nacreous tablet in *Nautilus*; arrow marks the boundary between two tablets; (i) Detail of the structure of the interlamellar membrane; (j) Acicular crystals of nacre from the septa of *Nautilus* (compare this with Fig. 3.30c); (k) Acicular structure of septal nacre in *Spirula*; (l) Detail of the preceding, illustrating the granular structure of the acicular crystals. i: Dauphin (2006b); h, k, l: Dauphin (2001c).

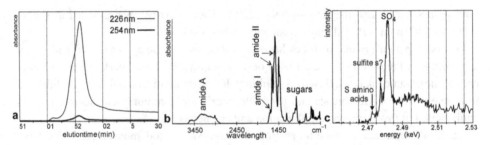

Fig. 3.31. General physical characterization of the organic phase in nacreous units. (a) Chromatographic profile of the soluble organic phase extracted from *Nautilus* nacre; (b) Infrared spectrum of the same, showing its mixed composition in sugars and proteins; (c) XANES spectra of nacre from *Haliotis*, showing the dominance of organic sulfates and the much smaller amount of sulfated amino acids. c: Dauphin *et al.* (2005).

in some specimens (Gotliv *et al.* 2003). The absence of well-marked bands was attributed at first to there being a weak association of disulfur bridges in the proteins. The massive use of detergents, urea, guanidine and other dissociating materials, combined with heating, and following classic procedures in denaturing electrophoresis, have not improved results. The inefficiency of these methods of dissociation of disulfur bridges is quite normal, since no peak corresponding to disulfur bridges (cystine) has been detected in XANES spectra carried out *in situ* or on lyophylized organic matrices. By contrast, the presence of organic sulfates is confirmed by both XANES spectra and mapping. Sulfated amino acids are likewise present in small quantities, but disulfur bridges could not be detected (Fig. 3.31c). Separations based on their isoelectric points are always very sharp.

It now seems that structural modifications intervene rapidly to play a role here, with a visible difference even between a dried shell and a very fresh one and, in other cases, between "completed" nacre and nacre that was in the process of being mineralized. Moreover, these two types of data "modification" are very closely associated. Such observations led to developing the model shown in Fig. 3.32.

The form of nacre tablets has major taxonomic implications; it is typical at the species level, in addition to the major characteristics shown by the three major groups. Observation of "the nacre" at diverse scales shows its variety, and it would be more appropriate to use as terminology, "the nacres." In truth, this layer, rare in occurrence, but the most studied of all types of microstructure, exhibits numerous variations. These depend on the systematic position of samples, such as the arrangement in columns (gastropods, cephalopods) or in steps (bivalves), as much as on the internal organization of tablets (when it exists). Thus, the particular form of tablets appears to be a specific character.

3.4.2 Growth mode of lamellar units forming the internal layer of the bivalve shell: evidence of control on crystallization at the level of the whole shell

The first data to demonstrate an overall control of shell formation is that of Wada (1961), who carried out a series of X-ray diffraction studies on the relationship between the

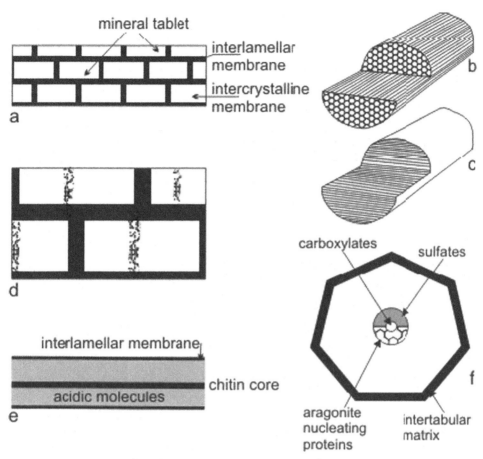

Fig. 3.32. Structure of nacre. (a) The "brick wall" nacre motif; (b) Acicular structure comprising sectors of Type 1 nacre; (c) Acicular structure comprising sectors of Type 2 nacre; (d) Detail of a tablet structure that displays a differentiated central zone; (e) Detail of the structure of interlamellar membranes of gastropods; (f) Structure of the interlamellar membrane.

crystallographic orientation of microstructural units and the overall geometry of shell margins. By using a focused X-ray beam projected onto small areas of shell surfaces from various species, Wada was able to determine precisely the orientation of crystallographic axes of microstructural units and compare these to growth directions of the shells. A basic and interesting conclusion resulted from this. In summary, it appears that in all cases a consistent relationship was observed between the overall geometry of shell margins and crystallographic orientations of components of the shell layers. Regardless of the sinuosity of these margins and their variations during ontogenesis, a constant correlation between the two is maintained. His conclusion was that the mineralizing cell layer of the mantle exerts the predominant influence on crystallization of microstructural units, not only at the level of individual units but also at the scale of the complete shell.

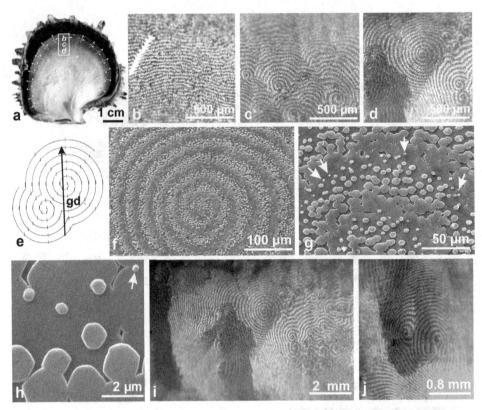

Fig. 3.33. Control of crystallization at the overall level. (a) Crystallographic orientation of microstructural units that follow the geometry of shell growth. Within the aragonite layer, the *c*-axis is perpendicular to the internal shell surface whereas the *a*- and *b*-axes of aragonite tablets are parallel and perpendicular respectively to the shell growth margin; crystal orientation depends on the direction of mantle extension; (b–d) Spatial arrangement of the nacreous tablets varies from (b) parallel to the shell border, to (c) irregular, or (d) largely spiraled; (e) Whatever their spatial arrangement, tablets share common crystallographic directions depending on the growth direction of the shell ("gd" arrow: growth direction); (f–h) Development of individual tablets from very early stages (arrows) up to lateral contact between tablets of the same level. Note irregularity of morphology and lateral contact between tablets, far from a purely crystallographic growth mode; (i–j) The nacreous surfaces show distinct areas having specific reflective orientations, irrespective of the figures made by tablet spatial arrangements. This optical pattern, not visible in the purely superficial mode of observation of the scanning electron microscope, suggests the existence of areas in which crystallization of tablets is controlled overall. e: Wada (1961).

As summarized in Fig. 3.33a, the diverging arrows at the periphery of the nacreous layer illustrate this relationship between growth directions of the whole shell and the crystallographic orientation of microstructural units.

This investigation of Wada was later extended to internal portions of the nacreous layer, resulting in even more surprising data. In *Pinctada* shells, for instance, the arrangement of nacre tablets varies as a result of the aging of the producing area of the mantle (Wada 1966).

At the periphery of the nacreous region (i.e., newly producing areas) the nacre tablets form growth lines that are practically parallel to the front of the mineralizing area (Fig. 3.33b: arrows). In more internal areas, tablets show a more irregular alignment, with formation of small spirals (Fig. 3.33c). Still more internally, large spirals become a very common pattern (Fig. 3.33d). Applying the focused spot X-ray diffraction method, Wada was able to establish that the crystallographic orientation of tablets is completely independent of their spiral arrangement. All of the nacre tablets (and eventually their internal subunits) have a common crystallographic orientation (Fig. 3.33e). This result is obviously of great importance with respect to the growth mode of nacreous units.

Closer study of the growth mode of nacreous crystals in the *Pinctada* shell shows the tiny central nucleus by which they are initiated (Figs. 3.33f–g: arrows). It has long been known that the nuclei of tablets are deposited on the surface of the previous layer of nacre as soon as it is completed, i.e., when the tablets of this layer are in close contact along most of their periphery, and not before. After this early stage, development of each crystal is essentially two dimensional, until making contact with neighboring tablets. In contrast to common statements, the hexagonal morphology is not at all constant (Fig. 3.33h). In *Pinctada*, most nucleating centers seem to be deposited at the limit between two tablets of the previous layer, but in most other nacre-producing species this is not the case. In the shell of *Pinna*, for instance, new nacre tablets are produced prior to completion of the underlying level, and consequently, appear at any point on their surface. Such a style of nucleation and growth of nacre crystals makes their constancy of crystallographic orientation even more remarkable.

The constancy of such tablet orientation, where tablets are developed from scattered nucleation centers, suggests that their own crystallographic orientation was fixed and consistent at the time of their deposition. Whatever the regularity of growth of tablets by addition of lateral increments (as in Fig. 3.29h, for instance) and the high degree of diversity of their polycrystalline organization (Figs. 3.29a, d, k), initial orientation determined by the nucleation center is maintained. This represents a striking parallel with the development of initial crystallization as it occurs from the center of crystallization at the top of the prism (see Figs. 3.21–3.22).

Study with the atomic force microscope clearly indicates that the complete tablet is composed of organically coated grains. This structure yields AFM images that are very similar to those of other calcified materials of molluscs and corals, thus, we can assume that the addition of new material is based on a similar process. More information regarding crystallization of nacre is provided in Chapter 5 (models of Ca-carbonate crystallization).

Other examples of crystallization control at the overall shell level

Due to its economic importance, nacre has received much attention from researchers, but the overall control of shell crystallization by the mantle is demonstrated just as well by components of numerous other molluscan shells. In shells of the commercial oyster, for instance, the observation of growth surfaces of several adjacent prisms shows an obvious conformity between the orientations of the intraprismatic mineralization frameworks of neighboring prisms (Figs. 3.34a–d). This type of common growth direction, shared by

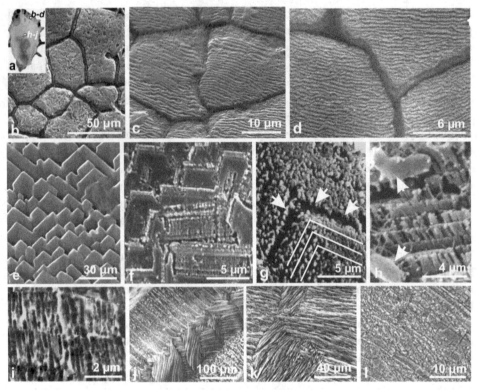

Fig. 3.34. Examples of control on the formation of microstructural units on large mineralizing surfaces in *Crassostrea gigas*. (a–d) Growth surface of calcitic prisms. The identical aspect of orientation of the axial organic framework between adjacent prisms suggests an overall control of prism formation; (e–i) The laminar layer. In spite of crystal-like terminations (e), the internal organization made visible by bacterial decay (f–h) clearly shows a linear arrangement of the submicrometer-sized units. (i) Equivalent data was obtained by Watabe (1965) using transmission electron microscopy; (j–l) In numerous other Ca-carbonate microstructures, such as cross-lamellar (j: *Phalium* and k: *Tridacna*), or irregularly layered (l: *Concholepas*, internal layer), no obvious indication of a localized control of crystallization has been found yet. h: Frérotte *et al.* (1983); i: Watabe (1965).

completely independent mineral units (after their origin), but separated by compact organic envelopes, clearly suggests that the crystalline features of each mineral growth surface are controlled by the mineralizing layer of the mantle.

The common, crystal-like morphology of nacre tablets is also seen in other types of microstructural units, such as the units of foliated calcite that form the internal layer of *Pecten* (Fig. 3.7a), and oyster shells (Figs. 3.33a–b). All terminate with angular faces, the angle of which has long been regarded as analogous to those of calcite rhombohedra (Tsuji 1958). In oysters, this internal layer is generally pure white and somewhat iridescent. For this reason, it is commonly – and wrongly – called "nacre" (as true nacre is always aragonite). In spite of the crystal-like aspect of their growth terminations, these laminar units are not at all monocrystalline.

Figures 3.34f–h show the results of bacterial hydrolyzing activity on these lamellae when fragments of oyster shells were left in artificial seawater for long periods (30 to 90 days). Bacteria isolated from the marine sediments surrounding these oysters were put into laboratory-made seawater lacking any organic nutrients. Fragments of oyster shells were thus the only sources of organic materials for bacteria introduced. Bacterial enzymatic digestion was complete on the upper organic envelopes of the lamellae, and this reveals the submicrometer units that make up the internal structure of the lamellae. It is shown here that there is a closely controlled arrangement of these micrograins, readily seen to have a stepping growth alignment (Figs. 3.34f–g), with their orientation almost perpendicular to the base of the lamella (Fig. 3.34h). These images conform to those obtained in thin sections perpendicular to the shell surface, prepared for transmission electron microscopy (Watabe 1965).

The crystal-like angular patterns commonly shown by the foliated lamellar units in oysters may be the result of this specific mode of growth. Here the stepping mineralization process produces growth increments composed of linear series of submicrometer units. As usual, the latter are made of organically coated grains. The organic envelopes of internal shell components are always much more fragile than those of the prisms in the outer layer, apparently enabling the final step in crystallization to develop morphologies that resemble crystal terminations. Nacre tablets themselves frequently show crystal-like growth patterns along with the layered mode of granular growth. Examples that clearly demonstrate this have been provided by H. Mutvei (1977; see Fig. 3.29h in this volume). The rectangular nacre tablets in the bivalve *Pinna nobilis* also display a four-sector organization, apparently in agreement with their internal crystallographic organization.

It should be emphasized that this type of crystallization occurs over broad surfaces of the shell. When evidence is seen of crystal orientations being aligned with the growth direction of shells, this is a good indication that the complete surface of the mineralizing area of the mantle has the ability to control the complete organic framework on which mineralization occurs. In most cases, however, access to such evidence is not available, in spite of the well-controlled organization of microstructural units. Such is the case in the cross-lamellar structure (Figs. 3.34j–k) where the method of control remains unexplored, and also, for most of the poorly organized shell structures, such as the internal coating of the gastropod *Concholepas* (Fig. 3.34l).

3.5 Skeletal development as a continuous process: evidence of transitional changes between molluscan shell layers and during coral ontogeny

Standard reference works (e.g., Bøggild 1930; Taylor *et al.* 1969, 1973), describe the microstructure of the molluscan shell as having two (sometimes three) superposed layers that are distinct and separate. However, observations that are more precise now allow us to recognize intermediate stages between layers, suggesting that a concept of *ontogenetic transition* provides a much better way to account for some degree of continuity in the biomineralization process. In this approach, detailed study suggests that changes occurring throughout shell formation reflect a modification of biochemical activity of the pallial epithelium in its function of secreting the organic components that generate microstructural units. This continues from the very early beginning of growth (even in pre-mineralized stages) to the formation of very late growth layers.

As we have seen in the pteriomorph bivalves, calcite prisms of the external layer acquire their crystallographic characteristics at the growth margin of the shell, and subsequent growth cycles essentially insure the continuation and reproduction of these characteristics. The prisms of *Pinna* provide good illustrations of crystallographic continuity that is maintained throughout their growth (up to 5–6 mm in length); a dedicated organic phase is seen playing an important role in this crystallographic activity. However, structural diversity can also be introduced during this elongation phase by modification of crystallization processes.

3.5.1 Microstructural change within the calcite prisms of Pinctada: transition from a monocrystalline to a polycrystalline organization

No biochemical framework comparable to that of *Pinna* prisms exists in *Pinctada margaritifera*. Conversely, prisms here show visible changes of both morphology and crystallography during their growth. In one flake (growth lamella) from the external surface of a shell (Fig. 3.35a), this change in prism morphology is clearly shown (Fig. 3.35b: arrows). An enlarged view (Fig. 3.35c) indicates that parallel, mineralized growth layers are present,

Fig. 3.35. Morphologic and crystallographic changes during growth of prisms in *Pinctada margaritifera*. (a–c) A growth flake (a) showing the lateral side of the prisms with a synchronous change of morphology as pointed out (b) by arrows, and (c) this horizon enlarged; (d–e) Monocrystalline prisms in the juvenile growth lamellae margins; (f–g) Polycrystalline constitution (f) of prisms after this change. Note (g) the organic material (arrows) forming internal limits between the prism subunits within the principal polygonal envelope.

regularly superposed in the upper part of prisms, whereas in their lower part, their external surface is more rough and irregular. There is a close correlation between this morphological change and the crystallographic organization of the mineral phase. During the first growth phase, prisms exhibit monocrystalline microstructure (Figs. 3.35d–e), but after approximately 130–150 µm of very regular growth, the prisms in effect assume a polycrystalline structure, formed by several distinct crystallographic units (Fig. 3.35f) that are separated by easily visible limiting organic matter (Fig. 3.35g: arrows).

This new state, occurring simultaneously in neighboring prisms, testifies to an isochronous metabolic modification in the entire surface of the pallial epithelium.

Validity of the theory of "crystal growth competition"

This example is also significant with respect to the commonly accepted hypothesis of "crystal growth competition" as the origin of microstructural organization in molluscan shells (e.g., Checa *et al.* 2009a). According to this hypothesis, the most rapidly growing crystals have a determinant advantage in the occupation of open or free mineralizing space, resulting in a reduction of the number of crystal units developed during shell growth.

The development of *Pinctada* prisms through time does not support this theory. It clearly demonstrates that not only is the number of prismatic units *not* reduced during shell growth but, considering the subdivision of crystallized elements within the prisms envelopes, the number of distinct crystalline units is actually increased, with a marked reduction in their transverse dimensions. Additionally, the regularity of this change, occurring repeatedly at the same stage of development in successive shell flakes, indicates that there is true biological control of this process and not merely "crystal growth competition."

3.5.2 Occurrence and diversity of transitional stages between shell layers

In some instances, as in the bivalves of the Family Trigoniidae, the transition from prisms to nacre is accomplished with remarkable simplicity. In this family, prisms are aragonitic and the internal thickening layer is nacreous aragonite. Very regular and parallel prisms (Figs. 3.36a–b) adjacent to one another are characterized by an internal microstructure formed of elongate, contiguous elements oriented perpendicular to the surface of the shell (see Figs. 3.26e–f). The base of the nacreous layer is marked by the progressive appearance of transverse lamellae that cut the axial fibrous structure into tablets of nacre. The transitional character of the contact between prisms and nacre tablets is shown here by the progressive emplacement of organic lamellae that separate and define the layers of nacre (Fig. 3.36d: arrows).

From a metabolic viewpoint, the formation process of nacre tablets in trigoniids is very similar to that observed in the early stages of development of the aragonite layer in *Pinctada margaritifera* (see Figs. 1.09d–f; also Figs. 3.38f–g). In both cases, elongate aragonite fibers are cut transversely by organic lamellae, first partially and then at regular intervals when the secretory metabolism that produces the parallel membrane is fully established and regularly cyclical.

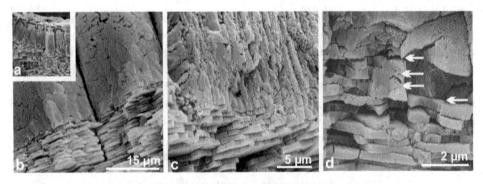

Fig. 3.36. The prism to nacre transition in the Family Trigoniidae (Bivalvia). (a–b) Aragonitic prisms and initial layers of nacre; (c–d) Progressive deposition of organic lamellae (arrows) that separate nacre tablets.

In *Haliotis* (see Figs. 1.14 and 1.15 for general views), a typical transition is present both at the summit and at the base of the prismatic layer. The external aragonite layer, formed of small units that have a spherulitic microstructure, is followed by large calcitic prisms. This is an example (Figs. 3.37a–c) that demonstrates a metabolic transition by simultaneous production of organic components responsible for determining calcite and aragonite deposition. It is remarkable that isolated aragonite spherulites are included completely within the large calcite crystals. NanoSIMS mapping of strontium distribution in these indicates that here, strontium concentration is not affected by concomitant calcite crystallization (Fig. 3.27c). This implies that bubbles of aragonite-generating organic materials are isolated within a mineralizing hydrogel that is in great part composed of calcite-generating substances.

At the other end of the calcite prisms, their morphology again exhibits a readily visible modification (Figs. 3.37d–e). During most of their growth, the calcite is formed as elongate fibers (Fig. 3.37e: "fib cal"). At their base, however, we can see that biogenic control of this calcite is changing (and probably weakened) because crystal growth occurs along planes (Figs. 3.37e–g: arrows) that result in the formation of an irregular inner surface to the calcitic layer. The passage to nacre is carried out first by the emplacement of fibrous aragonite in the form of spherulitic bundles. Following this, the first organic lamellae appear, and when the changeover has been completed, the remarkably regular nacre typical of *Haliotis* is produced (Fig. 3.37h).

With respect to the conditions that would allow this type of crystallographic transition to be carried out, the concomitant occurrence of spherulitic aragonite and elongate calcite crystal fibers cannot be obtained by simple crystallization within a common volume of liquid by any classical saturation–precipitation process. On the opposite side of this layer, the replacement of calcite by nacreous aragonite also results from progressive metabolic changes within relevant parts of the mantle.

3.5.3 *Biochemical transition in the sequence of prisms to nacre in* Pinctada

In addition to microstructural evidence regarding the transitional passage from calcite prisms to aragonite nacre (see Chapter 1), biochemical mapping provides converging

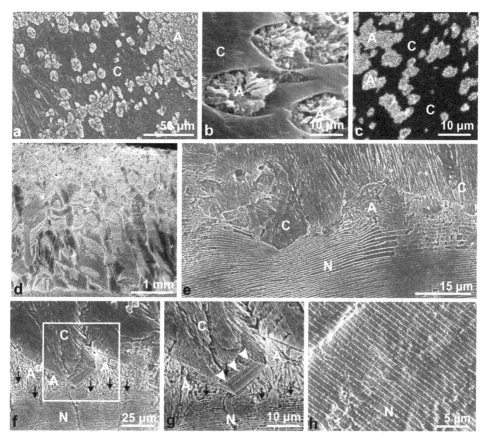

Fig. 3.37. Transition of external prismatic aragonite and prismatic nacre in *Haliotis*. (a–b) Transition of external aragonite and calcite; the two microstructural components are produced simultaneously; (c) Map of Sr distribution (mapping by A. Meibom, MNHN, Paris; CAMECA-NanoSIMS); (d) Irregularity of prisms that form the base of the calcitic layer; (e–g) Smoothing at this surface by the aragonite layer which is first produced in the form of spherulitic mineral fibers of aragonite ("A"), succeeded by the appearance of nacreous lamination (g: arrows); (h) The exceptional regularity of the aragonite of *Haliotis*, responsible for its vivid iridescence (image: H. Mutvei, Museum of Natural History, Stockholm). a, d: Cuif *et al.* (1989).

evidence that the transitional phase is more general than usually indicated by microstructural information. Figure 3.38a reminds us that prisms are covered at their internal end by an organic layer that is apparently formed by simple expansion of organic envelopes. Although the morphologic aspect of prisms does not show any modification, element mapping on the synchrotron at the European Synchrotron Radiation Facility (ESRF) reveals that the concentration of sulfated polysaccharides diminishes during final stages of secretion (approximately five to six days previous to this; Fig. 3.38b: arrows). Above this organic covering, the mineral material deposited is fibrous aragonite. Use of a time-of-flight (TOF) mass spectrometer allows study and identification of the organic fragments detached from the

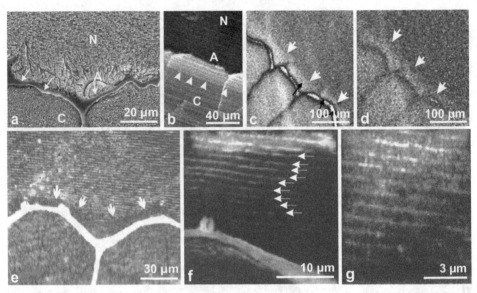

Fig. 3.38. Chemical and biochemical changes during the prism to nacre transition in *Pinctada margaritifera*. (a) SEM view of the transitional zone between prisms (below) and nacre; (b) Chemical mapping (synchrotron-based XANES mapping) focusing on the distribution of sulfated polysaccharides. A metabolic change occurs (lessening of S-polysaccharide concentration) before the end of calcite formation at the plane here marked by arrows. Based on counting layers, this change occurs five to six days prior to the covering of the prism surfaces; (c–d) Chemical maps (TOF, SIMS) showing that amino acid production is not fully synchronized with the beginning of aragonite production. For some of them (c: glycine, d: alanine) it is produced later, when formation of the nacreous membranes commences (images: A. Brunelle, ICSM-CNRS, Gif/Yvette); (e–g) Distribution of nitrogen during the prism–nacre transition (Nano-SIMS mapping). Arrows in (e) underline the distance (and time) between the thick organic sheet covering the prism and the appearance of nacreous layering. High-resolution mapping (f–g) shows precisely the development of nitrogen-rich laminae between nacreous tablets (images: A. Meibom, MNHN, Paris). These images provide chemical evidence that production of nacreous membranes is not synchronized with aragonite production but, rather, it reflects progressive modification of metabolism in epithelial cells of the mantle. a, b, e–g: Dauphin *et al.* (2008).

substrate by a beam of ions projected onto the specimen surface. Applying this methodology, it is possible to establish changes in the distribution of amino acids (Figs. 3.38c–d). Results indicate the specific composition of this covering membrane and identify an increase in concentrations of alanine and glycine (data: Farre and Dauphin 2009). This is an additional indication that the membrane is not simply an expansion of the prism envelopes, but a specific product related to the changes occurring in preparation for nacre formation.

The high-resolution mass spectrometer NanoSIMS allows extremely precise determination of the distribution of chemical elements during this transition process. Mapping nitrogen, as an example, establishes the progressive appearance of organic membranes that eventually separate the layers of nacre (Figs. 3.38e–g; data by A. Meibom). At this time

(2010), *in situ* data that show the distribution of biochemical components involved in the formation of the mineralizing organic phase are rare. However, these data, even though rare, add more evidence to support adjusting the concept of microstructural sequence in the Mollusca.

In contrast to the static description of superposed mineral layers, the microstructural sequence results from a progressive introduction of a series of biochemical modifications. The regularity of these testifies to the regulation of functions by those genes implicated in producing mineralized structures. A very specific practical application, the production of cultured pearls, has allowed us to make numerous significant observations on perturbations that are introduced into this well-studied biomineralization process; that is, the formation of nacre in *Pinctada margaritifera* (see Chapter 5).

3.5.4 Coral skeletogenesis as a transitional process: taxonomic and phylogenetic implications

Comparable analysis of skeletal changes during ontogenesis can be carried out on scleractinian corallites. The mechanism that assures taxonomic control of septal architecture is clarified by examination of initial stages in the formation of the coral polyp, and in particular, in its very early post-larval stages. Research on these first stages of mineralization was carried out by Vandermeulen and Watabe (1973), focusing on development of the coral species *Pocillopora damicornis*. In their study of the formation of mineralized structures during the initial stage of corallite formation in *P. damicornis*, Vandermeulen and Watabe showed that initial mineralization first appears on the undersurface of the polyp during very early stages of its development, resulting in a roundish mineralized platelet formed entirely of carbonate micrograins. Thirty-six hours after larval settling, the mineralizing surface then was restricted to radial zones that eventually form embryonic septa. These embryonic septa are also composed of "small crystals or granules" (Vandermeulen and Watabe 1973, Fig. 10). Their research also showed that a second process of mineralization developed subsequently, 72 hours after larval settling, to form a fibrous layer on each side of the primary (granular) septum, and that thereafter this layer continuously and progressively thickened each septum.

This description corresponds remarkably well to a hypothesis of the process of skeletogenesis that can be synthesized from observations on the overall organization of septal growth margins in adult corallites (see, for example, Figs. 1.22 and 1.23). Whatever the three-dimensional arrangement of zones of distal mineralization, the basic sequence of tiny granules followed by layers of aragonite fibers is always present. During ontogeny, changes that modify the function of distal portions of the ectoderm (the zone of distal septal mineralization) determine important changes in the arrangement of septal fiber bundles and, thus, septal microstructure (see various examples in Fig. 1.24).

It has long been suggested that these various three-dimensional arrangements are taxonomically linked (Ogilvie 1896). In order to test whether these modifications truly reflect evolutionary processes undergone by scleractinian corals since their Triassic origin, a study focused on 42 species of modern corals in which phylogenetic relationships inferred from comparison of the molecular structure of their ARN 28S were compared to classification based on microstructural analysis (Cuif *et al.* 2003b).

The objective of this research was to evaluate the families proposed in the classification of the Scleractinia by J. W. Wells (1956), the most widely accepted of the twentieth century. Research focused on the type species of genera (except for a small number of individuals having an increased importance resulting from historical taxonomic problems), and as far as possible, belonging to families important in the classification of the Scleractinia (e.g., *Favia fragum*, the type genus of the Family Faviidae and the Suborder Faviida). Twenty-five genera representing 12 families were therefore included in this comparison. The specimens utilized were collected alive in their original biogeographic areas (Caribbean, Polynesia, New Caledonia, etc.). For each, a microstructural analysis was carried out on the same aragonitic corallum whose tissues were utilized simultaneously to form the molecular phylogenetic tree. Thus, no sampling bias was introduced between the two types of analysis. The 5th branch of Segment 26S of ribosomal DNA was selected as the indicator of long-term evolution. Additionally, the method used for extraction of DNA permitted very long segments (more than 700 base pairs) to be obtained, thus conferring great significance to findings regarding relationships.

Results were as follows: comparisons of molecular data (using MUST software) indicate that several major taxa in the modern classification should be divided (e.g., Family Faviidae), while others are confirmed as is (i.e., Dendrophyllidae). From an overall viewpoint, this compares well with previous studies that employed other genes or molecular techniques, but also compares favorably with conclusions based on septal microstructures. Clearly, groups established on the basis of the specific patterns and arrangements of their zones of distal mineralization ("centers of calcification") provide a reliable foundation for taxonomy.

These results also support the use of the configuration of the distal mineralization zone as a marker reflecting scleractinian evolution (refer also to Chapter 7). Each specific group of "centers of calcification" at the growth edge of each corallite is the result of branching that is continuously maintained during corallite ontogeny. Evolution of the Scleractinia has superimposed various modifications on the basal microgranular plate and linear microgranular septa. One of these is the fragmentation of the continuous central line into shorter segments and eventually into even more restricted, point-like areas of distal mineralization. The dispersed centers of the Poritidae are the clearest illustration of this process.

This conclusion indicates some productive directions for more incisive study of fossil material, beginning with the rare specimens protected from complete diagenetic change (or recrystallization) by circumstances of fossilization. Beginning with these materials, whose frequency of occurrence varies a great deal according to geologic period, a realistic classification can be established, based on, and accompanied by a consistent evolutionary scheme (see Chapter 7).

It also appears that every corallite has a central septal system that, in part, reflects a portion of the evolutionary processes of its lineage. Beginning from a simple straight-line arrangement (that many modern Scleractinia have preserved), the fragmentation of this arrangement continues in various ways, each useful for defining phylogenetic lineages. Additionally, the hierarchical organization of septa allows direct observation of the end result of the evolutionary sequence, if not its beginning.

3.5.5 The cyclic organization of septa in a single corallite: microstructural transitions within and between septa

In a corallite, the first septa appearing between the six original septa are intercalated between them, and depending on the species, this process, when repeated in cycles, leads to the formation of a calice similar to that illustrated by Figs. 3.39a and 3.39b. This corallite of *Cynarina*, a member of the Family Mussidae, provides an excellent illustration of "Milne-Edwards's law," based on repeated alternations of successive septal cycles. Additionally, one should note that some of these "laws" are so rarely verified that some question arises concerning their actual validity.

Fig. 3.39. Microstructural development of septa in *Cynarina*. (a) Calicinal view of *Cynarina*. Four successive cycles of septa are regularly intercalated here; six first-order (1), six second-order (2), 12 third-order and 24 fourth-order septa. "Milne-Edwards's law" is rarely so perfectly recognizable; (b) One group of septa that illustrate different morphologies, progressing from first- to fourth-order septa; (c–e) Microstructure of adult first-order septa, where, at their summit, the pronounced teeth of the border within zones of distal mineralization ("dmz") are concentrated in circular groups (mussoid teeth); (f) In contrast, the growth edges of fourth-order septa display a continuous line of "dmz" (arrows).

Looking more closely at septal morphology, it is clear that the morphology of a given septum (its length, thickness, profile of its early growth line) also depends on its position in an ontogenetic sequence. Not only does septal morphology change from one cycle to the next but, in addition, the septal microstructure itself becomes modified. In *Cynarina*, for instance, the large spines of the first-order septa are characterized by concentrations of multiple early mineralization zones – the so-called "centers of calcification" (Figs. 3.39c–e), whereas septa appearing later (and thus ranked lower) exhibit a continuous "dmz" made of juxtaposed, very small "dmz" centers (Fig. 3.39f: arrows).

Some genera display more complex microstructural changes during their septal develop-ment. This is the case in the genus *Favia*, and more particularly, in the type species of this genus, *Favia fragum* from the Caribbean region (Fig. 3.40). In this coral, corallites usually have a cerioid arrangement (densely packed polygonal calices separated by complete walls; Fig. 3.40a), but their septa are greatly differentiated. At the upper margin of first-order septa, zones of distal mineralization are seen to be arranged in rows perpendicular to the median plane of the septum (Fig. 3.40b: arrows).

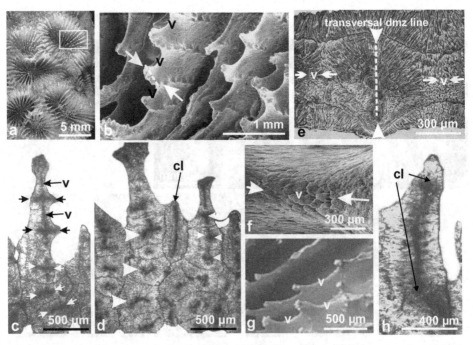

Fig. 3.40. Septal microstructure in *Favia fragum*. (a) Cerioid corallites; (b) First-, second-, and third-order septa with well-developed septal ornamentation in the major (first-order) septum shown here, formed of rows of "dmz" (arrows) perpendicular to the long axis of the septum; thus, between resulting septal platelets, the growth margin exhibits valleys ("v"); (c–d) Thin section views where transverse lines of "dmz" (arrows) and valleys ("v") are readily recognizable, but greater enlargement is necessary to see fiber orientations; (e–f) Fibers diverging from the transverse "dmz" line, the result of fiber growth in opposite directions in valleys; (g) Third-order septa exhibiting only weak septal differentiation; (h) Latest septa with a continuous "dmz" line ("cl").

Given their role in initiating calcification, these rows now are seen to occur in an elevated and prominent position with respect to the "valleys" (Fig. 3.40b: "v"). This arrangement is also readily visible in thin sections (Fig. 3.40c). Such a microstructural organization is typical of the first-order septa, but is attenuated in septa of succeeding orders and within septa of the latest (newest) order, the zone of distal mineralization is simply linear and continuous (Fig. 3.40d: "cl"). Thus, this sequence of successive cycles of septa provides an approximation of the microstructural evolution in this group, from very early stages (continuous linear zone of distal mineralization) to the pattern that is furthest evolved, shown by older septa belonging to the first cycle. Reconstruction of the microstructural evolution of major septa during development of corallites also leads to reconstruction of this feature for the species (see Chapter 7). This type of study is easiest done with solitary corallites, but for the great majority of coral colonies, it may also be of interest to note that the minor-order septa illustrate the earliest stages of early growth of septa, prior to development of their later and final stages of differentiation.

3.5.6 A century-long controversy: do "centers of calcification" in corals play the role of "centers of crystallization"?

Study of the development of prisms in the pteriomorph bivalves has provided an excellent example of what truly is a center of crystallization; a structure that determines the crystallographic characteristics of a unit (here, each prism) at its starting point, characteristics that will be continued in succeeding growth layers of each individual (prism). It should also be emphasized that the well-characterized centers of crystallization of *Pinctada* or *Pinna* prisms are not themselves mineral in nature, but are purely organic (see Fig. 3.22).

The concept of "centers of calcification" in scleractinian corals was initially described, in the first microstructure-based taxonomic investigation of Ogilvie, as being "the points from which fibres diverge" (1896). Half a century later, Bryan and Hill (1941) proposed that spherulitic crystallization could be the basic skeletogenetic process. Thus, they paved the way for a mechanistic representation of skeletogenesis in the Scleractinia as a purely physiochemical process, as stated by Constantz (1986): "The morphology and arrangement of the aragonite fibers are controlled by inorganic kinetic factors of crystal growth, akin to inorganically precipitated marine cements." Some years previously, Barnes (1970) had presented a roughly equivalent scheme, hypothesizing an "uplift" of the aboral polyp tissue, which would provide a supersaturated cavity in which aragonite crystals could grow.

Concerning the everpresent organic component, the interpretation of Constantz was that this was debris consisting of ectodermal tissue "pinched" between fibers during their growth. Present knowledge of the distribution of these organic compounds at a submicrometer scale completely refutes any such concept of accidental inclusion of these organic compounds within the coral skeleton.

Afterward, the then-hypothesized role of "centers of calcification" in crystallization of coral fibers was reinforced by the evidence of tiny crystals present and forming the first crystallized elements at the distal growth edges of septa. Fibers were believed to start from these small crystals and then to develop by following the mechanism of "crystal growth competition." Closer observation of the contact between centers of calcification and fibers in *Favia stelligera* (see Fig. 1.25) has already shown that fibers do not develop freely from centers of calcification, as they should if a liquid layer were present. Even when the centers of calcification are formed, the beginning of fiber growth depends on biochemical activity of the relevant mineralizing area of the basal ectoderm. Skeletal formation in the genus *Lophelia* presents an example of fibrous skeletal material produced without any contribution from "centers of calcification." Corallites of *Lophelia* are characterized by a thick wall (Figs. 3.41a–b; also Fig. 1.22), which is actually composed of several superposed concentric layers. The polyp occupies only the upper portion of the calice (Fig. 3.41c) and constructs a corallite in which a wall is formed that begins with a continuous circular zone of distal mineralization, thickened by the usual fibrous bundles (Fig. 3.41d). Because the basal layer covers only the top of the initial wall, these fibrous bundles remain moderately developed.

Much more important from a morphological viewpoint, this initial wall is thickened by calcification occurring in gel or mucus that flows from the polyp and covers all of the colonial skeleton, and eventually covers even organic tubes present, the last formed by a commensal worm (*Eunice norvegica*), as described by Freiwald and Wilson (1998; Figs. 3.38e–f). In this gel, aragonitic fibers grow and form a series of concentric layers whose thickness may even exceed the thickness of the initial wall that was directly constructed by the polyp (Figs. 3.41g–h).

These large fiber bundles, which sometimes also include sedimentary particles, since they are not protected from seawater and sediment, are clearly isolated from any influence of "centers of calcification." This is even more obvious when the fibers do not grow on the corallite tube itself, but develop in the gel that is being flushed onto the worm (Fig. 3.41f: worm tube, "wt"). These freely growing fiber bundles are excellent examples of biochemically induced crystallization.

On a polished surface (Figs. 3.41g–h), the two parts of the cylindrical skeleton are clearly distinct: the corallite-controlled and the gel-grown fibers together forming the thick outer enclosure comprising the corallite. These fiber bundles growing in gel are radially oriented, although slightly diverging from one another (Figs. 3.41h–j). Microprobe mapping provides evidence of the regular variation in composition of fiber bundles, the best illustration being that shown by the oscillation in magnesium concentration (Fig. 3.41k). The periodicity of this variation is not yet definitely established, but based on the slow growth rate of *Lophelia* corallites, the periodicity would seem to be on a scale approximating a month. Observing the internal cavity of the *Lophelia* corallite (Figs. 3.41o–p), we note that the mineralizing gel also gets flushed inside the corallite tubes. Additional thick layers of fibers result in the formation of a compact corallite lumen, as no regular dissepiments exist in this family. In both cases (internal or external formation), it is clearly understood that these successive layers of fibers are each formed without any centers of calcification.

It is conclusive that, although the expression "centers of calcification" is supported by long usage, such structures are in no way equivalent to the "centers of crystallization"

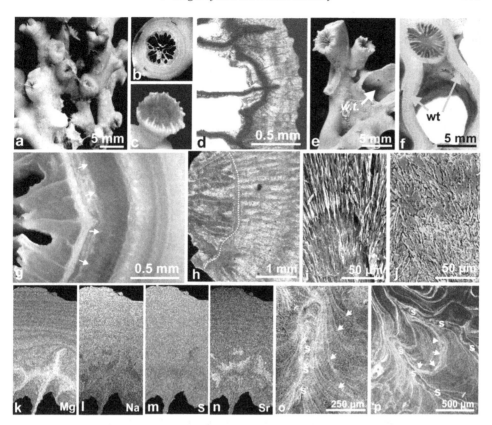

Fig. 3.41. Fibrous bundles crystallized within gel or mucus external to the polyp of *Lophelia pertusa*. (a) Example of a strongly incrusted corallite; (b–d) Section of a corallite (b). View of the polyp at the top of a branch (c). The thin section observed with the optical microscope (Alcian Blue staining) where zones of distal mineralization (septa and wall) are intensely darkened, and only a small part of the fibrous tissue is controlled by the polyp (compare to g); (e–f) Example of a worm tube incrusted by gel-produced fibers, where most of the fibrous layer is not covering the corallite tube but lying directly on the organic worm tube; (g–h) A polished surface in a *Lophelia* branch showing the limit of polyp-controlled structures (arrows). In (h), the dotted line also separates outer fibrous crystallization from corallite-based crystallization; (i–j) Fibers developed within the outer gel, where they directly crystallize as diverging bundles without any "centers of calcification"; (k–n) Distribution of minor elements. Mg shows visible zonation, others are weaker, and sometimes not visible (S) (images from T. Williams, NHM, London); (o–p) UV fluorescence of interseptal deposits (arrows) in the tubular corallite of *Lophelia* ("s": septa). This internal crystallization is equivalent to development of the mucus fibers outside the corallite wall.

identified in bivalve prisms, because the crystallization of coral fibers depends only on the composition of the organic gel that is secreted by the basal surface of the polyp. This conclusion justifies the position adopted by Wells (1956, part F). As a coral taxonomist, it was perhaps not completely acceptable to him that a simple spherulitic process of crystallization could produce obviously well-controlled and unique skeletal structures. Therefore,

Wells always used quotation marks around the printed expression "centers of calcification." In order to avoid confusion with true "centers of crystallization," a designation based on the position of these microstructural components has been suggested: early mineralization zones ("emz": Cuif *et al.* 2003), or still more clearly, distal mineralization zones ("dmz"), as used in this book. To emphasize their true function at the growth edges of the coral skeleton and their organizational role in forming the architecture of septa and walls, the pioneering work of W. Volz (1896) must be recalled, who named these regions "*urseptum*" (i.e., primitive septum), whose importance has long been neglected. Due to problems in establishing the temporal distinction between the "centers" and partially time-equivalent fiber bundles, we are now utilizing the term distal mineralization zone rather than referring to "initial" or "early" mineralization.

3.6 Conclusion

Beginning with a new structural approach to the mineral units produced by corals and molluscs, this chapter extends our view of the scale at which the mineralizing epithelium exerts its control on the formation of calcareous structures. From the analysis of the nano-structure of elementary growth layers in corals and molluscs to the overall organization of the very obviously distinct calcareous structures in the two groups, a time-based analysis of the functions of the mineralizing layers explains the different aspects of these calcareous structures, suggesting an in-depth similarity of biomineralization mechanisms in the two phyla.

3.6.1 Similarity of the organic–mineral interplay within the elementary growth layer of corals and molluscs

Comparison of physical and chemical characterization of growth layers provides us with similar evidence that the stepping growth mode of formation of mineral units, which has been found common to both corals and molluscs from a structural viewpoint, also operates in an identical way. In both cases, when we include biochemical information, mineral and organic components of the growth layers are seen to be associated in similar modes in both corals and molluscs. At the present-day highest-resolution level for microstructural analysis, no difference can be seen in the method by which biochemical control is exerted over a molluscan or coralline growth layer. It is understood, of course, that the organic materials involved have distinct and different biochemical compositions, but the molecular mechanisms of biomineralization are basically similar.

3.6.2 Species-specific shape versus crystal-like morphology of biocrystals: role of organic envelopes

When skeletal organization is regarded from a more integrative viewpoint, striking differences occur in the method of forming and controlling the sizes, shapes and three-dimensional arrangements of skeletal units. Many of them exhibit an overall shape that corresponds quite well with the normal morphology of the Ca-carbonate polymorph involved (calcite or aragonite). Therefore, the nacre tablets of *Pinna* are regularly quadrangular due to their composition being orthorhombic aragonite, and the foliated calcite forming the internal layer of oysters (Fig. 3.28b), *Pecten*, or even the fibrous prisms of *Mytilus* (Fig. 3.7a), all have angular terminations that compare well to those of calcite rhombohedra. However, it is now well established by observation with the atomic force microscope that all of these crystal-shaped units are built of organically coated grains, in ways that are similar to those in other units, but that form true species-specific morphologies. Envelopes of microstructural units may provide an explanation to this morphologic paradox.

The *Pinna* prisms of Fig. 3.12 show strong cohesive envelopes surrounding the gel layer when it is ready to crystallize. When crystallization occurs within the gel layer, only a few micrometers thick, the newly formed reticulate crystal spreads throughout all available space until encountering the resistant polygonal envelopes. The limit of the crystalline unit (i.e., the final shape of the crystal) is thus imposed by the organic polygon. Some observations made of the most distal portion of a growth lamella in the *Pinna* shell have shown that this polygonal network is already completed immediately prior to the beginning of crystallization (Figs. 3.20e–f). In the particular case illustrated here, further infilling of the polygonal compartments was not completed, possibly due to retraction of the mantle.

Conversely, when the chemical composition of the secretions of mineralizing epithelium does not permit formation of a strong and morphologically well-defined envelope, as is generally the case within the internal layers of bivalves, crystallization of the mineral occurring in the final step in formation of the growth unit can impose a shape conforming to the carbonate polymorph (calcite or aragonite). Thus calcite or aragonite biocrystals having limits reflecting relevant crystal angles may occur. However, it is definite that the internal organization of these crystals is typically reticulate, with organic material remaining within them. This paradox has been repeatedly illustrated by H. Mutvei, who has carefully studied nacreous structures among extant and fossil cephalopods. He has established concentric, stepping growth in the generally polygonal tablets of nacre, and noted that some even have a multicrystalline organization.

It is clear that, at this organizational level, the concept of "envelope" as a factor determining the morphology of skeletal units remains imprecise. Additional information will be provided by specific case studies (Chapter 5).

3.6.3 Formation of microstructural units: a prerequisite to their designation and comparative application

Focusing attention on the initial stages of the formation of skeletal structures enables us to make a distinction between the centers of crystallization that are the dominant control of crystal specifications in microstructural units of mollusc shells, and the widely used term of "centers of calcification" proposed by Ogilvie (1896) within coral skeletons.

Biominerals and Fossils Through Time

From a biological viewpoint, the so-called "centers of calcification" represent distinctly mineralized domains of coral skeletons. For every coral species, this arrangement is established in the earliest stage of polyp mineralization in the form of a circular mineral plate built by the young polyp directly after metamorphosis. During ontogeny of the polyps, and regardless of morphological complexity resulting from colonial development and modification of septal structure, microstructural continuity exists between the basal plate deposited by the juvenile polyp and the distal mineralization zone as situated at the extreme tips of septa and walls in an adult corallite. Every coral, from the simplest solitary coral to the most complicated and ramified colonies (such as *Porites* colonies with 1–2 m diameter) is truly built by fibrous bundles developed and overgrowing this extremely ramified micro-granular, tree-like structure that forms the microstructural framework.

Clearly, centers of crystallization, such as those visible on the uppermost surface of calcite or aragonite prisms in bivalve shells, have nothing in common with coral "centers of calcification." First, these are not themselves mineralized, and second, they play a key role in crystallization of each prism by fixing the orientation of its crystal lattice.

From an evolutionary point of view, the skeletal organization of molluscs and corals indicates that a cyclic biomineralization process is at the same time a permanent feature of calcified structures formed in these two major groups, and additionally, it is an extremely adaptable mode of biomineralization. It appears to be closely comparable in the calcareous structures built by phylogenetically distant organisms, such as fibers in corals and pillars of *Sepia* shells. However, there is a considerable difference in the ability of different phyla to create innovative structures. In corals, diversity essentially lies in the geometry of mineral-izing areas (e.g. Sorauf 1972). In the Acroporidae, the most abundant and diversified extant coral family, a truly innovative mode of mineralization is seen (Figs. 1.29–1.30). In contrast, the long separated classes of the Phylum Mollusca have shown an exceptional ability to create diversified skeletal units, all using the layered mode of growth, but with highly specific, taxonomy-linked patterns. The present status of their nomenclature needs to be updated to take full advantage of the potential of these patterns to help understand their phylogeny.

During the last few decades, microstructural analysis has proven to be an efficient tool in understanding the phylogeny of corals. This ability to trace evolutionary lineages in the Scleractinia is probably due to a general homogeneity of skeletal microstructures, as the skeleton-forming process has remained consistently based on the distal mineralization zone ("dmz") providing a framework for septal development.

In the Mollusca, numerous phylogenetic attempts based on shell microstructure have been made in the past. Kobayashi (1980), for example, established a microstructural classification with possible evolutionary relationships between the diverse microstructures of the Mollusca. The value of these attempts is directly tied to their degree of precision in the definition of microstructural categories. The usage of the terms fiber and prism is an excellent example of the ambiguity that can result from using names for microstructural units on the simple basis of their general or external aspect, without clearly understanding their mode of formation. As has been shown here, an identical name for a feature does not at all imply its having an identical structural or biochemical makeup, as can be recognized by the diversity of types of "cross-lamellar" structures, and diverse types of nacre. This emphasizes the importance of an accurate understanding of the origin and developmental specificity of skeletal units to insure the reliability of phylogenetic reconstructions.

3.6.4 Shell and corallite development as continuous processes

At a still higher integration level, the relationship between superposed layers deserves to be investigated. At present, most shell descriptions consider their microstructural sequences as superposed, clearly distinct layers of skeletal units. However, we have seen several examples of a transition phase between the superposed shell layers. This transition comprises both biochemical and structural modifications. This transitional stage, in contrast to the concept of distinct chemical crystallization, is fully understandable when we consider it as a consequence of a diachronous growth process. Progression from one type to the next in the microstructural sequence requires variations in composition of the organic materials directing crystallization. The mechanisms that regulate the functioning of those genes involved in producing these organic compounds proceed by a series of adjustments that occur in stages during time.

It is not necessary to emphasize that all of the properties illustrated in this chapter reinforce the concept of very precise controls being exerted by the mineralizing organ, whether the basal ectoderm of coral polyps or the external cell layer of the molluscan mantle. In all instances, a precise geometric control is required in order to form such coordinated assemblages of chemical, biochemical and mineralogical properties. From this viewpoint, detailed studies of the transitional stages seem particularly profitable.

To control such complex assemblages of organic and mineral components requires a direct link between the mineralizing organ and the growing mineral surface. The Crenshaw concept has been continuously supported by all of the new data that have resulted from improvement in our analytical instruments. The layered growth mode of biocrystals, a direct illustration of this concept, provides a perspective for explaining the continuity of the biomineralization process. Whether it is a question of explaining the distribution of isotopes and their fractionation process during formation of the coral skeleton, or of making shell materials based on the structure of the Bivalvia, the usual models propose that biocrystallization proceeds with the presence of an assemblage of chemical elements, either as ions or organomineral molecule groupings, *directly* producing calcareous structures without the presence of any structural mechanism ensuring control of the deposition. It is therefore a surprise that none of these models explains how, in the hypothesized liquid-filled spaces (with properties that would evidently be homogeneous from one edge to the other), microstructural types can be deposited that are at the same time sharply distinct, but possess contrasting mineralogical, chemical, and isotopic properties, and exhibit transitional stages between them. This suggests that a series of gene-on and gene-off states alternately occur in the cellular areas, and control progressive change in structure of skeleton components through modification in biochemical composition of the few-micron-thick mineralizing gel layer in which crystallization occurs.

4

Diversity of structural patterns and growth modes in skeletal Ca-carbonate of some plants and animals

The contrast between architectural and microstructural diversity at the macroscale and similarities at the submicrometer scale

In addition to molluscs and corals, two groups that contain highly diversified contributors to biogenic carbonate sediments in present-day shallow seawaters, many other organisms are also important Ca-carbonate producers. For some of them, such as the Echinodermata, their present role only poorly reflects their former importance during older geological periods. This chapter, by no means exhaustive, aims to present some of their diverse three-dimensional organizations, focusing on the fine structure of calcareous hard parts. The underlying objective is first, to provide some insight into recognition of the main categories of Ca-carbonate producers of carbonate-dominated sedimentary facies. Additionally, since knowledge of diagenetic change is an obvious prerequisite for obtaining reliable evaluation of characteristics in fossilized biogenic materials, knowledge of the initial state of these Ca-carbonate materials provides the reader with a basis for comparison.

4.1 Benthic algae

Very few algae have developed calcareous biomineralization, but they play an important geological role: only two families of the Chlorophyceae and two others among the Rhodophyceae. From a viewpoint of methods, this calcareous mineralization provides remarkable contrasts between the two groups, as follows:

(1) They are mineralogical opposites, as calcifying Chlorophyceae virtually always produce aragonite, while the majority of the mineralizing Rhodophyceae form calcite (although *Peysnella* and related types utilize aragonite);

(2) They also are structural opposites, as the Chlorophyceae produce carbonate external to their tissues, while the Rhodophyceae develop calcification at the level of cellular membranes;

(3) There is an additional ecologic contrast; the calcifying Rhodophyceae are developed in all seas, from the tropics to the Arctic, while the calcifying Chlorophyceae are warm water algae only. On the north coast of the Mediterranean, for example, their mineralization is reduced as a result, compared to that in individuals of the same species living on the south coast. These contrasting modes of mineralization, when associated with different biological ability, mostly depending on the wavelengths of the solar radiations utilized, determine their distinct but equally important geological roles.

4.1.1 The calcifying Chlorophyceae

In the shallow littoral areas of warm-water oceans, two groups of green algae are among the most important carbonate producers. Their respective contributions differ greatly. At the present time, the predominant genera belong to the Family Udoteacea (Order Bryopsidales); of these the most representative is the genus *Halimeda* (Fig. 4.1). The other producers of carbonates among green algae belong to the Family Dasycladaceae, a group having such a unique biological plan that its taxonomic position is still disputed. The group is generally considered to form an order by itself and has been proposed as the sole representative of a separate class (Dasycladophyceae) by Hoek *et al.* (1995). In effect, the Order Dasycladales, containing the well-known genus *Acetabularia*, is characterized by the peculiarity of only having a single nucleus for each complete thallus.

The historical development of these two groups is a good example of ecological sub-stitution through time. At present, the Dasycladales form a group with much reduced importance as compared to that of *Halimeda* and its related forms, a phenomenon of the Recent only. In contrast, in carbonate sediments of tropical seawater in older sedimentary basins it was the Dasycladales that played the principal role now played by *Halimeda*. Moreover, this predominance was maintained during an extremely long period, as the Dasycladales were important sediment producers during late Paleozoic, Mesozoic and Cenozoic times.

Calcification in Halimeda

Halimeda builds a ramified thallus that can attain a height of about 20 centimeters. Each branch forms a series of articulated segments (Fig. 4.1a), but in fact is constituted of a great number of tubes that are peculiar in that they are not formed of proper cells, but of siphons. Here, cytoplasm is contained in a continuous space between two walls of cellulose, and nuclei are distributed throughout the whole ensemble, which lacks true cell walls. These siphons are readily seen beneath the carbonate incrustation developed at the surface of each articulated segment (Figs. 4.1d–g). It is here that calcification develops, between these organic siphons. In a glucidic gel formed between the distal extremities of the siphons, the precipitation of aragonite here occurs in the form of acicular crystals forming a disordered felt (Figs. 4.1h–i).

Calcification decreases from the periphery towards the interior and within a single local-ity; specimens living at depth are less thoroughly calcified than are those living in shallower water. The calcareous structure of *Halimeda*, thus, is remarkably fragile, as the mineral elements are practically separate. They are maintained in their position only by being between the organic siphons, and dissociate after death of the plant. Geographic areas where *Halimeda* proliferates are characterized by very fine bottom sediments (Figs. 4.1l–m). These accumu-lated very fine acicular particles are easily transported, and thus can contribute to carbonate sedimentation in areas far from where they were produced. The very fine mud that they produce also can filter into residual space between sedimentary particles and constitute a major factor in cementation of sediments into rocks.

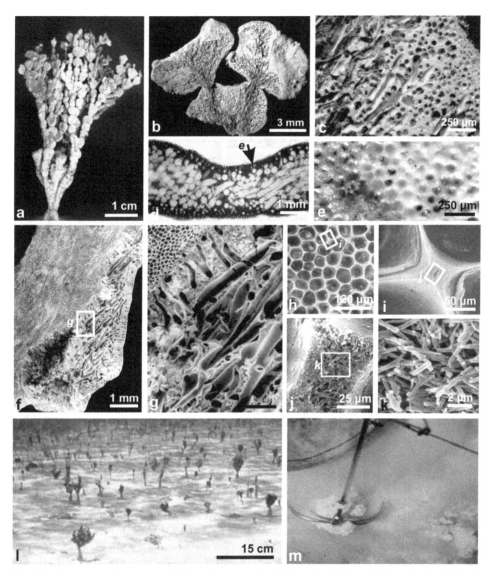

Fig. 4.1. Structure and occurrence of the calcareous alga *Halimeda*. (a) Typical view of an individual thallus; (b–c) The internal structure, an irregular network of filaments (siphons), with their size decreasing from central to peripheral parts of segments ("e": external surface). Note that the linkage between segments depends on the organic siphons only; (d–e) Enlargement of the external zone of one segment; calcification occurs first in the intervals between the terminal ramifications of the siphons; (f–k) Sequence of images ranging from the overall morphology of a segment (f: with organic siphons removed), to morphology of the micrometer-sized aragonite needles (k). Note that practically no contact exists between needles, leading to them being separated after decay of the organic extra-siphonal matrix; (l–m) Typical view of an *Halimeda* plantation in very shallow seawater (l) and fine mud resulting from accumulations of these fine-sized needles (m).

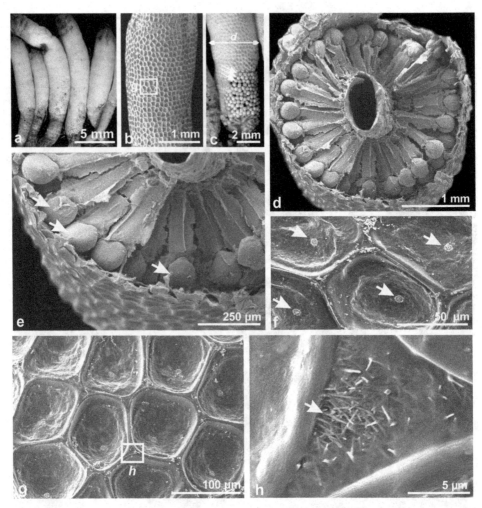

Fig. 4.2. General architecture of the thallus of a modern dasycladacian alga. (a–c) Morphology of some individuals. Note the regular arrangement of the distal parts of the siphons, already visible from the exterior; (d–f) Internal organization of dasyclades, here appearing as a single series of radiating branches. The terminal spheres are conceptacles, producing reproductive spores. Their apertures are visible (f: arrows) at the center of the outer layer; (g–h) Calcareous mineralization in this modern sample is reduced to a thin layer of irregularly arranged aragonite needles (arrow).

The thalli of the Dasycladales, formed of siphons just as those of *Halimeda*, are distinguished by a very different three-dimensional arrangement. Instead of irregular or dichotomous ramifications, the thallus of the Dasycladales displays an organization in verticils, which is to say that these ramifications are synchronous radial branches of the central bundle.

As their central siphon is generally unique, the Dasycladaceae have an architecture that is geometrically very strictly controlled. This property is of great importance in the identification of the hundreds of species that have been described within fossil floras. Even if diagenesis has completely destroyed the original calcareous material, the geometry of their arrangement is commonly preserved, permitting precise reconstruction of the verticils. The calcareous component of the thallus appears between the distal extremities of the verticils, similar to that in *Halimeda*. In addition, the carbonate is formed here of acicular micro-crystals of aragonite, arranged without any preferential order, except in the genus *Acetabularia* (below). The thickness of the calcareous incrustation varies according to species. The verticils are composed of a variable number of successive articulated segments, among them some that are transformed into sporangia. The sequence of ramification and the position of spore-bearing capsules are essential taxonomic criteria. Here, a difficult problem commonly results that bears on the identification of fossil forms because calcification develops from the exterior of the thallus towards its interior (and more or less deeply as a function of environmental factors, such as sunlight intensity). More internally placed branches are often nonmineralized, thus numerous fossil forms are only imperfectly known.

Acetabularia is a representative of the Dasycladales in which the vegetative (reproduc-tive) part is reduced to a single verticil, located at the distal end of a long vertical siphon (Fig. 4.3a). In this peculiar morphology, the single verticil in which spores are developed is likewise the only one calcified (Figs. 4.3b–c). A vegetative verticil, initially developed at the summit of a raised stalk, often remains visible at the center of the calcified disc (Fig. 4.3b: "st vert"). A fibrous aragonitic coating develops on the reproductive disc. This mineral

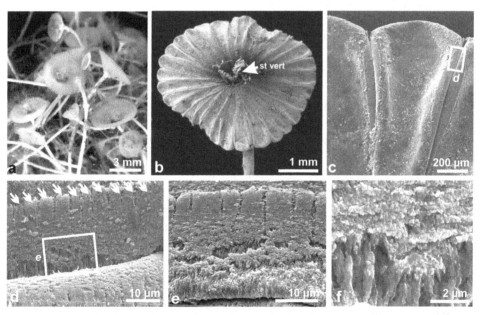

Fig. 4.3. *Acetabularia* morphology and microstructure. (a–c) *Acetabularia* at the reproductive stage: the calcified verticil (b: "st vert") is formed; (d–f) In contiguous radial elements of the disk, calcification is well developed. Elongation of branches occurs in a stepping growth mode (d: arrows). In contrast to other calcifying green algae, aragonite needles in this genus show some regularity in their arrangement on organic membranes of siphons (f).

material clearly shows the cyclic growth of the disc elements (Figs. 4.3d–e: arrows). Contrary to the other Dasycladales, the short fibers of aragonite that form this calcareous envelope are relatively well-oriented perpendicular to the cellulose membrane on which they develop.

Specimens of *Acetabularia* are well known, as they served as experimental material for the research of Hämmerling (1931, 1963) regarding the role of the cell nucleus in mechanisms of morphogenesis. The interest in this species for experimental purposes is due to each individual being unicellular. During the vegetative phase, only a single nucleus exists, localized at the base of the stalk. Experiments have shown that this nucleus produces organic compounds with the potential for morphogenetic control, thus establishing a role for the cell nucleus that had not yet been realized in the 1930s. Nuclear multiplication only comes into play in the formation of spores within radial elements that form the calcified disc.

4.1.2 Calcifying Rhodophyceae (Rhodophyta)

The Rhodophyta form a large and important group within the eucaryotes, with approximately 6000 species distributed in 670 genera (Woelkerling 1990). Their principal photosynthetic pigment is A-chlorophyll, with which are associated accessory pigments (phycobillines, phycocyanines, phycoerythrin) that are remarkably similar biochemically to the photosynthetic pigments in the cyanobacteria (Glazer *et al.* 1976). These varied pigments give the Rhodophyta the capacity to exploit the light spectrum in a way that is superior to the algae of the Chlorophyceae. Rhodophytes occur in marine water, from the shallowest zones to depths of several tens of meters when permitted to do so by water transparency.

In this highly diverse group, only a very few species calcify. Some of them construct fragile articulated thalli of centimeter-scale dimensions, the most typical of which is *Corallina officinalis* (Figs. 4.4a–d), whereas some others develop solid structures with varied morphologies, either ramified or incrusting (Figs. 4.4e–g). This capacity to incrust allows the construction of very solid structures capable of resisting the highest wave energies that can be generated by tropical storms.

Mineralization in the Rhodophyta is almost exclusively calcitic. Only one family produces aragonite: the Peysonellaceae. The calcite found in this group is always high in magnesium, and depending on environmental conditions, its Mg content can reach some of the highest values found in living organisms. These singular features were well established in the 1950s. This was the time when atomic absorption measurements first provided precise and easily reproducible values, drawing attention to mineralogical differences that are produced by organisms belonging to distinct taxonomic groups, even while living in the same environment and producing the same crystallographic form of calcium carbonate. These were some of the first demonstrations that the concept of vital effect, proposed with respect to isotopic fractionation, was also valid from a chemical viewpoint.

Another common aspect of mineralization in the Rhodophyta is that carbonate deposition is accomplished at the level of the cellular membranes themselves (Fig. 4.5). In contrast to the thallus of calcifying Chlorophyceae, formed of siphons that are tubular noncellular

Fig. 4.4. Diversity of morphological patterns in the calcifying Rhodophyta. (a–c) *Corallina*, the type genus of this family of calcifying rhodophytes that produce a branching thallus, with each articulated segment built by rows of cells that are only partly calcified in the central region; (d) Presence of noncalcified sections that allow the thallus to remain flexible; (e–f) Dendroid, spheroidal or incrusting habit of the Corallinaceae, partly dependent on environmental parameters (such as wave energy); (h–j) Incrusting Rhodophyta, able to build extremely solid structures on the front of barrier reefs (Marutea-South, Polynesia).

structures, the thallus in the Class Rhodophyceae is built by normal-sized cells. Calcification develops exterior to the plasma membrane (Figs. 4.5b–c). Calcification developed by two adjacent cells forms a symmetrical double-walled structure (Fig. 4.5e: aligned arrows, and Fig. 4.5f: dotted line). In early stages of calcification, communication between neighboring cells is maintained through the circular pores where plasma membranes remain in contact with each other, due to a lack of calcification there (Fig. 4.5f). Thereafter, with the growth of calcareous deposits, the cells die and no longer play more than a support role for the living superficial layer.

The contribution of the two joined cells to formation of a "calcified case" is clearly attested to by observation of the orientations of fibers (Fig. 4.6). This figure also illustrates

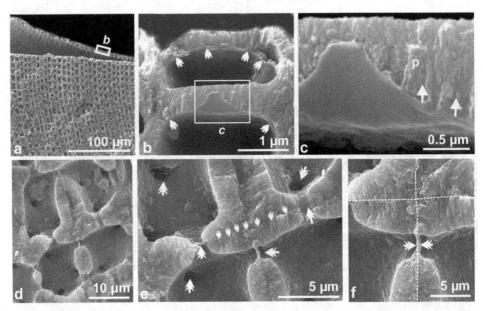

Fig. 4.5. Development of mineralized thalli in the Rhodophyceae (coralline algae), seen at the cellular level. (a) Typical aspect of the thallus of calcifying Rhodophyceae: rows of cells, most of them heavily calcified; (b–c) Cell membranes (b: arrows) and densely packed parallel high-magnesian calcite fibers (c: arrows); (d–f) Symmetrical development of fibrous calcite by adjacent cells. Note the circular, nonmineralized zones (e: double arrows).

the presence of the polysaccharide framework that separates adjacent cells. On untreated samples, the mineral phase appears as two opposing groups of continuous fibers, each of them depending on the activity of a separate cell membrane (Fig. 4.6a: arrows up and down). The polysaccharide framework is more easily visible after light etching (formic acid, 1/1000 for 30 seconds). Differential dissolution clearly establishes the layered structure of the calcified walls, Figs. 4.6b–e thus illustrating the stepping mineralization process of fibers that are formed by five or six stages of successive crystallization, separated by more easily soluble layers.

The mode of crystallization of calcite in the growth lamellae is readily visible by SEM observation (Fig. 4.6f), revealing the granular structure of the last (most internal) mineral layer. Atomic force microscopy is even more informative on this, in particular concerning the distribution of both organic compounds and mineral materials, and the three-dimensional aspects of units of mineralization.

Figures 4.7a–c illlustrate the typical sequence of height, amplitude and phase images on a 0.5 μm square area of an untreated surface of a calcified wall (e.g., Fig. 4.5c). The paired amplitude–phase images (Figs. 4.7d–e, f–g, h–i), ranging from a broader field of view (1 μm square area) up to higher resolutions (200 nm field of view), confirm that the organomineral interplay during calcite deposition in the walls of these red algae is basically similar to what we have observed in coral and mollusc skeletons.

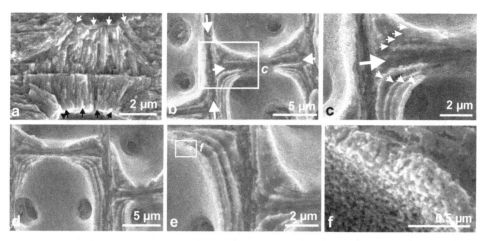

Fig. 4.6. The layered growth mode of calcite fibers of the red-algae skeleton. (a) Morphology of the fibers and symmetrical organization of the fibrous walls (SEM; arrows indicate remains of cellular membranes); (b–e) Growth lamellae of fibers, showing the symmetry of the process of mineralization by neighboring cells; (f) Granular ultrastructure of the latest growth lamella (see also Fig. 4.7).

Is the submicrometer structure in growth stages of the calcareous wall in corallinaceans then directly comparable to images obtained at the same scale in the organisms examined up to now? Certainly, the biochemical compositions of the organic phases are distinct, and the chemical or isotopic properties of the carbonates produced are likewise distinct. However, the similarity of their ultrastructure is truly remarkable and strongly suggests that comparable molecular mechanisms have come into play.

At both macro- and microstructural scales, calcification methods developed by the two groups of algae present a striking contrast, as is usually emphasized. However, the difference is not so great when the calcification modes are examined in relation to their cell membranes. The essential point depends on this: in both green and red algae, calcification is developed at the exterior margin of the cellular membrane itself. This results in a clear difference. Calcification by green algae is developed within the glucidic mucus that occurs between siphons, which does not provide a framework for any spatial organization of aragonite needles. Needles are disordered, and this occurrence is a good example of "biologically induced" calcification. In the calcifying Rhodophyceae, mineralizing material enters into contact with the polysaccharide wall that is common to two adjacent cells. This structurally rigid framework is made of insoluble polysaccharides, but also contains other polysaccharides that are more reactive, and able to bind cations. It is remarkable that sulfur is always associated with the glucides, the position of the sulfate groups and their number determining group structural properties and solubility. By contrast, differences in the structure and/or composition between parietal polysaccharides of noncalcifying red algae and those that produce Ca-carbonate in the Corallinaceae and Corallineae have not yet been determined.

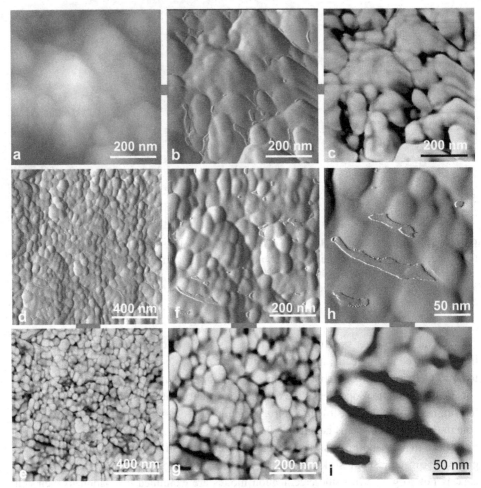

Fig. 4.7. Granular structure of the calcified layers in the Corallinaceae. (a–c) Height, amplitude and phase view of the cell wall of a calcifying red alga. High-magnification images show the granular structure of the material, whereas amplitude and phase images distinguish the geometrical relationships of organomineral-rich areas within submicrometer microdomains; (d–e), (f–g) and (h–i) A series of paired views at increasing enlargement (amplitude and phase in each case). Note the close similarity at the nanoscale between this calcite produced by algae and the calcite or aragonite of invertebrate animals.

4.2 Ca-carbonate production in sponges

In the Phylum Porifera (the sponges), production of a supporting structure is a vital necessity. These organisms get their nourishment by filtration and capture of organic particles suspended in water. In order to carry out this operation, they create a flow of water into the sponge through its entire external surface, and this water then circulates via a network of canals leading to cavities where the capture of nutritive particles is

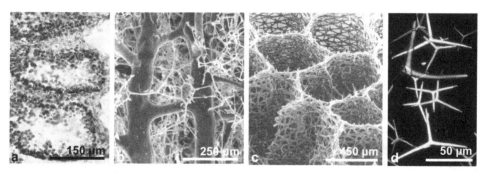

Fig. 4.8. Choanocyte chamber (a) and three examples of structures utilized by the sponges to support them. (b) Organic structure of the collagenous type; (c) Siliceous network; (d) Calcareous spicules (always calcite).

accomplished. The walls of these cavities are constituted of specialized cells, the choano-cytes, in which the cell body is prolonged by a small collar (in reality a group of fine lamellae arranged in an open cone). Every choanocyte carries a flagellum located in the axis of their collar, and it is the movement of these flagella that generates water movement. The particles captured are selected by the spacing of lamellae forming the collars. The total water flux now passes into this network of cavities that occupy the major part of the internal space of the sponge.

Following this, water is evacuated through one or several principal openings (= osculum). This canal system has no rigidity by itself, no more than do the walls of the choanocyte chambers. The essential function of the skeletal network, then, is to support the choanocyte system, permitting procurement of nutrient particles.

The Porifera produce three types of materials to construct such skeletal networks. The network can be entirely organic, or more precisely, proteinaceous, composed of a particular form of collagen, spongin. The principal structures (skeleton or scaffolding) support the whole of the organism, as well as the finest ramifications that form the individual baskets to support choanocyte chambers (Figs. 4.8a–b).

Sponges can equally produce siliceous elements of various forms, called spicules. These spicules can be free, simply soldered together by spongin, or associated in net-works. Along with this organic or siliceous scaffolding, a third group of sponges has been recognized for a long time. These are characterized by mineralization of calcareous spicules (Fig. 4.8d). These spicules are formed of calcite and are associated with a number of other histological and cytological characteristics. This mineralogical peculiarity defines this group of sponges, the Calcarea (created as the Calcispongia by de Blainville in 1834). In the sponges, it was long widely accepted that carbonate production was a character exclusive to the sponges of the Calcarea based on the mineral composition of their skeletal network components.

This was the situation at the time of discovery of a group of new sponge species that commenced in the 1960s. These have modified profoundly classical views regarding calcareous sponges. Not only have new data been gained regarding Ca-carbonate

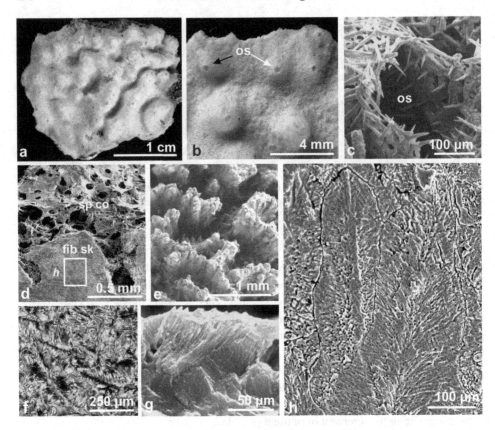

Fig. 4.9. The calcareous sponge *Petrobiona massiliana*, the first living sponge discovered with a nonspicular calcitic skeleton. (a–c) Overall morphology of the skeleton, with its nodular surface covered with spicules, and osculum (c: "os"), the opening for exit of water after its passage through the choanocyte chambers; (d–e) Superposition of spicular cortex ("sp co") and the fibrous mass ("fib sk"). The choanocyte chambers are localized in deep valleys of the upper part of the fibrous mass (e); (f–h) The fibrous mass becomes compact farther into the skeleton by additional mineralization that fills the valleys. Section viewed in polarized light (f), in an equivalent SEM view of a natural fracture surface (g), and in a polished and etched surface, SEM (h).

metabolism among the sponges, but also the role of sponges as reef builders has been fundamentally re-evaluated, making major modifications of paleoecological interpretations of reefs through geological time, and especially during the Paleozoic.

4.2.1 Nonspicular mineralization in the calcareous sponge Petrobiona massiliana

In 1958, Vacelet and Lévi discovered a new species of sponge (*Petrobiona massiliana*) in the Mediterranean Sea, which belongs to the Class Calcarea according to all its histological, cytological and spicular characteristics (Figs. 4.9a–c). This species also constructs thick, nodular incrusting laminae, with thicknesses in the centimeter range. In addition, it possesses a peculiarity recognized only here, in Porifera known at that time. The main part of the skeleton is a fibrous calcitic mass, *which is external*, i.e., directly fixed to the substrate and supporting the living part of the organism. Therefore, the choanocyte chambers are no longer braced by a spicular structure, but localized within cavities on the surface of this fibrous carbonate skeleton. The spicules, which are calcitic as in all other Calcarea, have the usual morphology of this class, but no longer serve as a system for support; instead, they only form a cortex covering the fibrous calcareous mass.

This new sponge was first interpreted as a "living fossil" and related to the pharetronids, a group of the Porifera created by Zittel (1878) for exclusively fossil species having a calcareous skeleton. From a taxonomic point of view, the discovery of this new method of supporting choanocyte chambers and concomitant functional modifications did not constitute sufficient basic features to warrant an important change in sponge classification. Cytological characters are considered fundamental in the classification of modern Porifera, as begun by Bidder (1898). However, this species represented the first evidence of a sponge with the capacity to construct mineralized supports in forms other than spicules.

The usage of the group name pharetronid for possible fossil predecessors of the modern *Petrobiona* also revealed the difficulty of developing a common system of classification uniting living and fossil sponges. This question of the classification of fossil forms began to be resolved in the years following the discovery of *Petrobiona*, so that in many respects it constitutes a chronological reference point.

4.2.2 The discovery of calcareous structures produced by the Demospongia

Since *Petrobiona massiliana* was discovered in submarine caves of the Mediterranean coast, further investigation focused on these very specialized biotopes thoughout the world. These had been inaccessible to traditional methods for investigation of marine features. In a few years, numerous entirely new species were collected, bringing unexpected, very fundamental data regarding capacities for mineralization in the Phylum Porifera.

When the spicules produced by these sponges became available for study, for the greatest part, it was then realized that these species had siliceous spicules of morphologic types already identified in numerous known sponges of the Class Demospongia. Additionally, all of the histologic and cytologic characters of these newly discovered calcifying sponges were clearly those normal for demosponges. The new sponges producing calcareous structures were thus recognized as belonging to the Demospongia, until then not known to produce anything similar. It is important to clearly state here that *the calcareous structures produced by these demosponges are never deposited in the form of spicules*.

Beginning in the 1960s, methodical exploration of overhangs and submarine caves of tropical reefs first permitted the collection of numerous specimens of a species whose analysis and interpretation had remained doubtful for a long time, and its importance unsuspected: *Astrosclera willeyana* Lister. Discovered in the Pacific at Lifu (New Caledonia) and nearby at Funafuti, this species builds a massive skeleton formed of spherulites of aragonite. It was found again at Christmas Island by Kirkpatrick (1910), who recognized the presence of siliceous spicules of the acanthostyle type, and he shortly thereafter observed the intracellular origin of the aragonitic spherolites. Limited by the concepts accepted at that time, he noted that this calcareous structure could not have been produced by a sponge. Thus, Kirkpatrick interpreted his specimen as an association of a sponge and a calcareous alga to which he attributed formation of the aragonite. An analogous hypothesis was formulated for another assemblage of the same type, siliceous spicules and a calcareous structure, already observed in another species (*Merlia normani*) discovered by Kirkpatrick himself in 1908.

It was only in 1965 that specimens collected by Vacelet and Vasseur at Tulear, Madagascar, permitted establishing that this association of siliceous spicules and carbonate skeleton was truly a fact in these rather unique organisms, not even belonging to the calcareous sponges (the Calcarea), although carbonate preponderates in its skeleton. They belong to the Demospongia; this led to abandoning the concept that demosponges were unable to produce carbonate materials.

Since that time, more species of the calcifying demosponges have been discovered during exploration of the biotope that apparently is one of their commonalities, the deep zones of reef slopes, the overhangs, and deep submarine caves of coral reefs. Hartman and Goreau (1975), who discovered several of them in the reefs of Jamaica between 1970 and 1975, proposed uniting them under the term of "sclerosponges" (or "coralline sponges"), terms lacking true taxonomic value. Actually, sponge classsification primarily depends on their larval development (Lévi 1973) and various histological, cytological and molecular criteria (Bergquist 1978; Hartmann 1982; Vacelet 1983; Hooper and Van Soest 2002), as recently summarized (Boury-Esnault 2006). As a result, these various groups of calcifying demo-sponges have to be classified in different families.

Astrosclera

Astrosclera builds irregularly nodular calcareous structures that can reach dimensions in the order of decimeters (Fig. 4.10). Their surface is marked by the presence of canals arranged in star-shaped clusters (= astrorhizae; Fig. 4.10: "as"). A section perpendicular to the surface passing through these astrorhizae (Fig. 4.10d) indicates that they constitute openings to a network of internal spaces developed at several millimeters beneath the surface. These spaces are the functioning choanocyte cavities of the sponge. This arrangement of mineral structures provides an innovation for *Astrosclera* alone. Here, there is no spicular network, but instead, a skeleton that is a porous calcareous mass in its upper part becoming compact in its lower part. The calcareous structure is built of well-defined spherulitic masses that are remarkable themselves, as also is their mode of emplacement. These skeletal units are formed within specialized cells that are differentiated in the deeper parts of the sponge, and are successively displaced towards more surficial areas (Figs. 4.10e–f). In the course of their trajectory, the size of the intracellular spherulite grows, and the carrier cell ends its migration by installing the spherulite at the summit of the skeletal growth region (Fig. 4.10g). In sponges, the production of spicules by specialized cells (sclerocytes) is

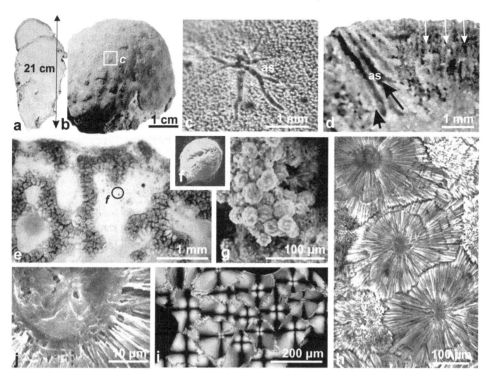

Fig. 4.10. Morphology and microstructure of modern *Astrosclera*. (a) Longitudinal section of a very old specimen from the Great Barrier Reef (courtesy J. Reitner, Göttingen University); (b–c) Surface of another specimen, showing numerous astrorhizae, and (c) an enlarged view; (d) Longitudinal section of an astrorhiza ("as"); (e) Microscopic view of a section in the upper part of the skeleton. The calcareous structures are formed of densely packed spherulites; (f) Individual spherulite moving to the upper part of the skeleton; (g) Accumulation of spherulites at the growing edge of the skeleton; (h) Radial arrangement of the fibers in an SEM view of a polished and etched surface; (i) Thin section of the spherulitic skeleton under cross-polarized light; (j) High concentration of organic matter in the center of a spherulite. e, g: Gautret (1985), specimens from J. Vacelet, Endoume Biological Station, France; f: Cuif *et al.* (1987); h: Dauphin (2002a); j: Cuif and Gautret (1991).

the rule. The sclerocytes of the calcifying demosponges carry out the same operation during construction of their initial calcareous spherulites; reflecting a considerable metabolic change. The summit elements of the skeleton thus are formed of an accumulation of spherulites, each of which has been transported by a single cell. In the following phase of their development, the spherulites continue to grow, until coming into contact with their neighbors. They then become polyhedral elements (Fig. 4.10h), whose fibroradial organization has been illustrated by study in thin section with polarized light (Fig. 4.10i).

We must remember that the aragonitic spherulites are produced by these sponges at the same time that all their cytological and biological characteristics place them incontestably within the Demospongia. Additionally, *Astrosclera* samples from the Pacific are distinguished from those of the Indian Ocean by the fact that the former still produce siliceous spicules conforming to other demosponges in their group, while the Indian Ocean *Astrosclera* no longer produce them. In these, nothing in their skeleton suggests that they are demosponges, and thus, it is not surprising that uncertainties in the identification of fossil demosponges persisted right up until discovery of their modern descendants. The discovery of these living Porifera has also brought an appreciation of their slow growth. Measurements carried out on the oldest parts of specimens measuring approximately 20 cm high indicate life spans of several centuries (Fig. 4.10a, personal communication, J. Reitner, Göttingen), and there is nothing to suggest that fossil forms had growth rates that were any different.

The Ceratoporellidae: tubular structures, siliceous spicules and a calcareous microstructure of fiber bundles

In 1911, another species producing siliceous spicules and a calcareous skeleton at the same time was discovered by Hickson: *Ceratoporella nicholsoni*. Using the same logic as Lister and Kirkpatrick, Hickson did not classify this species among the sponges, but rather in the Alcyonaria (a surprising point of view in view of the total absence of siliceous spicules in the Cnidaria). Very poorly documented prior to new collections amassed by Hartman and Goreau in Jamaica, this aspect is even more spectacularly shown than in *Astrosclera*; that of the simultaneous production of siliceous spicules and a calcareous skeleton in a poriferan.

Modern ceratoporellids build massive, generally hemispherical structures (Fig. 4.11a), but a very thin functional layer at the surface alone constitutes their biologically active region. This zone, 3 to 4 mm thick, is essentially formed of cylindrical tubes with joined walls (Fig. 4.11b), which contain the choanocyte chambers. Differing from *Astrosclera*, the ceratoporellids always form siliceous spicules that can be seen easily in functioning tubes of this superficial layer (Figs. 4.11c–d, f). The spicules are morphological types that are well known in the Demospongia (Fig. 4.11e). The functional layer of this sponge is always the same thickness, regardless of the overall size of the skeleton, because calcareous material is progressively deposited further to the interior that blocks the tubes, and grows at the same rate as the upper margin (Figs. 4.11g–h). *Ceratoporella* thus forms compact carbonate masses with dimensions of several decimeters in deep portions of Caribbean reefs, although only the external 3 mm is functional. The complete structure is formed of fibrous aragonite skeletal material where construction occurs in two phases. First, the walls of tubes are formed, with internal elements arranged in pointed bundles directed towards above and, second, large bundles of fibers continue the progressive infilling of functional space.

Spirastrella (Acanthochaetetes): an example of a "living fossil"

This genus was first established for a Lower Cretaceous fossil species, *Acanthochaetetes seunesi*, by Fischer (1970). Subsequently, a very similar living species (*A. wellsi*) was discovered in 1975 by Hartman and Goreau at Guam in the western Pacific, then by K. Mori (*Tabulospongia*, 1976). In spite of the great chronological gap from the Cretaceous to the Recent, the remarkable equivalence of mineralogical and microstructural characteristics of the calcareous structures built by these species at first led to them being classed as the same genus. This interpretation was chosen again by Reitner (1991), who placed the

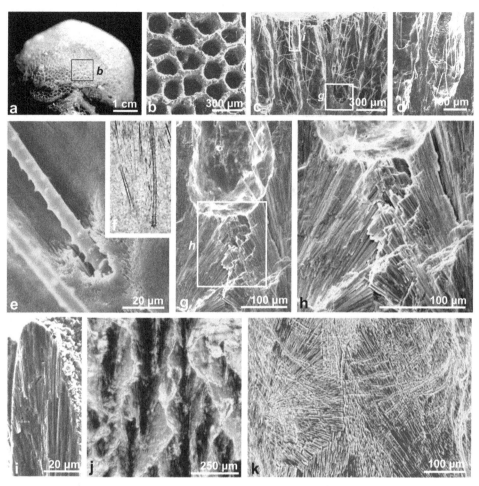

Fig. 4.11. Microstructure of the calcareous part of *Ceratoporella* (Demospongia). (a–d) Morphology of a small colony (a), geometry of the tubes at the surface (b), and a longitudinal section of tubes (c–d); (e–f) Insertion of a siliceous spicule into fibrous aragonitic skeleton; (g–h) Orientation of fiber bundles at the base of a tube with a compact structure formed of rectilinear fiber bundles; (i) Bundle of subparallel fibers at the summit of a tube (longitudinal section); (j–k) Fibrous wall in low magnification (crossed-nicols polarized light) and fibers with growth steps forming the infilling of tubes (k).

species in the subgenus *Acanthochaetetes*, regarding the latter as belonging in the preexisting genus, *Spirastrella*.

Their calcareous skeleton is generally hemispherical (Figs. 4.12a–c), with a tubular structure (Figs. 4.12d–f), partially interrupted by spines with a radial growth direction, perpendicular to the wall from which they originate. Spicules are comprised of long linear megascleres and a coating of very numerous small, massive, spiny spicules (Fig. 4.12g). The skeleton is formed of magnesian calcite (Reitner and Engeser 1987). The skeletal microstructure is characteristic and very different from the fibrous forms, whether calcitic

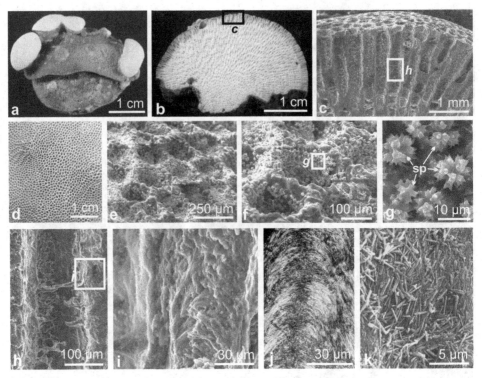

Fig. 4.12. *Spirastrella* (*Acanthochaetetes*). (a–c) General morphology and tubular structure with transverse plates; (d–g) Growth face and polygonal sections of tubes, and (g) the siliceous spicules ("sp") included in the living tissue; (h–k) Microstructure of the calcified walls with convex superposed lamellae formed of cylindrical elements 2 to 4 µm in length, irregularly arranged. Specimen: courtesy J. Reitner (Göttingen).

(*Petrobiona*), or aragonitic (*Astrosclera, Ceratoporella*). Here magnesian calcite is organized into fine superposed layers, whose general geometry is convex upwards, the same shape as the growth border of the tubes. These mineralized lamellae are made up of very small rectilinear elements 3 to 4 µm in length, with a submicrometer diameter (Figs. 4.12a–c) formed in a preliminary organic layer (Reitner 1991).

Vaceletia: *a demosponge with a sphinctozoan-like calcareous skeleton*

Establishing the presence of the calcifying demosponge *Vaceletia* in modern reefs represents an even more spectacular discovery than the preceding, because it allows a detailed examination of an architectural type recognized for a long time in ancient sedimentary rocks, and sufficiently abundant that a taxonomic category was created for it (and based on it): the Sphinctozoa (Steinmann 1882). The species, all fossils, which are united in this group, have a characteristic architecture of superposed domes, giving the whole of the calcareous structure, as seen from the exterior, a cylindrical organization

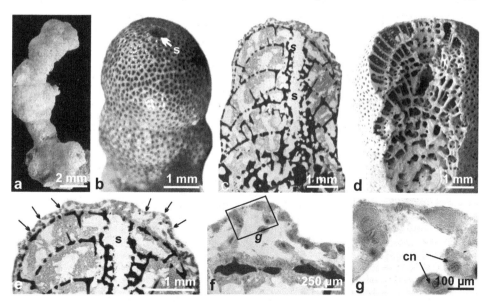

Fig. 4.13. Morphology and architecture of *Vaceletia crypta*. (a) A branch of *Vaceletia*. This species builds massive ramified complexes, formed of independent branches; (b) Extremity of a living branch: entry of water through pores distributed on the whole surface, then exiting through the axial siphon opening ("s"); (c) Longitudinal section. Tissues are localized in the superposed hemispherical chambers. Note the long, continuous axial siphon ("s"); (d) View of the skeleton only; the pillars supporting the hemispherical domes are easily visible; (e–g) At the summit of a branch, preparation of a future cupola is shown (e: arrows). Here it is still entirely organic, formed of distinct nodules ("cn") that are calcified after the cupola is in place. b, c, d: Vacelet (1979); e, f, g: specimens and histological sections from J. Vacelet (Endoume, France) and Gautret (1985).

marked by annular constrictions. This is exactly the case in the living *Vaceletia crypta*, discovered by Vacelet (1979) in sediments of the deep reef slope in New Caledonia (Fig. 4.13).

This sponge forms its skeleton as superposed carbonate domes, perforated by multiple openings (Figs. 4.13a–c). The series of hemispherical spaces is connected by an axial canal, providing the means for evacuation of water that has previously traversed the functioning flagellated chambers of the sponge located in successive domes. Here the choanocyte system is located in internal spaces, whose limits are defined by the superposed hemispherical layers forming the support structure of the sponge. It is noted that, compared to the common organization of the sponges, in which supporting structures are located within the body of the animal, the relative positioning of the soft tissue and skeleton are exactly the opposite in the Sphinctozoan sponges. However, the functioning of each sponges is comparable to the classical mode: water enters by numerous perforations at the periphery and exits via a central canal. This function is clearly reflected in the organization of the calcareous skeleton (Figs. 4.13c and 4.13e).

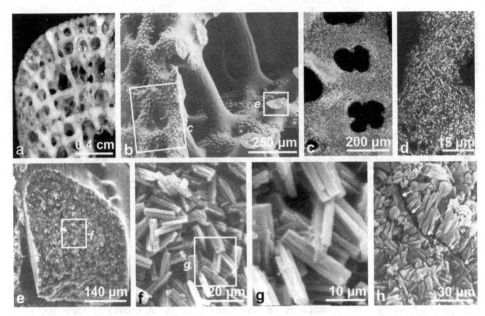

Fig. 4.14. Microstructure of the skeleton in *Vaceletia*. (a–b) Pillars and cupulae. Note the detailed ornamentation of the inhalent pores; (c–d) Microstructure with a granular appearance in optical microscopy; (e) Section of a pillar; (f–g) Morphology of composite aragonite rods with three or four of them clustered; (h) Boundary between a cupola (below) and the base of a pillar, separated by the organic membrane that limits the cupola (arrows). b–h: Gautret (1985).

Domes are formed successively, at first as an entirely organic framework (Fig. 4.13e: arrows) in which the central nodules are the units of mineralization (Figs. 4.13f–g). In the final phase, the dome takes its characteristic form, and mineralization is carried out in a single stage, concomitant with occupancy by functional tissue. The microstructure of the skeleton also contrasts with the usual fibrous calcitic or aragonitic structures. It is formed of very small, associated but parallel calcareous units in small groups that form elongate granules (Fig. 4.14).

4.2.3 Overview of present knowledge of the organic components associated with the calcareous skeletons of calcifying demosponges

These four examples illustrate the diversity of calcareous skeletal materials produced by the Demospongia. To date, 16 species of demosponges (Chombard *et al.* 1997) have been identified as producing a solid calcareous skeleton in addition to a spicular one (which at times is no longer produced). In spite of the interest in this calcification that is so unusual in the vast group of the Demospongia, biochemical data are fragmentary. Actually, it is a consistent feature of these calcifying species that other taxonomic criteria (larval

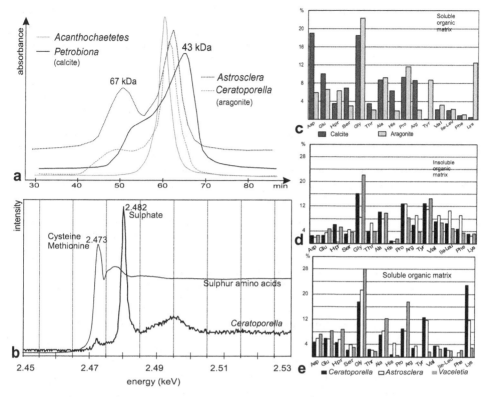

Fig. 4.15. Biochemical characterization of some organic components from the calcareous skeleton of living demosponges. (a) HPLC profiles of soluble matrices from calcitic and aragonitic nonspicular skeletons; (b) XANES profiles of a calcareous nonspicular skeletons (Ceratoporella); (c) Amino acids in soluble organic matrices of calcitic and aragonitic demosponges; (d–e) Amino acids in insoluble (d) and soluble (e) matrices of *Ceratoporella*, *Astrosclera*, and *Vaceletia*.

development, spiculation, etc.) can be applied to identify them in the normal fashion. As a result, research is not carried out on the calcareous parts, leading to the conclusion that, regardless of what its diversity may be, it cannot form a useful contribution to their classification.

After standard decalcification and purification processes, the chromatographic profiles obtained on the soluble organic phase (Fig. 4.15a) show characteristics of their molecular masses that are closely comparable in their orders of size to results produced from fibrous coral materials, for example, when prepared under the same conditions. In the analysis of these compounds, the presence of sulfated polysaccharides is important, and must be taken into account. The presence of these has been evaluated by use of the synchrotron X-ray absorption near edge spectrum method (XANES) (Fig. 4.15b: *Ceratoporella*). In chromatographic profiles, their contribution to formation of products with large molecular masses is highly probable, but their presence has only been established qualitatively at present.

Calcitic skeletons usually show higher concentrations of amino acids than do aragonitic ones (Fig. 4.15c). Concentrations in amino acids among the three principal aragonitic forms (*Astrosclera, Ceratoporella, Vaceletia*) are relatively homogeneous and comparable, with respect to both soluble and insoluble organic phases (Figs. 4.15d–e).

4.2.4 Taxonomic value of microstructural characteristics in calcifying sponges

Molecular data acquired in recent years (e.g., Chombard *et al.* 1997) have confirmed the conclusions of previous research founded on biological and histological criteria: calcareous nonspicular skeletons have been repeatedly produced in the Phylum Porifera. A calcareous, nonspicular skeleton can be produced by the Calcarea and has appeared on multiple occasions in the Demospongia. Grouping all these forms in a single taxon (such as Sclerospongia: Hartman and Goreau 1970; or Ischyrospongia: Termier and Termier 1973) should be abandoned, based on biological evidence. As a result of the taxonomy accepted at present, it is commonly written that the calcareous skeleton has no taxonomic value (Reitner 1991; Wörheide 1997). This simply means that the microstructural and mineralogic features of calcareous structures built by these calcifying demosponges cannot be used as criteria for defining taxonomic groups of species, as some species of the same unit do not have calcareous structures. This does not mean at all that skeletal features may strongly vary within a given species. Just the opposite; Reitner and Worheide, dealing with fossil representatives of calcifying demosponges, emphasized the extremely conservative character of the skeletal patterns (see Section 7.5). In addition to *Acanthochaetetes*, *Astrosclera* is another case where the similarities are so precise between living and fossil specimens (Triassic) that the same genus name has been used for both, although no fossil of intermediate age are known (Reitner and Worheide 2002).

Therefore it is especially necessary to carefully describe skeletal units (their mode of emplacement, sizes, shapes, and three-dimensional arrangements) in order to compare extant and fossil calcareous features of these sponges. The terms spherulite, or spherulitic, has often been used to refer to various types of fibrous structures, virtually wherever a more or less radial arrangement of bundles of fibers can be observed, even occasionally. This has led to both the fibrous skeleton of *Petrobiona* and that of *Astrosclera* being described as "spherulitic" (Vacelet 1964), in spite of obvious differences between the mode of skeletal development in the two species (Figs. 4.9 and 4.10). This emphasizes the necessity of creating biologically based descriptive terminology that can insure that no false comparisons are made due to confusion caused by use of the same term to apply to separate and different skeletal features.

4.2.5 Modes of biomineralization at the nanometer scale

The results of the examination of nonspicular calcareous structures produced by sponges at the submicrometer scale are similar to the surprising ones previously obtained by descriptions of molluscs, corals, and benthic algae. They have very sharply distinct

Fig. 4.16. Ultrastructure (SEM and AFM, tapping mode) of the aragonitic fibers of *Astrosclera*. (a) Scanning electron micrograph of a sectioned fibrous spherulite (aragonite); (b–c) Granular structure of fibers (b) and distribution of intracrystalline organic matter (c); (d–f) Submicrometer grains seen in the height mode (actual relief), in the phase mode (showing strong interactivity of the cortex), and in the amplitude mode (based on slope values).

microstructures at the optical scale, and even more so at the scale of the transmission electron microscope, but these differences are lost when they are studied in the atomic force microscope.

Astrosclera

In the well-crystallized skeleton of *Astrosclera* (see Fig. 4.10), the units observed in the atomic force microscope (tapping mode) suggest a distribution of interactive compounds that is very comparable to those that have previously been seen in the molluscs and corals. With diameters in the range of 50 to 100 nm, the characteristic structures are irregularly anastomosing nodular elements (Figs. 4.16b–c). Phase images show that they still contain, at least at their surface, interactive organic compounds (Figs. 4.16d–e). The similarity of ultrastructure shown by the relationship between organic compounds and minerals would indicate that the process of crystallization is comparable to the groups studied previously, in spite of obvious differences in the method of formation of the organomineral assemblage prior to crystallization. In spite of the perfectly monocrystalline appearance of fibers, each of them are composed of the now-familiar reticular crystals.

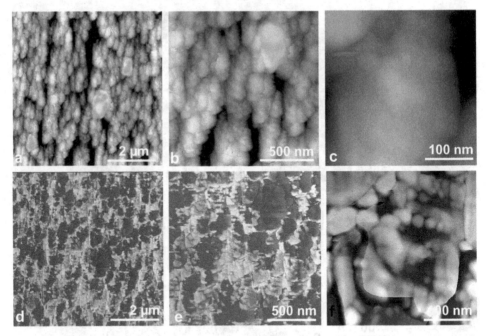

Fig. 4.17. Nanostructure, and organics and mineral distributions in fibers of *Ceratoporella*. (a–d) Granular structure of fibers. The alignment of grains corresponding to the elongation of the fibers is probably correlated to the presence of a highly organized organic component; (e–f) The components of the fibers are themselves complex structures, formed of elements with dimensions and organomineral structure that conform to general models.

Ceratoporella

Ceratoporella is also characterized by a well-crystallized, fibrous structure (see Fig. 4.11h). Within these fibers, atomic force microscopy establishes the linear arrangement of relatively large grains, that seems, at first glance, to be an exceptionally distinctive feature (Figs. 4.17a–b). These grains are included within an organic component (with surfaces that are strongly interactive) that is continuous, and covers the entire surface (Figs. 4.17d–e). At higher resolution, these grains appear constituted of microdomains (Fig. 4.17f) that also exhibit a distinction between an interactive cortex and a central region.

Vaceletia

The mineral portion of *Vaceletia* is formed up on a continuous organic substrate that provides a preformed structure dictating the cupola shape (Figs. 4.13e–g), with mineralization later providing the rigid nature of the cupula (Fig. 4.14). The groups of aragonitic units that crystallize within this organic phase are also seen to consist of granular subunits that are remarkably aligned (Figs. 4.18a–c) and well controlled, as suggested by their regular organic cortex (Figs. 4.18d–f).

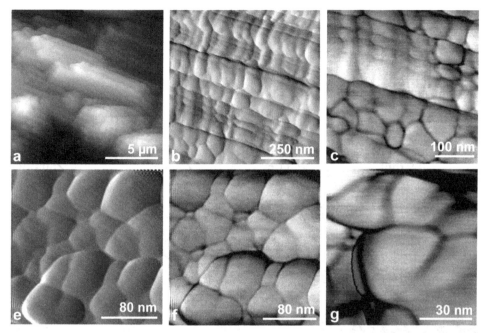

Fig. 4.18. Skeletal units of *Vaceletia* at the nanoscale. (a) Scanning electron microscope image of the fibrous composite units of the *Vaceletia* cupola; (b–c) AFM views showing alignment of granular subunits in fibers; (e–g) Height, amplitude, and phase images of granules within fibers.

In conclusion, regardless of architectural and functional differences between these calcified skeletons, the basic mechanism of calcification, that is, the interplay between organic and mineral components, appears to be similar. Taking into account the position of the Phylum Porifera as compared to other invertebrates, this overall similarity of calcification patterns suggests that methods of controlling crystallization involve strict dimensional limits to sizes of the crystallized units.

Data obtained during the last three decades concerning the formation of calcareous skeletons by modern demosponges have not significantly changed their classification, based on biological approaches. The true importance of these carbonate skeletons built by such "siliceous" sponges is apparent only with a paleontological perspective. The discovery of carbonate biomineralization in the Demospongia completely overturned our understanding of the role of this group during geological time, especially during the Paleozoic Era. From this perspective, their proper taxonomic position is essential for assessing the interpretation of the roles that different groups played as reef builders (see Chapter 7).

4.3 Hydrozoa and Alcyonaria

Within the Phylum Cnidaria, the Hydrozoa constitute a class that is highly differentiated in morphology and anatomy, but most of their modern representatives do not produce any

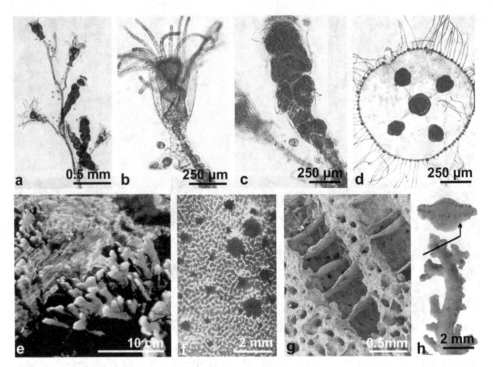

Fig. 4.19. Hydrozoans: *Obelia* (a–d), *Millepora* (e–g), and *Stylaster* (h). (a–b) Colony and vegetative polyps; (c–d) Polyps producing medusae and free medusa; (e) Site with colonies of living milleporines; (f–g) Colonial network and placement of polyps; (h) Fragment of a colony of *Stylaster* and cross section.

calcareous skeleton. The Alcyonaria on the other hand, have considerable capacity for mineralization, but this is also highly diverse, with some groups producing spicules with micrometer-scale dimensions, and others producing ramified calcareous axes attaining lengths of several meters.

4.3.1 Hydrozoa

The hydrozoans typically grow following a cycle in which two very anatomically distinct phases alternate. One phase lives as polyps, fixed to a substrate, forming colonies (Fig. 4.19a) with extremely variable architecture.

The basic colonial unit is the asexual polyp (Fig. 4.19b), which is most important in that it captures nutriments. The colonies also bear polyps that produce medusae (Fig. 4.19c), the free-swimming reproductive phase of the biological cycle (Fig. 4.19d). The strongly calcified colonies of *Millepora* appear as exceptions among the dominantly nonmineralized hydrozoa. Although they may be little diversified taxonomically, the milleporines are recognized as an order (Order Milleporina, Hickson 1901). They play a significant role in

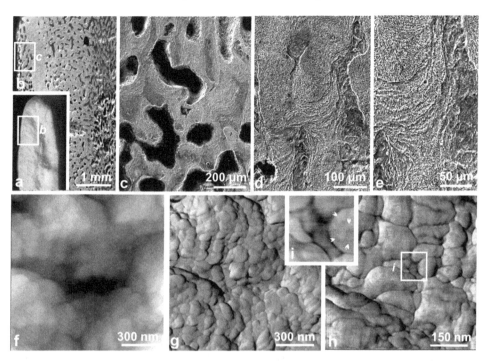

Fig. 4.20. Microstructure of the milleporine skeleton (*Millepora* sp.). (a–c) The two areas of growth in the colonial skeletons are visible as a result of their difference in compactness, but not by specific microstructural arrangements of the short fibers; (d) Both zones exhibit distinct growth layering (e); (f–g) Submicrometer-size grains are visible in height (f) and phase (g) images in the atomic force microscope; (h–i) Close-up view indicates that these are composed of partially merged smaller units (i: arrows).

shallow-water areas of carbonate deposition developed along tropical coasts (water depths up to 25 to 30 meters). Their colonies, either erect or incrusting, depending on environmental dynamics, are recognizable by the very homogeneous calcareous network that they construct (Figs. 4.19e–f). Polyps, often differentiated into two or three functionally distinct types, are lodged in the cylindrical cavities of this network (Fig. 4.19g).

The skeletal network of the milleporids is built of short aragonite fibers. The mineralizing epithelium produces two generations of fibers; the first constituting a lightweight, reticulated structure that is later reinforced by an accumulation of layers leading to the formation of a compact section (Figs. 4.20a–d). The coordination of crystallization between the fibers formed by each growth layer is poorly developed; long fiber bundles of several tens or hundreds of micrometers, commonly seen in the scleractinian corals, are not present here. At greater magnification and resolution, these fibrous layers are resolved into nodular grains (Figs. 4.20e–h), themselves formed of elements having diameters of 30 to 40 nm that are almost perfectly fused together (Fig. 4.20h). The model for crystallization of fibers by formation of reticulate crystals incorporating residual organic phases is well shown by *Millepora*.

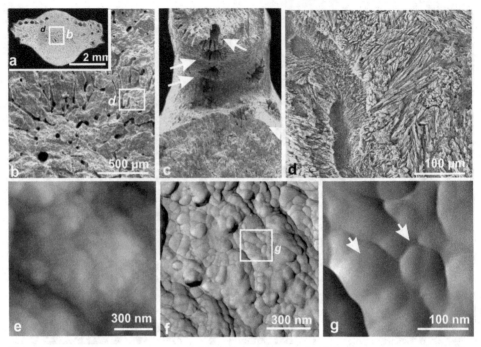

Fig. 4.21. Micro- and nanostructure of the *Stylaster* skeleton. (a–c) Sections perpendicular to the direction of growth, showing polyp cavities (c: arrows); (d) The massive part of the skeleton, built by dense sclerenchyme formed of disordered fiber bundles. Distal mineralization axes are visible from place to place in the skeletal sections, but lack any characteristic arrangement into specific mineralization patterns (d); (e–g) *Stylaster* skeleton seen at the nanoscale (in the AFM), displaying nodular constructional units in the 10 nm range (arrows).

The stylasterines, a second order of the hydrozoans, form arborescent branching skeletons with generally reduced dimensions, in which polyps are located in the alveoli of a fairly compact skeleton (Figs. 4.21a–c), with spaces utilized there for habitation by rare polyps (Fig. 4.21c: arrows). Opposite to the milleporines, the fibrous structure here is well formed, developed in zones of distal mineralization following a scheme quite comparable to that of the Scleractinia (Fig. 4.21d). As in the Scleractinia, the atomic force microscope shows that the calcareous structure is built by similar, densely packed nodular elements.

4.3.2 *Alcyonaria*

The Alcyonaria (formerly named the Octocoralla), with polyps that are characterized by having eight tentacles (Figs. 4.22a–b), show mineralizing activity that is apparently much less important than the constructional abilities of the Scleractinia with their large colonies in tropical areas. Two factors contribute to this reduction in their apparent volumetric importance (Carey 1918): first, in the majority of alcyonarians mineralization is in the form of

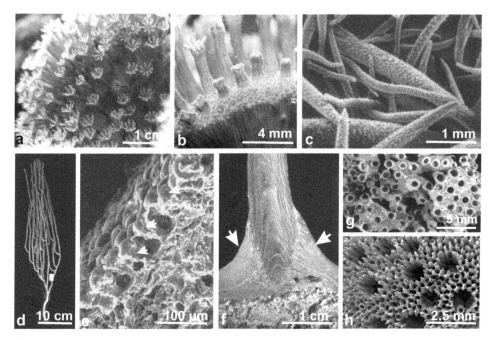

Fig. 4.22. Some examples of calcareous structures in the Alcyonaria. (a–c) *Sarcophyton*, an example of a "soft coral" (Alcyonacea), producing only spicules (c); (d–f) Gorgonacea, sea-fans with calcareous cortex formed of sclerites (or spicules: arrows) and skeletal axes variously calcified, from wholly organic to entirely calcareous (Scleraxonia); (g–h) Two atypical alcyonaceans: *Tubipora musica* ("pipe organ coral"), which builds distinct parallel tubes united by transverse lamellae (g), and *Heliopora coerulea* producing a fibrous aragonite skeleton (h).

mineralized spicules (or sclerites) that remain separated from each other within the tissue of the animal and are dispersed after its death and decomposition (Fig. 4.22c).

Second, the calcified structures of alcyonarians are built in the most part of high-magnesium calcite, the unstable form of calcite, and are very liable to dissolution. *Heliopora coerulea* (the "blue coral"), with its skeleton of aragonite, is a long-recognized exception to the rule. Recent data (Macintyre *et al.* 2000) established that hydroxyapatite is also present, at least in small amounts in various Gorgonacea. The overall sedimentological role of the Alcyonaria (or octocorals) should not be underestimated, as these organisms are quite tolerant with respect to both depth and temperature, and are essentially dependent only on nutrient resources. Therefore they are much more widespread than the Scleractinia.

In addition to their spicules, typically concentrated in ectodermal tissues, some have calcified axial skeletons that can attain appreciable size (in the order of meters). The axis of *Melithea*, as an example, can reach a diameter of 8 to 10 cm at its base, corresponding to a colony height of 2 to 3 meters. However, their principal characteristic is diversity, as much with respect to microstructures developed by different families, as by the degree of calcification that they carry out, and the degree of their crystallinity. In this very large group, the best-known representative is the red coral (*Corallium rubrum*) that builds a calcareous axis

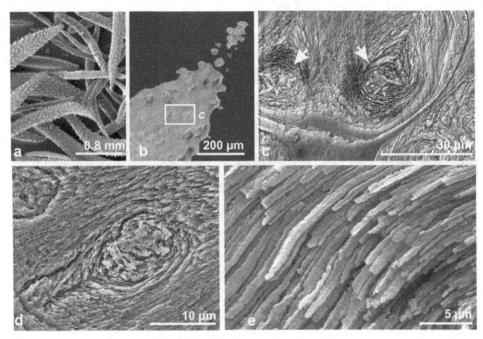

Fig. 4.23. Structure of spicules in *Dendronephthya*. (a–b) Morphology of spicules (a) and a tangential section of one (b). Note the long dimension of these elements. These can attain several millimeters in length (very different from the cortical spicules of *Corallium* or of the gorgonians, whose sizes are in the 20 to 50 μm range); (c–d) Microstructure of spicules. They are constituted of lengths formed of fibers grouped into cylinders of approximately 10 μm diameter (c), themselves united by sclerenchyme or fibrous colonial skeleton (d); (e) Fibers are sinuous, closely grouped and have homogeneous diameters in this sclerenchyme. Nothing in their morphology indicates their mode of mineralization.

with a precisely controlled microstructure. In contrast to this, numerous gorgonians form axes that are exclusively organic, or others in which only their most basal portions develop calcification, and this only very late in life (Fig. 4.22f: arrows). The Alcyonaria are distributed into five orders in which the production of mineralized structures is very unevenly distributed.

Alcyonacea

The Alcyonacea only produce isolated spicules, but these attain macroscopic dimensions, up to several millimeters in length. These spicules generally do not result in true rigidity of polyps, except in *Sarcophyton*, where they are particularly dense. These "soft corals" are broadly distributed in all oceans, and to great depths. In the Pacific, where they can even predominate in reef zones (unlike in the Atlantic), their spicules form a noticeable component of sediment volume. Spicules are illustrated here by those of *Dendronephthya* (Fig. 4.23), a typical representative of the group, forming colonies that can exceed a meter in height. At present, the genus numbers 248 described species. In the Red Sea alone, 32 species have been clearly delineated (Benayahu 1985). The taxonomy of such a vast group is evidently very difficult, in part resulting from the morphological diversity of their spicules, and also due to imprecise identifications based on these mineralized elements, still a risky task (due to the difficulty of analyzing assemblages, even modern ones, and much more so in fossils).

These spicules (or sclerites) are formed of long, joined fibers whose formation begins in the wart-like spots visible on the surface, commonly more abundant at the lengthwise extremities of the spicules. The groups of fibers that are thus constituted, and whose distribution on the surface of the spicule determines specific spicular morphologies, are included in a mass of parallel carbonate fibers, closely joined to form a compact structure (Figs. 4.23c–d). This type of fibrous calcification is very distinct.

Atomic force microscopy indicates that these calcareous fibers are formed of nodular microdomains that are very much analogous to those described above in other groups. In particular, phase images with sufficient enlargement to discriminate 10 nm objects allows observation of interactive compounds, distributed in rather compact layers between grains, but also present in the more clearly mineralized sectors. The fibers constituting the spicules thus are not purely mineral structures, and as seen in other fibrous materials, the formation of the fibers involves organic components (Figs. 4.24a–f).

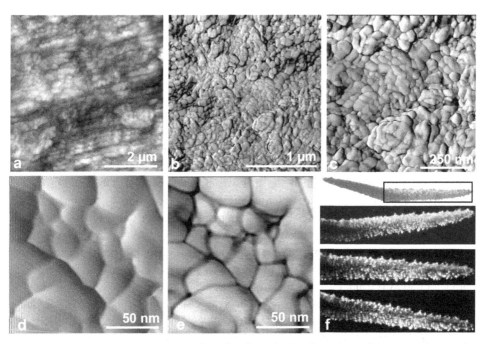

Fig. 4.24. Nanostructure of the carbonate fibers forming spicules of *Dendronephthya*. (a) Height mode: the field of view of 5 µm allows the fibrous structure of the spicules to be recognized; (b–c) Increasing enlargement shows the granular structure (phase mode); (d–e) Grains are seen separately by irregularly distributed interactive matter (= organic). No trace of crystal growth faces is visible; (f) However, spicules seen in transmitted polarized light show the consistent polarizing behavior of the parallel crystalline fibers, as shown by their progressive extinction as the spicule is rotated.

Gorgonacea

The Order Gorgonacea includes alcyonarians whose colonies are supported by resistant and ramified structures that form the growth axis of the edifice. Two groups are distinguished, based on the degree of calcification of their axis. The first are those that can be entirely mineralized (the Scleraxonia), and the second are the opposite (the Holaxonia), in which the majority are organic (sometimes completely), but also commonly have a mineral phase developed.

In *Melithaea* (Scleraxonia), the axial skeleton is formed of compact anastomosing lamellae in a dense network (Figs. 4.25a–c). This network is formed of fibers arranged in cylinders with a diameter in the order of 10 μm (Figs. 4.25d–f), with the fibers themselves having diameters of approximately a micrometer. The margins of calcified lamellae clearly show the early development of these fibrous cylinders (Fig. 4.25g: arrow) occurring in tissues of the basal ectoderm. Their diameter is increased by the addition of additional later fibers up to a median diameter of approximately 10 μm. These are successively incorporated into the compact structure of the lamella.

No histological or cytological data have been obtained concerning the process of formation of the fibers of sclerites or on their method of grouping prior to being incorporated in skeletal lamellae. Observation of the transverse section of an axis at a horizon where a lateral ramification is being extruded, shows the zone of formation of sclerites (Figs. 4.26a–b), and their increasing mineralization during approach to the massive structure in which they are finally incorporated (Fig. 4.26c). Their incorporation into the axial skeleton produces very dense "tissue" having great resistance to breakage, due to the jumbled distribution of sclerites (Figs. 4.26d–g). Under cross-polarized light, it can be seen that each sclerite is a single unit, with continuous polarization, with fibers that are all arranged in a homogeneous fashion. However, in groups of the nearly parallel sclerites that form skeletal lamellae, each sclerite is distinct and independent (Figs. 4.26f–g). Therefore, it can be concluded that their individual crystallographic properties were acquired prior to incorporation in the massive skeletal material. Observation in the scanning electron microscope of sclerites during their formation in the intermediate zone of the fleshy ramifications (Figs. 4.26h–j) adds little in the way of indications regarding their process of mineralization.

In contrast, the atomic force microscope (Figs. 4.26k–n) clearly shows the structure of the elemental fibers, formed by deposition of calcareous grains with dimensions in the neighborhood of 50 to 100 nm, the same order of size as the microdomains observed in mollusc shells and scleractinian skeleton. There is remarkable similarity at the micro- and nano-structural scales between the spicules of *Dendronephthya* and those of the subunits forming lamellae of the axial skeleton of *Melithea*. In the two instances, the basic unit is a long calcareous fiber that appears as a polarized unit in thin section, attesting to the similar crystallographic orientations of grains within the fibers.

The presence of collagen that contributes to forming an insoluble organic matrix in these mesogloeal spicules has been recognized in other gorgonids (Kingsley *et al.* 1990). This

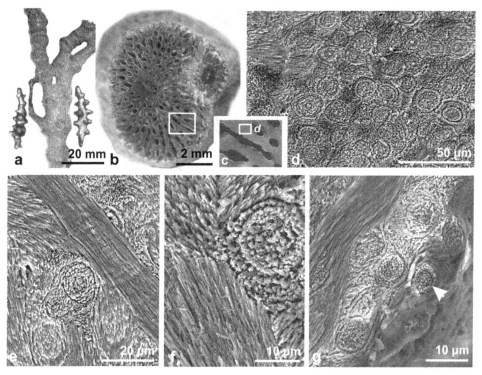

Fig. 4.25. The fibrospicular skeleton of *Melithea*. (a–c) The external epithelium produces isolated spicules (dimensions 200 µm), which remain isolated in the organic tissue (a). The axial skeleton is formed of lamellae (b–c) oriented in generally radial directions; (d–f) Microstructure of lamellae in the axial skeleton of *Melithea*, where the basic units (sclerites) can be seen, here cut transversely (d). In (e) and (f), the fused fibers are readily visible in longitudinal sections of the sclerites; (g) Here, on the margin of a radial lamella, a section is shown of a sclerite during its formation within soft tissue (arrow).

biochemical component of the gorgonid octocorals may also be involved in the organization of the axial skeletal system through mineralization of the long, parallel fibers. This would be in accord with mineralization in other cnidarians.

Corallium rubrum is the Mediterranean species whose populations have been exploited since antiquity as "red coral." Its polyps (Fig. 4.27a) are located within a thick cortical layer containing numerous spicules, and are supported by a calcitic axis (Figs. 4.27b–d). It is from this compact and very hard, branching, vividly red-colored axial carbonate that jewelry is made. The formation of these calcareous axes was first interpreted (Lacaze-Duthier 1864) as the result of a progressive agglomeration of cortical spicules. This interpretation was justified by the observation of the margins of the axial skeleton (Fig. 4.27h). At the tip of the branches calcification occurs through the accumulation of distinct units with poorly controlled morphologies (Fig. 4.27e). These also

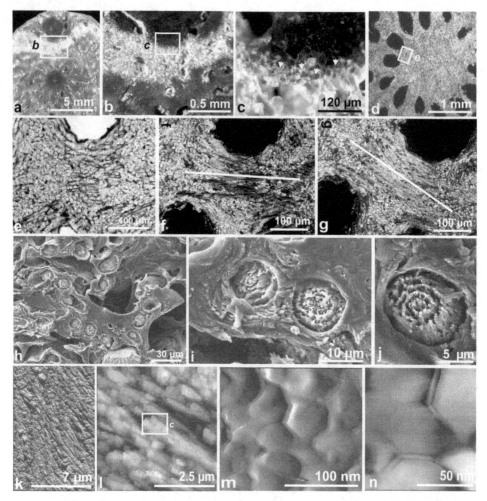

Fig. 4.26. Sclerites of the axial skeleton of *Melithea*. (a–c) Optical views of the zone of formation of sclerites and their formation within tissue separating two branches (c: arrows); (d–g) Arrangement and polarization of sclerites. Note the transverse (circular) or longitudinal sections of the sclerites (e), illustrating their jumbled arrangement, and their monocrystalline behavior (e–g), clearly visible by their individual extinction in cross-polarized light (crossed nicols); (h–j) SEM views of the formation of sclerites (regions indicated in c). In transverse section, the structure of the sclerites is clearly visible, with distinct fibers and a spiral arrangement shown (j); (k–n) Nanostructure (AFM) of the sclerites constituting the skeletal lamellae of the axis of *Melithea*: fibrous structure of a sclerite cut longitudinally (k), alignment of granules within fibers (l–m), and detail of the granule cortex (n).

had structures clearly distinct from those of spicules formed within the mesogloea (Fig. 4.27f), which are just the opposite, having very characteristic morphologies. Microstructural differentiation at the summit of the axial structure is common (see also *Corallium johnsoni*, Fig. 4.29). Here, an element appears as a result of the configuration

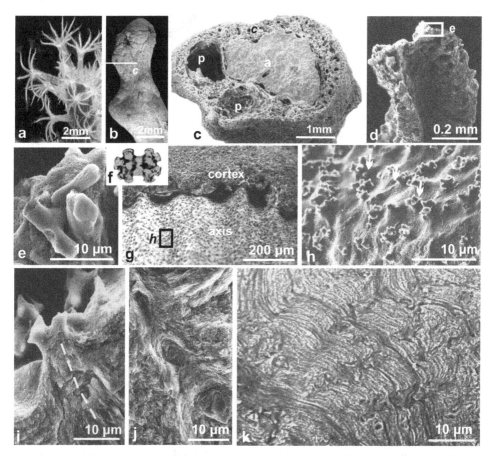

Fig. 4.27. *Corallium rubrum*. (a–b) A branch of *C. rubrum* with polyps extended (a) and a dried branch covered with a cortical layer of spicules, with tissue having been removed (b); (c) Transverse section of a branch. Dwelling spaces for polyps ("p") and the axis ("a") covered by spicular cortex are readily visible; (d–f) Growth tip of the skeletal axis (d) with agglomerated spicule-like bodies (e). These do not exhibit the typical spicule morphology for the species *C. rubrum* (f); (g–h) Enlargement of the axis surface, showing a spiny aspect that was previously interpreted as a spicular organization; (i–k) The layered structure of the axis. In the protruding bodies of the axis surface (i–j) there is a microstructural continuity with the circular layering, indicating that they do not result from the incorporation of spicules from the cortex.

of the external surface of the axis with numerous spiny projections that perfectly simulate spicules partially buried within the massive skeleton (Figs. 4.27g–h). It was this aspect that was the basis of Lacaze-Duthier's interpretation of the axial skeleton as being spicular.

However, transverse sections of the calcareous axis in *C. rubrum* show that beneath this summit, the calcareous axis is formed of concentric and continuous growth layers. SEM study of sections perpendicular to the growth axis indicate that the spiny projections at the

Fig. 4.28. Crystallization and nanostructure in the skeletal axis of *Corallium rubrum*. (a–b) External surface of the axis and UV fluorescence of the organic layer in contact with the crystallized structure; (c–d) Distribution of crystalline units in columns perpendicular to the layer of growth. Within the columns, crystals are oblique in relation to the growth direction; (e–g) Nanostructure of the crystals. Not a single face of crystalline growth is present in these nanometer-sized units, although polarized light shows coherent orientations of the mineral component.

surface are formed of calcareous lamellae in microstructural continuity with those forming the carbonate axis (Figs. 4.27i–j). The dynamic of calcification in *C. rubrum* has been analyzed (Allemand and Benazet 1996) by marking with [45]Ca and [14]C. This clearly confirms the existence of two distinct regions of calcification within the axis of *Corallium*, as well as the presence of a calcifying system of an epithelial type that forms growth lamellae with micrometer thicknesses (Fig. 4.27k). More recently, field observation and marking experiments have shown that the growth in the diameter of axes is very slow but highly variable, with a mean of 0.35 mm/year, plus or minus 0.15 mm (Marschal *et al.* 2004).

The calcareous axis of *Corallium rubrum* (Figs. 4.28a–c) has a very characteristic microstructure, with crystals of magnesian calcite organized into radial rows having orientations that alternate regularly from one row to the next (Fig. 4.28d). A staining method that uses Toluidine Blue following gentle decalcification with 2% acidic acid (Marschall *et al.* 2004) allows the distribution of mineralizing organic material in annual growth zones to be established, a temporal estimate validated by calceine staining experiments on living specimens. This procedure led to observation that the concentration in magnesium also is zoned (Vielzeuf *et al.* 2008) and this alternates very regularly with the organic zonation.

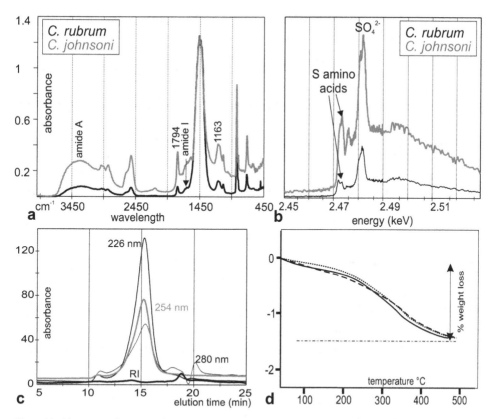

Fig. 4.29. Character of the organic phase of the calcareous axis of *Corallium rubrum* and *C. johnsoni*. (a) Identifying characters of the organic phase by infrared absorption; (b) Establishment of the presence of sulfated polysaccharides by XANES fluorescence at the ESRF synchrotron at Grenoble, France; (c) HPLC chromatographic profile; (d) Measurement of the quantity of nonmineral components in the axial skeleton of *C. rubrum*. The weight loss before reaching 500 °C is approximately 1.5%. a–c: Dauphin (2006a).

Data relating to the organic phase included in the axis of *Corallium rubrum* are scarce. Thermogravimetric measurements indicate that the proportion of nonmineral components there is lower than in the scleractinian coral skeleton. The loss of weight prior to carbonate decomposition has a mean of 1.5%. The presence of proteins and glucides is directly established by infrared absorption (Fig. 4.29a), and sulfated polysaccharides are clearly detectable by X-ray fluorescence (Fig. 4.29b). Extraction by decalcification obtains an organic phase with protein masses distributed in a practically continuous fashion, centered on a maximum of approximately 150 kDa, with indications of above average amounts present in the region of 10^6 (Dauphin 2006a). A biochemical estimate of their relative importance (Allemand *et al.* 1994) suggests a relationship of 80% for polysaccharides to 20% for proteins.

These associated microstructural and biochemical data show that the growth of the calcareous axis of *C. rubrum* takes place as a biomineralization sequence in two stages, consisting of an initial part (Figs. 4.27d–e), formed by the accumulation of disjointed units with varied morphologies (Fig. 4.28e), and a second step in which concentrically stratified growth occurs (Fig. 4.28i). Utilizing chemical marking methods involving ^{45}Ca and ^{13}Ca, Allemand and Benazet-Tambutté (1996) were able to describe metabolic differences tied to these two phases of axis growth, thus confirming the biological specificity of the biomineralization mechanisms corresponding to these two phases. In the second phase of this form of skeletogenesis, the crystallization process shows a great similarity here to the model of biocrystallization in two stages described for scleractinian corals. In transverse section, fluorescence UV allows the evaluation *in situ* of the initial biochemical layer of the biomineralization cycle then current (Fig. 4.28b). The complete surface of the axis is effectively covered by a fine, strongly fluorescent lamina situated under the basal ectoderm. It thus reproduces exactly the situation of the basal glucidic lamina observed in the Scleractinia by Goreau early in the research on biomineralization in the latter (Goreau 1956).

Corallium johnsoni builds an irregularly branching axial rod, always with an asymmetrical cross section resulting from the presence of a laterally situated commensal worm (Figs. 4.30a–b). The mineral material deposited here is also high-magnesium calcite, but lacking pigmentation. The initial stages of mineralization, observed at the center of the axis (laterally displaced) are fibrous, and the cyclic process of mineralization is easily seen. The result of this is the formation of a structure some hundreds of micrometers in diameter (Figs. 4.30c–d). The process of mineralization thus has developed under rather similar conditions to those in *C. rubrum* (or apparently so). The external surface does show similar spiny features, but lacks the longitudinal canals of *C. rubrum* (Fig. 4.30e). Moreover, the axial microstructure is more highly differentiated here than in *C. rubrum*.

The concentric layers of magnesian calcite in *C. johnsoni* are traversed by cylindrical features with generally radial orientations (Fig. 4.30f: arrows). These are seen to be made of thin concentric lamellae (Figs. 4.30g–i) that exhibit a monocrystalline behavior and therefore represent a unique case of crystallization lacking any known equivalent among other extant Cnidaria. Comparable structures that correspond to this singular mode of crystallization do exist, and are well known in fossils whose taxonomic position is still somewhat uncertain.

Corallium johnsoni also offers good examples of the granular organization of skeletal fibers when examined at the submicrometer scale with the AFM (Fig. 4.31). Growth lamellae (Fig. 4.31a), fibers (Fig. 4.31b), and granular carbonate subunits are visible.

Gorgonacea, Holaxonia

The Holaxonia are best represented by the "bamboo corals," which are attracting considerable interest for purposes of paleoclimate reconstruction because they are widely distributed in cold seas. They are characterized by calcification in distinct segments, between which the axis remains wholly organic. A typical genus is *Isidella* (Fig. 4.32a), where calcified transverse sections of the axis (Figs. 4.32b–c) show a characteristic arrangement in divergent curved zones extending out from a center. Bright zones, lacking organic material, directly exhibit relatively irregular growth zonation, with superposed layers that are 4 to

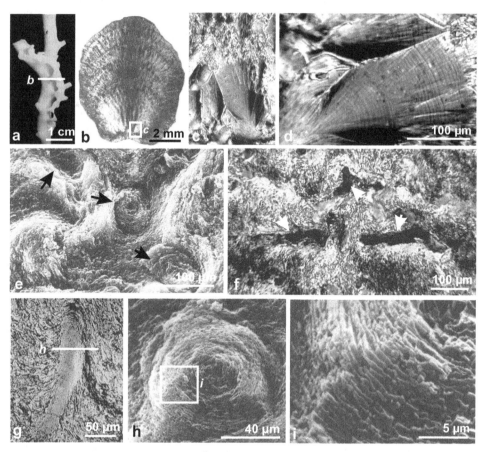

Fig. 4.30. *Corallium johnsoni*, the "white coral." (a–d) Morphology and section through a branch. Note asymmetrical development due to the presence of a commensal worm, and the fibroradial microstructure of the axial portion of the branch; (e) The spinose surface of the branch (comparable to the surface of the axis in *C. rubrum*); (f–i) The layered growth visible in the axis and the remarkable concentric lamellar microstructure of the cylindrical structures. b, d, i: Cuif *et al.* (1985); g–h: Cuif *et al.* (1987).

10 μm in thickness (Fig. 4.32d). Darker zones owe their coloration to higher concentrations of organic materials that are sensitive to collagen dyes. This suggests a parallel relationship to biochemical data of other gorgonians (Kingsley 1990), whereas in *Isidella*, good results are also obtained by staining with Alcian Blue (Fig. 4.32e).

The calcareous component is organized as short fibers gathered into parallel layers with an overall spiral arrangement. Cross-polarized light allows recognition of an overall control on crystal orientation (Fig. 4.32h).

Observation in the AFM provides images which clearly illustrate the similarity of mineralization patterns in this group to those reported above. The few-micrometer-thick growth layers are uniformly formed by densely packed grains (Figs. 4.32i–j). Residual

Fig. 4.31. The reticulate structure of the calcite crystals in *Corallium johnsoni*. (a–b) At relatively low enlargement (field of view 2.5 micrometers), growth layering (b) and fibers are visible; (c) Phase image: crystallization produces an irregular deposit of organic materials; (d–g) Typical sequence of AFM views: height (d) and phase (e) images. Note the irregular nature of the organic accumulation subsequent to crystallization of the Ca-carbonate. f: Dauphin (2006a).

organic components show an irregular accumulation, and provide typical features supporting hypothesizing the identical process of segregation as in stepping crystallization models (see Fig. 3.11).

Along with the great diversity of size and morphology, their functional organization makes them one of those groups in which taxonomy is still uncertain. The alcyonarians produce some structures that are extremely original, whether on an architectural basis (*Tubipora*) or on a mineralogical basis (*Heliopora*). Thus, in spite of their possessing a small number of species, families and even orders have been created for each of them.

Tubipora musica is the "pipe organ" coral, an easily recognizable skeleton formed of vertical parallel tubes joined by horizontal platforms (Fig. 4.33a). Tube walls are apparently made of distinct spicules, but examination of growth lines indicates that mineralization occurs as a continuous stepping process that is well synchronized at the growth edges of tubes (Figs. 4.33d–g).

Short calcite fibers are repeatedly produced, and are made of typically reticulated material as seen at the 10 nm scale. Crystallographic continuity between growth steps appears to be quite weak, and does not produce elongated fibers as scleractinians do. This last observation contrasts strongly with the obvious microstructural organization of the most singular of the Alcyonaria, the fibroradial aragonite skeleton of *Heliopora coerulea*.

Heliopora coerulea, the "blue coral," produces massive and nodular colonies several decimeters in size. The architecture of the colonies is remarkably homogeneous and simple.

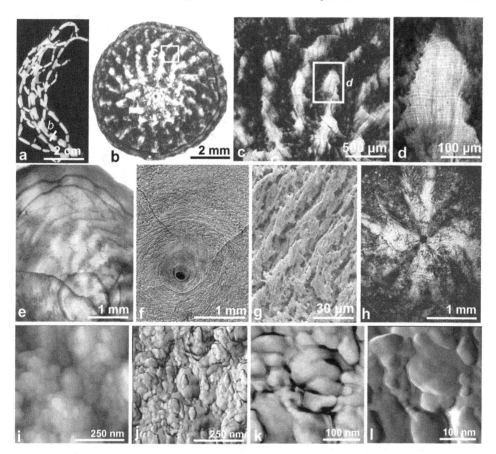

Fig. 4.32. Microstructure of the calcified segments of a specimen of *Isidella* sp. (a) Typical "bamboo coral," in which series of calcified sectors are joined by short organic segments; (b–d) Optical view of a section perpendicular to the growth axis. Note the specific spiral organization of alternating sectors. In light sectors (weakly collagenous), growth layers are easily visible (d); (e–h) Microstructure of a calcified axis. Fibrous bundles with a spiral arrangement: (e) Alcian Blue staining, (f–g) SEM views of short calcite fibers, and (h) view of thin section under crossed nicols showing evidence for overall control of fiber orientation; (i–l) AFM views; (i–j) Height and phase images illustrating the granular structure; (k–l) Enlarged view showing the residual organic component subsequent to crystallization.

Polyps lodge in parallel but distinct tubes of approximately 300 μm diameter (Figs. 4.34a–c). Tube walls, as well as the mineralized structure surrounding the tubes, are made of vertical elements that are polygonal in section and in partial contact with each other. This results in a regular but unique skeleton. The contrast between these two species extends both to mineralogy and microstructural organization (Figs. 4.34d–h). Instead of calcite, the usual mineral of the Alcyonaria, *H. coerulea* produces an *aragonite* skeleton. Instead of undifferentiated skeletal "tissue," made of superposed short fibers, it builds vertical units with well-developed axial centers, having fibers regularly radiating from a center. Such structures can be compared

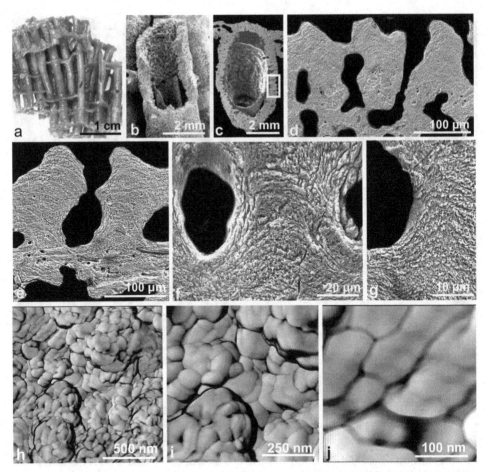

Fig. 4.33. Microstructure of the skeleton of *Tubipora musica*. (a–c) Morphology of the colony and individual tube; (d–g) Layered growth mode of the tube wall. No visible crystallographic continuity exists between successive layers: calcite fibers remain short, without specific microstructural patterns; (h–j) Sequence of AFM views showing typical reticulate composite carbonate material.

to a similar skeletal organization described in septae of scleractinians as "trabeculae." The stepping growth mode is clearly shown (Fig. 4.34f), as is the particular microstructure of these continuous "centers of calcification" (Figs. 4.34g–h: arrows). The mineral skeletons of *Tubipora* and *Heliopora* differ in many respects (mineralogy, overall disposition, and microstructure), so that their similarity at the nanoscale deserves to be emphasized. Figures 4.33h–j and 4.34i–k differ from both mineralogical and biochemical viewpoints, but the distribution and spatial arrangement of organic and mineral components are so alike when compared at enlargements in the 100 nm range, that one can conclude there are identical mineralizing mechanisms operating at the molecular level in both. These are what control crystallization, whatever the differences in geometry of the basic growth layer in each species (which depends

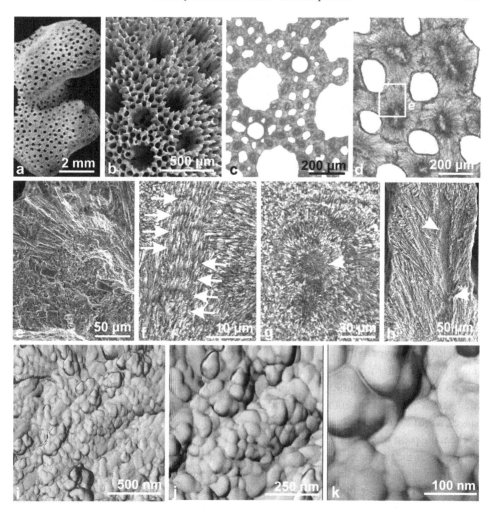

Fig. 4.34. General morphology and enlarged view of the surface of *Heliopora coerulea*. (a–b) Microstructure and nanostructure; (c–d) Optical view of a thin section (natural light); (e–h) Radiating fibers (e) built by the stepping growth mode (growth increments marked by arrows in f), surrounding a central axis (g) made of minute crystals organized into a continuous central cylinder (h): typical "trabecular" arrangement; (i–k) Sequence of AFM images showing the reticular nodular arrangement of the submicrometer-scale structure of fibers.

on the anatomy of the polyp), and the details of crystallization resulting (or not) in continuity of crystallization between successive growth layers, and thus end in forming different degrees of elongation of calcite or aragonite fibers.

In the Cnidaria, a highly diversified group with respect to their skeletal organization, no exception has yet been found to this common mechanism of crystallization, allowing us to extend it throughout the phylum.

4.4 Brachiopods

4.4.1 General description

The brachiopods are a zoological group lacking great importance in the modern world; they only contain about 330 living species. Distributed in all oceans at present, these species, generally living in somewhat deeper water, pass practically unnoticed. However, they are actually the last, dispersed representatives of a very large taxonomic group that formed a major component of benthic marine faunas during the Paleozoic. They played a significant role during the Mesozoic, but have truly regressed in the Cenozoic. More than twelve thousand fossil species have been identified, which is numerically comparable to the modern species of the Bivalvia.

For a long time, the brachiopods were divided into two groups of unequal importance: the Inarticulata and the Articulata. This division was based on the presence or absence of a hinge between two valves and an internal skeletal element, the brachidium. As a result of the group of works by Williams *et al.* (1996), a new classification has been adopted in Part H of the Treatise on Invertebrate Paleontology: the Inarticulata was provided with two branches, the Linguliformea and the Craniformea, while the Articulata, by far the most abundant in the number of species, are now named the Rhynchonelliformea.

These animals construct bivalved shells, and due to this fact are at times confused with the Bivalvia (or Pelecypoda), with which they differ on all essential levels of organization. Their anatomical peculiarities, which clearly differentiate them from the Bivalvia, have been recognized for a long time (the taxon Brachiopoda was created by Duméril in 1806, although he placed them in the Mollusca), and these are manifested at early developmental stages, as for example, in their lack of a trochophore larva.

From the point of view of general shell architecture, the difference in organization is equally very sharp. In the Bivalvia, the plane of separation of the valves also forms the plane of symmetry of the shell. In principle, therefore, the two valves are identical except at their apparatus for articulation. They differ from this generality only in that some families follow a differential development, which determines and develops great dissimilarities between valves. In the Brachiopoda, the two valves are only identical in some of the most archaic forms of the group, in which the two valves lack a tooth and socket apparatus and only maintain their shells in place by muscles. These are grouped in the former inarticulate brachiopods. In the majority of brachiopods, the two valves are articulated, and remain together due to their very effective tooth and socket system. Another difference with the bivalves, whose valves open when ligaments and adductor muscles are destroyed after death, is that the two valves of the articulate brachiopods remain closed after death, thus also during fossilization. The two valves are very clearly distinct, morphologically, as one of them bears an opening, the foramen (Figs. 4.35a–d), which allows free passage for an important organ, the pedicle, by which the animal fixes itself to a substrate. The other carries the system that assures nutrition, the lophophore and a lophophore support, provided that it has one.

At the level of their nutrition system, the Brachiopoda are easily compared with the Bivalvia. The brachiopoda are "suspension feeders," in which the capture of food particles is

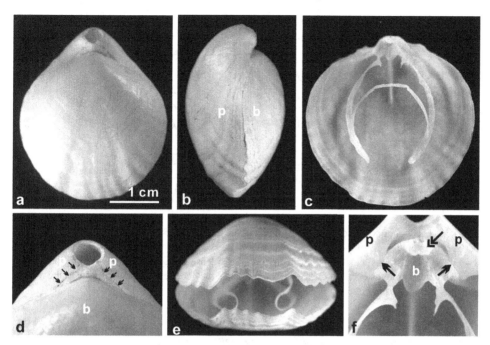

Fig. 4.35. Shell of a modern brachiopod (*Magellania*,Terebratulida). (a–b) Upper and lateral views ("p": pedicle valve; "b": brachial valve); (c) Internal view of the brachial valve, showing the brachidium, the calcified support for the lophophore; (d) Closer view of the foramen. Here, the circular foramen is mostly formed by the pedicle opening, with the addition of specific plates (arrows); (e) Maximum opening of the shell for filter feeding; (f) The interior of the brachial valve ("b"), with muscle attachment (double arrow), and sockets in which remains of the pedicle valve ("p") are still in place (arrows).

carried out by a highly differentiated organ, the lophophore, which occupies the greatest part of the internal space of the brachial valve (Fig. 4.35f). This organ represents the functional equivalent of the gills (ctenidia, branchia) of the Bivalvia (or lamellibranchs). The analogy holds for the method of transporting trapped particles towards the mouth. Mucus cells in specialized gutters in the brachiopod lophophore insure the movement of particles to the mouth by using cilia in the lophophore (Figs. 4.36c–d).

Along with the articulate brachiopods, the smaller component of the Phylum Brachiopoda is the group of former "Inarticulates," whose most visible characteristic is the structural independence of the two valves. However, on a biological plane, and in particular, with respect to developmental biology, very profound multiple differences are present in the early stages of their embryonic development (Yatsu 1902). All major organizational components of these two subdivisions of the phylum are different: the organization of the muscular system, that of the alimentary tract, the origin of the pedicle, etc. This justifies the new basic taxonomy for the group, and is even true when taking into account the chemical compositions and microstructures of their shells.

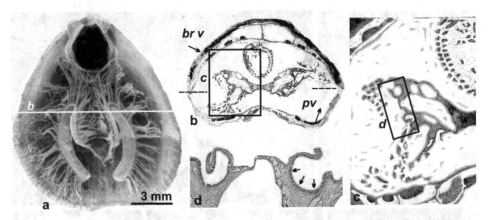

Fig. 4.36. The lophophore occupies the main part of the internal space in the shell (*Terebratulina* sp.). (a) Overall internal view of the brachial valve; (b–c) Transverse section of a shell after decalcification. Two main lateral loops and a smaller median one bear dense tentacles that play a nutritional role; (d) Two mucus grooves occur in one of the lateral loops: they convey nutrient particles to the mouth and thus to the digestive tract.

4.4.2 Rhynchonelliformea

The shells of Brachiopoda are produced by an epithelial organ, the mantle, which covers the interior of the valves as in the Bivalvia. One might then assume that identical mechanisms are developed to generate shell in them. This is not so, however, because it is here that the two groups differ at the basic level of the process of biomineralization itself. This establishes very important differences between shells of the two groups. Rather than being a process of well-synchronized secretion that generates mineralized growth layers that are continuous over large surfaces, as is the case in the Bivalvia, the mantle of the brachiopods functions primarily at the cellular level.

It is due to Alwyn Williams (1956 and later) that we have a workable model for the microstructural organization of brachiopod shells, beginning with that in the modern Terebratulidae. At its distal margin, the mantle is subdivided into two lobes by a very deep groove. Here there is produced a thin, continuous organic layer, the periostracum, which extends to cover the complete exterior of the shell. Beneath the periostracum, the calcified shell has a structure made up of two superposed layers that differ not only in their thickness, generally unequal, but *above all* by their method of formation. Immediately beneath the periostracum, the primary layer is formed, produced by the most peripheral part of the mantle. It is generally only a minor component of the shell (Figs. 4.37b, d and j; see also Fig. 4.38).

Its microstructural organization varies according to genus. It can appear to be irregularly prismatic (Fig. 4.37j), and can be constituted of units with a monocrystalline appearance, also assembled in a very irregular fashion (Figs. 4.38e–f).

It is the internal layer of the brachiopod shell (the secondary layer) that forms its major component. This is generated by a narrow mineralizing zone at the peripheral margin of

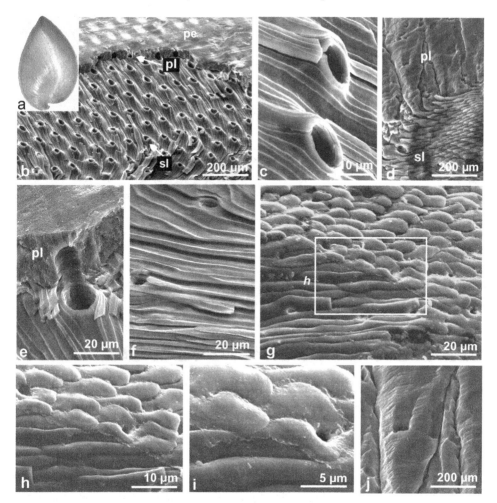

Fig. 4.37. Microstructure of a modern terebratulid brachiopod (*Gryphus* sp.). (a–b) The shell (a) and its three shell layers: periostracum ("pe"), the primary layer ("pl"), and the secondary layer ("sl"). The secondary layer forms the principal mass of the shell; (c–e) Shell fibers oriented parallel to the exterior and longitudinally, turning around the *punctae*, which allow passage of tubular expansions of the mantle. These expansions penetrate into the primary layer, but do not reach to the shell exterior; (f–i) In spite of having irregular cross sections and nonrectilinear morphologies (f–g), the fibers are positioned tightly alongside each other. At their ends, the cells that produced them can be seen (h–i) at the internal surface of the shell; (j) The primary layer is composed of very irregular units, sometimes oriented perpendicular to the surface. Their growth by superposed lamellae (that can include several neighboring units) presents something close to an analogy with the prismatic layer of the Bivalvia.

the mantle, which appears to be made of elongated sinuous units that are tightly joined (Figs. 4.37c–f). These are calcitic fibers, whose organization and method of genesis are truly remarkable. Their principal characteristics are, first, they remain perfectly continuous in spite of their great elongation and, second, in spite of their sinuosity, to have a

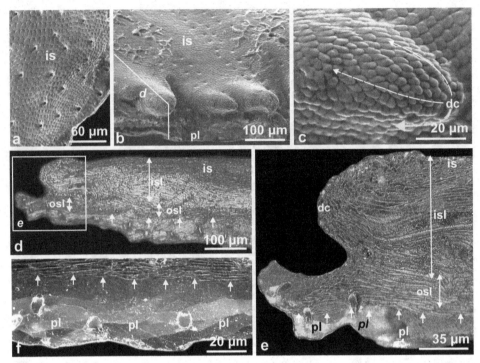

Fig. 4.38. Distal zone of the shell of a terebratulid brachiopod (*Dallina* sp.). (a) Internal surface of the test ("is") in a lateral region with reduced growth; (b–c) Morphology of the growth border. The external part of the shell is reinforced on its internal surface by the development of conical elements that fuse laterally. These conical structures form centers of divergence ("dc") determined by the spatial arrangement of cells that produce fibers; (d–e) Radial section of the anterior border of the shell. The cone of divergence of fibers ("dc") in (e) forms the internal secondary layer ("isl"). A fibrous zone ("osl": outer secondary layer) continues to cover the internal surface of the primary layer ("pl"); (f) Radial section of the primary layer: the microstructure shows large and irregular units, each of them with single crystal behavior (reflected phase-contrast microscopy).

monocrystalline behavior. The latter property evidently indicates continuous control during their formation. At the distal extremity of the fibers, the explanation of this remarkable peculiarity is furnished by observation of the cells comprising the mantle (Figs. 4.37g–i); as it can be seen here that the extremity of each fiber is covered by a mantle cell.

This is an essential difference with the mode of shell generation shown by bivalve molluscs. We have seen that in the latter (Chapter 1), there is no actual relationship between cells producing growth layers and the morphology that each microstructural unit assumes (prisms, nacre tablets, etc.). In contrast, here *each fiber results from the activity of a single cell*, and its growth is stopped when the cell concerned reaches its final position resulting from ontogeny of the mantle during the process of shell construction. Differing from the prismatic structures in the Mollusca, fibers here are essentially parallel to the surface of the valve. This arrangement in continuous longitudinal units, joined but distinct, explains the surprising solidity of brachiopod shells, given their limited thickness. The organization of the principal layer of the shell (the secondary layer) is opposed to that of the external layer (primary layer) in that one can discern a bedded structure (Fig. 4.37j), noticeably parallel to the shell surface. This indicates that in this external layer, mineralization is carried out by the collective secretory function of the cells forming the most peripheral zone of the mantle.

The internal surface of the brachiopod shell is covered by epithelium formed of cells that have finished their mineralizing activity once they reached their final position within the shell. Therefore, the shell does not increase its thickness between the region of the beak and the periphery. It is at the distal border of the mantle that the essential features of shell structure are defined. Traversing the primary and secondary layers, the mantle also branches into tubes extending perpendicular to the surface of the shell (through punctae) and terminating in a closed ampule, not opening out to the shell exterior as its terminal face is mineralized and covered by periostracum (Figs. 4.38b, c, e). These shell perforations are characteristic features in the large group of punctate brachiopods. The physiological role of punctae has not been precisely established. The growth margin of a terebratulid brachiopod clearly shows the method of shell formation. In *Dallina*, this border has a characteristic row of conical structures formed in the fiber generating zones (Fig. 4.38b). A related observation (Fig. 4.38c) would indicate that at the summits of these cones, the fibers are exactly oriented in the direction of growth; the cross section of cells here is practically circular. On the flanks of these same structures, the cellular surfaces become oval because the direction of growth then is oblique with respect to the surface. These conical structures result from differentiation of the internal zone of the cellular layer that produces the fibers. The summit of each cone is a zone of cellular multiplication that forms a center of divergence for the mineralizing cells (Fig. 4.38c: "dc", arrows).

In the most exterior part of this generative zone, the fibers form a simple layer that is the most distal portion of the secondary layer (Fig. 4.38: arrows), the conical structures being in the most internal position. This fibrous layer is covered on its exterior surface by the primary layer, which has an irregular structure, as shown by Figs. 4.38e–f. Phase-contrast images (in reflected light) outline the organization of this primary layer as a series of blocks having irregular morphologies. This view corresponds to SEM images (Fig. 4.37j) that indicate that these blocks with irregular morphologies result from general stratification and reflect the synchronous collective functioning of peripheral cells of the mantle, as opposed to individual cellular functioning in forming the primary layer.

Compared to the microstructural units that we have observed in corals and molluscs, the fibers of the Brachiopoda are objects of true originality. They are densely assembled and perfectly joined along their complete length (Figs. 4.39a–b), in spite of their relative irregularity as individuals. They evidently owe this property to their formation by distinct

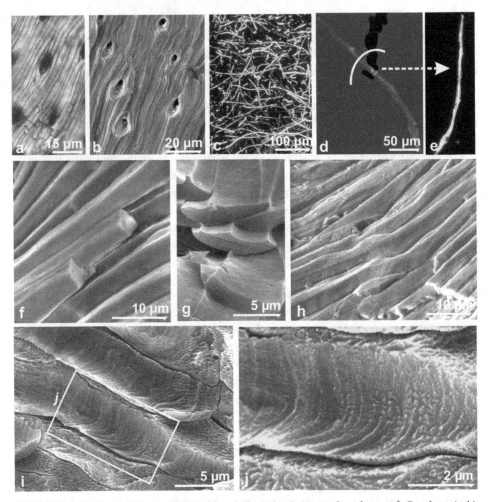

Fig. 4.39. Properties and mode of growth of fibers in the secondary layer of *Gryphus*. (a–b) Arrangement of the fibers *in situ*; (c) Felt of dissociated fibers; (d–e) Two different orientations of a single fiber in a polarizing microscope with crossed nicols. The monocrystalline nature of each fiber is clearly indicated by the uniform extinction throughout the entire length of this curved fiber; (f–h) Different transverse sections in a single shell (*Gryphus*), and variability of fiber morphologies and arrangements during growth; (i–j) Fiber growth by repeated addition of submicrometer-thick growth layers that reflect successive positions of the secreting cell.

cells whose positions respective to one another change continuously during growth of the shell. Their most remarkable character is their elongation, and the ratio between their length and transverse dimensions is extremely high. By dissociating the secondary layer in a solution of sodium hypochlorite, a felted layer of isolated fibers is obtained (Fig. 4.39c) whose fiber continuity can be followed for several hundreds of micrometers, although its transverse dimension can be only 4 to 6 µm. In polarized light, an even more surprising characteristic appears; each fiber is monocrystalline with a constant crystallographic orientation throughout its entire length (Figs. 4.39d–e) regardless of morphological variation occurring during the hundreds of micrometers of length commonly attained. In particular, when fibers bend around punctae the crystallographic directions remain unchanged.

Taking into account their mode of formation by distinct cells with positions that vary during time, transverse sections of fibers show considerable variability, even within the same shell (Figs. 4.39f–h). Through etching of a polished surface on which fibers are sectioned longitudinally, their individual mode of growth is easily observed. Fibers are elongated by the repeated addition of curved growth layers, whose curvature reflects the surface morphology of the generating cell. It can be seen that on two neighboring fibers (Fig. 4.39i) the morphologies of the growth surfaces differ, thus bearing witness to their deposition by cells functioning independently. With a thickness that is much less than that of the secondary layer, in general only some tens of micrometers, the primary layer is notable in its completely distinct microstructure (see also Cusack and Williams 2001). A growth surface (Figs. 4.40a–c) reveals its granular nature, with grain dimensions in the neighborhood of 0.1 μm (Figs. 4.40d–e). The fundamentally stratified nature of the primary layer is also clearly seen here (Fig. 4.40f), and it is at this level that the distal extremities of punctae occur. In the older portion of the primary layer, microstructure becomes more compact (Fig. 4.40g), suggesting that the grains that are distinctly visible on the growth surface continue to be cemented by action of the organic material. This modification of grains, however, does not cause the complete disappearance of the fundamentally stratified structure of the primary layer.

Figures 4.39i and 4.39j illustrate the stratified structure of the fibers that form the secondary layer. These very long units, each with a monocrystalline optical appearance, are formed by the superposition of successive growth mineralization units, established here by their differential reaction to etching. Selective dissolution is an obvious paradox in a feature that appears to be monocrystalline. Once again, the morphology of the surface of growth units here has no direct relationship to crystal symmetry planes. Atomic force microscopy reveals that this striking discrepancy between the monocrystalline behavior of fibers and the curved surface of their growth units can be transposed to the building blocks of the growth units themselves. A longitudinal section of a *Terebratulina* shell with a fiber segment cut on its median plane shows once again the mode of growth by superposition of incurved units (Figs. 4.41a and b).

As clearly shown on the image formed by the "height" signal, the grains forming growth units do not at all appear as faceted calcite microcrystals. Rather, they are irregularly ovoid and, as shown by the "phase-lag" images, they are surrounded by fine, interactive, organic or amorphous material. Thus, in spite of the great difference in the type of mineralization here, with monocellular formation of brachiopod fibers and their long-recognized perfect crystalline individual behavior, no trace of crystallinity can be found at the submicrometer level of observation.

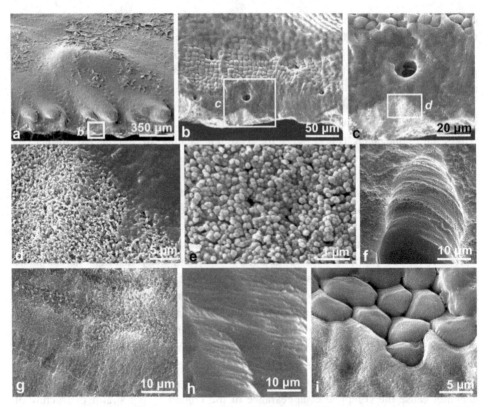

Fig. 4.40. Microstructure of the primary layer in the shell of *Gryphus* sp. (a–c) Growth margin of the primary layer showing the organic cover over the mineralized phase; (d–e) Submicrometer grains constituting the primary layer immediately beneath the organic cover; (f) The distal end of a puncta (pl. punctae) where the superposition of mineralized laminae forming the primary layer is clearly visible; (g–h) The mineralization of the primary layer shows variations in the mineralization process that determine a laminar, more or less compact aspect in successive stages (h: from the same area as Fig. 4.37j); (i) The contact between the internal portion of the primary layer and the first fibers of the outer secondary layer reflects the sudden change from an organic mineralizing secretion forming the successive laminae of the primary layer to the individual cellular mineralization typical of the secondary layer.

There is no need to emphasize how important it is to find this basic similarity in the organomineral interplay at the nanometer level between brachiopod fibers produced by individual cells and the epithelial products previously examined. The clear suggestion is that the mechanism of Ca-carbonate transportation to the mineralization site is essentially similar, regardless of the mechanism of calcification at the tissue (or supracellular) level.

Obviously, there must be some coordination existing between the secretory activities of cells in the brachiopod mantle in order to result in species-specific morphologies. Compared to the mechanisms that insure coordinated activity in the

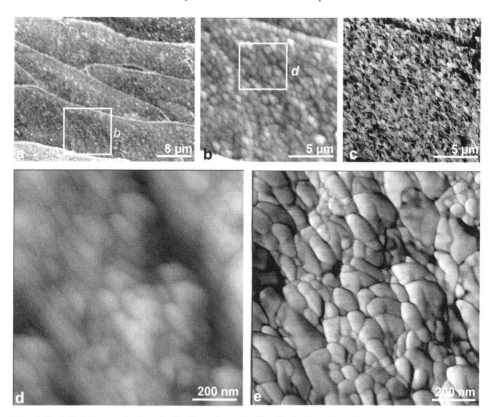

Fig. 4.41. AFM views of a longitudinal section of a fiber in the brachiopod species *Terebratulina* sp. (a–b) In this species, whose fibers are very thick, growth stratifications are clearly visible within fibers. The height signal illustrates the morphology of the successive growth steps (compare to Fig. 4.39j); (c) A phase view of the same area as (b) produces a less-distinct picture of lamination due to dispersion on independent, small, interactive surfaces; (d–e) Enlarged height and phase images. The composite organization of the granular units is individually visible. a, b: Cusack *et al.* (2008).

molluscan mantle, there are important differences found here. However, at the molecular level, a basic similarity between these two types of mineralization is revealed by the common organization of the reticulate crystals, which is the result of the submicrometer-scale interaction between the Ca-carbonate and closely associated organic materials.

Analytical data

Clarke and Wheeler (1915) established the calcitic nature of fibers in the shell of the rhynchonelliform brachiopods, along with an organic component present in an amount on the order of 1% (then evaluated by thermal breakdown). Improved chemical characterization demonstrates the taxonomic specificity of its amino acids (Jope 1971, 1973).

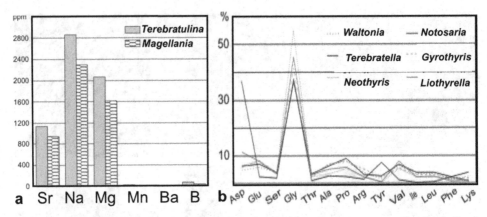

Fig. 4.42. Chemical data on modern brachiopod shells. (a) Minor element concentrations in shells of modern brachiopods; (b) Amino acid composition of intracrystalline organic matrix in shells of some modern rhynchonellid brachiopods, with data of primary and secondary layers combined (data from Walton *et al.* 1993).

This property permitted Collins *et al.* (1991) to develop a sero-taxonomic analysis of living terebratulids (see also Walton *et al.* 1993; Endo *et al.* 1994). Complex proteins and/or polysaccharides in brachiopod periostracum were also identified by Jope (1973). The presence (in more than 70% of cases) of histidine in the N-terminal position suggested to Jope that it plays a significant role in calcification. More recently, the biochemical complexity of the organic materials associated with fibers has been notably increased by the discovery of lipids (Curry *et al.* 1991) and carbohydrates (Collins *et al.* 1991). Cusack and Williams (2001) characterized as many as 21 distinct proteins in a single soluble organic sample, having masses distributed between 16 and 109 kDa.

This series of biochemical characterizations does not reflect the crystallographic and structural specifics of the Brachiopoda. Cusack and Williams clearly expressed this difficulty. No single protein could be identified as involved in calcification of one or the other of the principal shell layers (Cusack and Williams 2001). More recently, synchrotron-based data yielded additional chemical support to the three-zone model that had been proposed by Williams (primary layer, and inner and outer secondary layers) through mapping the distribution of organic sulfates (Cusack *et al.* 2008a). After noting the presence of organic sulfur in sulfated polysaccharides (Fig. 4.43a), the differential distribution of these compounds was established at a submillimeter scale. Concentration of this glucidic acid phase is noticeably higher in the primary layer than in fibrous parts of the shell (Fig. 4.43b: arrows). Additionally, a difference is seen in the concentration of polysaccharides between the "inner secondary layer" (the fibrous layer directly next to the internal surface of the primary layer), and the "outer secondary layer" that apparently originates by the joining of conical structures (Fig. 4.43b: dotted line).

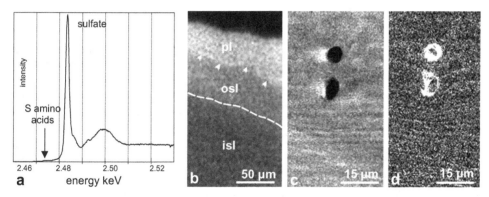

Fig. 4.43. Synchrotron-based characterization of sulfated polysaccharides in *Terebratulina* shells (Brachiopoda, Rhynchonellida). (a) Peak of organic sulfates. The very weak concentration of sulfur is associated with sulfated amino acids (almost absent); (b) Distribution of polysaccharides in the three shell layers ("pl", primary layer; "osl", outer secondary layer; "isl", inner secondary layer); (c–d) High-resolution mapping of sulfated polysaccharides (c) and sulfated proteins (d) in the fiber layer. Concentrations of polysaccharides are higher in the fiber envelopes, but they are also present within envelopes.

At higher resolutions, mapping of the fibrous region (Figs. 4.43c and 4.43d) illustrates differences in composition between the organic envelope of fibers and that of the organic phase associated within the mineral portion, as the contents in sulfated polysaccharides are higher in the fiber envelopes than in the mineral phase itself. Mapping of sulfur tied to proteins is of equal interest, because it shows (Fig. 4.43c) that an unequal distribution of sulfur in fiber envelopes is clearly visible while the fibers themselves are practically sulfur impoverished.

In parallel with characterization of this type, physically located on organic components, research on improvement of mineralogical and crystallographic descriptions continues progressively to define numerous and new, specific patterns. More precise knowledge of fiber crystallinity is important, not only to understanding mechanisms of biomineralization, but also its consequences. In effect, since the first studies carried out by Lowenstam (1961), the brachiopod shell, formed of low or moderate magnesian calcite, has been considered as an ideal source of chemical and/or isotopic proxies. However, newer, more detailed studies that take into account which valve is used, its microstructure, and other features (hingement, muscle attachments, etc.) of the samples, has led to fine-tuning nuances and enhancing their potential for serving as proxies. Great variability has been determined in diverse species from diverse localities, indicating the presence of a measureable "vital effect" (Carpenter and Lohmann 1995); more recent data on other taxa confirm this. Additionally, the two layers of the same shell of *Caioria* have ^{18}O and ^{13}C compositions that are distinctly different (Fig. 4.44a; Parkinson *et al.* 2005).

Crystallization mode of calcareous fibers in the Rhynchonelliformea

The precision of crystallographic controls exerted by mineralizing cells of the secondary layer has been established by utilizing the analysis of crystallographic orientation in electron

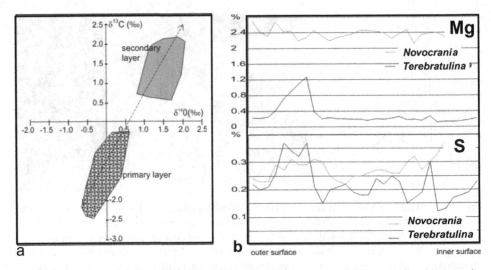

Fig. 4.44. Isotopic and chemical data on rhynchonelliform and craniiform shells. (a) Isotopic composition of the two layers of the shell of *Calloria* (data from Cusack *et al.* 2008b); (b) Chemical compositions along a transect from outside to inside of *Terebratulina* (rhynchonelliform) and *Novocrania* (craniiform) shells (data from England *et al.* 2006).

backscattered diffraction (EBSD) (Schmahl *et al.* 2004, 2008; Cusack *et al.* 2007). Each cell maintains a coherent orientation of its crystallographic axes, regardless of geometric inflections during growth, and most notably where they bend around the numerous punctae (see also, Griesshaber *et al.* 2007). The statistical study of crystallographic orientation likewise shows that this behavior in groups of such fibers simultaneously changing their growth direction results in developing a complex structure, one that is very effective at dispersing forces exerted on the shell. This structure explains very well the overall observation of their surprising resistance to fracturing in inarticulate brachiopods, even more surprising as they are always much thinner than bivalve shells. To these properties of overall crystal state, modern researchers are striving to locate the exact positions of minor elements, such as magnesium or strontium with respect to the crystal lattice of the calcite. Concerning magnesium, the existence of a biological control over its concentration has been known since the work of Lowenstam (1961), but its position still remains uncertain. Pérez-Huerta *et al.* (2008) established that while there are observable differences in Mg concentration between the primary and secondary layers, the amount in the outer secondary layer remains larger than the amount in the inner secondary layer. It also appears that the fibers forming the inner secondary layer can be used to evaluate paleotemperature, based on the changes in the Mg/Ca ratio in them. The question of the position of Mg with respect to the crystallization process still remains unresolved, magnesium appearing to be tied to the sulfated fraction of the organic matrix in *Terebratulina retusa*, but concentrated in the mineral crystal lattice in *Novocrania anomala* (England *et al.* 2006).

The specific changes in the process of mineralization between different portions of the brachiopod shell are also apparent in terms of isotopic fractionation. Cusack *et al.* (2008c)

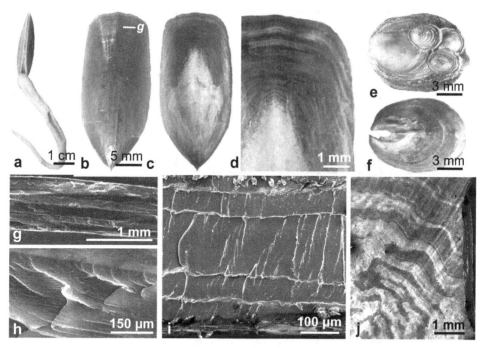

Fig. 4.45. Shells of lingulid brachiopods. (a–d) Morphology of the *Lingula* shell. Animal with its pedicle (a): external (b) and internal (c) views, enlarged view (d) showing the calcareous deposits; (e–f) Discinids; (g) Stratified structure of the shell; (h–i) Appearance of the phosphatic material on a fractured surface; (j) Stepping growth deposition of calcareous material on the internal side of the shell.

demonstrated the separation by layers of the parameters $\delta^{13}C$ and $\delta^{18}O$ into two sharply distinct groups within *Calloria inconspicua*, confirming the necessity of selecting microstructural areas utilized in isotope studies, and more precisely still, the necessity for a fuller understanding of the biocrystallization mechanism. The last must also take into account the mineralizing activities of the organic materials closely associated with the formation of the mineralized shell.

4.4.3 The shell of the Inarticulata

The mineral portion of shells of the linguliform type is composed of a form of apatite (calcium phosphate) and thus differs from shells of the craniform type that are calcitic. SEM images of fractured surfaces of the shell of *Lingula* show a series of stratified thick and compact layers (Figs. 4.45g–i). According to X-ray diffraction analysis, the shell of *Lingula* is a fluoroapatite similar to francolite (McConnell 1963), with impurities that notably include $CaCO_3$ and $CaSO_4$ (Iwata 1981). However, Le Geros *et al.* (1985) and Watabe (1990) suggest that the dominant mineral is dahllite. Thus, this apatite differs markedly from that in vertebrates (Puura and Nemliher 2001).

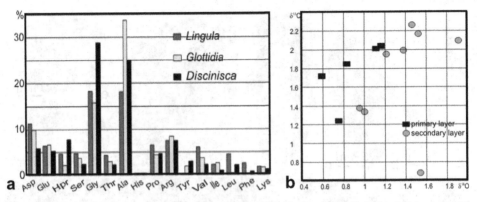

Fig. 4.46. Amino acids and specific isotope fractionation in brachiopod shells. (a) Amino acid compositions of the shells of three modern "inarticulate" brachiopods (data from Iwata 1982, normalized identification); (b) Isotopic composition of the two shell layers of a single species, *Crania*, showing the great disparity between layers.

According to Iwata (1981), the shell of *Lingula* is composed of a punctate alternation of organic and mineralized layers. The acidophile organic layers seem to be structureless when unstained, in both optical and electron microscopes. After staining, fibrils with a diameter of less than 10 nm become visible. These fibrils are themselves composed of bundles of filaments with 3 nm diameters. The structure of the mineralized layers is equally difficult to see (Iwata 1981). Following treatment with sodium hypochlorite to remove organic material, acicular crystallites appear, with orientations and dimensions that differ in different zones of the mineralized layer.

In lingulids, seven compounds have been separated by electrophoresis, each having different molecular weights that range from 6 to 46 kDa. Their amino acid compositions differ, but their contents of serine, glycine, alanine and valine are elevated when taken as a whole (Williams *et al.* 1994). Amino acid analyses of other genera show that the principal components are similar, namely alanine and glycine (Fig. 4.46a). While the presence of hydroxyproline has been determined in scattered genera, fibers of collagen as yet have only been observed in the lingulid shell.

The secondary layer of *Discinisca* is made up of a succession of thin, phosphatized organic layers. The basic unit of these layers is a granule of apatite, from 4 to 8 nm in diameter, wrapped in a chitinous, proteinaceous matrix (Williams *et al.* 1992). Although the adult shell can be phosphatized, the shell of juvenile *Discinisca* is composed of a mosaic of siliceous tablets, of micrometer size, themselves composed of granules of crystalline silica, exclusively silica of biogenic origin (Williams *et al.* 2001). Under the periostracum, Cusack *et al.* (1999) described a thin, external primary layer, with a relatively constant thickness, and an internal layer that is secondarily laminated.

The shells of craniform inarticulates, although calcitic like those of the articulates, have a different structure. The Mg content of this calcite is particularly high, up to 8.6% (Williams *et al.* 1999). The secondary layer of *Neocrania* is composed of regular layers with an alternation of mineral and organic layers, and composed of tablets with spiral growth (Williams and Wright 1970; Williams *et al.* 1999). A fibrous, glycosylated protein of 60 kDa occurs within the lamellae. Taken together, this structure is sometimes called semi-nacre (England *et al.* 2007) because of some morphological resemblance to the nacre of molluscan shells (which is always aragonitic).

Isotopic data on inarticulates are not abundant because the generally thin and weakly mineralized shells are usually poorly fossilized. However, available results seem to be similar to those obtained from the Rhynchonelliformae, and exhibit a large amount of intraspecific variability, and are even highly variable within the same shell (Carpenter and Lohmann 1995).

Even though this is a very brief survey of their mineralization, the Brachiopoda apparently provide us with a spectacularly different type of biocrystallization compared to the epithelial mode of growth in mollusc shells. The cellular origin of shell fibers, taken with their crystallographic continuity in their entire length (irrespective of sinuosity), makes these microstructural components unique among calcareous biominerals. Additionally, it should be added that the brachiopods are a group with roots deep in the Paleozoic Era, and most of the highly diversified groups of families had disappeared by the end of the Permian Period. The examination of Paleozoic brachiopods (see Chapter 7), reinforces this particular and unique aspect of the phylum.

However, it is also true that this cellular formation of fibers involves stepping growth, based on production of organically coated units when seen at the submicrometer scale. At this level, the calcareous biomineralization of extant brachiopods presents some similarities (both structural and biochemical) with the process as developed by the other groups examined thus far.

4.5 Echinoderms

In several respects, echinoderms form a unique group among the invertebrates in terms of their skeleton. Modern species, estimated at 7000 total, are distributed in five classes: the Echinoidea (sea urchins, Figs. 4.47a–c); Crinoidea (sea lilies, Figs. 4.47d–f); Asteroidea (starfish, Figs. 4.47g–h); Ophiuroidea (brittle stars, Figs. 4.47i–j); and Holothuroidea (sea cucumbers, Figs. 4.47k–l). All possess skeletons that differ greatly in architectural aspect, but are very similar in their mineralogy. They are all composed of magnesian calcite. Actually, they are also easily recognizable based on their morphology, but in addition to their nearly identical mineralogy, all of the animals in this phylum build skeletal structures using a method of biomineralization that is surprisingly homogeneous throughout. If we include the very numerous fossil classes (15 following Ubaghs 1967, or 18 according to Paul 1979, 1988), the phylum provides an example of exceptional stability in their mode of skeletal biomineralization during geologic time, including both their mineralogy and chemical composition.

Echinoderms are often noted as having five-fold symmetry, a unique characteristic in the animal world. In reality, however, the primary symmetry is bilateral, and pentamerous

Fig. 4.47. The five extant classes of the Phylum Echinodermata. (a–c) *Diadema*, sea urchin (a) with radioles (spines) in longitudinal (b) and broken surface (c), a remarkable example of spatial arrangement of the calcite stereome; (d–f) Stem and articulated arms of a fixed crinoid; (g–h) Starfish, overview and complex appendices; (i–j) Ophiuroid (brittle star), dorsal view (i) and skeletal unit (vertebra) from an arm (j); (k–l) Holothurian, overview (k) and dermal spicules (l). j: Dauphin (2005).

symmetry appears later, being superposed on the bilateral symmetry. Paradoxically, in the course of their evolution a secondary bilateral symmetry also appears, more or less masking the five-fold symmetry. The five major classes exhibit great diversity in their mode of life, and this is always reflected in their skeletal organization, thus easily recognizable.

The unifying characteristic of the entire phylum is the mode of skeletal growth and its resultant microstructure. In contrast to most invertebrates, the Ca-carbonate skeletal structure of the Echinodermata forms within, and is produced by, mesodermal tissue. Regardless of their particular skeletal architecture, all skeletons are formed of distinct skeletal units that are free or are closely joined (the plates or ossicles, or at times, spines), their articulation determining the degree of rigidity of the skeleton. All of these skeletal elements are reticulate: the highly perforate stereome is so specific that even a small shapeless fragment can be recognized as belonging to the Echinodermata throughout the geological column

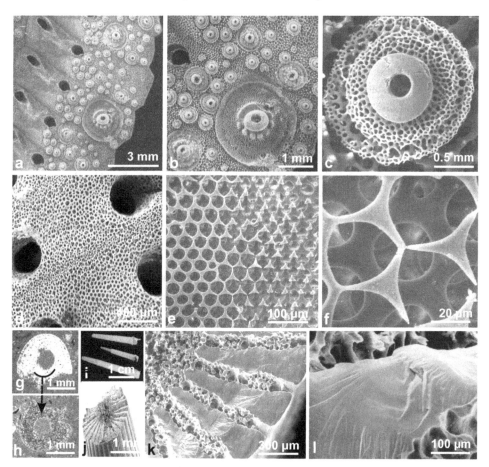

Fig.4.48. The reticulate structure of the components of the echinoderm skeleton. (a–c) At a submillimeter scale, the skeletal units of echinoderms show their reticulate organization: here are shown three sea-urchin plates with pores and basal articulation for a spine; (d–f) Usually reticulate without preferential direction (d), echinoderm stereome at times shows sectors in which crystal-like features become obvious (e–f); (g–h) Evidence of monocrystalline behavior for an ophiuroid armpiece in cross-polarized light (h: with crossed nicols); (i–j) Broken surfaces of an urchin spine, with a conchoidal surface contrasting with the numerous crystalline features. e: Dauphin (2002a).

(Figs. 4.48a–f). An individual plate of an echinoid contains, by volume, approximately 40% mineral material (Raup 1962). The most remarkable thing about them is that each plate, urchin spine or holothurian spicule, whatever its complexity, is a crystal-like structure. This is another property that makes recognition of echinoderm skeletons simple, even when only based on fragmented material. As illustrated in Figs. 4.48g–h, rotation between crossed nicols shows the monocrystalline behavior of the entire piece (Schmidt 1924), as did early X-ray diffraction analyses (West 1937). Urchin teeth are exceptions, due to their being the result of fusion of several elements during their embryological formation, they are polycrystalline.

The *c*-axes of these calcite crystals have a preferred orientation with respect to the morphology of the individual, but significant differences in this crystallographic orientation can be correlated to taxonomic positions. In *Strongylocentrotus* (urchin), the *c*-axes of the plates of the test are perpendicular to the surface of each plate, and therefore to the test in general. However, in *Eucidaris*, the *c*-axes are almost tangential to the surface of each plate, and more nearly parallel to the columns formed by the plates (Raup 1962).

Yet, in spite of such spectacular and universal behavior under cross-polarized light and other evidence of well-controlled crystallization, broken surfaces of skeletal elements do not show crystal-lattice developmental plans typical for true crystals (Figs. 4.48k–l). Owing to the sizes attainable by their skeletal units (the long spines of *Diadema* reach more than 30 cm in length), the calcite of echinoderm skeletons illustrates at best the paradox of calcareous biocrystals: precise control exists over morphology of calcite units (e.g., Fig. 4.48b), ordered crystallization over a long distance, and the lack of many features typical of inorganic crystals. Additionally, owing to the size of these pseudo-monocrystalline units, fossilized echinoderms provide spectacular evidence of some of the changes that occur at a molecular level during diagenesis (see Chapter 7).

4.5.1 Composition: mineralogy and crystallinity of intraskeletal organic components

During the last few decades, authors have argued against the monocrystalline nature of echinoderm skeletal units (Nissen 1963; Towe 1967; Tsipursky and Buseck 1993). Observations by transmission electron microscopy and electron diffraction analyses of plates and radioles of sea urchins reveal the presence of structural features with approximately 10 nm diameters. In fact, the hypothesis of Towe (1967), proposing that echinoid (sea-urchin) radioles (spines) are a combination of polycrystalline aggregates and simple crystals, seems most probable, as this observation has been confirmed by Berman *et al.* (1990) and Aizenberg *et al.* (1997). These structures would thus be composed of microcrystallites aligned in such a fashion that they act as a monocrystal (Magdans and Gies 2004). More recently, Beniash *et al.* (1997) have shown that the first larval stages of echinoid skeletons are dominated by amorphous calcium carbonate that is progressively transformed into calcite during growth.

Mineral and organic components

The content of $MgCO_3$ of echinoids varies from 1.4 to 4.5 mol % for radioles (spines), and from 4.2 to 7.8 mol % in plates of their test (Tsipursky and Buseck 1993), but these contents vary for each species (Magdans and Gies 2004). Thus, Mg quantities can attain a maximum of 43.5 mol % (Schroeder *et al.* 1969). Such elevated contents are also detectable in infrared spectroscopy, based on the positions of identifiable bands (Dauphin 1999a) (Fig. 4.49a). In theory, such calcite is unstable under conditions of normal temperature and pressure. Additionally, within a single radiole, the Mg contents are not uniform; they rise about 2%

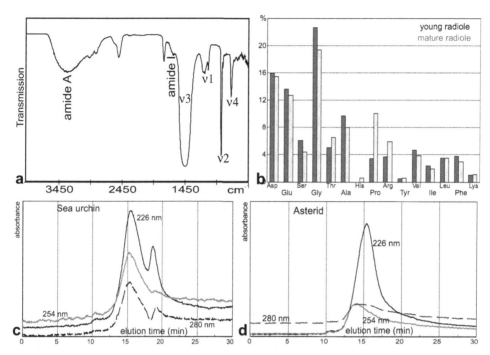

Fig. 4.49. Infrared and chromatographic evidence of organic compounds within echinoderm skeletal carbonate. (a) Infrared spectrum of the echinoid (sea-urchin) skeleton, showing its composition in calcite, rich in Mg (*v* bands) and organic compounds (amide bands); (b) Amino acid composition of soluble matrices from young and mature radioles of *Paracentrotus*; (c–d) Elution profiles of molecular masses extracted from the skeleton of a sea urchin (c) and ossicles of a starfish (d).

from the summit to the base. In general, Mg content shows a linear relationship with water temperature (Magdans and Gies 2004). Experiments are presently in progress on starfish that suggest that there may also be a correlation between some trace elements, the amounts of the isotopes O^{18} and C^{13}, and temperature (Ranner *et al.* 2005).

At first, no organic matrix was detected in the echinoderm skeleton (Currey and Nichols 1967), and this was confirmed by observations in both transmission electron microscopy and electron diffraction (Tsipursky *et al.* 1993). However, after demineralization of the initially formed skeleton (often called spicular), a fibrillar organic matrix was observed (Benson *et al.* 1983). Since then, organic compounds have been extracted and their quantity estimated at 0.01% by weight (Weiner 1985; Berman *et al.* 1990). The majority of data has been obtained on echinoids. Infrared spectrometry (DRIFT method on skeletal powder) allows the direct determination of skeletal organic compounds (Dauphin 1999).

Chromatographic elution profiles at various wavelengths (Figs. 4.49c–d) indicate that soluble matrices extracted from an echinoid skeleton and ossicles of a starfish, are complex. Several dozen proteins and glycoproteins have been identified (Killian and Wilt 1996), and matrix composition is apparently characteristic at the species level.

Very little data is available on the amino acid composition of echinoderm proteins. The two fractions of the soluble skeletal matrix of *Paracentrotus* are rich in acidic amino acids (>25%) and in glycine (>20%). Alanine and serine are equally abundant (Weiner 1984). Radioles of both juvenile and mature urchins belonging to *Paracentrotus* show a high content of glycine, and the acidic amino acids (aspartic and glutamic acids) (Fig. 4.49). Mature radioles are much richer in proline and arginine, but less rich in glycine than are those of juvenile stages (Aizenberg *et al.* 1997).

Glycoproteins of the N and O types are present in the insoluble matrix (Benson *et al.* 1986; Ameye *et al.* 2001), however, their distribution within the skeleton is variable. The activity of certain of these compounds in the secretion of skeleton has been demonstrated in the echinoids (Ameye *et al.* 1999). In the same way, ossicles of starfish (asteroids) contain soluble intracrystalline organic matrices, so that, in *Pisaster*, nonpolar amino acids (glycine, alanine and proline) represent 65% of the mass of the proteins, but sugars apparently are only weakly present (Gayathri *et al.* 2007).

Echinid stereome at the nanoscale

Figure 4.50 presents examples of the nanometer-scale structure of calcite in the Echinodermata. A striking observation is that this skeleton, known to exhibit such easily recognizable features with the use of a simple polarizing optical microscope, cannot be separated from other invertebrate calcareous structures when observed at the 10 nm scale.

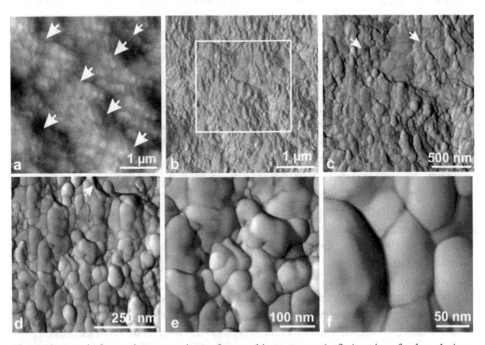

Fig. 4.50. Atomic-force-microscope views of sea-urchin stereome. (a–f) A series of enlarged views showing the typical reticulate assemblage common to all calcareous skeletal materials examined to date.

The densely packed nodular elements form a reticulate calcareous structure that closely resembles each of the examples shown above. The main specific character of the Echinodermata is the very large dimension and complexity of the surface over which crystallographic consistency is sufficiently maintained to produce the classical aspect of a monocrystalline unit. Figures 4.48 and 4.49 provide illustrations of this complex growth surface.

Some observations suggest that materials with this degree of similarity could have resulted from a rather comparable mechanism. In the spines of *Heterocentrotus*, spherical particles with diameters of 80 nm observed in the transmission electron microscope have been interpreted as islands of proteins within the interior of the mineral material (Su *et al.* 2000), while others have demonstrated the existence of microcavities in ossicles (Robach *et al.* 2005). The diameter of these cavities varies from 10 to 225 nm, with a mean diameter of 50 to 60 nm. The total volume of these cavities approximates 20% of the ossicle, and they are filled by organic compounds in the form of hydrated gels.

From decimeter-long urchin spicules to ten-micrometer-wide holothurian sclerites, simple observation with the polarizing microscope demonstrates the efficiency of the process controlling crystallization of all skeletal elements of the Echinodermata. Without a doubt, this phylum provides the most spectacular examples of consistently oriented crystallization, forming huge crystal-like units, such as large urchin spines (Figs. 4.51a–b), and yet also producing the minute and delicately shaped holothurian sclerites (Figs. 4.51c–e). Remember that these elements were all grown within the mesoderm of the animals.

The combination of an exact morphological control at scales of up to centimeter's in length, and their complex reticulate structure seen at a 10 nm scale, makes any model inadequate that has skeletal units produced as single-crystals shaped by adsorption of free

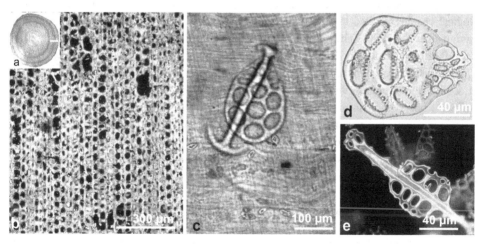

Fig. 4.51. Two extremes in size of mineralized structures among the Echinoderma. (a–b) *Cidaris* spine, up to 12 cm long, magnified view showing the crystal-like network (b: as seen under crossed nicols); (c–e) Dermal sclerites in the skin of *Holothuria*.

molecules within a liquid-filled compartment. Modulation of localized growth by addition of amorphous organomineral material followed by crystallization and the accompanying expulsion of organic compounds out of the developing mineral lattice provides us with a more plausible mechanism. This allows up-to-date data to be gathered. According to this viw, growth of skeletal units of echinoderms is in accord with the layered growth and crystallization model. One prominent difference is that, owing to the great porosity of these skeletan units, coordination mechanisms required for this type of growth process must operate on a three-dimensional level. Data are still lacking regarding the mechanism that enables localized application of such mineralizing secretions over these broad and complex surfaces, thus making the skeleton of echinoderms still somewhat enigmatic.

4.6 Foraminifera

In the Foraminifera, calcareous tests (calcitic with only a few exceptions) are produced intracellularly, thus demonstrably different with regards to calcification examined so far, which is generally epithelial. Owing to the importance of this group as geological tools, for both chronologic and environmental research, the microstructure of the skeleton has been thoroughly studied for decades. Not only the geometry of the calcareous part was described with the optical microscope, but also the fabric of the calcareous material itself was investigated, an approach facilitated by geological applications. For decades, microscopic study of petrographic sections was a major working tool for geologists, especially sedimentary geologists. As a consequence, the Foraminifera is the first group for which taxonomic criteria were based on skeletal microstructures and these incorporated into diagnoses of major taxa, to the point of being more important than morphological or architectural characteristics. As pointed out by Loeblich and Tappan (1974), this methodological change resulted in a dramatic set of changes in the Foraminifera as a taxonomic group since the 1930s. The classification used by Cushman (1925) recognized ten families in this group, as did de Blainville a century previously. Based on shell microstructure, 72 families were characterized by Rauzer-Chernousova and Fursenko (1959), whereas Loeblich and Tappan (1964) employed 94 taxa of family rank. Moreover, the beginning of SEM observation in the 1970s led them to introduce significant changes due to their access to a new observational scale (Loeblich and Tappan 1974). At that time, recognition of the presence of organic components at a submicrometer scale was already integrated into the representation of skeletal formation, although the resolution accessible at that time did not allow the precise description of its location and composition. A major distinction regarding the wall structure of the Foraminifera, and more precisely the relationship between microstructure and mode of growth, has been recognized for more than a century. In Foraminifera producing microgranular walls (including the "porcellaneous" wall), each chamber is added individually, without contributing to the overall skeletal thickness. Under the microscope at standard petrographic thicknesses, these species exhibit dark coloration. In other families (such as Rotaliidae, for instance), the addition of a new chamber results in an additional layer added to the external side of previous chambers. Various forms of this laminar growth process have been recognized (i.e., Smout 1955). From a microstructural and biomineralization viewpoint, it is remarkable that these successively added layers result in a well-ordered microstructure for the wall, the mineralized units forming radiating prisms with an overall orientation direction perpendicular to the shell surface. Additionally, a complex system of canals is developed within septa and chamber walls that demonstrate the very high degree of structural control on the skeleton of these unicellular animals. It is important to mention the ability shown by Foraminifera of some families, generally regarded as primitive, to build composite skeletons by including sedimentary particles and cementing them together with intracellular calcareous materials.

Four examples among the groups of extant Foraminifera illustrate the diversity of their skeletal organization, diversity that reaches a much higher level when fossil forms are also taken into account. At the same time, these examples reinforce the concept of their possessing common constraints on the calcification mechanism operating at the nanoscale. In *Globigerina*, for instance, the regular arrangement of elongate calcareous units with their orientation perpendicular to the rounded shell surface (Figs. 4.52a–b), justifies a general description of a wall built by crystals with *c*-axes perpendicular to the growth direction. Closer observation reveals the layered growth mode of these fibrous crystals (Fig. 4.52c: arrows), and AFM observation illustrates the submicrometer organization of growth layers themselves (Figs. 4.52d–e). In contrast to the polarizing properties of skeletal units that built the chamber walls, so easily seen with the optical microscope, AFM does not reveal any trace of crystallization in the morphology of the elements contributing to the formation of the skeleton. Here the height images suggest the existence of densely packed cylindrical units; this much more precise phase imaging (Figs. 4.52e–f) is very sensitive to the presence of the organic coating here surrounding skeletal units. Additionally, this mapping mode reveals that the chamber walls are formed of numerous smaller units having variable morphologies. This was not shown by height imaging. Surprisingly, except for the morphologies of the submicrometer-sized units, the overall organization of the skeleton in *Globigerina* suggests a growth and crystallization mechanism rather comparable to those of metazoans. This conclusion is reinforced by study of the nature and distribution of associated organic compounds.

Figure 4.53 presents a series of synchrotron-based fluorescence maps made on the chamber wall of an individual of the genus *Globorotalia*. The distributions of proteins and polysaccharides are separated and identified here by the two distinct coordination states of sulfur, producing specific XANES peaks at 2.473 keV and 2.46825 keV. Additionally, the X-ray fluorescence method (EXAF) reveals profiles specific for sulfated polysaccharides that are clearly distinct from those of mineral sulfates, allowing the unambiguous recognition of sugars among organic components of these calcareous skeletons (see Fig. 2.42).

As in mollusc shells, a permanent correlation is observed between the distribution of organic contents (both proteins and sugars) and the calcareous structures. Repeated

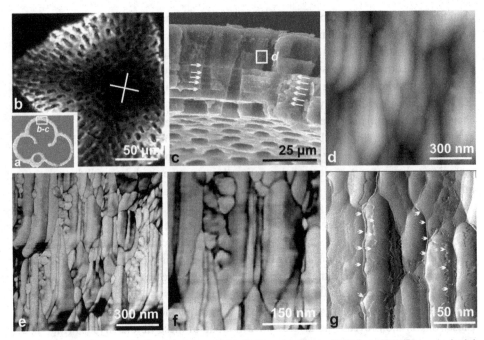

Fig. 4.52. Ultrastructure of the chamber wall in *Globigerina* sp. (a–b) Polarization figure (uniaxial cross) obtained in viewing a fragment of wall under crossed nicols (b), illustrating the classical description of the wall built by crystals with their *c*-axes perpendicular to the chamber surface; (c) SEM view of the same part: the approximately 20 μm thick wall is built by superposition of micrometer-thick growth layers (arrows); (d) AFM view (mode height imaging) of a fracture surface perpendicular to the limits of the wall; (e–f) AFM (mode phase imaging): the organic components of the wall are located at the periphery of mineral units; (g) Amplitude imaging emphasizes the limits between the mineral core and organic coating (arrows) of the skeletal units.

observation of this indicates that the interplay between the mineral and organic components occurs at the submicrometer level. It also contradicts the common concept of using a template model for skeletal formation in Foraminifera (as proposed by Hansen 1999, also by Kunioka *et al.* 2006), wherein skeleton deposition occurs by the deposition of mineral material on both sides of a thin organic membrane. Variations in organic concentration are visible on biochemical maps of sections through walls (Figs. 4.53d–g), as commonly seen in the long-accepted layered growth mode of wall chambers, but an alternation of pure calcite crystals developed on an organic layer is not compatible with the results of the biochemical mapping shown here.

Additionally, and again conforming to what was observed in *Globigerina* (Fig. 4.52), the layering visible in Fig. 4.53d does not represent an alternation between organic and mineral layers. Mapping of organic components shows that they are permanently incorporated within the calcified structure, with changes in density and possibly in accompanying proportions of protein/sugar components. These variations may be related to variations in metabolism or nutriments available (as seen in Fig. 2.7 within *Pinna* prisms). Continuity of

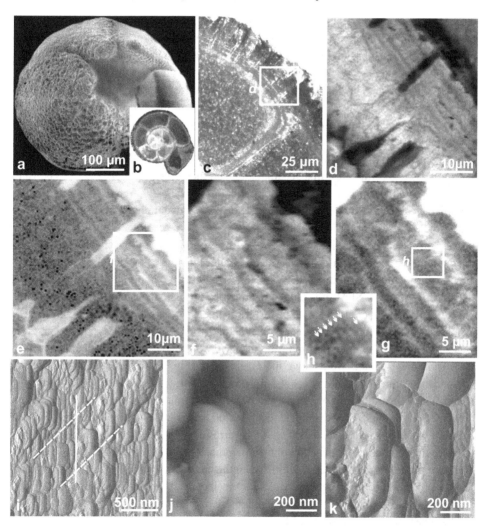

Fig. 4.53. Distribution of organic components within the wall chamber of *Globorotalia* sp. (a–b) Morphology and section of a specimen; (c) Polished section of the chamber wall; (d–e) XANES mapping of sulfur distribution showing distribution of polysaccharides (d) and proteins (e). Note the opposite concentrations, notably in wall pores (infilled with cell cytoplasm), and also between the outer and internal parts of the wall; (f–h) Closer view of the mapped sectors. Image (h) shows that the major layers are subdivided into micrometer-thick growth layers (arrows), the thickness of which is rather comparable to the mineralization cycle observed in invertebrate shells; (i–k) AFM view of the mineral components. The calcite units do not exhibit any crystalline habits. Their arrangement is rather parallel. The overall direction (white dashed lines in i) are oblique to the layered growth process (compare to Fig. 3.5g) d, f: Cuif *et al.* (2008b).

the organic component at the nanometer scale provides an important feature that contradicts the template model of biocrystallization, in which alternation between purely organic and mineral layers is the basic scheme. As in previous examples, this location of mineral and organic material together is well illustrated by AFM observations of their spatial interaction at the submicrometer level. Figures 4.53i–k show the remarkably large subunits of the skeletal system within the wall of *Globorotalia*, but they are still generally less than the thickness of the elementary growth layer shown in Fig. 4.53h. It is not surprising that, when using mapping tools with a resolution of one micrometer at best, organic and mineral components appear as having identical distributions. Thus, the two-step crystallization scheme for the growth layer (first accumulation and then crystallization) seems to be valid for the Foraminifera also. Other examples belonging to the extant representatives of those commonly called the major Foraminifera (Soritidae, Alveolinidae, and Nummulitidae) permit the extension of this conclusion.

The Soritidae and Alveolinidae belong to the broad group Miliolina, characterized by the microgranular microstructure of their wall and septa (Fig 4.54). Characteristically, observation of the elementary units of the skeleton is practically impossible with the optical microscope. With the SEM, it is only at the micrometer scale that individual units become visible (Figs. 4.54a–c and f–i). Rather well ordered in *Marginopora* (Fig. 4.54d), these elementary components seem not oriented at all in *Alveolina*. In both cases, however, these very small units are apparently composites themselves, built by subunits with dimensions in the 10 nm range. This is well shown in *Marginopora* (Fig. 4.54e) and also in the skeletal components of the alveolines (Figs. 4.54j–l). In the latter, it is noteworthy that there is an abundance of accumulated nanograins, still being distinct and not yet been incorporated into definite skeletal units. They cannot be identified on the AFM height image (Fig. 4.54j) but are readily visible in phase images (Fig. 4.54k: arrows). This AFM phase mode also allows us to see the composite organization of the micrometer-sized units in microcrystalline tests. The remnants of organic material left between the coalescent grains within units indicates that the process of forming composite units from elementary grains has begun, but remains at a very low level of general organization in these organisms.

The other major group of extant Foraminifera, the Suborder Rotalina, offers a striking contrast with the previous groups based on shell microstructure. The basic difference is easily seen with the optical microscope in standard petrographic sections. Here, instead of opaque sections resulting from microgranular material of the test, which transmits light poorly, the shell of the Rotalina appears highly translucent.

As an example of the historical preeminence of foraminiferal studies in micro- and ultrastructural analysis, the long-recognized contrast between the Miliolina and Rotalina was first explored by electron microscopy (as carbon replicas) in the early 1960s (Hay *et al.* 1963). These authors were able to show that the skeleton of Miliolids is composed of grains 0.5 to 2 µm in length and 250 nm in diameter, with their surface covered with organic matter. The random orientation of these tiny skeletal units was then considered as "unique in biological systems" (Towe and Cifelli 1967) and was related to the presence of "an incoherent colloidal active organic ground mass." As summarized by Loeblich and

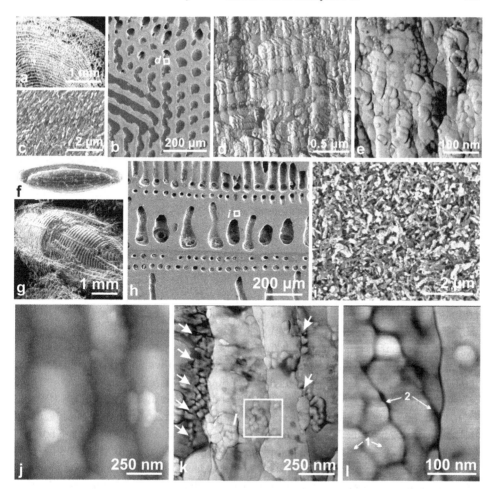

Fig. 4.54. Ultrastructure of calcareous units in Soritidae and Alveolinidae. (a–b) *Marginopora* sp. (New Caledonia). Partial view and polished surface; (c–e) Ultrastructure: submicrometer grains (c) showing rather well-organized arrangement (d). Phase imaging (e) makes their composite structure clearly visible, with incompletely fused nanograins and remains of organic material appearing as coating on composite units; (f–h) Alveolinidae: morphology (f–g) and polished section parallel to coiling axis; (i) Granular aspect of the skeleton viewed with the SEM, medium enlargement; (j–l) AFM height (j) and phase (k–l) imaging; submicrometer units are also made of grains. Note the irregular distribution of organic residues: from traces between fused grains (arrows "1") to denser concentration after formation of reticulate crystals (arrows "2").

Tappan (1974), "... the crystals fill spaces in the passive membranes that give shape to the test." In contrast, optical studies carried out at the same time led to considering the tests of the Rotalina as built of elongate crystals consistently arranged perpendicular to the shell surface (radial microstructure). Multiple demonstrations of the presence of both radial and random microstructures (Bé and Ericson 1963, and others) reduce the basic contrast between these two microstructural types. Additionally, the demonstration of the "mosaic" structure of the apparently coherent crystals observed with the polarizing microscope also led to a consensus that "when the terms radial and granular are restricted to use in a crystallographic sense in respect to optical effects, a sharp distinction can be made," but that "apparently uniform crystals as seen in the polarizing microscope, are shown by electron microscopy to represent mosaics of smaller crystals" (Loeblich and Tappan 1974). It is unnecessary to emphasize the excellent agreement between the in-depth analyses carried out in the 1960s and present-day AFM observations, where the latter allow easy access to the 10 nm range and simultaneous imaging of the distribution of organic components. Figure 4.55 illustrates the nanoscale view of the skeleton of an operculinid, a last representative of the family Nummulitidae. This family exemplifies the complexity of the structural organization of the foraminiferal test. The walls of chambers are formed by fibroradiate crystals and marked by the development of canal systems that allow communication between the endothecal and exothecal parts of cellular cytoplasm (Figs. 4.55a–e). Observing the ultrastructure of fibers in the wall (Figs. 4.55g–i) and in the septa (Figs. 4.55j–l) provides substantially identical images. In both cases, the crystal-like fibers exhibit the reticulate ultrastructure as it results from the coalescence of grains during crystallization with concomitant exclusion of organic components towards the exterior of the newly formed crystal lattice. As usual, but well illustrated here, fusion between neighboring grains is imperfect; the gray colors reflecting various densities show that some portion of the organic compounds remains entrapped within the mineralized granules (Figs. 4.55h, k–l: arrows), included within the skeleton, *but not in the lattice*. These remaining organic components explain the crystal defects already detected in the 1960s in crystal-like fibers by investigators using X-ray diffraction. In addition, it should be noted that the crystallization models accepted at that time were based on the "template model," based on an epitaxial development of crystalline calcite with the c-axis as a preferential growth direction (Towe and Cifelli 1967). In this model, the polysaccharide components of the organic matrix were considered as a "passive substance which provides the form of the chamber, whereas proteins were the active substances providing sites for mineralization" (Loeblich and Tappan 1974).

Gathering synchrotron-based maps and AFM phase imaging provides more nuanced information regarding the distribution of organic components in the Foraminifera. Consistent with the principle of superposition of mineral and organic components within a layered mode of growth and formation of crystallized layers with reticulate ultrastructures, we now understand that, after crystallization and formation of the reticulate crystal, most of the observed "intracrystalline" matrices are not within the crystals but around them (with minute residues that can be located within crystal lattice defects). However, tribute must be paid to the research carried out in this seminal decade of the 1960s, when taxonomists of the Foraminifera investigated their ultrastructure and organomineral relationships at the submicrometer scale. In practice, most of the observations made at that time can easily be integrated into the overall scheme that results from many decades of progress in physical instrumentation.

Fig. 4.55. Wall and septal ultrastructure in operculines (Nummunilitidae). (a–c) Morphology (a), and X-ray radiography (b–c) of operculine tests; (d–e) Radially oriented pores in the wall: enlarged view in X-ray radiography (d); microscope view (thin section, natural light) of the wall (e); (f) Location of the AFM images; (g–i) Ultrastructure of the wall. Phase image (g–h), amplitude image (i). Note the very irregular distribution of organic remains; (j–l) Ultrastructure of the septum; generally similar to the wall. Note the variety of organic occurrence, from weak traces (arrows) within the mineral features to very dense, highly interactive (but irregular) coating of the skeleton.

4.7 Calcareous structures of Vertebrates: otoliths and eggs

It has long been known that Ca-carbonate is formed as a minor component of vertebrate bone (with its proportion increasing during aging), but two nonskeletal mineralized structures in vertebrates are purely calcareous: (1) otoliths and otoconia, essential structures for equilibrium, and (2) the shells of bird and reptile eggs. It is impressive that, although constructed under very different anatomical conditions, and with different modes of growth, these two structures exhibit fine-scale organizations that are comparable in many

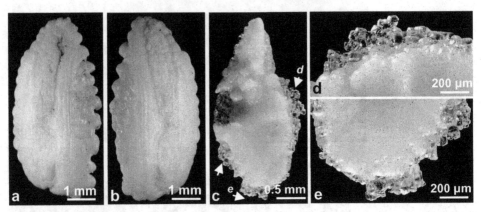

Fig. 4.56. Otoliths. (a–b) Two faces of an otolith (sagitta) of *Gadus morhua*; (c–e) Otolith with vaterite crystallization abnormalities (arrows). a: Dauphin and Dufour (2008).

respects to the common patterns observed among previous examples, all of which are invertebrates.

4.7.1 Otoliths

Otoliths (ear stones) are small carbonate secretions present in the internal ear of bony fish. They participate in the hearing process, but their principal role is to insure the animal's equilibrium. Three pairs of otoliths are present in each ear: sagitta, lapillus, and asteriscus. The sagitta is the largest of the three elements (Figs. 4.56a–b).

Each otolith is contained in an otic capsule and is suspended in liquid that fills the capsule, the otolymph, without being in contact with the epithelium that covers the internal side of the otic capsule. However, otoliths have a species-specific form. It has also been demonstrated that otolith structure records a sequence of daily, lunar and seasonal variations. Additionally, as variation in composition of the Ca-carbonate can be an indicator of fish migration, otoliths are much studied by fishery institutes. Otoliths are most commonly composed of aragonite. Interestingly, from a biomineralization viewpoint, they frequently show abnormal crystallization (Figs. 4.56c–e).

Recent studies (Tomàs and Geffen 2003) have established that juvenile populations of herrings can have vaterite crystallization up to a percentage of 7–8% in the area of the mouth of the Clyde River, whereas the proportion of abnormality rises to about 14% in the Irish Sea. By studying the reactivity of the nervous receptive system, Oxman *et al.* (2007) have demonstrated a significant loss of sensitivity to vibrations among specimens producing significant amounts of vaterite (2.5 to .5 dB).

Microstructure

Sectioned otoliths exhibit columnar elements that radiate from the otolith center (Figs. 4.57a–b). The nucleus (center) corresponds to a first phase of mineralization,

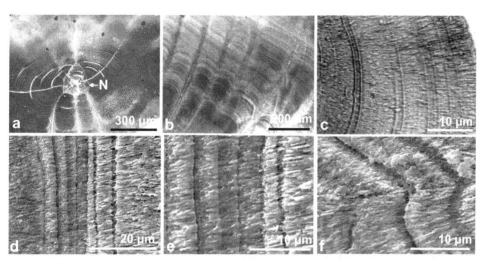

Fig. 4.57. Sections of a sagitta. (a) Polished section showing the nucleus ("N"); (b) Same, external zone with growth striations and sectors; (c) Polished and etched, zone with regular growth striations and lacking sectors; (d) Same, detail of growth zones; (e–f) Same, fibrous aragonite crystals in growth layers. The acicular crystals are themselves composite elements with a typical nanometer-scale reticulate structure. a, c: Dufour *et al.* (2000); e: Dauphin and Dufour (2008).

commonly referred to as premetamorphose (a term used by analysts dealing with fish otoliths), the sole remnant of the original otolith (Fig. 4.57a). Afterwards, growth continues from several centers of calcification. The columnar elements (Fig. 4.57b) exhibit growth striations on a micrometer scale, producing a time series that can be analyzed to reconstruct the life history of the fish (Fig. 4.57c). In fact, these growth striations occur as diurnal doublets corresponding to days and nights, a pattern quite similar to that visible in the nacreous layer of *Pinctada* shells (see Fig. 1.5). Microscopic observation of thin sections shows narrow dark lines alternating with thicker, translucent layers. The usual interpretation of this is that each mineral layer (translucent) was precipitated on a thin organic layer (dark), thus providing another illustration of the "template model" for biocrystallization (e.g., Wright *et al.* 2002). However, previous authors (Mugiya 1965; Ans *et al.* 1982) have shown that this alternation of dark and clear layers is instead due to differing proportions of mineral and organic components.

Chemical composition

Contents of chemical elements such as Sr are very low in otoliths, in spite of their aragonitic composition, and quantities of Mg and S are low as well (Fig. 4.58a). Mapping of Sr has shown that the topographic growth zonation corresponds to a chemical zonation (Tzeng *et al.* 1999). The first organic compounds identified were those of the insoluble phase when Degens *et al.* (1969) described a collagenous protein here. Since then, the presence of a soluble matrix composed of numerous proteins, glycoproteins, and proteoglycans has also been described (Tadashi and Mugiya 1996; Borelli *et al.* 2001; Dauphin and Dufour 2003).

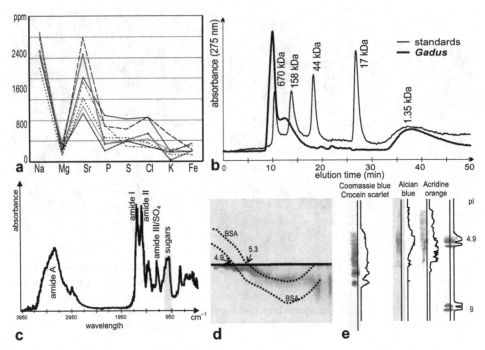

Fig. 4.58. Chemical composition of seven modern otoliths belonging to different genera. (a) Diagrams of minor-element concentration (EDS microprobe measurements); (b) Elution profile with liquid chromatography indicating the presence of very large molecules in the soluble organic matrix of *Gadus*; (c) Infrared spectrum of soluble matrix extracted from otoliths of *Gadus* that indicates the presence of proteins and sugars; (d) Electrophoresis of soluble matrices of *Gadus*: here a titration curve shows the acidic nature of the proteins; (e) Isoelectrofocusing indicating the presence of proteins and acid sulfated polysaccharides. c–e: Dauphin and Dufour (2003).

Molecular masses of matrix proteins extracted from *Gadus* range through a wide spectrum (Fig. 4.58b). The most abundant masses are not yet detectable by classic electrophoresis, but infrared spectrometry clearly indicates the additional presence of sugars, and perhaps organic sulfates (Fig. 4.58c). Titration curves with isoelectrofocusing (Fig. 4.58d) illustrate the acidity of these compounds. Coomassie Blue staining additionally confirms the presence of proteins or glycoproteins, while Alcian Blue and Acridine Orange staining indicate the presence of sulfated polysaccharides. Comparison of the two staining results establishes that the polysaccharides are more acid than are the proteins.

Clearly, beyond the taxonomy-linked diversity and variability due to individual life conditions, the mineralizing matrices of fish otoliths demonstrate considerable resemblance to equivalent biochemical assemblages present as the mineralizing agents of various groups of invertebrates.

Structure of otoliths at the submicrometer scale

Observation with the AFM shows that the fibers of the otolith growth layers exhibit the reticulate structure typical for crystals formed by coalescence of granules of variable

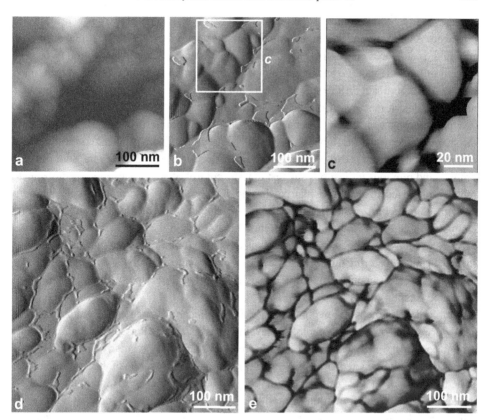

Fig. 4.59. AFM images of aragonite in an otolith of *Gadus* sp. (a) Height image, here at low magnification, illustrates granules grouped in linear assemblages; (b–c) Amplitude image, demonstrating the interactive (= organic and/or amorphous) nature of the (c) envelopes; (d–e) Complementary images – amplitude (d) and phase (e) – summarizing the structural pattern of otolith calcification, and emphasizing similarities with invertebrate biological carbonates. c, e: Dauphin and Dufour (2008).

dimensions during the crystallization process. More or less aligned (Figs. 4.59a–b), the densely packed granules are surrounded by a strongly interactive, irregular coating (Fig. 4.59c). Because of the similarity of this to nanostructures that have previously had their atomic forces measured (see Fig. 3.2), we can definitely state that this envelope, 5 to 10 nm thick, is primarily organic material.

In the overall layered structure of otoliths and the presence of reticulate crystals forming fibers in the 10 nm range, the similarity of this vertebrate calcification to invertebrate calcifications previously discussed is obvious, and needs no further emphasis. It is logical to assume that control of the crystallization of these fibers is regulated in the same way, thus producing biocrystals with identical internal structures. The aberrant development of vaterite (or sometimes calcite) reflects the suppression of control on crystallization. Not only does mineralogy change, but in the "translucent ring of vaterite" (Tomàs *et al.* 2004, Fig. 1;

see also Fig. 4.56) surrounding the aragonite core, crystals with planar growth faces are produced. As a result, the structure of the otolith then can no longer be identified as having columnar sectors and concentric growth layers in columns. In their investigation of protein content of both the vaterite and aragonitic portions of herring otoliths, Tomàs *et al.* (2004) did not find significant differences in the amount of soluble proteins. It is worth noting that no precise determinations of differences in the nature of the proteins associated with the two polymorphs were made. Focusing discussion on soluble proteins tends to neglect the shaping role of the insoluble phase, usually interpreted as simply here controlling "the otolith overall shape." It can be seen that the overall shape is nothing but the sum of local growth, and previous indications suggest that the control of shape is exerted by envelopes, mostly composed of insoluble matrix, throughout the area of micrometer-scale stepping growth. A reduction in the production of insoluble matrix (or any modification of its composition that leads to losing effectiveness) may explain the correlation between transparency and the appearance of faceted crystals. Examination of mineralogical and morphological changes occurring during pearl development provides additional examples of similar micrometer-scale stepping growth processes in dictating the shape of the microstructural units (see Chapter 5).

From a developmental viewpoint, occasionally a linear arrangement of grains has been seen during the first hours of otolith development in *Dano* (Pisa *et al.* 2002). This is reminiscent of the early stages of otolith formation, marked by regularly aligned granules of glycogen with diameters of approximately 16.5 nm. The existence of this specific microstructural pattern at this early stage of development represents an additional point of similarity between the formation of otoliths and the formation of the more common shell structures. This in turn would *suggest that there are common rules that cross the highest taxonomic levels* that must be present and operating at the nanometer scale in order for animals to succeed in producing controlled Ca-carbonate structures.

4.7.2 *Vertebrate eggs*

Calcified eggs exist in the reptiles and birds (eggs of monotreme mammals are covered by a resistant but noncalcified membrane). For obvious economic reasons, the shells of chickens have been extensively studied, and the chicken egg provides a common example of extremely rapid calcification, one of the most rapid among living organisms. A medium-sized egg corresponds to about 6 g of shell material, and is formed in 20 hours, following a very precise sequence of calcification events during its transit through the oviduct. Both the structure and composition of eggs produced by birds and reptiles vary according to taxonomy.

Microstructure of eggshells

Eggshells of reptiles and of birds are composed of different layers whose nomenclature varies somewhat according to author. The structure of reptilian eggs is simple (Figs. 4.60a–b, Fig. 4.61b), with an internal surface composed of organic fibers, more or less composed of

Fig. 4.60. Egg microstructures. (a) Vertical section (perpendicular to shell surface) in a tortoise egg; (b) Vertical section in a crocodile egg; (c) Vertical section in a ratite (terrestrial flightless bird) egg, showing the well-developed external layer; (d–e) Thin section of a ratite shell in natural (d) and polarized light (e) showing the coarsely crystallized mammillary layer ("mml"); (f) Organic fibers and insertion of a mammillary cone within a bird's egg; (g) Crystals at the base of the mammillary cone; (h) Detail of a palisade (or prismatic) layer, showing the chevron structure; (i) Prismatic units and pore opening ("po") at the external surface of a bird's egg. f: Dauphin (1990).

collagen and rich in sulfur. On this organic layer (*membrana testacea*) are anchored mineralized mammillae whose centers are composed of crystals that are loosely assembled, and somewhat similar to the "centers of calcification" of corals. This is not always considered as a completely separate layer, and some authors integrate these "centers of calcification" with the layer that follows it: the mammillary layer that has radial fibers organized around the cores of mammillae. Growth lines are generally visible, whether dealing with tortoises, crocodiles or geckos. The mammillary layer, composed of a series of cones, is discontinuous at its base, with

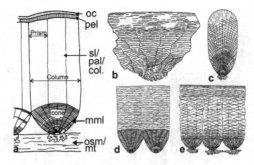

Fig. 4.61. Vertical sections of major structural types of eggs having a mineralized shell. (a) Terminology: "oc": organic cuticle; "pel": prismatic external layer; "sl": spongy layer / "pal": palisade layer / "col": column layer (depending on authors); "mml": mammillary layer; "osm": organic shell membrane / "mt": membrana testacea (depending on authors); (b) Schematic section for a crocodilian egg; (c) Reptile (dinosaur); (d) Neognathid birds; (e) Ratites (flightless birds).

the cones progressively enlarged and joined to each other in what is called the prismatic layer (Figs. 4.60a–c: "pl"). The junction between these prisms is not perfect because there are pores present here, extending through the total shell thickness, with their openings visible on its external surface (Fig. 4.60i: "po"). The distal extremity of these prisms is at times flat and at times swollen. The surface of the egg can then be ornamented by bosses or by more or less elongate crests. A very fine organic pellicle coats the external surface of the shell.

The structure of birds' eggs is more complex: they resemble a reptilian egg with supplementary external layers (Figs. 4.60e–f). The mammillary layer is prolonged into a structure in which the divergence of the crystal units is no longer visible, but in thin sections under polarized light, large prismatic units can be recognized (Figs. 4.60e and g). Each prism is composed of small crystals whose arrangement forms a chevron structure (herring-bone pattern) (Fig. 4.60h). This layer carries different names (Fig. 4.61a). The thickness ratios between the mammillary layer and prismatic layer differ according to taxa.

As in the reptiles, the most internal layer is organic and composed of fibrils (Fig. 4.60f). The most external mineralized layer is prismatic. In the ratites (ostriches, emus, etc.), it is relatively thick, as was early recognized and described. It also seems to be always present in the neognathid birds, but on a much finer scale, thus its structure is less easily interpreted. An organic pellicule covers the external surface. Thickness proportions of different layers, as well as height-to-diameter relationships of cones in the mammillary layer have been utilized as phyletic criteria (Zelenitsky and Modesto 2003). However, this is difficult to do, because in a bird's egg, the thickness of the shell varies between its poles and equator; moreover, these thicknesses vary with the age of the animal. Apparently, no one has yet studied reptilian eggs from this point of view.

Preferential orientation has also been described, not only for the aragonite of tortoise shells, but also for calcite crystals in other reptile egg shells (Silyn-Roberts and Sharp 1985); their (001) plane tends to be parallel to the surface of the shell. Generally, the maximum degree of such preferred orientation is situated near the external surface. Detailed analysis of

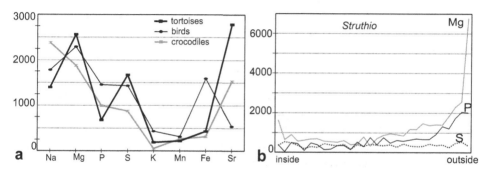

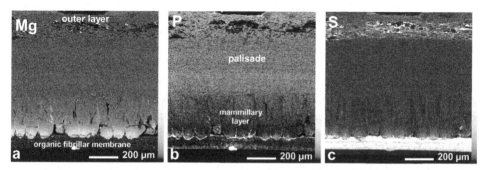

Fig. 4.62. Minor element concentrations in eggshells. (a) Median chemical composition of aragonitic eggs of the tortoise, and calcitic eggs of reptiles and birds; (b) Profile showing gradients through ostrich shells in Mg, P and S (the fibrillar organic layers are excluded). b: Dauphin *et al.* (2006).

Fig. 4.63. Mapping of a vertical section through an eggshell to illustrate the distribution of Mg (a), P (b) and S (c), showing gradients of composition and the enrichment of the internal organic membranes in sulfur (S). (Electron microprobe, C. T. Williams, Mineralogical Dept., NHM, London). a–c: Dauphin *et al.* (2006).

crystallinity in the chicken egg has shown that the orientation of crystallites varies between various microstructural layers (Lamnie *et al.* 2005). Thus, the small crystallites in fibrillar organic membranes are aligned parallel to the surface of the shell. The size of the crystallites in the mammillary zones is smaller than in the other layers, and the crystals are not regularly oriented. In the palisade layer, they are much larger and regularly oriented, with the (0001) axis of the calcite perpendicular to the surface of the egg (Dalbeck and Cusack 2006).

Composition of eggshells

All eggs have a calcitic shell, except those of tortoises, which are aragonitic. Contents of Mg are not much different between calcitic and aragonitic shells, and contents of Sr are small, even in the tortoise shell (Fig. 4.62a). Contents of P vary according to taxa, and S content differs greatly, according to whether one takes into account the internal organic membranes or not. Contents of Fe are not uniform throughout the shell thickness (Cusack *et al.* 2003; Dauphin *et al.* 2006, as shown by quantitative analyses (Fig. 4.62b) and maps of chemical distributions (Fig. 4.63). As yet, only a few species have been studied from this point of view.

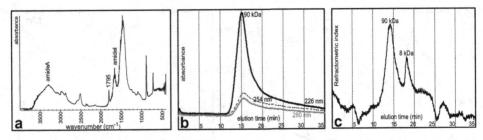

Fig. 4.64. Overall characterization of eggshell matrix. (a) Infrared absorption; (b) High-pressure liquid chromatography of soluble organic matrices extracted from a chicken eggshell: UV detection showing the presence of proteins and glycoproteins; (c) During the same elution, refractometer detection indicates the presence of nonprotein compounds, very probably polysaccharides.

However, in both calcitic and aragonitic shells, they are not purely mineral, as shown by infrared spectra (Fig. 4.64a). In addition to the internal organic fibrillar membranes and the external cuticle, they contain proteins and associated sugars (Figs. 4.64b–c). The internal membranes are composed of sugars and proteins rich in disulfide bridges. The organic matrix associated with crystalline units contains sulfated polysaccharides (chondroitin sulfate), proteins (here comprised of collagen) and proteoglycans (Arias *et al.* 1992, 2003; Fernandez *et al.* 1997). The palisade layer is notably rich in proteoglycans having high molecular weights. Intracrystalline proteins are presently classified in three categories; certain proteins of eggwhite (ovalbumin, ovostransferrin, and lysozyme) are present (Hincke 1995; Hincke *et al.* 2000; Gautron *et al.* 2001), forming the first category. The second is composed of nonspecific proteins; for example, the osteopontine found in the egg. Third, proteins known exclusively from the shell occur: ovocleidines and ovocalyxines (Hincke *et al.* 1995, 1999). It is worth noting that the soluble matrix of the chicken eggshell may contain more than 520 proteins (Mann *et al.* 2006). However, their composition, structure, and role in formation of the shell have not yet been determined for the greatest number of eggs, in spite of intensive research related to controls on industrial production of about 850 billion eggs each year.

Ultrastructure

Inserted onto the basal membrane, and more precisely, onto the crystals of the mammillary layers, are small linear units (in the micrometer range), formed of well-ordered fibers, with their long axis perpendicular to the shell surface (Figs. 4.65a–b). The crystalline behavior of these units, obvious in thin section under cross-polarized light at the microstructural level (Figs. 4.65d, g), is also very clearly visible at low magnification in the atomic force microscope. The broken ends of fibers are roughly parallel (and oblique) to the growth axis (compare with the prisms of *Pinna*, for instance, Fig. 3.5).

These figures provide us with evidence that the crystalline units so clearly visible under cross-polarized light in the mammillary layer are actually typical reticulate crystals. The overall crystalline organization continues within the prismatic layer, with a noticeable

Fig. 4.65. Ultrastructure of the mammillary layer in the eggshell of *Numida* sp. (a–b) Overall view (field of view 10 μm) as height (a) and phase (b) images, with short units having axes perpendicular to shell surface; (c–d) Enlarged view of fracture surfaces of the fibers, which clearly show their crystalline structure; height (c) and phase (d) images; (e–f) Surfaces of fibers showing traces of fused granules of the same composition as fibers. Some of these are still visible at the margin of the fiber; (g) Phase image of partly fused grains, with excluded organic material present as an irregular coating (white arrows). Within fused grains, note that the AFM tip is sensitive (gray density variation) to some lower concentrations of organic remains (black arrows).

increase in the irregularity of crystallization (see Fig. 4.60e). Specifically in birds' eggs, a "chevron" ultrastructure epitomizes the reticulate crystals. At the SEM scale, the spongy – or palisade – structure indicates that roughly 10–20% of the layer is not composed of mineral material, and when looking at the mineralized part itself, AFM observations reveal the importance of this interactive component of the shell layer. Additionally, in this part of the bird's egg, apparently added during the evolution of birds from reptiles, controls on crystallization are weaker than in the more primitive (i.e., internal) layer, based on increasing irregularity observed at the ultrastructural level (Fig. 4.66).

4.8 The calcified cuticle of crustacean Arthropods

Although organisms belonging to the Phylum Arthropoda represent about 80% of living animal species, their role as sedimentological contributors is limited; only a restricted number of groups have developed a mineralized skeleton. Arthropods are characterized by their rigid exoskeleton, mainly made of chitin. Owing to the rigidity of this material, the skeleton is made of distinct articulated units: plates on the body, rings and tubes around the appendices. Taking advantage of the multiple high-level mechanical properties of the

Fig. 4.66. AFM views of eggshell (*Numida* sp.). (a–b) Height and phase images of the constructional units in the spongy layer, aligned in the direction of growth; (c–d) Enlarged views illustrating the composite structure of fibrous units, already especially visible in the phase image (d), here with a notable abundance of interactive (organic) material; (e–f) Granular composite structure that exhibits marked irregularity in the size of building units; (g) Amplitude image that is barely decipherable due to the abundance of organic material in the palisade layer.

chitin-based skeleton was a key point in the evolutionary success of the phylum. As a counterpart, the necessity of periodically renewing the skeleton led to the elaboration of a sophisticated molting process, typical for the whole phylum. All the pieces of the cuticle (or exoskeleton) are simultaneously shed and a new cuticle is secreted, enabling both the size and the organization of the animal body to be progressively modified, until reaching an adult stage. Generally considered as a subphylum, the crustaceans comprise six distinct classes (Martin and Davis 2001), of which the Class Malacostraca unites the most heavily calcified arthropods (e.g., crabs, lobsters, etc.).

Observations of the structural complexity of these calcified structures were part of the early stages of microstructural research (Williamson 1860). The presence of four layers has been identified by Travis (1963, 1965), of which three are easily visible (Figs. 4.67a–b): exocuticle, endocuticle, and, in an internal position, the membranaceous layer. Underlying the endocuticle, the nonmineralized membrane forms an intermediate layer between the base of the calcareous structure and the cellular layers of the epidermis. The thickness of the fourth and more external layer (epicuticle) is in the micrometer range (Fig. 4.67c).

Chemical identification indicates that the mineralized layers contain up to 40 to 50% by weight of organic materials, depending upon the species. Clarifying the distribution and organization of such a considerable amount of the organic component has been a major

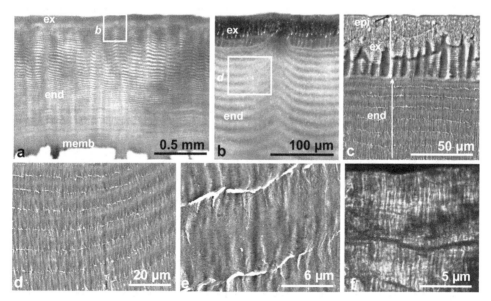

Fig. 4.67. Structure of mineralized crab cuticle. (a–c) Two distinct mineralized layers are visible on a polished surface: the exocuticle ("ex"), a few tens of micrometers thick, and the endocuticle ("end"), a few millimeters thick depending on species (optical view). The basal layer is the membranaceous nonmineralized layer, directly overlying the cellular mineralizing tissue. Thin epicuticle ("epi") is visible at the top of the microstructural sequence (c); (d–e) Etched endocuticle shows that the superposed layers are composed of mineralized units that are oriented perpendicular to the cuticle surface (d). This is irrespective of inflections of layers (e); (f) A thin section of the endocuticle layers (crossed-nicols observation) shows the small, parallel, densely packed rod-shaped units within each growth layer.

issue in understanding the relationship between the molting process and biomineralization mechanisms in the arthropod families that calcify (Travis 1963).

Perpendicular to the layering in the cuticle, a closely packed series of parallel canals has been established by Richards (1951). These canals are cylindrical expansions of the membranes of the cells of the basal epidermis. Passing through the three superposed layers of the cuticle, they connect the cells and the outermost cuticle layer.

Ultrastructural investigations carried out with the transmission electron microscope (Bouligand 1972) have shown that the calcified structures forming the main part of the carapace (i.e., the exocuticle and the endocuticle) include a dense series of chitinoprotein fibers, organized into distinct, superposed, and parallel layers. Within a given layer, fibers are parallel, but their orientation changes from a given layer to the next. Thus, a regular "plywood" structure is created. TEM observations suggest that each fiber is composed of a central core of chitin covered by a layer of proteins (Rudall 1963; Bouligand 1966; Neville 1975). In the exocuticle, fiber diameters are 100 to 250 angstroms. A comparable organic framework, with similar spatial arrangements, is present in the underlying endocuticle, but there the diameter of chitinoprotein fibers is greatly increased, to as much as 300 to 500

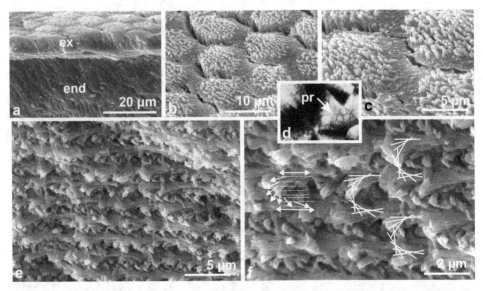

Fig. 4.68. Organic networks within the calcareous cuticle of a spiny lobster. (a) The thin exocuticle ("ex": only a few micrometers thick) shows a polygonal organization; (b–c) The polygonal network is characterized by dense mineralization, the spaces between the vertical canals also occupied by the mineral phase; (d) Portion of a published image by Giraud (1977) showing the prisms at the uppermost part of the exocuticle during the early crystallization phase in *Carcinus* sp.; (e–f) SEM view of the endocuticle showing the regular changes in orientation of the mineralized chitinoprotein fibers (see also Fig. 4.69).

angstroms (Giraud 1977). The regularity of directional changes between successive fiber layers leads to the appearance of a specific visual pattern, and is readily visible on sections perpendicular to the cuticle surface (Figs. 4.68f and 4.69c). This "arched" pattern has long been described as an original structure of the organic framework, but three-dimensional reconstructions of the fibrous "plywood" organization have clearly shown that this appearance is due to the regular changes in orientation between the parallel fibrous layers of chitinoprotein.

It is worth noting that the two-layered organization of the chitinoprotein framework is related to the two-step process of molting. The organic framework of the exocuticle is prepared *before* shedding the old cuticle, whereas the main part of the new cuticle (i.e., the endocuticle) is developed *after* molting. Of course, in both layers, mineralization occurs only after molting: the preexisting framework of the exocuticle thus appears as a preparative phase of the post-ecdysis stage (= after removal of previous cuticle), allowing the calcification process to occur almost immediately, making this critical portion of the animal's life as brief as possible.

From the biomineralization viewpoint, transmission electron microscopy and optical observations (polarized light) show that mineral deposition first occurs as radiating laminar spherulites just below the very thin epicuticle (Giraud 1977). In early stages of

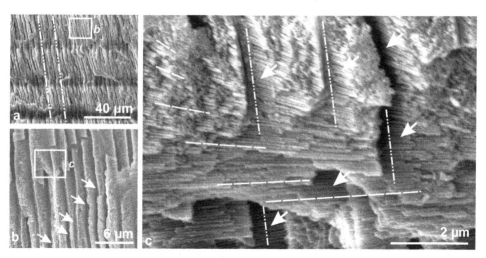

Fig. 4.69. Endocuticle layering, parallel pore canals, and orientation of fibers visible on a fracture surface perpendicular to the cuticle surface. (a–b) Longitudinal sections of the parallel pore canals that join the mineralizing cells to the epicuticle; (c) A magnified view reveals the fibrous pattern of the mineralized structure, suggesting a close relationship between mineral deposition and chitinoprotein fibers.

biomineralization, the "vertical canals" joining the cells of the mineralizing epidermis to the upper part of the mineralizing layers play a major role. In the outer part of the cuticle (the exocuticle layer), the units of mineralization form a polygonal network, with each unit cell of the network being about 5 μm wide (Figs. 68a–d). It was observed (Giraud 1977, p. 19) that this polygonal pattern corresponds to the dimensions of the cells of the basal mineralizing membrane. According to this observation, the mineralized polygonal framework seen within the polygons (Figs. 4.68b–c) is a direct expression of the cellular organization of the mineralizing epithelium.

In the deeper layers of the carapace (the endocuticle) this correlation no longer exists (Figs. 4.68e–f and 4.69). When mineralized, the structure of the cuticle is markedly influenced by the closely juxtaposed, parallel pore canals (Fig. 4.69a–b: arrows). With increased magnification the chitinoprotein fibers are seen to have a typically twisted arrangement, here impregnated by calcite (Fig. 4.69c).

Chemical data

The mineralogy of the cuticle is neither simple nor uniform The presence of calcium phosphate was noted earlier by Huxley (1879) in crayfish cuticle. Calcite is common; aragonite has also been reported (Prenant 1925; Vinogradov 1953) along with amorphous Ca-carbonate and phosphate. Comparison of recent X-ray diffraction patterns (Al-Sawalmih 2007) shows the distinctiveness of crabs and lobsters with respect to Ca-carbonate crystallization (Fig. 4.70a). Crab cuticle displays well-crystallized calcite, whereas the diffraction pattern of the lobster suggests a sizable proportion of amorphous Ca-carbonate present.

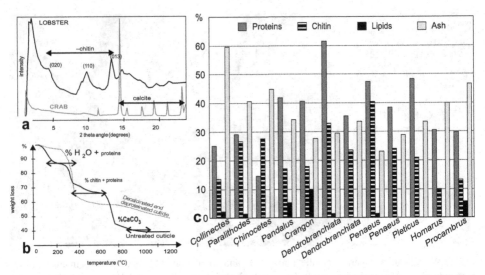

Fig. 4.70. Compositional data for some calcified arthropod cuticles. (a) Illustration of different crystallization state between crabs and lobsters; (b) Thermogravimetric profiles for a complete untreated carapace, and for the same sample after decalcification and deproteinization; (c) Diverse overall compositions in various crustacean cuticles. a, c: redrawn from Al-Sawalmih (2007); b: redrawn from Romano *et al.* (2007).

Biochemical information also varies greatly. Usually, β-chitin (the most widely distributed form of chitin in nature) is also the common form of chitin in crustacean cuticle. However, Raabe *et al.* (2006) found that in the lobster's endo- and exocuticle, chitin is the α type. Pratoomchat *et al.* (2002) reported the presence of α-chitin and β-keratin as the major organic phases in the crab *Scylla serrata*. Along with these organic structural frameworks, proteins with medium isoelectric points and 5 to 30 kDa molecular weights have also been recognized (Kragh *et al.* 1997). In crab cuticles, numerous proteins have likewise been found (Compère *et al.* 2002), with molecular weights of 10 and 110 kDa. They are glycosylated: O-linked oligosaccharides and N-linked mannose rich glycans.

The compositional data gathered thus far indicate great diversity with respect to both mineral and biochemical compositions. It is noteworthy that additionally mineralized cuticle is an essentially metastable structure. Between two molting processes, the mineralogy and crystallography are submitted to a cyclic process that first, increases its hardness and crystallinity (with variations in different body parts), and second, enables premolting remobilization of the innermost cuticle in the final part of the cycle (Drach 1939). Thus, even for a given species, a single and definite mineralogic and crystallographic status of minerals involved is difficult to assess.

Ultrastructural patterns of the mineral phase and the calcification process

Physiological events linked to the molting process are begun by the partial remobilization of endocuticle, whereas hardening of new cuticle begins within the first hours after ecdysis (shedding of the old cuticle). This involves first, the biochemical modification (tanning) of the proteins (Dendinger and Alterman 1983), followed by the calcification process. A five-step sequence has long been recognized, leading to a fully mineralized status (Drach 1939). As pointed out by Pratoomchat *et al.* (2002), data concerning deposition of the mineral phase are "very sparse." In an early stage (A), a complex melange of dicalcium phosphate and calcium octophosphate has been reported in the form of spherules (Pratoomchat *et al.* 2002, in *Scylla serrata*). Comparable spherules have been shown to be calcareous in other species (Ziegler and Miller 1997). Phosphate spherules are rapidly replaced by calcite and Mg-calcite in the following stage (B).

With respect to formation of the cuticle itself, Giraud (1977) characterized carbonic anhydrase activity in the cuticle, using the Haüsler-Hansson method paired with the Stoelzner method for imaging Ca concentration. Traces of the membranes of mineralizing cells could be recognized within the polygonal structures of the exocuticle layer. This suggests a leading role for these cells in the early phases of calcification. Accordingly, development of pore canals in the exocuticle layers (Figs. 4.68b–c) indicates a role for them in ion transportation. Further observations (Giraud-Guille *et al.* 2004) confirmed the presence of glycoproteins that have a possible calcium binding function.

These observations are significant with respect to the images obtained by atomic force microscopy (Fig. 4.71), showing the overall orientation of structures in the 10 nm range perpendicular to the layering of the cuticle (i.e., parallel to the direction of polygonal structures and pore canals). Rather limited information exists regarding the crystallization processes and factors influencing deposition of the mineral phase in cuticle. Travis (1963) reported mineral elements present in the vertical canals (actually cellular membrane expansions) but, as pointed out by Giraud (1977), the presence of solid mineral material within these tubules is improbable. Concluding a detailed physiological study of ion and molecule fluxes during the pre- and post-ecdysis periods, Cameron (1989, Fig. 11) suggested a purely chemical scheme of crystallization, without indicating which factors can drive the deposition of newly formed mineral units. Transverse and tangential sections (Giraud-Guille *et al.* 2004) fully confirm the role of the canal system in providing mineral elements to calcification sites and, in addition, the importance of the polygonal network formed by epidermal cells within the chitinoprotein framework. There is little doubt that the polygonal network provides preferential calcification sites for early deposition of the mineral phase. The fibrous framework also acts in guiding calcification. Giraud-Guille *et al.* (2004) noted that calcification can occur in free spaces between the fibers of the chitinoprotein framework as long as the space between fibers exceeds 300 angstroms. This conclusion is consistent with Fig. 4.69c that images mineralized surfaces of the fibers. On the other hand, microscopic observation of thin sections of cuticle under crossed nicols indicates that polarization of the submicrometer-sized mineral units is consistently oriented to their overall orientation perpendicular to the cuticle surface. Such a feature is consistent with AFM imaging of the units of micrometer-sized mineral materials (Figs. 4.71a–b).

With respect to the crystallization process, it is worth noting that no indication has been found of a mechanism for a saturation method of growth of grains, at any scale of

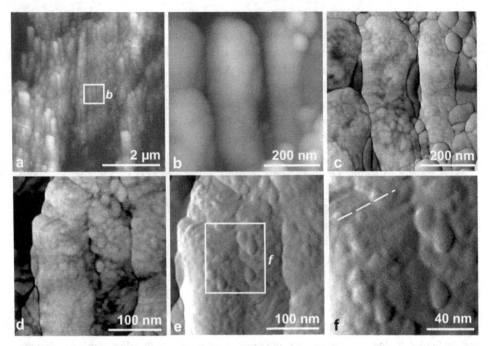

Fig. 4.71. AFM images of calcite units in the crab endocuticle. (a) Seen at a micrometer scale, the mineral units exhibit an overall orientation of their long axes perpendicular to the cuticle surface; (b–c) The elongate units have a soft, rounded surface (b: height image), but phase-contrast imaging reveals their formation by accretion of small mineral granules; (d–e) A closer view confirms the nanogranular structure of the individual mineral units; (f) Sometimes, weak indication of alignment of the nanogranules suggests that chitinoprotein fibers are also involved in the deposition of the mineral phase.

observation. The densely packed, nanometer-sized grains, basic units of the elongated micrometer units (see Fig. 4.71), do not show any trace of crystallization from saturated solutions, a process that always produces faceted crystals as they occur in polymer-driven crystallization of ions from solution (Grassmann *et al.* 2003). Consistent with the previous case studies in this chapter, this observation suggests that a purely chemical crystallization mechanism for the calcite in cuticle is improbable. Additionally, the abundance of amorphous Ca-carbonate in arthropod skeletons, and the long-recognized use of such organically stabilized forms of Ca-carbonate for mineral storage (Luquet *et al.* 1996), suggests that such a method is used in the transportation of mineralizing materials to calcification sites. The transfer of mineralizing material by tubular expansions of the cell membranes (the pore canals) in an organically stabilized form seems more probable than free ion transit of calcium ions in the animal's blood, as suggested by Cameron (1989). The overall orientation of these calcareous structures indicates that further evolution of this material after its passage to extracellular spaces is controlled by the polygonal cell membranes and pore canal bundles. As a first approximation, this could be considered as a "direct transfer" from mineralizing cells to surfaces of mineral growth, as suggested by Crenshaw (1980), here adapted to the particular structural conditions resulting from the molting process of the calcifying Arthropods. Note, however, that such control does not exist during formation of the endocuticle, the main part of the cuticle. As a result, no precisely defined microstructural units can be observed.

Microstructures in the carapace of barnacles

From a microstructural standpoint, this conclusion highlights the striking contrast between the undifferentiated cuticle of the crabs which, although highly crystallized, does not form distinct microstructural patterns, and that of the barnacles, in which five types of micro-structures (lamellar, granular, microcrystalline, radiating crystalline, and disoriented prismatic) have been characterized (Bourget 1987). Remarkably, histologists have been able to identify three major cell types within the mineralizing tissue which forms the barnacle shell plates, most notably chitin-producing cells and matrix-secreting cells (Costlow 1956; Bocquet-Védrines 1965; Bubel 1975; Klepal and Barnes 1975a). This heterogeneity suggests that formation of variously structured shells in the barnacles is dependent on cellular metabolism and biochemical secretions.

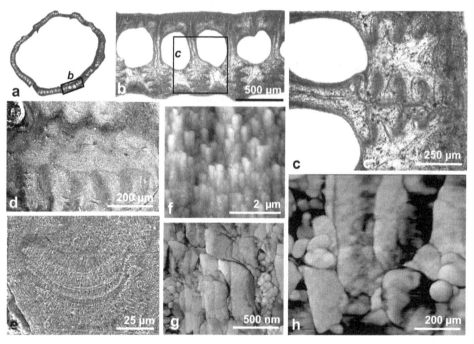

Fig. 4.72. Layered growth of barnacle shells. (a–c) Microstructure of the plates. The distinct microstructural patterns in the internal portions of the shell ring are taxonomy-linked; (d–e) SEM views of the fibrous microstructure, growing by cyclic processes that produce micrometer-thick growth layers; (f–h) In AFM views, higher magnification indicates that the skeletal crystallites are similar to equivalents in crab cuticle, both being constructed of fused grains.

In contrast to the weakly differentiated crab and lobster carapaces, barnacles produce long-recognized microstructures with taxonomy-linked three-dimensional arrangements of the microstructural units (Fig. 4.72). Closer proximity between the mineralizing tissues and the developing calcified structures in the barnacles, as shown by Bourget (1977), probably allows formation of these well-defined microstructures, thus differing from what can be observed in the main part of crab cuticles. Unexpectedly, microstructural differences between the calcareous structures produced by these two groups of Arthropods once more exemplify the importance of Crenshaw's concept of a "direct transfer" in the formation of well-defined skeletal units.

4.9 Conclusion: the overall distribution of a calcification process based on layered control of carbonate crystallization at a micrometer scale and the resultant formation of organomineral reticulate structure at the nanometer scale

Although not exhaustive, this survey of various calcareous materials produced by taxonomically distant organisms shows that comparable organomineral patterns are visible within these skeletal structures when observed at a 10 nm scale. As interpreted in terms of the calcification process, this suggests that maintaining controls over Ca-carbonate crystallization requires strongly convergent methods. Fish otoliths commonly show features that reflect the consequences of control deficiency, as this immediately results in visible development of aberrant crystal-faceted units (see Figs. 4.56c–e). In contrast, atttention must be drawn to the fact that, whatever the scale of observations, no traces of "classical" crystallization have been observed. In spite of an overall polarizing behavior for all this calcareous mineralization, some of them are clearly organized in well-ordered microstructural units, whereas others are essentially made of disordered granules. In all cases, the crystallized elements always grow by formation of very specific reticulated material in which reticulated rounded units and irregular organic coating are indicated at the 10 nm scale. Bearing in mind differences in modes of secretion and the resulting diversity of these skeletal architectures, this suggests that interpretations resulting from the detailed comparison between mollusc and coral skeletons can be extended to any biogenic calcareous structure.

In moving onward in this review, it is now reasonable to compare the conclusions suggested above by this broadly based, "top-down" method of investigation, to results that have been obtained by research developed following an opposite, "bottom-up" approach: chemical and biochemical methods that strive to understand the relationships between organic and mineral components of structures through analysis and in vitro experimentation.

5

Connecting the Layered Growth and Crystallization model to chemical and physiological approaches

Ongoing conceptual changes in biocalcification

Research on the formation and properties of mineralized structures produced by living organisms is characterized by the diverse origins, methods, and aims of scientific investigators. This has resulted primarily from the rapid development of analytical methods in the second half of the twentieth century. As has been emphasized above in preceding chapters, the contrast between shell structures was never thought greater than at the beginning of this period. At that time, structures were still described with the optical microscope in the same way as Bowerbank had, yet, at the same time, were analyzed with the latest developments in theoretical and applied physics. These provided new data on isotopic fractionation in the calcium carbonate of the same shells.

Here we should emphasize that underestimating optical studies is a serious error. The logic of interpretations concerning the formation and method of growth of calcareous biominerals was influenced by their crystalline appearance in microstructural units, as observed with the polarizing microscope. It is important to acknowledge the great accuracy and precision of descriptions carried out at the end of the nineteenth and early twentieth centuries, and acknowledge our debt to these early researchers for the development of many of the major concepts in the field of biomineralization. Bourne (1899), in a seminal paper dealing with the formation of skeleton in the Anthozoa, carefully described the crucial zone of contact between the calicoblastic ectoderm and the corallite skeleton. He was the first to make the formal statement (p. 534) that "no calcium carbonate crystals can be seen within the calicoblastic cells themselves." This was controversial at the time, as Ogilvie supported the hypothesis of calcification within cells. Then, later in the same paper (p. 538), he noted that "the spine [= fiber] grows by the addition of minute particles of carbonate of lime which crystallize out of an organic matrix – or colloid" (p. 539) secreted by the calicoblastic cell layer, "and attach themselves to the crystalline structure already present and become oriented conformably to it." To Bourne, the carbonate of lime "must pass through the membrane" to crystallize, a view that led him to name the translucent layer between the calicoblastic ectoderm and the crystallized Ca-carbonate as "matrix."

In this outstanding example of microscopic study, Bourne observed that the translucent "membrane" is not continuous but rather, "better visible at the growing tips of the spines," exactly as has been recently confirmed, more than a century later, by labeling coral fibers by growing them in seawater enriched with [86]Sr. Mapping the distribution of the incorporated isotope by using a secondary ion mass spectrometer with the highest resolution presently available (CAMECA NanoSIMS) permits the demonstration of the high concentrations of [86]Sr at the tips of fibers, revealing their zone of maximum growth and confirming Bourne's observation completely (Houlbrèque *et al.* 2009). In addition to the first occurrence of the term "matrix" used as the name for the crystallization medium, it is also worth noting that in Bourne's view, the translucent "colloid layer" observed between the calicoblastic epithelium and the skeletal fibers clearly had no relation to the liquid layers "close to seawater" imagined by some later authors.

Further progress in analyzing what occurs physically and chemically within that micrometer-thick translucent layer was simply not possible at that time. Half a century later, when Goreau (1956) characterized the polysaccharide layer external to the calicoblastic epithelium, he still experienced the same difficulty that Bourne did: to demonstrate clearly its role in skeletogenesis. Therefore, during the long period of time between the peak of optical studies and the diversification of physical and chemical methods that now enable *in situ* characterization (e.g., the first TEM observation of the ultrastructure of the matrix layer made by Johnston in 1977, see Fig. 2.40), most investigators aiming at reconstruction of the biomineralization process had no alternative to using indirect approaches. This could only be achieved by influencing the complete live organism by changing its living conditions, or by in vitro chemical experimentation conducted with chemical compounds extracted from biological carbonates. One could then infer the role of these compounds from their specific properties, or use them as part of crystallization experiments.

Such multiple approaches resulted in a huge accumulation of data covering a wide range of areas, from physiology and cellular biology to the physical study of interactions between mineral surfaces and organic components or, reciprocally, between complex organic molecules and mineral ions. Apart from a series of reports issued at regularly spaced meetings dedicated to biomineralization that allowed direct access to the results of this broad spectrum of approaches, data are dispersed through diverse journals that are as varied as the methods used by investigators. The observation has to be made that only a very restricted number of general concepts or reliable models emerged from this continuous flux of publications, surprisingly based on ever more restricted categories of materials studied. For example, there is a striking contrast between the generalized use of the "nacre-prism" model, largely considered as the prime reference for discussing biomineralization processes on the one hand, and the actual proportion of shells constructed on this microstructural scheme on the other hand.

5.1 Summary of chemical research on organic compounds involved in calcareous biomineralization and relevant models for biocrystallization

By the early part of the twentieth century, it had become obvious that the organic compounds permanently associated with biologically produced calcareous crystals must constitute essential agents in their emplacement. Thus, it followed naturally that, in parallel with microstructural studies with a "top-down" approach, researchers developed chemical approaches that sought first to analyze the way by which these organic compounds might influence the process of crystallization, and, more ambitious, attempted to reconstruct these organomineral assemblages in vitro. Initially motivated by the analysis of structures of biological origin, this biogeochemistry of crystallization has been extended to the analysis of all combinations that involve an organic compound and a mineral ion. A typical example is provided by the ferritin molecule, a nanocage protein having a diameter of approximately 80 angstroms, formed of 24 associated subunits. Understanding the conditions and mechanisms controlling the reversible association of iron to this complex organic assemblage is a typical example of biomineralization research conducted at the molecular level. From occurrences of a natural type, a rapidly expanding research area developed, bringing with it numerous potential practical applications. In this research, organomineral combinations studied are often very far from those in naturally occurring biological products and mineralized structures. Long ago, Chave, who participated in pioneering research on natural biomineralization as an environmental tool, was irritated (seemingly) by this research direction when he noted "There are many strange experiments and measurements carried out in the name of biomineralization" (1984). Such an overly broad concept of biomineralization research is exemplified by a special issue in 2008 of the *Chemical Review* dedicated to this topic.

From a simple quantitative point of view, the surprising power of organisms to generate huge amounts of crystallized calcareous material (commonly in environmental conditions inimical to such crystallization) led early investigators to focus on the role of biologically secreted molecules that could potentially influence the process of crystallization. The discovery of carbonic anhydrase by Meldrum and Roughton (1933) provided the first example of a possible *calcification-enhancing* molecule. This enzyme accelerates the conversion of carbon dioxide to bicarbonate, thus enabling formation of Ca-carbonate. Their pioneering paper, clearly emphasizing the potential of this enzyme for assisting in shell formation, resulted in a series of investigations aimed at an exact definition of this function and its potential for explaining deposition of calcite or aragonite. Freeman and Wilbur (1948) made a detailed study of twelve bivalves and eight gastropods, and their conclusions supported this function of carbonic anhydrase. However, they also noted that in some species with well-developed shells, such as *Atrina* or *Crepidula*, concentrations of carbonic anhydrase are hardly detectable. Stolkowski (1951a, b), while showing that anhydrase concentration is higher during periods of maximal shell growth in *Helix*, further supported this interpretation. An additional observation was made that anhydrase inhibitors prevent shell regeneration. Actually, carbonic anhydrases form a family of molecules widely distributed among living organisms, where they are involved in various biological functions unrelated to biocalcification. For example, in their extensive study of the sea urchin *Strongylocentrotus purpuratus*, Livingston *et al.* (2006) identified 19 genes encoding for carbonic anhydrases, but only one of them included the signal for secreted molecules (i.e., potentially active outside the cell).

5.1.1 Early development of a template model for biocrystallization

In parallel with investigations of the physiological mechanisms affecting Ca-carbonate precipitation, the early 1960s were also marked by more structural approaches to the biomineralization process. The concept of crystallization by ionic saturation could explain crystal formation, but the visible three-dimensional ordering of skeletal units obviously demands some additional factors acting on a broad scale to create specific arrangements for each type of shell layer. At that time, the long-recognized membranous structures containing elevated amounts of nitrogen (Schlossenberger 1856) were selected to serve as potential agents of this function. Based on observations using the transmission electron microscope, Grégoire (1960) suggested that these membranes might have played the role of "template" for the development of calcareous crystals. Following this concept of crystallization template, the polycyclic microstructural units would have had to be built by an alternation of mineral materials crystallizing on organic substrates having properties that could simultaneously influence their mineralogy and the crystallographic orientation of mineral components. It is noteworthy that the matrix layer identified by Bourne (1899) was not regarded as a template. Bourne described the mineral particles assembling themselves *within* the matrix layer, not at its surface.

Although simple alternations between purely organic substrates and purely crystal layers are not confirmed by modern submicrometer description of biologically produced carbonates, the "template" scheme still exists nearly in its original sense, and is used in experiments. The concept has been considerably expanded by experiments utilizing artificial organic membranes, with the objective of forming "nanocomposites," materials with vast possibilities for application (see Mann *et al.* 1989, for a review). An extension of this approach has been to replicate, in vitro, the formation of these template membranes by a process of self-assembly. This is obviously an important step in the biomimetic goal of seeking to duplicate the crystallization process. Thus, Langmuir layers are used as substrates to determine the characteristics of mineral materials (for review, see Fricke and Volkmer 2007). The essential advantage of this method is that of procuring chemical substrates with a known structure at the molecular scale. Volkmer (2007) clearly pointed out that if this research had as its objective the formation of structures identical to natural materials, such monomolecular layers have an "oversimplified character."

5.1.2 Discovering the biochemical complexity of skeletal matrices

The complexity of molecular assemblages found in biogenic calcified structures became rapidly obvious because of progress in analytical biochemistry. This has led investigators to search for accessible organs in which crystallization could occur, so that they could directly analyze the chemical properties of these specific localities. The liquid phase present in some places between the internal shell surface and the outer mantle in the Bivalvia was long considered a nearly ideal experimental site. Easily accessible, this liquid was analyzed by numerous authors. The role of the outer cell layer of the mantle in determining the compositions of this liquid compartment has been thoroughly explored, from Manigault (1939), to Stolkowski (1951a, b), to Crenshaw (1972a), and to Wada and Fujinuki (1976). These authors, and also Simkiss (1965), outlined an experimental approach to identifying the function of organic matrices in calcareous biocrystals, emphasizing the association between carbonic anhydrase as the *precipitating agent*, and the role of the organic phase as the *substrate for crystallization*, thus determining the specific mineralogy found in different layers in a given shell. This conclusion was drawn from experiments showing that different results can be obtained using extrapallial fluid or organic matrices extracted from shells. This approach was developed during the following decades after the discovery of the "soluble portion" of organic shell matrix.

By the end of the 1960s, it was shown (Voss-Foucart 1968) that a part of the organic matrix was easily separated from the more dense fibrillar or lamellar organic components observed thus far by use of the transmission electron microscope (e.g., Grégoire *et al.* 1955). Crenshaw (1972b) showed that in the supernatant fraction after centrifugation, organic molecules existed and that, after separation, they continued to show an elevated capability for fixing calcium ions. This discovery has proven to be of fundamental importance. Since this discovery, the ability to separately analyze the two phases that contribute to the mineralization process allows us to determine their distinct compositions (Weiner and Hood 1975; Krampitz *et al.* 1976). The specific and taxonomically diverse compositions of such soluble matrices have been well established, in most part based on amino acid analysis, but also by use of electrophoresis (Weiner *et al.* 1977). In several respects, the evidence of the two types of organic compounds in matrix provided a crucial point in the evolution of research on biomineralization mechanisms, and resulted in somewhat unexpected findings.

The high percentage of acidic amino acids in the soluble part of matrices determines the presence of proteins having very low isoelectric points, as compared to "mean intracellular proteins." This exceptional composition suggests their potential mineralizing role as a possible ligand of calcium ions, and thus perhaps initiating the crystallization process. This was also illustrated, for instance, by Mitterer (1978) for the skeletal matrix in corals. Thus, the concept of "organic template" with its epitaxial function of nucleating crystallization was transferred from the membranous substrates to molecular assemblages that were not necessarily forming "solid" or even visible structures.

Evidence of a chemical interaction between soluble matrices and mineral ions in solution was provided by De Jong *et al.* (1976), who used a marked solution containing radioactive calcium ions. When a chromatographic elution of organic molecules (previously having been placed in the radioactive solution) is carried out, an increase in radioactivity indicates the passage of organic molecules, which were thus identified as carriers of radioactive Ca. A similar investigation, with measurement of pH change in a solution where Ca-carbonate precipitates, provided Wheeler, George and Evans (1981) with comparative diagrams to illustrate this interaction. When calcium chloride is introduced into a solution of bicarbonate of soda, sodium chloride is formed and Ca-carbonate precipitates, immediately lowering the pH of the solution. When soluble organic materials obtained from

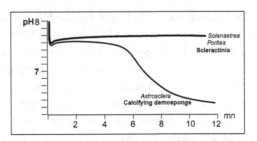

Fig. 5.1. Diagrams of lowering of pH during Ca-carbonate precipitation in the presence of soluble matrix material from scleractinian corals and from a calcifiying demosponge, utilizing the Wheeler experimental process.

calcareous biominerals were added to this initial solution of bicarbonate, it was observed that the lowering of pH is slowed to a greater or lesser degree. Figure 5.1 presents typical examples of this retardation, as produced by several organic matrices, originating both from corals and from a calcifying demosponge. The difference in the profiles expresses the clear and distinct ability of organic matrices from corals and sponges, respectively, to capture Ca ions from solution, thus resulting in slowing Ca-carbonate formation and pH decrease.

It is important to note that, due to the capability of these "soluble" molecular assemblages to interact with ions, they are able to play two opposite roles. They can capture free ions in solution and promote crystallization, or conversely, they are able to fix themselves on solid surfaces of growing crystals, thus inhibiting further growth of the mineral unit. This is the first step in developing a broad area of investigation: the study of the origin and results of varying atomic or molecular forces at the interface between mineral surfaces and potential mineralizing compounds.

In a concept of a microstructural unit developing as one or more growing crystals, some shaping process must be applied in order to attain species-specific morphologies. Soluble matrix molecules offer the possibility of providing this crucial function. One can reasonably hypothesize that species-specific combinations of molecules would be able to produce the necessary morphology of microstructural units in a given shell layer. This is the beginning of the "biomimetic" approach (see below, Section 5.4).

However, Crenshaw (1980), after having repeatedly analyzed the liquid compartment located between the mantle and the shell of the mollusc, was led to the conclusion that these investigations "have contributed little to understanding of the shell formation." Having made the observation that most analysts previously took extrapallial liquid from the central portion of the shells and not from the marginal parts of shells (where most of the calcification occurs), he formulated the statement (counter to accepted opinion of the time) that between the mantle and the growing surface of the shell, "the transfer of material is essentially direct" (1980, p. 120). The importance of this statement has not been generally recognized, as indicated by the number of anatomical models still being published that show broad spaces containing crystallization fluid between the mantle epithelium and the growing shell. This conclusion was formulated the same year as a paper entitled "Compartment and envelope formation," by Bevelander and Nakahara (1980), in which a general biomineralization hypothesis was proposed.

5.1.3 The "compartment and envelope" hypothesis

A first "compartment" hypothesis had been proposed earlier by Bevelander and Nakahara (1969), based on a transmission electron microscope study of several nacreous layers. As later noted by Crenshaw (1980, p. 120), it has "generally fallen into disfavor." At the time of Crenshaw's remark, numerous biochemical identifications (most of them amino acid analyses) had shown that a significant compositional difference exists between the "soluble" part of the organic component, admittedly more closely linked to the mineral phase, and the less acidic external envelopes surrounding microstructural units. Having obtained spectacular images of entire skeletal matrices of prisms and nacre by transmission electron microscopy, Bevelander and Nakahara suggested that this compositional difference correlates to the very distinct appearance of the two types of skeletal matrix. This led them to formulate a new, more complete version of their previous hypothesis: the "compartment and envelope" concept (Bevelander and Nakahara 1980).

To summarize this, transmission electron microscopy observations indicate that the mineralized part of the shell comprises a continuous network composed of small circular or ovoid "envelopes" (e.g., Bevelander and Nakahara 1980, p. 22, Fig. 3), interpreted by the authors as the "organic layers covering the successive initiation of new crystals." An important point is that, in their terminology, "envelopes" are described as the intracrystalline organic component. Thus they indicated that the "soluble part" of the matrix is not composed of molecules dispersed within the mineral matter, as they would necessarily be in the "shaping by adsorption" hypothesis, but rather, they are organized into readily visible shapes (although surprisingly irregular). Conversely, the solid external features surrounding mineralized microstructural units, commonly referred to as "envelopes," were called "wall" by Bevelander and Nakahara. In their hypothesis, the "envelopes" were actually active in the calcification process, although no further information was provided by the authors on the origin of the envelope structures and the mechanisms leading to carrying out the role hypothesized. Whatever the further influence of this theory, the basis of the experiment was suspected of errors, such as inadequate fixation. Our attention focuses on two points that are not interpretations, but are directly shown on electron micrographs provided by Bevelander and Nakahara. These are as follows:

(1) In contrast to any suspected inadequate fixation, one should observe that the sizes and shapes of the circular or ovoid "envelopes" visible on the TEM preparations illustrated by Bevelander and Nakahara (e.g., 1980, p. 22, Fig. 3), exactly match the dimensions of the organic coating of the nodular elements that form the reticulate crystals now established as the basic components of all skeletal units examined by AFM microscopy. As we have seen through a comparison of AFM and TEM imaging (see Figs. 3.1 to 3.10, for example), the organic features visible there, coating the mineral elements of the reticulate crystals, have shapes that approach irregular cylinders that could

be regarded as the "circular or ovoid envelopes" of Bevelander and Nakahara when they are seen in ultrathin TEM sections.

Doubtless, from a purely structural viewpoint, the TEM images of Bevelander and Nakahara provide strong and unexpected support for a new model, here termed "layered growth and crystallization." This conclusion is reinforced by the second and more detailed version of their model, in which Bevelander and Nakahara (1980) added the observation that identical types of "envelopes" are observed within mineralized parts of both nacreous layers and calcite prisms. This is exactly what is shown in Fig. 3.10, where the nodular units of calcite prisms and nacreous aragonite are clearly indistinguishable. Naturally, this geometric similarity leads to emphasizing the difference in their origin. In the Bevelander and Nakahara hypothesis, the nanometer-sized "envelopes" *exist prior to calcification*, whereas in the layered growth and crystallization model, they are simple coatings of the crystallized part of the growth layer, segregated from the mineral lattice during the crystallization step and concentrated at the periphery of the reticulate crystals (also, see Chapter 4 conclusion).

(2) Demonstrating a spatial three-dimensional organization for the "intracrystalline" soluble matrix disproves the concept of crystal shaping by continuous adsorption of freely moving organic molecules interacting with specific mineral sites on growing crystals. Such a mechanism, derived from the Wheeler experiments, requires crystal growth in a liquid medium, as is actually the case in in vitro "biomimetic" research on biocrystallization.

Studies of the fine structure of biological Ca-carbonate and the resulting extensive sets of data reviewed in previous chapters strongly suggest the general distribution of the reticulate aspect of mineral structures, along with their organic coatings (and consequently Bevelander and Nakahara's "envelopes"). These correlate well to the TEM structural data obtained by Bevelander and Nakahara, thus supporting the conclusion that these are not artifacts of preparation, as suggested immediately after the images were published. Now, AFM observations have enabled us to check their hypothesis concerning the mineralization process itself, and one part, "compartment" formation by resistant "walls," cannot be generally applied to calcareous microstructures.

In some instances, such as in the prismatic envelopes of *Pinna* (Figs. 2.50c–d), observations indicate that the strong lateral envelopes (i.e., the "walls" in Bevelander and Nakahara's terminology) actually provide morphological limits to lateral growth, and thus are clearly distinct from areas of mineralization. Such cases illustrate best the "compartment and envelopes theory": the overall shape of the microstructural unit is related to taxonomy through the specific morphology of the "wall" (*sensu* Bevelander and Nakahara).

Envelopes can also be limiting membranes, but apparently lack sufficient influence to determine the shape of microstructural units, as the formation of reticulate crystals occurs during the final steps of forming the growth layer. Here for example, in the foliated units of internal layers of *Pecten* or *Ostrea*, the morphology of these calcareous units is in large part dictated by that of the Ca-carbonate polymorph (see Figs. 3.37 and 3.38). The insoluble part of the coral skeleton has been characterized by *in situ* fluorescence and by immunological labeling (see Fig. 2.38). These examples in corals show that the insoluble materials surround the fibers; they are also less acidic than the "soluble part" of the skeleton organic matrix. In agreement with the terminology of the "compartment and envelopes" theory, these insoluble matrices are forming "compartments" but their shaping influence is weak, or nonexistent.

It can also be noted that these envelopes are not very resistant to decalcification by acidic solutions. The observation has commonly been made that, during the final step of the preparation process, considerable amounts of their mass become soluble.

A third case is exemplified by the Levi-Kalisman *et al.* (2001) model for formation of nacreous crystals. Here, an insoluble substrate composed of chitin and proteins also contains an additional layer containing numerous acidic amino acids. This assemblage forms an insoluble but active substrate for crystallization. In addition, it is noteworthy that nacre itself is not an homogeneous material. Several important points are yet unanswered in the debate regarding the initiation of growth of nacreous tablets, for example, the positioning of centers with respect to underlying structures, and whether the apparent crystallographic continuity between superposed layers is overly simplified.

5.1.4 The recent decades: progress in biochemical models, development of physical studies of organomineral interactions, and importance of the genomic approach

Although the presence of sugars has long been noted in the biochemical composition of skeletal matrices (e.g., Grégoire 1967, 1972b), Crenshaw (1972b) determined that soluble matrix from *Mercenaria* contains about 20% carbohydrates, emphasizing the role of organic matrices, and what is rarely taken into account: the duality of its biochemical composition. The Weiner-Hood model (1975) was based on distances between β-carboxyl groups of aspartic acid in the peptide chain (a pure protein model for nucleation). As an alternative possibility, Crenshaw and Ristedt (1975) identified a sulfated polysaccharide calcium binding site present in the center of nacreous tablets. A few years later, the role of the sulfated portion of the skeletal matrix in determining the formation of Ca-carbonate units was stressed by Greenfield, Wilson and Crenshaw (1984), on the basis of the previous observation of Thiele and Awad (1969), who favored the formation of well-ordered crystals by "ionotropy" (i.e., formation of crystals around dispersed sites). This model requires a very precise spatial arrangement of calcium binding sites (also required by the peptide chain model).

Addadi *et al.* (1987) have also proposed a "cooperative" model for crystal nucleation, with a separate role provided for the sulfated polysaccharides (an "attraction" function, creating localized supersaturation), whereas the α-carboxylate chains play an essential role in fixing orientations of the crystallized units. Reporting research on the regulation of crystallization by matrices in oyster shells, Wheeler, Rusenko and Sikes (1988) summarized the different hypotheses suggested to that date, concerning ways by which mineralizing matrices have been hypothesized to regulate growth of microstructural units, "by initiation of crystal growth, determination of crystal polymorph, control of crystal shape, or termination of crystal growth." They also emphasized how difficult a comparative approach was, owing to "the lack of agreement on matrix composition." They noted that methods used at that time for decalcification of shells in order to separate included organic material into fractions, resulted in a variety of results that made reliable analyses of the control method and process "practically impossible." They were also of the opinion that the distinction between soluble and insoluble matrices itself had certainly been overestimated, because "a large part of insoluble matrix has an amino acid composition similar to soluble matrix," and after solution in basic liquids, it does not show any great difference in chromatographic profiles or "crystal growth inhibition."

This final expression summarizes the concept shared by numerous investigators that a given microstructural unit, *as a crystalline structure*, must be explained first by identifying a nucleation process. This is then followed by a growth process, during which normal growth by saturation of the medium is balanced by localized inhibition of growth by specialized molecules to reach the species-specific size and shape for each type of skeletal unit.

This sort of concept of the biocrystallization process led to a series of in vitro experiments using major organic components extracted from mollusc shells as reactive agents to control the development of crystals produced in saturated solutions. A mixture of α-chitin and silk fibroin for substrate and aspartic acid-rich soluble matrices were thus assembled in vitro (Falini *et al.* 1996). Scanning electron microscopy and Fourier Transform InfraRed (FTIR) characterization of minerals resulting from various combinations of these revealed a rather good correlation between artificially produced polymorphs and the original mineralogy of the shell that was the source of the matrix materials employed.

In the same year, Belcher *et al.* (1996) used matrices from *Haliotis* shells for comparable experiments, showing that the matrix proteins were able to induce sequences of calcite or aragonite, switching the orientation and the nature of the Ca-carbonate polymorph without a membrane substrate.

It is also through in vitro experimentation that Aizenberg *et al.* (1997) established the specific interaction of soluble matrix materials extracted from sea-urchin spines on growing crystals of calcite (synthetic or biogenic). Comparable experiments by Albeck *et al.* (1993) showed that it is possible to demonstrate multiple interactions between crystal surfaces and organic compounds in solution. This showed the diversity of adsorbed molecules, depending on the lattice surfaces exposed and the resultant modification of growth, the last leading to formation of curved surfaces. This point is important, as it corroborates the remarkable characteristic of broken surfaces in biogenic sea-urchin calcite, which do not have normal rhombohedral cleavage, characteristic of pure calcite crystals (see Figs. 4.48g–i). In a more detailed study that included single layer templates, Aizenberg (2000) tested a number of different ways it is possible to constrain the crystallization process. Atomic force microscopy now allows viewing surfaces while they are growing, and measuring forces with unprecedented accuracy and precision. This new capability leads to the re-examination of basic processes, such as structural consequences at a molecular scale, of ion interactions between crystal surfaces and mineral ions or small molecules. For instance, Davis *et al.* (2000) and Teng *et al.* (2000) have both investigated the local increases in solubility due to fixation of Mg ions on to crystal surfaces with resultant changes in morphology. For the first time, measurement of forces and observation of the lattice itself are possible at a relevant scale, and in addition, the influence of the presence of mineral ions on crystallization can be observed (Davis *et al.* 2004).

In addition to the appearance of such new instrumentation that permits direct observation of crystal surfaces, research on skeletal matrices has also been marked by the progressive use of methods of molecular biology. For instance, a recent re-examination of the nacreous-prismatic model (Nudelman *et al.* 2007) combines the classical approach of organic matrix extraction and amino acid analysis with *in situ* identification and information on the distribution of chitin through the use of specific markers obtained from molecular biology. They found that, in addition to *in situ* characterization, high performance SEM views provide images of prisms growing by accumulation of "densely packed particles," thus fully supporting the "layered growth mode" of the prism. In their discussion, Nudelman *et al.* (2007) emphasized the paradox between the mode of growth of the prisms and their final appearance as monocrystalline. They were thus led to the conclusion that "this paradox can be conceivably resolved if we assume that the *mineral crystallizes after deposition.*"

In the Nudelman *et al.* model (see Fig. 5.7), mineralization of nacreous tablets is presented as a single-step process that occurs between two sheets of "interspaced α-chitin," forming "a scaffold onto which the hydrated gel-like acidic proteins are adsorbed and form the crystal nucleation site." This description of nacre mineralization uses the common concept of crystal growth, except that a first step formed by "colloidal mineral particles" has been substituted to control development of purely crystalline units. This proposed method for growth of nacre was inspired perhaps by the famous "pyramidal" growth of nacre tablets between preexisting organic layers (the classical example being that of nacre in *Haliotis*). However, as pointed out by Nudelmann *et al.* (2007), nacreous material differs greatly between major taxa, and thus does not allow hypothesizing an identical mode of growth for nacre tablets that are structurally very different. In the case of the nacre-producing bivalves, the proposed description does not fully account for other classic micrographs showing the development of tablets from an initial site at the growth front of each layer, or for growth figures that commonly show the formation of polycrystalline tablets through the repeated addition of more or less concentric growth layers (Fig. 3.29). However, owing to the methods used and the inclusion of some innovations, this improved description of a basic model for biomineralization served as a good transition between previous syntheses of classical analytical biochemistry and the new, rapidly evolving approaches employing the innovative methods of molecular biology and genomics.

5.2 The genomic approach: a new scale of biological complexity and perspectives for understanding environmental influences on biomineral formation and growth

During the last few years, the slow progress in synthesizing data resulting from a half century of analytical studies on organic compounds extracted from calcified structures has been revolutionized by the occurrence of the genomic approach, as it is applied to cells forming the mineralizing epithelium. Suddenly, instead of a few tens of proteins isolated from matrices of different taxa – often incompletely described and identified – gene sequencing and molecular biology techniques reveal thousands of sites having a potential for involvement in the biomineralization process, and the amazing complexity of an intra-cellular matrix-preparation event or sequence of events. Rapidly, it has become obvious that the process producing the organic phase that controls calcification outside the cell cannot be inferred simply from the biochemical analysis of mineralizing matrices themselves. This had been the only approach available from the beginning of biomineralization studies.

As pointed out by Read and Wahlund (2007, p. 227), "It is clear that the traditional approaches directed towards identifying the molecular players in biomineralization pro-cesses do not possess the power of genomic and proteomic approaches for discovering multiple gene, protein, and metabolic pathway interactions that must occur during this complex process." This is doubtless true, and although genomic approaches had only been developed a few years earlier, research dealing with gene expression in mineralizing cells is already providing a molecular panorama that is amazingly complex. Four decades of biochemical investigation that focused on the organic phase included within minerals (whose growth they have driven) has largely shown their diversity, but the number of fully identified proteins remains so far very small; only some tens at best, distributed between a few different taxa. Hard data on their actual role and method of action in the process of forming microstructural units are even more scarce.

5.2.1 *The surprising result of a comparison of the proteins and genes involved in mineralization of different taxa*

An illustration of the huge difference between data resulting from "traditional approaches" and that coming from genomic analysis is provided by a paper from Jackson *et al.* (2006), dealing with mineralizing activity by the mantle of *Haliotis asinina*, a member of the group of gastropods remarkable for the microstructural and mineralogic diversification of their shells (see Figs. 1.14–1.15, 2.5). Summarizing the results, Jackson wrote, "25% of the genes expressed in the mantle encode secreted proteins, indicating that hundreds of proteins are likely to be contributing to shell fabrication and patterning." Additionally, these proteins are novel and species-specific in a very high proportion. There is no need to emphasize the correctness of Read and Wahlund's statement concerning the difference between this first result dealing with the *Haliotis* shell and analytical results provided by the "traditional approach" (see Chapter 2).

The availability of such a breadth of information allows for detailed and broadly based comparisons to be made, the results of which are frequently intriguing. As an example, phylogenetic proximity between vertebrates and echinoderms has long been suggested (Grobben 1908) and more recently developed (Jefferies 1986), mostly based on interpreta-tion of a very primitive group, the Homalozoa (Whitehouse 1941), formerly named the carpoids. However, this interpretation is not supported by recent data obtained from study of genomic expression during the development of *Amphioxus*, or by sequencing of the 18S-rRNA (McClintock *et al.* 1994). The hypothesis of the vertebrate–echinoderm phylo-genic relationship has also been tested by taking into account gene encoding for proteins of the skeletal matrix in *Strongylocentrotus purpureus* (Livingston *et al.* 2006). The result of this comparison is unambiguous: contrasting to the hypothesized phylogenic proximity, no similarity is found in the extracellular proteins involved with the skeletal mineralization of vertebrates (Wilt and Ettensohn 2007, p. 206).

Furthermore, a comparison of proteins involved in controlling Ca-carbonate deposition has been carried out between two invertebrate groups (molluscs and calcareous sponges) and the result likewise does not provide any evidence of significant similarity (Wilt *et al.* 2003). Perhaps there is a touch of disappointment in Wilt and Ettensohn's statement that "it is mostly premature to pursue these comparisons very deeply" (2007, p. 206). Interestingly, and in contrast to Read and Wahlund's earlier statement, Wilt and Ettensohn emphasized the need for "more detailed structural analyses and combined structure–function studies" to improve our understanding of the methodology of Ca-carbonate deposition: that is, the reconstruction of what occurs *outside the membrane* of the mineralizing cells.

Attention should be drawn to the clear contrast between the marked differences in the parts of the genomes involved in the biomineralization process in various major taxa, and the striking similarities shown by the resulting materials observed at the submicrometer scale. Surveying the skeletal ultrastructures illustrated in Chapters 3 and 4 establishes that the supramolecular control of Ca-carbonate deposition is carried out by the same type of organomineral frameworks that are often practically indistinguishable in their three-dimensional organization between distant taxa. It now becomes apparent that these similar organomineral assemblages are actually the results of distinct genomic activities and vary between major taxa.

As stressed by Wilt and Ettensohn (2007), the diversity of the genomic sectors involved in biomineralization, as deciphered in various taxa, demonstrates that mineralizing phyla have in common "only basic strategies" concerning the biochemical composition of the organic framework required to control Ca-carbonate deposition. These chemical requirements have resulted in long-recognized common properties (abundance of acidic proteins, presence of sulfated polysaccharides, etc.; see Chapter 2), but obviously very different intracellular pathways have been developed independently by living organisms to fulfill these chemical requirements.

From a fundamental perspective, improvement in deciphering the extracellular process of biocrystallization now requires close collaboration between analysts of genomic activity and investigators able to characterize the molecular architectures developed outside mineralizing cells. At the present time, it would seem that atomic force microscopy has the potential to identify molecules at an individual level and thus can serve as the major tool in deciphering the chemical assemblages involved in the control of crystallization within mineralizing growth layers. From an environmental viewpoint, deciphering the parts of the genomes involved in biomineralization has already demonstrated the potential of this type of study for such explanations, such as the case of the coiled planktonic foraminifera *Neogloboquadrina pachyderma*, widely used as a thermal indicator. Considered a single species, specimens attributed to this taxon are now thought to belong in two cryptic species having a difference of 0.5 per mil in their oxygen fractionation ratio, thus recording the same seawater temperature as being about 2 °C different (Bauch *et al.* 2003).

Results like this are pertinent to this type of study, and direct our attention to an additional area of genomic studies: sensitivity of genome expression to environmental factors. More than the "genetic blueprint" within the cell nucleus, other factors affect the activity of gene-regulators, enabling the expression of genes to be modified, thus leading to biological responses to external conditions, such as environmental stress. Seen from the perspective of biomineralization, knowledge of the potential of genomic "machinery" to produce the complete set of molecules involved in the process of Ca-carbonate deposition is a first step. The full understanding of this process must also include the recognition of, and explanation for, the influence of living conditions on the functions of this complex intracellular system and the resulting composition of the molecular assemblages secreted.

One cannot overemphasize the importance of the perspectives opened by this aspect of the genomic approach in both focused and broad applications of it. At the level of an individual specimen, this can explain a structural or compositional particularity. At a geological scale, different responses of various biological groups to changes that have occurred in the global environment can suggest causes for the overall evolution of faunas and floras, and thus change our resultant reconstruction and presentation of life history in geologic time.

Microstructural diversity observable in the culture of pearls provides a spectacular illustration of perturbations caused by stress to mineralizing epithelium. The following series of examples allow us to make numerous conclusions regarding various aspects of the biomineralization process, but we also need to keep in mind that all of the various features (each resulting from diverse organic and mineral conditions in what would generally be abnormal associations) have here been produced by a single type of mineralizing tissue: the nacreous producing portion of the mantle of *Pinctada margaritifera*.

5.3 An example of change in the mineralizing activity of epithelium after stress: microstructural variation during the early stages of mineralizing activity of pearl-producing epithelium

The method for producing cultured pearls was established in its definitive form by Mise and Nishikawa in the early 1900s, after earlier investigations by Saville-Kent. They established the possibility of carrying out mineralization starting with a spherical core material, surrounded by a fragment of mantle, the "pearl-sac." This method, still used today, consists of cutting off a fragment of epithelium, a few millimeters square – the graft – in the zone of nacre production of a bivalve (Figs. 5.2a–b). The graft is then deposited over a small sphere of carbonate – the nucleus – previously introduced into the visceral cavity of another bivalve of the same species (Fig. 5.2c: arrow). The graft, whose active mineralizing surface (the side of the mantle that produces nacre) has been positioned against the surface of the nucleus, spreads onto the nucleus, surface up, to cover it completely, thus forming the "pearl-sac." Actually, formation of the composite organic feature in which the pearl will grow (Figs. 5.2d–f: arrow) has long been a matter of debate (Machii 1968), but molecular studies now lead to the conclusion that pearl-sac cells result from the proliferation of grafted epithelium itself (Arnaud-Haond *et al.* 2007).

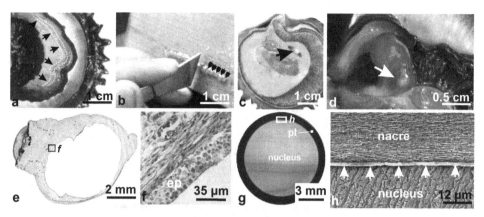

Fig. 5.2 Key steps in cultivating pearls. (a) Cutting a ribbon (arrows) from mantle tissue in the nacre-producing area; (b) Cutting the grafts; (c) Position of the nucleus (arrow) within the animal body (drawing); (d) Finished pearl still located within the pearl-sac (arrow), just before collecting; (e) Histological section of the pearl-sac (courtesy N. Cochennec, IFREMER-Taravao biological station); (f) Pearl-sac mineralizing epithelium ("ep"). This image shows the dual composition of the pearl-sac, with the internal epithelial layer and the external complex structure mostly made of host cells. This pictures explains why the actual origin of the pearl layer has long been a matter of debate; (g–h) Cross section of a pearl at its maximum diameter ("pl": pearl layer), and zone of contact (h) between the nacreous layer and the surface of the nucleus. Between the nucleus (lower part of the figure) and the nacre of the pearl layer, a continuous layer of organic material is produced (arrows) as the first step in the secretion activity of the completed pearl-sac. a: Dauphin *et al.* (2008); g: Cuif *et al.* (2008a).

After about one month, pearl-sac epithelium restarts its mineralizing activity. Newly formed minerals are not yet in direct contact with the nucleus surface. Secretory activity of the mineralizing epithelium first produces a continuous organic layer (Figs. 5.2g–h: arrows). Then, as would be expected, taking into account the origin of the mineralizing epithelium, nacreous layers are deposited (Fig. 5.2g).

5.3.1 First indications of delay in the recovery of the nacreous mineralization mechanism

The production of nacre by the pearl-sac is rarely immediate and regular. The production of non-nacreous material in the first phases of functioning of the pearl-sac is so frequent that Kawakami (1952) proposed a theory of "regeneration" of the graft, in order to explain the presence of non-nacreous material at the base of the pearly layer, which then forms the first stages of mineralizing activity of the pearl-sac. According to this theory, the mechanism for production of nacre, active when the graft was prepared, was perturbed by the process of cellular multiplication that led to forming the pearl-sac of mineralizing epithelium. When the latter recommences its activity, it returns to a complete sequence of the structure of the *Pinctada* shell, periostracum, then prisms and finally, nacre. Observational methods now available allow reinterpretation of the Kawakami hypothesis, but fully confirm the

Fig. 5.3. Thickening of the organic basal layer, leading to a radial organization of the initial organic compounds. (a–c) Section and overall view of the basal pearl layer, observed in natural (b) or UV light (c); (d–e) Optical view of the basal pearl layer in natural (d) and UV light (e). Note the fluorescence of the basal pearl layer and the weak indications of a radial organization of the structure, i.e., perpendicular to the normal layering of nacreous material; (f–g) Backscattered electron view of the basal pearl layer. BSE observation emphasizes the contrast between the organic compounds – black surfaces are not producing backscattered electrons – and the mineralized parts of the field of view. The abundant organic components secreted in the early phase of mineralization are obviously organized in radially oriented irregular structures. A closer look at the regions between the organic columns allows observation of the first "horizontal" lamellae (i.e., parallel to the nucleus surface). Note their irregular aspect when they first begin (arrows); (h) BSE view focused on the first appearance of the organic laminae parallel to the nucleus surface. These layers are at first discontinuous and irregularly spaced, but the cyclic process rapidly acquires regularity and continuity. In spite of a rather thick first step of radially oriented organic components, production of nacre can be considered as being quickly established. a: Cuif *et al.* (2008a).

frequency of nonimmediate nacreous production by the mineralizing cell layer, in spite of the demonstrated identity of the cells which were producing nacre just prior to cutting the graft.

In this first example, indications of the formation of a radial organization are weakly visible in optical observations (Figs. 5.3a–e), but much more evident in backscatter electron microscopy (BSE). In BSE observation, the contrast is proportional to the atomic number of elements present in the substrate, essentially the organic phase here, with C, H, O, and N, producing a less intense signal than areas containing calcium.

The organic components, instead of producing the so-called "bricks and mortar" system typical for the nacreous layer, are here organized into a layer of radially oriented units of purely organic material (Figs. 5.3f–h). Between these irregular pillars, with their general orientation perpendicular to the nucleus surface, a better mineralized phase is visible (lighter gray in BSE observation). It is worth observing that, within these mineralized areas, fine black lines parallel to the substrate mark the first occurrence of nacreous envelopes. At this early period of functioning, the pace of production of these organic layers is slow (Fig. 5.3h: arrows), therefore the "tablets" between the organic layers are thick, massive, and imperfectly formed. However, higher in the section (i.e., later in pearl layer development), the production of organic lamellae is rapidly accelerated. When production of the radial organic columns is interrupted, the typical nacreous layering is definitely in operation. From a practical viewpoint, the additional structure produced in the early stages of the pearl layer does not disturb the later development of nacre.

This case simply indicates that recovery of normal production of nacre can take much longer than producing a simple organic laminar substrate, as illustrated by Fig. 5.3h. The organic material produced by the pearl-sac epithelium also here *demonstrates its ability to develop its own structures*. The next example (Fig. 5.4) presents a more fully differentiated

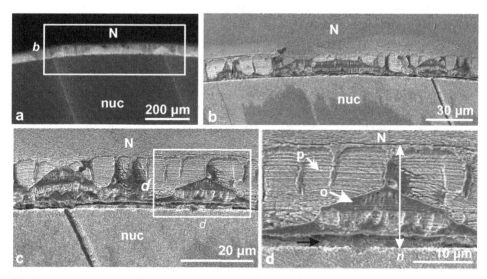

Fig. 5.4. Example of a complex pre-nacreous layer with composite organization. (a) UV fluorescence of the first part of the pearl layer between the nucleus ("nuc") and the nacre ("N"). Its structural complexity is already visible, seen in the heterogeneity of its fluorescence; (b–c) Microstructure of the pre-nacreous layer. The early organic components (brightest sectors under UV light) are very irregular: the upper part of the pre-nacreous layer is composed of a well-mineralized zone with subdivisions due to radial organic laminae; (d) This picture emphasizes the heterogeneity of secretions produced by pearl-sac epithelium during deposition of the first 15 micrometers of the pearl layer. a–c: Cuif *et al.* (2008a).

"pre-nacre" stage, in which obvious indications of prismatic structure are visible. Within this 15 μm thick initial layer, mineralizing epithelium immediately forms a stratified structure, but with a depositional rhythm and material composition that have nothing to do with nacre production. All of this material is UV fluorescent, thus differing from the nacreous layers (Fig. 5.4a: "N"). The heterogeneity of the initial layer is visible, with purely organic portions fluorescing intensely, and alternating with weakly fluorescent sectors. Etching allows discrimination between the fully organic materials (irregularly arranged in this case), and a well-layered deposit of mineralized material. Here the regular stratification is interrupted by radial organic laminae.

Comparing the early stages of the pearl layer to typical nacreous deposition indicates that a significant perturbation has occurred in the secretion mechanism of the pearl-sac epithelial cells after the grafting operation. The diversity of the materials that are deposited during these early stages of pearl-sac activity is remarkable, and more important, the length of the recovery period necessary, as well as the thickness of the intermediate zone likewise are. Figure 5.5, as an example, exhibits the structures that are formed as above, but in a 100 μm thick layer. Even more apparent than in the two previous examples, here the tendency to form radial structures is visible, as well as the distinct layering of mineralized areas. The geometry of the latter seemingly has no relationship to the spatial organization of the nacre tablets.

However, some similarities exist in this material with the previous examples. The initial step in recovery is marked by forming a UV fluorescing layer in which radial fluorescent lamellae emerge. In addition, the layered organization is also readily visible in all types of secretion. Obviously, the cyclic secretion process is the basic pattern of the biomineralization mechanism for the production both of purely organic and mineralized phases. Laser confocal fluorescence study of the organic components (Fig. 5.5c) provides values for the density of radial organic membranes, by utilizing the depth of laser penetration into the mineralized material, and in addition shows that the radial lamellae are built by superposition of micrometer-thick layers (Fig. 5.5d).

The importance of this data deserves underlining. Throughout this early step, the pearl-sac epithelium produces organic components that are able to assemble themselves and produce prism-like structures. In the first example, the prismatic envelopes are rather imperfect, but their mode of growth is quite similar to the envelopes of prisms in the outer layer of the *Pinctada* shell. Their production somewhat reinforces the Kawakami hypothesis of "rejuvenescence" of the pearl-sac epithelium during its phase of spreading to cover the nucleus surface. However, further biochemical and mineralogic determinations will undoubtedly lead to substantial modifications of this interpretation and will provide additional data on the nature of the perturbations undergone by the graft throughout the history of various individuals.

5.3.2 *Diversity of sequential timing and spatial distribution of non-nacreous materials produced by the pearl-sac epithelium*

The perturbations that apparently affect initial stages in the function of the pearl-sac epithelium thus result in this considerable variety of resultant structural sequences.

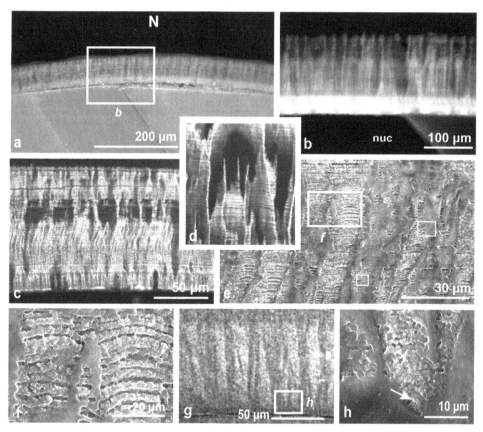

Fig. 5.5. Example of radial mineralized structures in the early stages of a pearl layer. (a–b) Optical views in natural (a) and UV (b) light; (c) Laser confocal fluorescence (488 nm, Mineralogical Dept., NHM, London). Due to laser penetration, the density of organic radial membranes is well demonstrated; (d) Laser confocal three-dimensional reconstruction of the growth mode of the organic radial lamellae. Their layered formation is well shown, as well as the imperfect polygonal structure that results from their development. They are nearly forming prisms in the basal part of the pearl layer; (e–f) SEM view of the mineralized layers; (g) XANES mapping of organic sulfates in the prismatic early stages of the pearl layer. Sulfate concentrations clearly separate membranes and mineralized sectors; (h) SEM view of the initial mineralization (arrow) just above the compact basal organic layer. The small granules forming the mineralized phase are distinctly visible. a, b, d: Cuif *et al.* (2008a).

Compared to Figs. 5.2g–h that present an ideal case with only a thin organic layer separating the nucleus and the initial nacreous layers, Figs. 5.3, 5.4, and 5.5 demonstrate alterations in function affecting the complete pearl-sac epithelium. Nevertheless, other specimens show much more highly differentiated patterns. Figure 5.6 shows that mineralization here can be distributed in distinct areas within which microstructural and even mineralogical character-istics differ markedly.

In this approximately 250/300 μm thick basal portion of the pearl layer, two distinct prismatic zones can be seen (Figs. 5.6a–c). Not only are they distinctly pigmented, but also

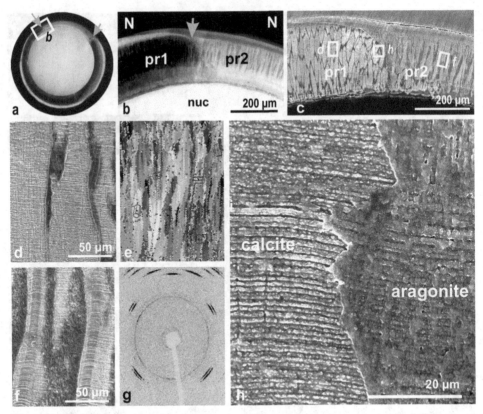

Fig. 5.6. Simultaneous occurrence of calcite and aragonite in a basal pearl layer. (a–c) Example of simultaneous doubled mineralogy in prismatic form within a basal pearl layer; (d–e) Growth layering of the calcite prisms (d) and (e) electron backscattered diffraction diagram showing their monocrystalline organization (imaging by A. Perez-Huerta); (f–g) Growth layering (f) and microdiffraction diagram (g) in the area of aragonite prisms. The positions of the diffraction spots indicate the near monocrystalline structure of the material (image: P. A. Albouy, S. Rouzières, LPS, Orsay); (h) SEM view of the contact between the calcite and aragonite areas. Note the continuity of stratification across the boundary. b, h: Cuif *et al.* (2008a).

the prisms have a distinctive morphology; however, the two regions are undoubtedly prismatic, having quite comparable, layered microstructure (Figs. 5.6d and f).

Crystallographic determinations lead us to a surprising result: the dark pigmented prisms here are composed of calcite, as shown by SEM-EBSD identification (Fig. 5.6e), whereas the prisms of the light zone are aragonitic. Microdiffraction diagrams (Fig. 5.6g) made on 30 μm diameter areas of this portion of a thin section leave no doubt about this mineralogy. In addition, both electron backscatter (EBSD) and microdiffraction diagrams indicate that these calcite and aragonite prisms each behave as single crystals.

This abnormal presence of aragonite prisms built by epithelium cut from a piece of the mantle of *Pinctada margaritifera* has to be underlined, as not only was this piece of mantle producing nacre tablets prior to being cut, but prisms composed of aragonite are *never* formed by bivalves belonging to this family.

Thus from a biological viewpoint, these prisms are true microstructural chimerae because they associate organic envelopes of prismatic types with a mineral phase usually deposited in association with lamellar envelopes (= nacre). This abnormal association demonstrates, above all, that perturbation of the biomineralization mechanism by the grafting process has been more fundamental than observed in the previous examples. Additionally, it indicates that, during the recovery process, unexpected biochemical associations may occur. When looking closely at the growth mode of these prisms, we can see the continuity of layering observed directly at the contact between the calcite and aragonite areas (Fig. 5.6h). Growth layering passes with complete continuity from the zone of calcite deposition to that of aragonite. It seems obvious that, when the mineralizing epithelium was in the process of producing this series of strata, *cells of the two distinct areas were then producing different mineralizing compounds*, but yet preserved perfect synchronism.

This image provides indisputable evidence of the independence between the sizes and shapes of mineralizing units and the identity of the Ca-carbonate polymorph deposited in one or the other of these two synchronized mineralizing areas.

The mineralogic properties of calcite and aragonite in these prisms have been established by localized measurements and mapping, using CAMECA electron microprobes. The results are summarized in Fig. 5.7. The overall composition of these mineralized units corresponds exactly to the properties of the calcite and aragonite layers respectively, in the shells of *Pinctada*, and generally to those of other bivalve shells.

As previously observed (Fig. 2.7), magnesium concentrations are much higher in calcite prisms than in aragonite, and the strong variation in concentration here is strictly correlated to growth layering (Fig. 5.7c). Conversely, sodium is more abundant in aragonite, permitting mapping that thus distinguishes it from neighboring calcite (Fig. 5.7d). As in all mollusc shells, strontium concentrations are much lower than those of sedimentary aragonite (or corals), and sulfur is more abundant in calcite (Fig. 5.7f).

These localized measurements support the conclusion that mineralizing activity of the matrices in the pearl-layer epithelium results in mineral phases (calcite or aragonite) comparable to the mineral phases of prisms and nacre of the original shell itself. The disturbance of the mineralization mechanism that results in producing calcite in some sectors (and sometimes over the whole surface of the pearl-sac epithelium) seems to be more linked to the process of regulation than to alteration of the mineralizing mechanism itself.

These results show conclusively that sizes and shapes of microstructural units do not result from any progressive shaping of a growing monocrystal. The stepping growth mode of mineral phases here results in crystal-like units with unexpected and highly variable sizes and shapes, simply due to the variation in composition of the portion of the organic matrix that produces envelopes (see also Fig. 5.9). Additional and significant data are also provided by synchrotron mapping of the organic phase itself. A basal pearl layer in which calcite and aragonite prisms are distributed in several distinct sectors (Figs. 5.8a–b) is seen here in optical views, and then additionally as a backscattered electron image in SEM (Fig. 5.8c), in

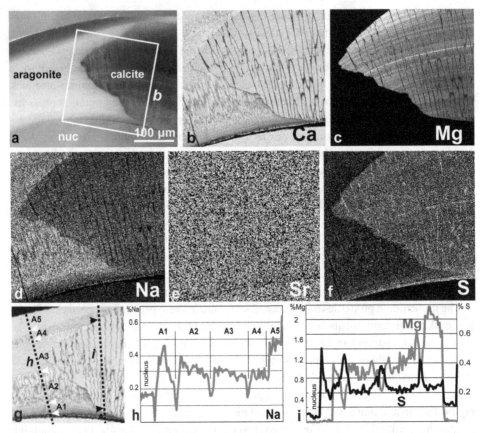

Fig. 5.7. Measurement and mapping of minor-element concentrations in the "abnormal" calcite and aragonite within basal pearl layers. (a) Image of the pearl layer studied, with indication of the areas mapped with the electron microprobe, and aragonite and calcite labeled; (b) Backscattered electron imaging of the studied surface. Black lines correspond to the main organic envelopes; (c–f) Distribution of the usual minor elements in Ca-carbonates; (g–i) Position (g) and profiles (h–i) of the microprobe measurements (mapping by T. Williams, Mineralogical Dept., NHM London; CAMECA SX50). Profile (h) is through the aragonitic part of the basal layer; profile (i) is mostly calcite, except for the two extreme ends. b–f: Cuif *et al.* (2008a).

order to allow the continuity of growth layers to be established. Figure 5.8d also emphasizes the contact between these two areas of radial prismatic microstructure. In addition, the distribution of sulfur has been mapped in the two sectors located on both sides of this contact: one in the median part of the thick prismatic layer, another at the base of the mineralization zone.

Figures 5.8e–f emphasize the contrast in distribution of sulfated polysaccharides between the calcitic area, in which they are present (with visible variations related to growth rhythms), and the opposite side (aragonitic) where their abundance is much lower or absent. Figures 5.8g–h are even more demonstrative of the dominant role of the sulfated

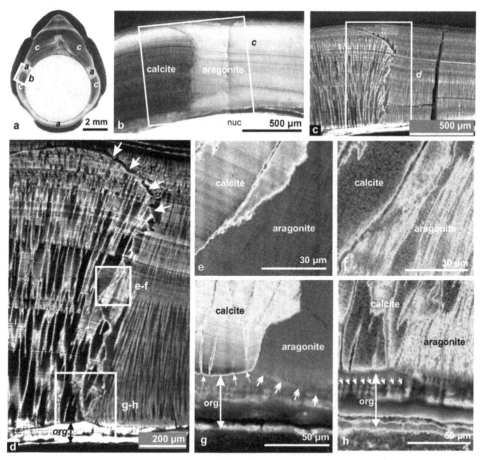

Fig. 5.8. Distribution of sulfur from polysaccharides and proteins as shown by synchrotron-based XANES mapping. (a) Transverse section of an irregular pearl. Below the nacreous layer (black), calcite ("c") and aragonite ("a") exhibit irregular development, leading to formation of a "baroque" pearl; (b–c) Zone at the contact between calcite and aragonite; optical view (b) and SEM, backscattered electron image (c). Continuity of layering between the calcite and aragonite sectors is clearly visible (see also Fig. 5.6h); (d) Close-up view showing the position of the two sectors mapped. Note the radial orientation of the organic envelopes on both sides of the field of view and the progressive regression of the calcite in the upper part (arrows); (e–f) Correlation between formation of calcite and preponderance of the sulfated polysaccharides (e) and formation of aragonite in areas where their concentration is lower; (g–h) Distribution of sulfated polysaccharides (g) and S-proteins (h) in the very early stages of formation of the pearl layer. Initially it is mostly proteinaceous and weakly differentiated ("org"), but then the basal organic layer abruptly splits into two adjacent portions according to dominance (single arrows) or absence (double arrows) of sulfated polysaccharides. Formation of calcite (g) or aragonite (h) occurs accordingly.

polysaccharides in the formation of calcite. After secretion of the nonmineralized layer that surrounds the nucleus (Figs. 5.8d and g–h), production of mineralized material begins with two different but adjacent phases simultaneously produced: calcite and aragonite on both sides of the contact between them, which oscillates during most of the formation of the

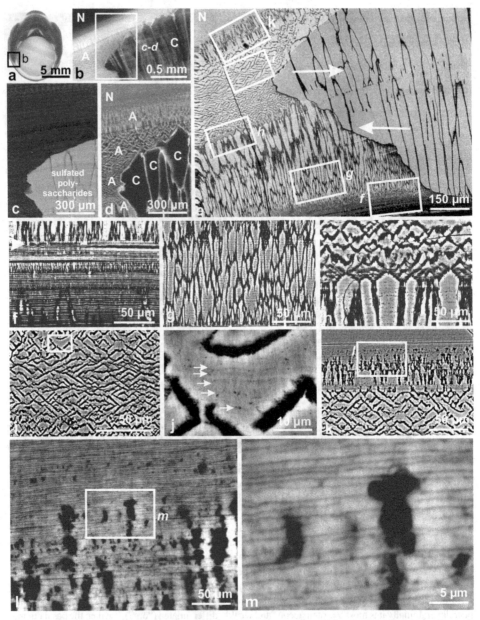

Fig. 5.9. Diversity of organic envelopes in the "pre-nacreous" pearl layer. (a–b) Pearl section and surface of the pearl layer in which calcite ("C") and aragonite ("A") have been simultaneously produced; (c–d) XANES mapping of sulfated polysaccharides (c) and sulfur in proteins (d) in the same area. The diversity of microstructures in the aragonite area is already visible; (e) The same sector, enlarged: this image has been obtained by backscattered electrons. Here, contrast is due to differences in atomic number, thus, mineralized surfaces (light gray) are clearly distinct from organic ones (black). Note the precision in the mapping of the envelopes shown here in the aragonitic areas. Arrows draw

non-nacreous layer. In the last part of the transition, the expansion of aragonite over calcite produces a major inflection of this contact (Fig. 5.8d: arrows). The distribution of calcite compared to aragonite is closely correlated to the presence or absence of sulfated polysaccharides.

It should be noted that this correlation is probably restricted to the Phylum Mollusca, and perhaps only to the Bivalvia themselves. In scleractinian corals, for example, sulfated polysaccharides are abundant, but the skeleton is entirely composed of aragonite. At the same time, this represents a good example of discrepancies between genomic activity related to the biomineralization process in different phyla, and even more clearly, an indication of the difficulty that exists in assigning a precise role to the organic compounds involved.

From a functional point of view, these three examples provide us with evidence that when the epithelium of the pearl-sac restarts its activity, the activity can be separated into different areas in which it functions synchronously, but independently with respect to the composition of the mineralizing matrices being produced.

5.3.3 The stepping growth mode of envelopes and their morphological diversity

Our microstructural observations on the pearl layer have already provided images of the stepping growth mode of the envelopes surrounding the calcite prisms (see Fig. 5.5d). The sizes and overall morphologies of these prismatic walls, although frequently imperfect, are comparable to the corresponding components of the external shell layer in *Pinctada*. This emphasizes the variability of the microstructural envelopes seen in the aragonite sector, a location where we should see the classical "bricks and mortar" arrangement. Figure 5.9 presents a series of such envelopes formed by the insoluble organic matrix components, from the beginning of mineralization of the pearl layer until beginning the formation of nacre, about 1 mm above, a thickness that corresponds approximately to one year's growth, based on their mean standard growth in favorable environmental conditions (Figs. 5.9a–b). It is noteworthy that, during this period, XANES mapping of sulfated polysaccharides and S-amino acids does not reveal any significant change in the composition of the mineralizing matrices in either calcite or aragonite sectors.

Caption for Fig. 5.9. (cont.)

attention to the extension of the calcite area, followed by its regression. Enlarged views of each of the selected sectors are shown below; (f) The first insoluble layers, parallel to the nucleus surface; (g) The prismatic structures (remember, this is an aragonitic area); (h) Transition from prismatic to the "chevron" portion of the sequence. Note the traces of stepping growth in the upper parts of the aragonite prisms; (i–j) Overall view and detail of the "chevron" structure. Whatever the morphology of envelopes produced by the insoluble matrices, the basic stepping mode of growth remains unperturbed (j: arrows); (k) Transition from "chevron" to the radial pre-nacreous stage; (l–m) At this stage, horizontal nacreous layering is produced, with some irregularity and persistent residues of insoluble prism-like matrices. The latter finally disappear in the upper part of sequence (m). After about 0.8 mm (i.e., about half of the thickness of a standard pearl layer) the non-nacreous aragonite phase is finished. a, e: Cuif *et al.* (2008a).

During this time, the pearl-sac epithelium has produced a series of at least four strikingly different structures before producing the thin organic layers between the aragonite tablets of nacre (Fig. 5.9e). At the base of the series (Fig. 5.9f), regularly spaced, thick organic layers were produced, suddenly changing (arrow) into a radial prismatic pattern (Fig. 5.9g). This prismatic organization (recall that this is an aragonite-producing area) also changes quite suddenly into a chevron pattern (Figs. 5.9h–i). Looking closely at these mineralized parts, we can see traces of the stepping mode of growth for this series (Fig. 5.9j: arrows), and this layering is also visible in the prismatic structure noted above. At the top of the chevron layer, there is a new occurrence of radial organization (Fig. 5.9k), but this is less well developed than previously. Simultaneously with this, the first indication of "nacreous-like" layering becomes visible (Fig. 5.9l). Finally, secretion of the compounds that produce the strong organic architectures definitely stops, and aragonite is included between the regularly spaced horizontal layers (Fig. 5.9m), comparable with the corresponding components within the external layer of the *Pinctada* shell. This underlines the variability of micro-structural envelopes visible in the aragonite sector, a place where we should see the classic "bricks and mortar" arrangement of nacre tablets and organic materials.

It is worth noting that this sequence of variable envelopes formed in the aragonite-producing area is temporally equivalent to the very regular construction of prismatic envelopes in the adjacent calcite-producing sector (Fig. 5.10). The major interest in this "nonclassical" mineralizing process is due to the fact that, in spite of the abnormal character of resultant microstructures, *the layered growth mode is always very well shown throughout.* It is even more easily seen by careful observation of the prismatic boundaries in both aragonite sectors (Figs. 5.10b–c) and in areas of calcite prisms (Figs. 5.10d–f). Here, the layered growth mode finds its most convincing expression.

Figures 5.10d–e must be examined bearing in mind that each calcite prism resulting from this stepping growth process appears as a monocrystal when seen in images generated by backscattered electrons (Fig. 5.6e). This is also the case when aragonite prisms are studied by microdiffraction (Fig. 5.6g). The layered mode of growth implies that in both the aragonite and calcite sectors, envelopes are formed first, and resultant polygons (with a different shape in each sector) are then filled in by its distinct mineralized carbonate polymorph.

Obviously, the sizes and shapes of the prisms are not determined by the crystallization process, which is the final step in forming each growth layer, but instead, by polymerization patterns resulting from the biochemical composition of the envelopes.

5.3.4 The shift to forming nacre

Throughout the stepping mineralization process in these adjacent areas, either producing calcite or aragonite, the limit between the two minerals varies. Commonly, even areas producing aragonite (i.e., the normal and expected mineral for this type of epithelium) can shrink and be laterally replaced by calcite. We can see this type of change in the margins of aragonite/calcite areas in Fig. 5.9e (arrows from right to left). In this particular instance, after

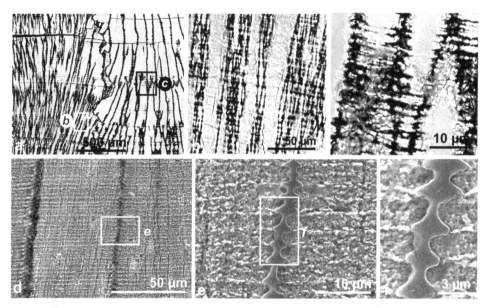

Fig. 5.10. The synchronized stepping growth mode in envelopes of calcite and aragonite prisms in the basal pearl layer. (a) Two adjacent sectors in a basal pearl layer: aragonite (right) and calcite (left); (b–c) Radial envelopes in the aragonitic area, providing a different polymerization pattern (compare to the series in Figs. 5.9f–k); (d–f) In the calcite-producing area, the prism envelopes clearly show the stepping growth mode, synchronized with the growth of the mineralized phase. Such perfect coordination in the respective thicknesses of the insoluble matrix forming the envelopes and the mineralizing matrix that generates the mineral phase clearly suggests closely controlled dual secretions.

a maximal expansion of the calcite-producing surface, a reverse change occurs, and aragonite replaces calcite (higher arrow of Fig. 5.9e). In some cases, this mineralogical change does not result in a complete halt of calcite production and it can happen that, at the time the pearl is harvested, a part of its surface is still composed of calcite. Such incomplete recovery of nacreous production is illustrated in Figs. 5.11a and b. In the transverse section of the pearl, the line marking the shift from calcite to nacre ("sl") is remarkably symmetrical with respect to the pearl axis (Fig. 5.11b). Above this line, almost the entire resultant conical portion of the pearl is made of calcite with strong prismatic microstructure and UV fluorescent envelopes (Figs. 5.11c–d).

It is worth noting again that the shift from calcite to nacreous production is not an immediate, or a one-step process (Fig. 5.11e). Backscattered electron imaging, emphasizing the contrast between calcium-rich areas and purely organic materials, illustrates this progressive metabolic change.

The intermediate status of the envelope matrices is well illustrated on UV fluorescence images (Fig. 5.11d), clearly seen due to the strong fluorescent response that these "intermediate" matrices exhibit, contrasting with the thin envelopes of the nacre. These

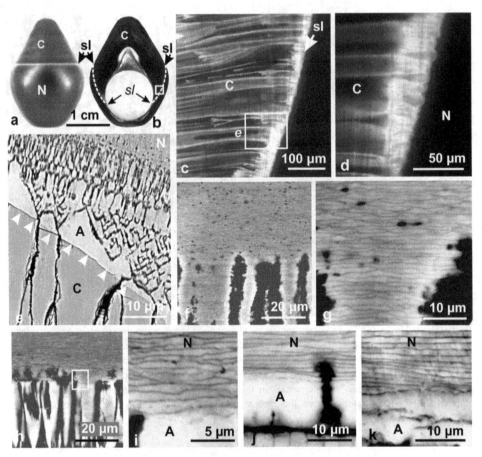

Fig. 5.11. Examples of recovery of nacreous layering after abnormal mineralization in the early pearl layer. (a–b) External view (a) and medial section (b) of a composite pearl with a continuous calcite compartment ("C") throughout the entire cultivation period. Note the line ("sl") separating the calcite and aragonite (and nacreous) areas, and the progressive shift of this boundary during pearl growth (b); (c–d); UV fluorescence view of the superposition of calcite prisms and aragonite areas. Note the absence of fluorescence in the nacreous sector; (e) BSE view of the contact zone between calcite prisms ("C") and aragonite areas; (f–g) Transition zone between the two types of envelopes; the distal parts of prism-like envelopes are concomitant with early secretion of the nacreous insoluble matrices; (h–k) Examples of irregularity in the early layers of nacreous envelopes. Note the thick layers of unstratified aragonite ("A"). (Sample: courtesy T. Bernicot, Moorea, Polynesia.)

images suggest the independence of the group of genes that form the envelope matrices and others that are responsible for the organic components that control mineralization. Add to this the fact that a clear difference exists between the two main types of insoluble matrices, one being strongly fluorescent, and the other not. Where the latter are present, so are the nacreous envelopes. This situation contrasts with what normally takes place in the shell mineralization process, where formation of prismatic envelopes is interrupted long before the appearance of regular nacreous layering. In these pearls, it appears clear that two mechanisms are active simultaneously. The latest strongly fluorescent radial matrices are still being formed at the same time as are regular, well-ordered (and nonfluorescent) nacreous layers.

5.3.5 The contrast between diversity of envelopes and similarity of mineralized materials magnified at the 10 nm scale

The observation has previously been made that the mineralized materials forming the calcite prisms and nacreous layers of *Pinctada margaritifera* show a striking similarity when observed at the 10 nm scale (see Fig. 3.9). The complete independence between the structural organization at the 10 nm range, and the morphology of microstructural units (in the 10 μm range, an order of magnitude larger), results from the organization of envelopes. This is even more clearly shown by comparing the mineral phase imaged within the superposed layers of the pearl (Fig. 5.12). Inside the compartments formed by the insoluble molecules assembled in various ways, the similarity of structural organization of mineralized parts at various stages of pearl growth allows us to confirm the impressive continuity of the mineralizing mechanism in them.

The occurrence of the nacreous envelopes themselves, very distinct from the preexisting prismatic or "chevron" microstructures, does not correlate to any change in nanostructure of the mineral phase (Figs. 5.12a–d). Likewise, this aragonite (Figs. 5.12i–l) is also indistinguishable from the calcite simultaneously deposited (Figs. 5.12b–d).

Thus, the mineralizing events observed on this surface of approximately 1 mm^2 provide us with an important summary of the major results reflected in the formation and mode of growth of the microstructural units building calcareous skeletons.

5.3.6 General information derived from the structures produced in the pearl layer

From an overall viewpoint, examination of these intermediate, non-nacreous structures that precede the nacreous layering of the pearl permits us to roughly classify them according to their diverse differences that occur between the ordinary and expected nacreous materials on one hand, and the mineralogical and microstructural patterns of this "unexpected" mineralization on the other (in this case, both mineralized material and organization of the associated envelopes). In general, this amount of microstructural diversity cannot simply be explained as being due to influences of environmental conditions. The pearls described here were produced in lagoons at Tuamotu, where temperature, nutriment distribution, and seawater characteristics are nearly homogeneous. Descriptions of *Pinctada* shells from these localities have never exhibited any comparable variability in their microstructure.

An important point is that the starting point for the grafting process is the cutting of the graft out of the mantle of a living animal (Fig. 5.2b). Owing to the usual separation of roles in functional pearl cultivation methods, the insertion of the graft itself is carried

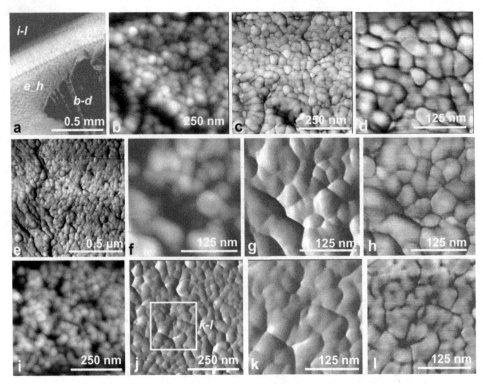

Fig. 5.12. Similarity of the 10 nm scale structures in the calcite, aragonite, and nacreous layers, as shown by the pearl layers illustrated in Fig. 5.9. (a) Location of the three sectors illustrated: calcite (b–d), non-nacreous aragonite (e–h), and nacre (i–j); (b–d) Calcite: typical granular appearance of calcite prisms, height image (b), and phase images (c–d); (e–h) Aragonite with "chevron" microstructure: phase image low enlargement (e), closer view with height (f), amplitude (g), and phase images (h); (i–l) Granular structure of a nacreous tablet (i): amplitude images (j–k), and phase image (l). No substantial difference can be seen in the submicrometer-scale nature or organization of the mineral materials, whatever their mineralogy or microstructural organization may be.

out by specialized operators who receive each graft pre-cut. Usually, the mantle of an adult *Pinctada* animal is large enough to be able to furnish 30 to 50 grafts (from both sides of the mantle). Depending on the efficiency of the grafting operator, the last graft of the series can wait at least one hour, in conditions very unusual for it, prior to being placed within the second animal's mantle cavity, and commonly this wait can last much longer than one hour. Even then, the small epithelial fragment is not in its usual physiological environment. Thus, the small, ten-square-millimeter fragment of epithelium is submitted to considerable biological stress. No systematic study has ever been published concerning possible correlations between the degree of disturbance of the mineralizing process and the waiting time the graft has been submitted to during the grafting operation. The hypothesis can reasonably be made that some relationship may exist between the duration of this biologically disturbing step and the intensity of the resulting metabolic perturbations. This disturbance is reflected by the production of abnormal mineralizing matrices and resulting combinations of envelopes and mineralized material.

From a functional viewpoint, the mineralizing epithelium of the pearl-sac typically is a self-controlled biological unit, artificially placed, and developed within the body of a second animal (close to the gonads). As a result, it is separated from the original, normal organic environment of the parent bivalve animal. Therefore, the secretory activity of the cells making up the mineralizing epithelium of this pearl-sac is not submitted to the overall regulation that generally maintained the usual functional relationships between the living tissues of the animals when they were in their original anatomic situation. However, after some amount of time, which varies, during which non-nacreous structures are produced by this mineralizing epithelium, the majority of these pearl-sacs shift to producing nacreous material until the end of the commercial process. Histology suggests that the mineralizing epithelium of the pearl-sac is not completely isolated from the bodily fluids by which regulating molecules are distributed within the second *Pinctada*. By diffusing through the complex tissues surrounding the mineralizing epithelium (see Figs. 5.3e–f), in the major part composed of cells from the host *Pinctada*, the regulatory factors find access to the mineralizing cells themselves. Thus, depending on the intensity of stresses resulting from individual details of stages in the operation, variable thicknesses of non-nacreous layers of the pearl can be observed that were deposited prior to the recovery of nacre deposition. We have seen therefore, that perturbation of the mineralizing mechanism results in strongly abnormal materials and structures, such as aragonite in the form of prisms.

With respect to ongoing research on understanding intracellular aspects of the biomineralization process, careful analysis of the microstructural assemblages provided in the examples above can provide data on the genomic activity responsible at each level of formation of these unusual microstructural sequences. As pointed out by Wilt and Ettensohn (2007), owing to the complexity and taxonomic diversity of the genomes involved in biomineralization, only cooperative research on intracellular analysis and detailed investigations of the resulting extracellular materials can enable us to formulate a reliable and consistent concept of the relationships between the process operating on the inner and outer sides of cellular membranes.

5.4 The two-step growth and crystallization model – a solution to formation of taxonomy-linked microstructures through biochemical control of crystallization at the submicrometer level: concluding remarks

Remarkable syntheses of mineral crystallization that appeared in the first half of the twentieth century emphasized the diversity of the calcareous structures produced in different groups of living organisms, and potential applications of these, based on characteristics of microstructural units forming distinct layers in shells (taxonomic, evolutionary, and environmental studies). All of the examples in Chapters 1 through 4 have shown the organization of skeletal components and their description. The applicable terminology was based on earlier optical observation. The top-down approach, investigating structures and their mode of growth in the multiplicity of very different calcareous objects, now seen at unprecedented magnification, results in surprising conclusions.

5.4.1 Similarity in the layered mode of growth

These distinct calcareous microstructural units (prisms, foliated, cross-lamellae, etc.) are not composed of pure crystals specifically nucleated and subsequently shaped during continuous growth by external influences, such as organic molecules adsorbed on their surfaces. The microstructural units earlier observed with the polarizing microscope as crystalline features are actually secondary, and thus must be considered three-dimensional structures formed by the superposition of micrometer-thick growth layers. The three-dimensional organization that leads to the formation of long-known microstructural units is due to the concomitant association of two distinct materials secreted within each growth layer: deposition of an organomineral assemblage that later evolves into crystallized material, and production of organic components specialized to form morphological networks, usually named "envelopes."

Utilizing the word "envelope" was consistent with the concept of a single monocrystalline unit, contained within an organic membrane. However, reconstructing the formation and crystallization of the elementary growth layer provides us with a different view of the relationship between the "envelopes" and the mineralized matter. In each of the biomineralization cycles, crystallization of the mineralized portion occurs independently of the concomitant polymerization of the material producing the envelopes. The final size, shape, and structure of microstructural units depends on the precision in the superposition of the repeated biomineralization cycles.

Since the beginning of microstructural studies, the use of the calcite prismatic layer of pteriomorph bivalves as an example to describe and explaine calcareous biomineralization is primarily due to the exact alignment of the successive micrometer-thick growth layers in the prisms of these species. The repeated secretion process results in two distinct structures, the "envelope" surrounding the "crystal." Just the opposite is shown by the structures observed in pearls (see Fig. 5.9), showing that differential polymerization of envelope material may occur during the growth of pearl layers, generating various (and, thus far, unclassified) morphologies as microstructural units. The diverse sizes, shapes, and spatial arrangements of microstructural units do not depend solely on the crystallization process itself.

5.4.2 Similarity of mineralized structures at the submicrometer scale

Among the different groups examined so far, the mineralized parts of their biogenic calcareous structures are formed of similar materials. One can regard this material as resulting from the interplay of two phases, one a mineral component, organized as a densely packed, reticulate structure, and the other an organic material coating the mineral elements. Atomic force measurement and mapping carried out on this organomineral feature allow us to assess the compositional differences between these two basic components of growth layers (Figs. 3.1 and 3.2). Regarding dimension, this organomineral intermixture develops at approximately the 10 nm scale throughout the complete range of known calcareous skeletons. From a descriptive viewpoint, the nanometer-scale interaction between organic and mineral components allows us to explain the similarity of distribution of biochemicals and the organization of mineral materials.

5.4.3 Reconstructing the two-step formation of the growth layer

Imaging in the transmission electron microscope (TEM), and accompanying diffraction studies led to the suggestion that the observed submicrometer interaction between the mineral network and its organic coating results from a two-step formation of the growth layer. Evidence of continuity between the layering of distinct microstructural shell layers (e.g., nacre and prism layering, see Fig. 1.5) indicates that the mineralization cycle involves the complete active surface of the mineralizing organ, resulting in the formation of an isochronous layer of hydro-organic gel comprising distinct mineralizing areas. With specific compositions that correspond to each of these areas, the mineralizing gel contains amorphous calcium carbonate (i.e., Baronnet *et al.* 2008), as well as organic compounds. This mineralizing gel is exocytosed during the first phase of formation of the growth layer, and accumulates between the preexisting mineral surface and the cell membrane (lacking any compartment with liquid "close to seawater"), creating a "pre-crystallization" layer having a thickness in the micrometer range. This is probably the translucent "colloid layer" observed by Bourne (1899) between the skeleton and the calicoblastic ectoderm of coral polyps. Crystallization occurs synchronously on the entire mineralizing surface. Synchronism of this crystallization step is shown in microprobe maps of minor elements that always show a close correlation between structural and chemical layering. However, no precise data exist regarding the chemical or biochemical mechanism that triggers the crystallization process. Magnesium zonation, consisting of parallel bands that closely correspond to sinuous skeletal growth, extending over broad areas of the growing *Porites* skeleton (see Fig. 3.13), may possibly be linked to such a triggering mechanism. At present it cannot be interpreted as causing crystallization itself. However, mapping of this element aids in the accurate assessment of the synchroneity of crystallization over the entire surface of a growth layer. Such crystallization obviously differs from models proposed as having pure crystals freely growing in a liquid layer and "competing for space."

To account for the uniform organomineral assemblage seen in all calcareous skeletons at the 10 nm range, we suggest that in the layer of organomineral gel, first described by Bourne for corals but at times directly observed here (Fig. 3.12), amorphous calcium carbonate is disassociated from the organic component that stabilizes it. Thus, crystallization occurs within this micrometer-thick layer of hydrated-organic gel having a species-specific composition. It is possible that the crystallization process is influenced first by the biochemical composition of this gel and additionally by local conditions (temperature, nutrients, etc.). The sensitivity of this process to conditions at that instant in time is clearly shown by high-resolution microprobe maps showing layer-to-layer variation in composition (in Fig. 2.7, for example). Formation of the Ca-carbonate lattice from mineral components that accumulate within the nodules of organic matter and amorphous carbonate (see sub-section 5.4.5 below) results in the formation of dense reticulated crystals. As a result of this, most of the organic components are pushed out to the periphery of the crystallizing material to form what appears as a "coating" on the reticulate crystal. Therefore, the mineral component retains the generally "nodular" pattern recognized in numerous observations of skeletal material in various taxa.

According to this model, the organic-mineral interaction visible in AFM observations is then related to the second step in the formation of the growth layer, the layered crystallization proper. It must also be added that AFM phase-mode observations invariably show the persistent "gray levels" of variable intensity in the mineralized portions of field images. This persistence permits the assumption that some interaction still exists between the cantilever tip and mineralized material. Low concentrations of nonexpulsed organic molecules surely remain within mineral lattices, causing the disruption of its crystallographic uniformity, and resulting in variable concentrations of localized lattice defects.

5.4.4. Controlling crystallization through stepping growth

An interesting and rather unexpected result of AFM study of a broad spectrum of Ca-carbonate skeletal materials is that a comparable method of crystallization now emerges from these high-resolution observations. Apparently, this similarity of the detailed arrangement of matrix and mineral phases with an approximate 10 nm size, can now be interpreted as the result of common controls of crystallization processes. Through limitation of both the lateral extension of the crystallizing surface and the thickness of the volume of material crystallized, the mineralizing system prevents development of purely physical crystallization that could lead to forming the normal crystal shapes characteristic of pure Ca-carbonate.

5.4.5 Agreement of the layered growth and crystallization model with emerging concepts of biocrystallization based on transient amorphous Ca-carbonate

The diverse roles of amorphous minerals in biology was reviewed by Simkiss (1994), but at that time it was difficult to identify and describe these noncrystallized materials, as has long been the case, which prevented the adequate appreciation of their broad distribution and importance. The recent development of physical methods – above all, infrared absorption spectroscopy and synchrotron-based extended X-ray fluorescence (EXAF) methods – now permits more efficient investigation of this. With respect to the mechanism of biocrystallization, the finding by Beniash *et al.* (1997) that amorphous calcium carbonate appears as a transient mineral material in the formation of larval spicules of sea urchins suggests a completely new impression of the methods by which Ca-carbonate can be concentrated in areas where calcareous structures are to be produced. Now, indications of the frequent and diverse occurrences of this mineral form are so common that, in a recent paper, Addadi, Raz and Weiner (2003) considered it as a "tantalizing possibility" that many other phyla accumulate their calcium carbonate previous to crystallization by the same method. Additional data, culminating in the recent report of amorphous Ca-carbonate in the brachio-pod shell mineralization process (Griesshaber *et al.* 2009), suggest that what was a possibility in 2003 is now a probability.

The importance of the widespread occurrence of this process of calcium-carbonate accumulation in the sequence of events leading to calcareous biomineralization of shell materials cannot be overestimated. Compared to the commonly accepted model of crystallization in a liquid-filled space, the "radical change" suggested by Addadi *et al.* (2003) relies on the fact that transformation of amorphous Ca-carbonate to one of several crystalline polymorphs can occur without any dissolution and recrystallization, thus making the concept of crystallization within a liquid layer irrelevant. In various aspects, the layered growth and crystallization model, produced by the top-down approach to shell structure illustrated in this book, appears to qualify as one of the "radical changes" to be expected with the introduction of amorphous calcium carbonate being recognized as the common transient phase during the formation of biogenic Ca-carbonates.

(1) This hypothesis is fully compatible with the micrometer-thick accumulation of material prior to crystallization as observed in microstructural and biochemical research, as had been suggested for the first step of the growth model. The concept of "direct transfer" of material from the outer surface of the mineralizing organ to the growing mineral surface, as hypothesized by Crenshaw (1980), is perfectly consistent with the layered growth and crystallization model. This has been repeatedly noted above, and Crenshaw's observation is likewise one of the "radical changes" suggested by Addadi *et al.* (2003).

(2) The similarity of the nanometer scale interaction between the mineralized material and its irregular organic coating that has been seen in all cases studied also indicates that formation of reticulate crystals by stepwise crystallization is apparently the general mode for crystallization of growth layers.

(3) In respect to the crystallization process itself, as it occurs within the basic growth layer, neither precise indications nor a reliable hypothesis exist at present concerning the molecular mechanism that causes the transformation of amorphous calcium carbonate into carbonates with well-organized crystal lattices.

Experimental research has long focused on the processes of crystallization and growth of Ca-carbonate in conditions other than in the hydrous environment. The use of gel as a crystallization medium would seem to provide an appropriate pre-adapted medium for investigations into understanding what occurs in biogenic carbonates, based on data derived from the new results of microstructural investigations. Research on the influence of poly-acrylamide gels (Henisch 1988; Putnis *et al.* 1995; and more recently Grassmann *et al.* 2003) has succeeded in forming crystals of calcite from $CaCl_2$ and $NaHCO_3$ solutions as the result of diffusion of each of these solutions from opposing sides of a cylinder composed of newly synthesized hydrogel. When the two solutions meet within the polyacrylamide medium, an aggregation of small calcite crystals is produced in the pores of the hydrogel.

This resultant carbonate occurs in the form of monocrystal-like pseudo-octahedra. The TEM view (Grassmann *et al.* 2003, p. 649, Fig. 3) illustrates that these faceted microcrystals are aligned into monocrystal-like aggregates. Furthermore, ATG indicates that a part of the gel has been incorporated between submicrometer-sized crystals (up to 0.7% of weight).

In spite of some similarities, such as the formation of very well ordered crystal-like units by clustering of single crystals, and having some acrylamide gel present within the octahedra, comparison with the reticulate mineral materials that are crystallized from gel as seen in every biogenic Ca-carbonate does not seem to have further implications. The faceted crystals observed by Grassmann *et al.* have never been found in any biogenic material. Additionally, the use of liquid solutions to produce calcite precipitation is the part of this research that reduces its relevance in comparing their results with the biological transformation of amorphous calcium carbonates.

Under the general term of "mesocrystals," objects formed through a "nonclassical" crystallization process have recently been produced (Cölfen and Antonietti 2008), with detailed analysis of the mechanisms by which the spontaneous assemblage of submicrometer crystalline units can lead to the formation of crystals with large dimensions (for a review, see Meldrum and Cölfen 2008). They emphasize two major points that are of interest with respect to the properties of biominerals: (1) the shape of the final objects grown through this process is infinitely variable, because addition of extra elements can occur without producing "classical" growth plans, or interfacial angles, and (2) mesocrystals are not necessarily compact. Actually, a number of biominerals (such as molluscan prisms, and coral fibers) are cited or illustrated as examples of materials resulting from "nonclassical crystallization."

There is little doubt that the concept of "nonclassical crystallization" *sensu lato* can include the formation of crystal-like microstructural units; each of the previous chapters has emphasized the impossibility of "classical crystallization" accounting for the properties of the materials that result from the biomineralization process. Just the reverse, the initial step in the process of nonclassical crystallization *sensu stricto* is the formation of "individual nanocrystals" (Cölfen and Antonietti 2008, p. 96). These, according to the authors, may be the result of diverse processes. Amorphous particles were also hypothesized by Cölfen and Antonietti as possible starting points in the formation of mesocrystals. Assuming that Ca-carbonate concentrations result from intracellular processes, this would furnish us with a broad new research area of biological study.

5.4.6 Implications for the biomimetic approach

The ultimate objective of the biomimetic approach is the creation of materials closely resembling natural structures, with these materials having properties that are commonly very striking from an engineering point of view. Research into the interactions between crystals and organic molecules began long ago, as a logical approach to investigating mineral ion capture by organic molecules or substrates. Kitano and Hood (1965) explored the consequences of amino acids, organic acids, or sugars being present in solutions where

Ca-carbonate is precipitating. Later, Kitano *et al.* (1969) added various concentrations of magnesium ions to their solutions. Surprisingly (here in 1969, when soluble matrices were yet unrecognized), their results indicated a reduction in crystallization rates and complexity of results in terms of both mineralogy and shaping. Wheeler *et al.* (1981) also reported that their matrices were composed of compounds with high molecular weights that were strongly reactive to PAS staining, which suggested to them that sugars were present. At the same time, the importance of organic sulfates in providing potential crystallization sites was also stressed by Greenfield *et al.* (1984). Wheeler and Sikes (1984) explored modification of crystal shapes when crystallization occurs in solutions containing organic molecules. Using solutions containing soluble shell matrices, Albeck *et al.* (1993) experimentally demonstrated that portions of these matrices could be fixed on crystal faces, thus providing certain of them with a curvilinear morphology. These studies were truly the first biomimetic approaches. The resultant change of crystal faces is very real, as shown by models of coccoliths or otoliths (otoconia) formed by Mukkamala *et al.* (2006), who utilized polycarboxylate ions to influence growing calcareous crystals. The results show that expertise in utilizing chemical shaping methods to form carbonate structures resembling the shapes of natural objects can be impressive.

The point to be emphasized is that coccoliths and otoconia are precisely the two cases in which a chemical shaping approach is relevant. Coccoliths are developed within Golgi vesicles (for a review, see Young and Henriksen 2003), whereas otoconia are formed as calcified protein matrices in the vertebrate hearing vestibular system (Steyger *et al.* 1995). Both associate curved and polyhedral morphological surfaces, a pattern that has never been observed among the numerous examples studied (and are never present in normally growing fish otoliths for instance; see Fig. 4.56). Therefore, the consistent series of data covering the wide range of calcareous structures previously shown suggests, *a contrario*, that developing a mimicking approach based on the concept of shaping a single crystal during its growth within a liquid would seem to be a misdirected effort to reproduce the microstructural units built through the stepping growth and crystallization mode observed in most of the natural biocalcifiations.

6

Microcrystalline and amorphous biominerals in bones, teeth, and siliceous structures

Data illustrating the nature of calcareous biocrystals

In contrast to the process of calcareous biocrystallization, for which different interpretations are still being debated in current literature, questions concerning the state of the mineral phase in bones and teeth on the one hand, and in siliceous structures on the other, are not particularly a matter of controversy at comparable scales. As in calcareous structures, the observation of thin sections in polarized light with the optical microscope has permitted establishing the crystal organization of phosphatic material in bones and teeth, as well as the amorphous state of the silica in sponge spicules (as well as in diatoms, unicellular algae, and the Radiolaria). Surprisingly, questions that do arise concerning the more precise character-ization of these two very different mineral phases have comparable causes. In bones and teeth, crystals are so small that, until electron microscopy, and now atomic force micro-scopy, direct observation was impossible. Yet bone and tooth constituents are by far the most studied of all mineralized biological structures, essentially in medical or dental research. The situation is still more frustrating concerning the mechanisms leading to the striking paradox in all siliceous biomineralized objects. In every case, morphology is seen clearly to be controlled, right up to the highest resolution of instrumentation, although this was first demonstrated in the spectacular drawings of microscopic observations of radio-larians made by Haeckel in the nineteenth century, and since completely confirmed by scanning electron microscopy. However, until development of molecular biology techni-ques, no data pertaining to their biomineralization mechanisms were available, and even now, the formation of these highly delicate architectures remains largely enigmatic.

In addition to the contrast between the crystallized and amorphous conditions, another difference makes these two biominerals opposites. In bones and teeth, the mineralized structural features are extracellular, and optical observation easily shows the mineralizing cells dispersed within the mineralized portion of bone (Figs. 6.1a–c). Conversely, siliceous structures are basically intracellular, even when, due to their extended dimensions (as in some siliceous sponge spicules several meters long), much of the mineralized material has subsequently become external.

From a geological perspective, bones and teeth are quantitatively insignificant, whereas siliceous biomineralization results in the formation of impressive sedimentary units. The contribution of siliceous biomineralization to marine sediments has long been known, from the early Antarctic exploration of Ross (1839–1842), to Lisitzin (1972), Nelson *et al.*

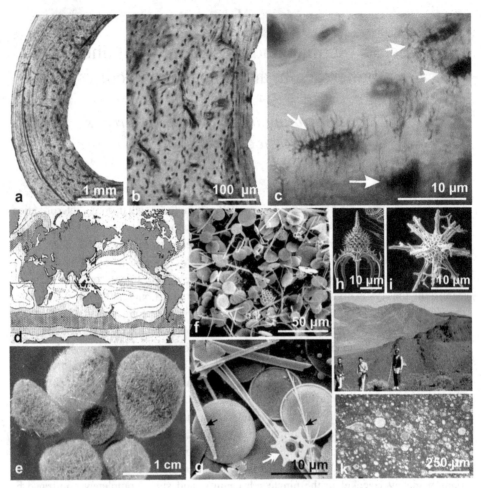

Fig. 6.1. Calcium phosphates and silica: two major mineralizing systems ranking only behind Ca-carbonates in abundance. (a–c) These images indicate how dispersed cells are within the mineralized structure, pointing to the essential difference in the overall organization of bone compared to calcareous biocrystallization. Note the specific aspect of the mineralizing cells: a dense network of thin cellular processes – the canaliculi – allows communication and exchanges between the neighboring cells to be maintained (c: arrows); (d–e) Distribution of silica on the world ocean floor (d).The high-latitude circumpolar areas are characterized by a high production of 500 g/m²/year (data from Lisitzin 1972). Nodules are made by agglutination of siliceous spicules. Such nodules cover wide surfaces of the Antarctic floor (collected by N. Boury-Esnault, Marseille); (f–g) Another ingredient in siliceous oceanic sediments: diatoms (unicellular algae), sponge spicules (g: arrows), and silicoflagellates, a minor group of silica-producing organisms (g: double arrow); (h–k) The third major contributor to silica concentration in the ocean: the Radiolaria. One of the famous drawings by Haeckel (h), SEM view of another species (i), arid mountainous landscape of the Zagros mountains, Iran (j), entirely formed by radiolarites, siliceous rocks made by the accumulation and cementation of Radiolaria (k).

(1995), and numerous others, making it obvious that although limited to a restricted number of biological groups (Figs. 6.1d–k), concentration of silica by marine organisms is a major geological factor. Additionally, in continental areas, the formation of solid concretions of opaline silica in tissues of higher plants is an important part of the geological cycle of silica.

Although there is general agreement regarding the structural and compositional patterns of these two distinct types of material, their use in providing information for various geological or paleontological applications is not so firmly established. Knowledge of their fine-scale organization is important, specifically for bones and teeth, in order to improve comparative microstructural studies between vertebrate fossils. Due to the importance of vertebrate fossils in the study of terrestrial environments, in modern geological studies this knowledge is vital. It is especially important to understand their structure and composition in order to carefully assess chemical data obtained from them.

6.1 The concept of *microstructure* in bones and teeth

Microscopic examination of a polished surface or thin section prepared in the compact cortical part of a long bone (e.g., the femur of a tetrapod vertebrate) permits the recognition of specific organizations, justifying the use of "bone microstructure" here. This is comparable, with respect to the scale of observation, to the microscopic approach to calcareous invertebrate exoskeletons or shells. It is important to realize how confusing the use of this term can be when applied to materials that are markedly different, not only with respect to their mineralogy but in every aspect of their formation, growth mode, and stability, etc.

The most significant of the differences between bone and the calcareous features built by invertebrates is the relationship between cells and mineralized parts. This remark may seem surprising, because in both cases the mineralizing material is extracellular. However, in considering the overall organization of mineral components, we have noted numerous times that in Ca-carbonate skeletons of most invertebrates, the sizes, shapes, and spatial arrangements of microstructural units are independent of the cells secreting the materials that form these mineralized components (the secondary layer of brachiopods is a notable exception). In bone, it is just the opposite: there, the spatial arrangement of the mineralizing cells determines the microstructural patterns.

The most characteristic part of bones in this respect is the organization of their external part, named "compact bone," although containing a dense network of canals called the "Haversian" canal system. These canals provide passageways for blood vessels and nerves. Primarily longitudinal Haversian canals are connected by transverse Wolkmann's canals (Fig. 6.2a). Micrographs in the SEM of simple fractured surfaces (Fig. 6.2b) show this organization in the 100 μm range, and even more so on polished surfaces (Fig. 6.2c). The concentric layering of mineral material around the axial canal is determined by the arrangement of the cells producing these superposed layers. Instead of a continuous epithelium that produces mineralized growth layers, as in calcareous shells, the bone-producing cells (the osteoblasts) are free and distinct (Fig. 6.1c shows their typical appearance, and their filamentous extensions). When commencing production of the first layer of mineral around

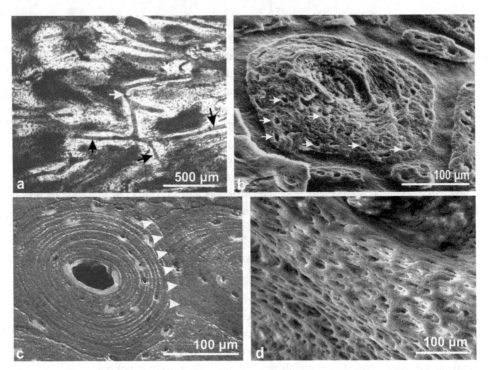

Fig. 6.2. Microstructure of "compact" and spongy bones. (a) Canals in bone structure; (b–c) Haversian system: a fractured surface (b) and a polished surface (c). Note the partial removal of an Haversian system (on the right by the system at left center). Arrows here mark the limit of the process of resorption; (d) Spongy bone with no visible or specific microstructure. b: Dauphin (2005).

a Haversian canal, the mineralizing cells (osteoblasts) remain distinct and never form a continuous epithelial-like cell layer that might be compared to the mantle epithelium of the molluscs or the basal ectoderm of corals. At the 100 μm scale, observation of bone reveals the regular superposition of microstructural layers. Although distinct, osteoblasts form a given layer and between superposed layers they remain in contact through a dense network of cellular processes. Coordinated secretion obviously exists, but the bone layering visible in the osteon (the basic microstructural unit surrounding the Haversian canal) does not result from a repetitive secretion process. Each of the superposed layers requires the grouping of new osteoblasts. More generally, in this respect the growth mode, as well as the mineralization mechanism of bone, differs basically from those in shells, corals, or any other calcareous skeleton. Accordingly, bones are structures that contain a number of cells: the mineralizing cells of the osteons, and the cells included in the Haversian canal system. Their presence provides the fossil nucleic acids found in bones (but not in enamel; see below), whereas shell structures are entirely made of extracellular material.

The primary difference between shell and bone microstructures is further accentuated by the consequences of the mechanism of resorption, which is specific to bone (it does not exist in enamel, see below, subsection 6.1.6). In parallel with the osteoblasts responsible for bone mineralization, another set of specialized cells exist in bone, whose role is to remove the bony areas that need renewing. Aging is certainly the major reason for this process; it has been estimated that approximately every five to six years, the complete skeleton of a human adult has been fully renewed. Thus, "Bone substance has one of the highest turnover rates of all the tissues or organs in vertebrates" (Glimcher 2006). Other factors, such as adaptation to new constraints or specific bone diseases, may also trigger the renewal process. Therefore, turnover occurs at highly variable rates among bones of an individual, and even within different areas of a given bone.

In every instance, the cells in charge of bone resorption, the "osteoclasts," utilize a biochemical process leading to dissolution of bone matrix. When an osteoclast commences the process of bone removal (Baron 1989), the cell membrane facing the bony surface develops a specific "ruffled" pattern that results in increasing the cell's chemically active area. Additionally, at the periphery of the cell, a ring seal is formed, creating a closed volume between the cell membrane and the bone surface. The cell then releases hydrolytic enzymes (cathepsin and metalloproteases) that remove organic bone matrix, while increase of acidity in this closed volume results in dissolution of the mineral phase. This process provides osteoblasts with space for formation of fresh bone, whose structure is thus adapted to the constraints operating at the moment of formation. Figure 6.2c illustrates the localized aspect of the removal/reconstruction process. The surface marked by white arrows shows a surface of partial resorption of an Haversian system, the space corresponding to the dissolved part having been replaced by similar bone, which here is part of a system with a different orientation.

It should not be necessary to emphasize the importance of such a continuing process with respect to the use of fossil bones as sources of biological or environmental information. This differs from the use of shells or coral skeletal material that continuously increases volume and/or features, although some rare examples of partial shell removal do exist among the gastropods, as an example. For this reason, a vertebrate skeleton can provide information on a certain period of time only. This is also true for tooth enamel.

Additionally, such well-differentiated formation of microstructure in bone can follow various pathways, based in part on two distinct origins for bone-producing cells. Bone can be produced by transformation of cartilaginous substances, especially during early growth stages of the individual, whereas membranous ossification results from the direct trans-formation of stem cells into osteoblasts.

6.1.1 Overall chemical characterization of bone: mineral and organic phases

In addition to the basic difference in growth mode, bone (but not enamel) is also distinct from calcareous biominerals with regard to the proportion of the organic components associated with the mineral phase. Bone contains approximately 30% by weight organic matrix (see infrared absorption profiles, Fig. 6.3a). Chemically, the mineral phase of bones is a calcium phosphate, in a mineral form belonging to the apatite group. Its composition is

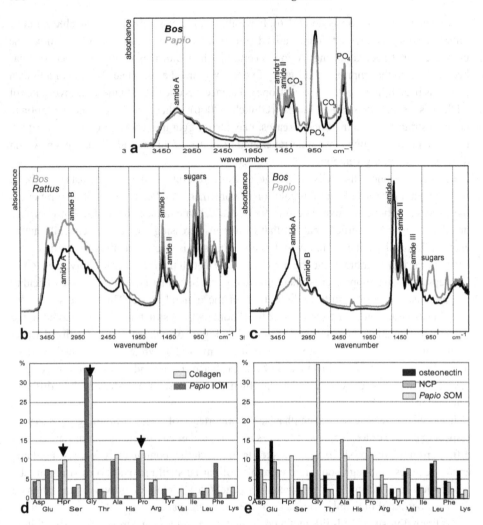

Fig. 6.3. Overall characteristics of bone. (a) Overall IR spectra of bones; (b–c) Infrared spectra of organic matrices extracted from two bones: insoluble matrices (b), and soluble matrices (c); (d) Comparison between amino acid concentrations in typical collagen and insoluble matrix from *Papio*; (e) Comparison of soluble matrices from *Papio* bones, noncollagen proteins (NCP), and osteonectin (specimen from B. Senut and M. Pickford, Paléontologie, MNHN, Paris). a–c: Dauphin (1998).

complex, and its structure at the molecular level is still a matter of debate. Hydroxy-apatite, carbonated apatite, carbonated hydroxyl-apatite, and dahllite have all been reported. The general formula of this bio-apatite is $10Ca[6(PO_4) -x(CO_3)] 2(OH)$. Phosphorus represents from 12 to 16% by weight and Ca from 30 to 35% by weight.

Carbonates are also present in bones, approximately 7% by weight, and are generally considered as substituted for phosphates. However, in spite of the large number of studies carried out in medical research, neither the exact nature of the carbonate substitution nor the state of hydroxylation of the mineral lattice are as yet well known (Wopenka and Pasteris 2005). The mean concentrations of minor elements in bone is provided in Fig. 6.3b. It is important to remember that, as pointed out by Glimcher (2006), the composition, structure, and other properties of the mineral phase in bone vary continually, depending on metabolic constraints and individual living circumstances.

The decalcification of bone liberates a soluble organic matrix and an insoluble one. Infrared spectrometry and the proportions of amino acids in these two organic phases illustrate their compositional differences (Figs. 6.3b–e). Structurally, the main component of bone matrix is collagen. It is a rigid and fibrous glycoprotein in the form of a braid with three strands, having a length of 300 nm and a diameter of 5 nm. It is rich in proline and hydroxyproline. The collagen fiber is an aggregate of fibrils in the form of a ribbon that can be seen with the optical microscope. In bone, the majority of collagen is Type 1 (numerous different collagens have been identified in various biological groups). Noncollagen proteins (NCP) represent from 10 to 15% of the protein in bone. There are more than 200 of them; the most studied are:

– Osteocalcin, specific to bony tissue; has a marked affinity for hydroxyapatite.
– Gla-protein of the matrix that is not specific to bone.
– Osteonectin, the most abundant glycoprotein of NCPs, having a molecular weight of approximately 40 kDa.
– The sialoproteins, with molecular weights that vary from 44 to 200 kDa. An example is osteopontin, BSP.
– The proteoglycans.

The occurrence and distribution of these proteins and proteoglycans are still poorly known, because very few species have been studied in detail. Furthermore, some proteins are present in the animals when young that disappear in adults. Additionally, the role of proteoglycans or noncollagenic proteins has not yet been clarified. They are possibly involved in controlling the localization and degree of mineralization.

Amino acid compositions of soluble and insoluble phases resulting from bone decalcification indicate the difficulties in comparing varying published results (Figs. 6.3d–e). Some analyses have been carried out after eliminating the cellular portion of the bone, while others take into account the complete bone. Furthermore, the chemicals used for decalcification directly influence the proportions of the included amino acids. Some types of collagen are soluble in acetic acid, and thus cause higher percentages of proline, hydroxyproline, and glycine. This type of discrepancy between analytical procedures casts doubts on comparisons and interpretation of the exact functions of the NCPs.

6.1.2 Bone structure at the micrometer level

In long bones, the first-order microstructural units (the osteons) are regularly arranged parallel to the bone axis (Fig. 6.4a: arrows). At higher resolution, they show a more three-dimensional organization. Each of the concentric layers that build the osteon is made of joined cylindrical fibers, parallel within each layer, but with alternating orientations from layer to layer. At medium enlargement (Fig. 6.4b), these fibers appear as the elementary

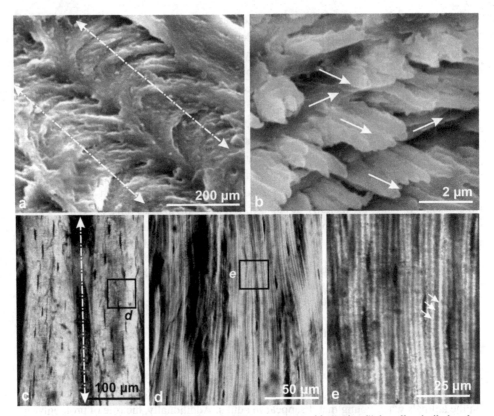

Fig. 6.4. Osteons in long bones. (a) Plywood-like structure with "twisted" lamellae built by the mineralized collagen fibers; (b) Lamellar bone layers with an alternation of orientations; (c) Thin section viewed in polarized light (crossed nicols) at low magnification of an osteon that was cut axially (c: arrow); (d–e) Enlargement of fibrous layers. Color uniformity indicates that, within a given layer, all microcrystals share identical orientations. Polarized light emphasizes the differences in crystallographic orientation of structures between successive layers of osteons. Arrows in (e) draw attention to three layers forming a sequence with maximal contrast. a, c: Dauphin (2005).

component of bone. Analyzing and reconstructing their structure was key to understanding the essentials of bone formation; that is, establishing a reliable model of the spatial relationships between the major organic component, the collagen molecules, and the tiny apatite crystals associated within this collagen framework.

Generally speaking, resolution in observations needed to be increased by three orders of magnitude prior to reaching the nanometer scale in order to view the arrangement of collagen molecules.

The formation of nanometer-sized mineral units in bone: an evolution of
interpretations, from chemical to biochemically controlled precipitation

Glimcher (2006) summarized the paradox of early interpretations of phosphate crystallization in bone by purely chemical precipitation. Based on the fact that supersaturation of body fluids is required to explain a chemical crystallization process, the main question still perplexing researchers in the 1920s to 1940s was how it could be that phosphate crystallization does not occur in the whole body. Clarifying the structure of the collagen molecule and the arrangement of densely packed collagen molecules took several decades, but finally resulted in the replacement of purely chemical models of precipitation by a newer, "heterogenous crystallization" model (see Glimcher 2006 for a detailed synthesis).

A readily understandable model of the relationship between the mineral phase and densely packed collagen molecules was obtained at the time when it was realized that phosphate nanocrystals were distributed within "holes," that is, in the space existing beween the C-terminal end of one molecule and the N-terminal end of an aligned neighboring molecule (Hodge and Petruska 1963; Hodge 1967). One first had to compare the volume created within collagen fibers by these holes, repeated at a distance of approximately 67 nm within each coaxial collagen fibril, and regularly placed between neighboring files. This volume then could be compared with the quantity of mineralized material within fully mineralized bone. The conclusion was that the crystals were added in the intermolecular spaces (see Glimcher 2006, Fig. 17d).

This impressive biochemical reconstruction of the bone mineralization process basically depended on the structural biochemistry of collagen as it forms a nucleation substrate. Another part of the process involves noncollagenous molecules that control nucleation and development of the crystals themselves in bone. Deciphering the activity of the complex set of molecules involved, and their respective role and relationship to gene activity, is the subject of ongoing research (e.g., Murshed *et al.* 2005).

Previously, images of bone apatite crystals were obtained through transmission electron microscopy. Different physical methods have been used to describe precisely the overall characteristics of the apatite crystals located within collagen fibers. Landis *et al.* (1993), carrying out a three-dimensional reconstruction using high-voltage electron microscopy, were able to identify the dimensions of these crystals: a length of as much as 170 nm, a width of 30–45 nm, and a thickness of 4–6 nm. These plate-shaped units have been found occurring as even thinner platelets (1.5–2 nm) in fish bone (Burger *et al.* 2008). In a given bone, crystal lengths and widths may vary, whereas their thickness appears to be the more stable dimension. Results obtained from mineralized tendons of domestic turkeys confirm the more general observations by electron microscopy (Weiner *et al.* 1986; Traub *et al.* 1989; Weiner and Traub 1991). A small-angle X-ray scattering method enabled Fratzl *et al.* (1996) to determine the orientations of both collagen molecules and hydroxyapatite crystals and their dimensions in mature bones of a horse. The transverse orientation of crystals was studied and found to vary according to the position of the sectors analyzed with respect to the antero-posterior orientation of the bone. Crystals are seen to be larger in the anterior part of the bone. It is only in the last decade that preparative processes have been developed to isolate these mineral components from their organic substrate (Kim *et al.* 1996), thus allowing physical identification and the description of free bone crystal units. According to these observations, the nanometer-scale units exhibit comparable patterns whatever the origin and type of bone.

As emphasized by Glimcher, the search for a definite shape of calcium phosphate crystals in bone must be subordinate to the consideration that bone cells are continually nucleating, growing, and dissolving, and this occurs both in the collagen substrate and in the mineral units. Concerning this, the only published statement of Glimcher notes that in vitro investigations cannot provide a reliable base for understanding biochemically driven growth mechanisms for crystallization within collagen fibers.

Diversity of bone histology

In addition to the bone microstructure illustrated in Figs. 6.2 and 6.4, which are specific to the cortical part of long bones, other types of organization of bone structures and modes of ossification exist within all individuals. The presence of Haversian systems is typical for the medial parts of long bones (diaphyses). They do not exist in the trabecular bone that forms the distal parts of the long bones (epiphyses). In these regions, orientation of the pillars and sheets of bony material are adjusted and continuously modified to match specific constraints that result from muscle insertions and body weight. Other bones (actually most of the bones of the human body, for instance) are built by yet another process: endochondral calcification, in which cartilage is transformed into bone. Bones (whatever their organization) and cartilage surfaces are covered by a dense layer of connective tissue, the periosteum, itself coated internally by a specialized layer, the endosteum. Cells of these two layers have osteogenic potential, in that they can transform themselves into osteoblasts. Diversity in bone microstructures results from a conjunction of these two main mechanisms. Variability within taxa (e.g., the particular light structure of a bird's bones) and the influence of major physiological processes (such as homeothermy) have produced convergent structural evolution resulting in a variety of structural patterns whose interpretations are still debated.

6.1.3 Microstructural patterns of dentine and enamel in teeth

Teeth are the mineralized elements inserted into the jawbones of numerous vertebrates. Directly linked to the mode of nutrition, they have acquired characteristic shapes and structures throughout vertebrate evolution, resulting in extreme diversity. The numerous generations of rather similar and morphologically simple teeth developed during the life of fishes and reptiles have been replaced in mammals by teeth characterized by highly differentiated morphologies adapted to their functions. Mammals possess only two generations of teeth (the numbers of which vary according to individual species), and these teeth are sharply differentiated within a given individual.

Teeth are formed of three components. Typically, the exterior layer (i.e., the only one normally visible) is the enamel, the hardest of all biominerals. Enamel covers the "crown" of the tooth. Beneath the enamel is dentine, a bony tissue that includes ramified nerve endings and blood vessels. It is sometimes called ivory. Not usually visible in healthy teeth, dentine normally can become exposed in teeth with continuous growth (Fig. 6.5e, h: white arrows). Enamel and the upper part of the dentine form the crown of the tooth, whereas in the lower

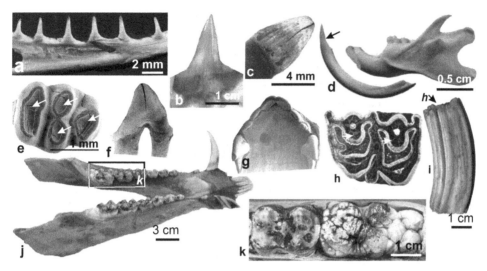

Fig. 6.5. Morphology of teeth. (a–c) Fish and reptile. Jaw and teeth (a) of a bony fish, lateral view (b) of a shark's tooth, and lateral view (c) of a crocodile tooth; (d–k) Mammalian teeth. Lower jaw and incisor (d) of a rodent, and occlusal view (e) of a rodent tooth (molar): note the visible dentine (d: arrows). Lateral view (f) of *Didelphis* tooth (insectivore), and view of the palate and teeth (g) of *Talpa* (insectivore). Upper surface (h) and lateral view (i) of a horse molar. Lower jaw (j) of *Sus scrofa*, with canines developed for defense and incisors forming a scoop, and occlusal view (k) of the molars of *Sus scrofa*. j: Dauphin *et al.* (2008b).

part of the tooth, dentine forms the roots (single or multiple). Cement, the third tooth component, is bony tissue covering the teeth's roots. Each of these layers possesses a particular structure and composition, according to the taxon.

This generalized outline of tooth organization cannot be universally applied. This is the result of differences in the biomineralization mechanism for enamel and dentine in different taxa. Dentine is a bony tissue, resulting from the activity of cells that are essentially similar to osteoblasts, but differ importantly regarding the rhythmicity of their activities such as remobilization. However, enamel differs in its basic makeup. It is secreted by ectodermal cells (ameloblasts) that function during a rather short period, after the upper part of the dentine of a tooth is fully developed and needs to be covered by a harder layer. This occurs just prior to the time when teeth erupt from flesh (gums). No ameloblast cells are incorporated into the enamel layer, thus it differs from bone. When a tooth becomes visible, the mineralizing activity of the enamel-producing cell layer is finished. Thus, enamel is only produced once for each tooth and is a "nonreparable" feature.

The method described above covers the formation of teeth that have limited growth. Other special growth modes have been developed by biological groups in which teeth are submitted to intense wear. Horses, for example, whose diet includes silica-rich plants (Graminacea), have molars that are structurally complicated by the formation of longitudinal infolds (Fig. 6.5h), involving both enamel and cement. Thus, a resistant abrading surface is created (Fig. 6.5i). Additionally, growth of these molars is continuous (although slow). The age of the animal that furnished these specimens can be estimated precisely by examining the infolds. Sometimes incisors also grow continuously (beavers, hares, etc.) and become adapted to cut even hard woody (thoroughly lignified) plants (Fig. 6.5d). In all cases, not only tooth morphology is functionally modified, but also highly specialized microstructures are developed according to special needs.

This applies fully only for the mammals, but can at times also occur in reptiles. In fishes this tooth material has been given a variety of names, such as enameloids, vitrodentine, placoid enamel, and durodentine, thus reflecting some uncertainty concerning its origin (see Poole 1967). It could be hypercalcified dentine, and thus not formed by ameloblasts. A certain amount of consensus has been established, so that the enamel of reptile and mammalian teeth is considered "true" enamel, based on its having been formed by ameloblasts.

6.1.4 Chemical and biochemical data

Enameloid, enamel and dentine are composed of apatite. Nevertheless, in spite of this common general mineralogy, not only do their internal structures differ, but also their chemical compositions vary considerably with respect to their minor and trace element contents. They vary according to tissue of origin, the mode of growth of teeth, and taxon. Similarly, the proportion varies between mineral and organic matter, and in amount of included water. Thus, mature enamel contains from 95 to 99% mineral matter, so that, from this viewpoint, dentine is quite similar to bone (approximately 75% mineral).

Enameloids are possibly very rich in F, but this is not true in all animals. In general, data are rare on this material, and no statistical study is available. However, it has been suggested that the contents of F and Fe in enameloids of bony fish are related to their phylogeny (Suga *et al.* 1992), but that the contents of F and Fe are independent of each other. In the same samples, dentine did not show any enrichment in either F or Fe. Although not abundant, analyses of reptilian enamel nevertheless suffice to allow comparison with mammalian enamel. In both groups, enamel is rich in Na and poor in S and Mg (Table 6.1 and Fig. 6.6); its contents of F are always low, and enrichment in Fe seems tied to orange coloration. The content of various elements can sometimes be used to deduce the animal's alimentary regime, but this is a complex (and risky) process.

Table 6.1 *Mean content of major and minor elements in dentine (D) and enamel (E) of reptilian and mammalian teeth – in ppm.*

		Na	Mg	S	Cl	Sr	Fe	K	Mn	Si	P	Ca
Reptiles	E	10682	3082	195	1377	710	83	416	57	81	175431	325945
	D	2990	10267	1942	1281	703	924	233	215	208	107379	164257
Mammals	E	8612	1923	269	2835	719	470	776	54	1538	176226	317534
	D	6608	10845	1779	1394	935	452	800	131	1957	139781	225494

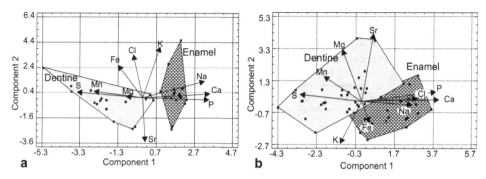

Fig. 6.6. Comparison of chemical composition in enamel and dentine of modern reptiles (a) and mammals (b) by Principal Components Analysis. Dauphin and Williams (2007).

The composition of dentine is the most variable of all tooth materials. It is largely dependent on the method of growth of a particular tooth, but is always poorer in P and Ca than enamel is. In teeth with continuous growth, as in rodent or rabbit incisors, Mg content can attain 2%, while in those teeth having closed roots (as do human teeth), Mg quantities are in the order of 2000 ppm (Dauphin and Denys 1994). No difference of this magnitude has been observed in the enamel of the same teeth. Dentine is also rich in S.

Two major categories of proteins have been identified in the enamel of adult amphibians, reptiles and mammals: amelogenins and enamelins. Enamelins have also been recognized in the enameloid of both bony and cartilaginous fish, and additionally, in larval stages of amphibians (Moss-Salentijn *et al*. 1997).

Amelogenins are hydrophobic proteins, rich in proline. In mature enamel, they represent 90% of the matrix, and are present in the form of nanospheres having diameters of approximately 30 nm. During maturation of enamel, these are progressively degraded and disappear, but in such a way that they modify the amino acid composition in adult enamel. According to Satchell *et al*. (2002), amelogenin is the most conservative protein organically associated with enamel. Enamelins are large, hydrophylic glycoproteins with molecular weights greater than 150 kDa, and are very acidic. They remain bound to crystallites of enamel and are also found in mature teeth (Inage *et al*. 1989). In pig or bear enamel, enamelin has a molecular mass of 186 kDa at the time it is secreted. Subsequently, it undergoes a series of enzymatic hydrolyses that finally result in forming polypeptides that remain within the enamel.

Ameloblastin apparently is the second most abundant component of the organic matrix in enamel, although only representing about 5 to 10% of the total proteins. It is locally concentrated at prism peripheries. As is common with other proteins, it is secreted with elevated molecular weights (> 65 kDa), and subsequently degraded to form polypeptides with smaller sizes. The distribution and organization of proteins and minerals are possibly taxon-dependent (Diekwisch *et al*. 2002). For a long time, enamel has been described as totally lacking collagen; lately, however, collagen has been reported in mature human enamel (Acil *et al*. 2005).

Table 6.2 *Example of amino acid compositions of some proteins in mammalian dentine (from Linde 1889).*

	Phosphoryn		Rattus	
	Rattus	*Bos*	95 kDa GP	60 kDa GP
Asp	993	366	165	114
Thr	8	7	65	58
Ser	530	514	111	69
Glu	20	11	180	118
Pro	1		43	115
Gly	24	34	156	64
Ala	7	8	66	105
Cys				3
Val			34	93
Met				1
Ile	2	1	26	30
Leu			36	78
Tyr	2	4	6	22
Phe			8	34
His	7	4	33	23
Lys	10	48	43	40
Arg		3	28	30

Collagen is abundant in dentine, comprising approximately 90% of the organic phase. As in bone, the majority of it is Type I collagen. The noncollagenic proteins (NCPs) are very closely tied to the mineral phase. The phosphoproteins make up part of these, and they are habitually highly phosphorylated (PP-H) (Table 6.2). These are the most acid proteins known, with a pH of 1.1 in rats (Jonsson *et al.* 1978), and one assumes that they are tied to collagen electrostatically.

The proteoglycans are one of the major noncollagenic compounds in dentine. These are macromolecules having a protein center on which are tied lateral chains of sugars in a covalent fashion. The chains of sugars are glycosaminoglycans. Molecules of the galactos-aminoglycan (chondroitin sulfate) have also been identified. As a group, these proteogly-cans are acidic. Other NCPs are present, and are often acidic (rich in aspartic, glutamic, and sialic acids), and also rich in sugars.

6.1.5 Microstructure and nanostructure of dentine

Numerous uncertainties still exist with respect to the dentine found in fish teeth. Diverse names and different attempts at their classification have been proposed, but none has received general approval. Thus, such terms are found as osteodentine, vasodentine,

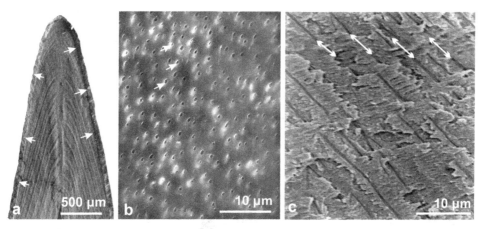

Fig. 6.7. Dentine in a crocodile tooth. (a) Longitudinal section of the apical zone, showing dominance of dentine and the thin enamel layer (arrows point to its base). Growth lamination in the dentine is readily visible as well; (b) Oblique sections of the tubules in crocodile dentine (arrows); (c) Longitudinal section in crocodile dentine, illustrating parallel tubules (arrows are within tubules). a, b: Dauphin (2005).

plicidentine, and orthodentine (see Bradford 1967). Osteodentine has a structure resembling that of bone, vasodentine possesses blood vessels, etc., while orthodentine, characteristic of higher vertebrates, is also present in some fish, such as the rays. It is characterized by the presence of numerous tubules having diameters of approximately 1 μm, oriented parallel to one another, perpendicular to the surface of the tooth. Tubules in orthodentine are occupied by prolongations of cellular tissue (Fig. 6.7).

In reptiles all dentine is orthodentine. Growth striations, parallel to the external surface of the tooth, are easily visible, most notably at the level of the zone covered by enamel. Dentine of the mammals is also orthodentine (Fig. 6.8), indistinguishable from the orthodentine of fish or reptiles. The number of tubules per square millimeter varies according to taxon, but statistical studies are sparse. Additionally, variations occur within a single tooth. Tubule density in dentine of the rat is approximately 15 000/mm^2 in the external zone, but 55 000/mm^2 near the pulp cavity. In some taxa, dentine can be divided into peritubular and intertubular dentine. The peritubular type is very hard and rind-like. In numerous situations, even within the same tooth, some zones reflect this subdivision while in other zones, peritubular dentine is absent. It is interesting that, in those rare edentates that possess teeth, they are completely composed of dentine.

6.1.6 Enamel microstructures

Differing from bone, *the enamel layer lacks cells*, and this is true, both for enameloids and enamels. In all cases, the structure at the micrometer scale closely depends on the taxon.

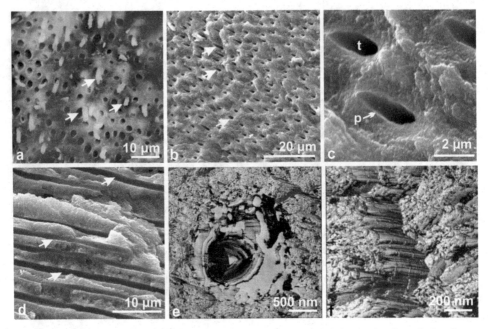

Fig. 6.8. Structure of mammalian dentine. (a) Section in the incisor of a horse showing tubules. Some are still occupied by cellular processes (arrows); (b–f) Wild boar molar. Oblique section (b) showing the regular distribution of tubules (arrows), detail (c) showing peritubular dentine ("p") associated with tubules ("t"), longitudinal section (d) showing parallel tubules (arrows), section (e) of one tubule showing peritubular dentine, and detail (f) of a tubule wall. b, c, e, f: Dauphin *et al.* (2008b).

Fishes

In fish teeth, the thinness of this outer layer makes its description difficult. The teeth of fishes are actually covered by a hard layer with a developmental process that differs from that forming other vertebrate enamels. This enameloid material has a mixed origin: in part ameloblastic (ectodermal cells) and in part odontoblastic (ectomesenchymal) (Gillis and Donoghue 2007). The enameloid in the elasmobranchs seems to be the most primitive of all hard tissues, and geologically, it appeared before enamel, to which it may be homologous (Moss 1977; Fearnhead 1979), or perhaps only analogous, according to Sasagawa and Ishiyama (1999).

The development of elasmobranch teeth comprises three major phases: (1) formation of the enameloid matrix, (2) mineralization of the enameloid and formation of the dentine matrix, and (3) the maturation of enameloid and mineralization of dentine. These stages have been described in detail by Sasagawa (2002). The organic matrix is produced by cells of the internal epithelium. Their organic fibrils are organized into bundles and apatite crystallites are deposited along these fibers. Radial and longitudinal crystallites have been described in several patterns within the teeth of both cartilaginous and bony fishes (Schmidt and Keil 1958). In *Lepisosteus*, the orientation of crystal long axes is practically parallel to the surface of the enameloid forming the tooth's external layer. In this external zone, the width of crystals only reaches 16–33 nm; in the internal layer they are even smaller (Sasagawa and Ishiyama 2005). Changes in the orientation of crystal bundles provide a prismatic aspect to the enameloid of pharyngeal teeth in the parrotfish. Near the dentine junction, crystallites have very irregular orientations (Carr *et al.* 2006). Three layers have been identified in the enameloid of butterflyfish: an external cuticle, a superficial layer that is very well organized crystallographically, and a nonoriented internal layer (Sparks *et al.* 1990).

Enameloid is also present on the teeth of actinopterygians, whereas true enamel has also been identified in some dipneustian lungfish (Satchell *et al.* 2000).

Reptiles

Reptilian teeth have simple forms; most often they are conical and they are commonly flattened, or at the least not round (Fig. 6.9). Their large pulp cavity is occupied by a replacement tooth during its formation; it will then follow the first. The enamel of reptilian teeth is true enamel, but it is always very thin. This is perhaps related to the facts that (1) most reptiles do not chew their food, and (2) the number of replacement teeth is large (from 40 to 50 generations for a crocodile). Thus, these teeth do not need to be particularly resistant to shock or to caries. The surfaces of crocodilian teeth are sprinkled with hollow tubes and narrow raised bands (Fig. 6.9a). The eventual role of these structures, described long ago, is still unknown. On some teeth, or even only on certain parts of teeth, prisms or pseudo-prisms are visible (Fig. 6.9b). In section, enamel is here composed of small elongate fibers that are more or less parallel to one another, but are perpendicular to the tooth surface (Fig. 6.9c). Sections also show growth striations paralleling the external surface of the tooth (Fig. 6.9d). The elongate fibers are themselves composed of smaller units (Fig. 6.9e). Growth striations are visible on the teeth of other reptiles (Fig. 6.9f), although "prisms" have only been described in the crocodilians. Examination of crocodilian enamel shows that in addition to the parallel crystalline elements, numerous irregular granules exist (Fig. 6.9g). In this poorly mineralized enamel, it is probable that the granules may be the remains of organic compounds. Additionally, the crystallites are not always regularly parallel (Fig. 6.9h), and their rectangular shape is shown in more detailed views (Fig. 6.9i).

Mammals

The development of mammalian enamel has been much studied because of implications for human dentistry. Several stages have been described in detail, notably the cellular events that intervene when a dentary epithelial cell changes into an ameloblast. At this moment, a marked increase occurs in the number of organelles within the cell, most notably those involved in protein synthesis. The newly differentiated ameloblasts commence deposition of the enamel organic matrix, and additionally, matrix of adjacent dentine begins to be mineralized. Several types of proteins are then secreted (Sasaki 1990). Amelogenesis is an appositional process (the formation of growth layers), and is influenced by external events. A rhythm of daily deposition has been observed in human teeth, adding a thickness of 2.7 to

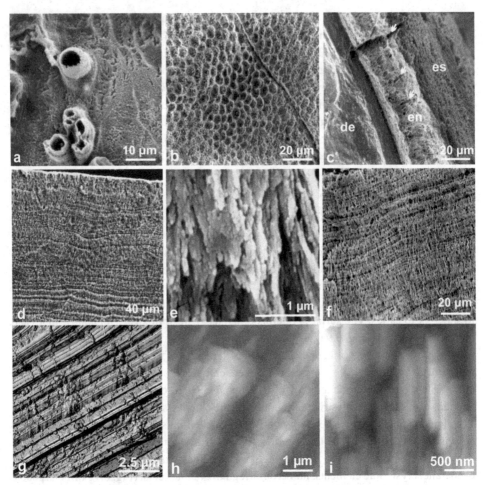

Fig. 6.9. Structure of the enamel in reptilian teeth: (a–f) SEM images, (g–i) AFM images. (a) Tubes and ridges of the external surface of a crocodile tooth; (b) "Prisms" of the external surface of a crocodile tooth; (c) Section showing the tubes (arrows) in the same sample ("es": external surface; "en" enamel; "de" dentine); (d–e) Enamel fibers (d) polished and etched, and (e) individual enamel fibers of the crocodile tooth; (f) Growth striations of the fibrous enamel of *Varanus*; (g) Parallel fibers of crocodile enamel. The irregular granules are perhaps the remains of organic compounds (phase image); (h) Fibers that are more or less grouped in packages in crocodiles; compare with (e) (height image); (i) Crystallites of crocodilian enamel (height image). Compare to Fig. 6.10 (mammals). a, d, e: Dauphin and Williams (2007); b: Dauphin (1984).

4.6 μm. Hunter-Schreger bands, visible in section, result from changes in the orientation of prisms between adjacent bands (Osborn 1965).

Mammalian enamel generally has a greater thickness and more complex structure than reptilian enamel. Since mammals only have two generations of teeth of very unequal durations, their teeth and more specifically their enamel, must resist the shocks imposed

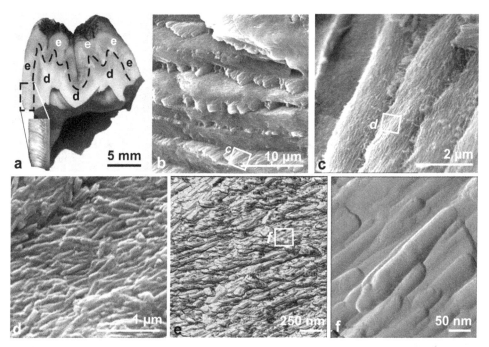

Fig. 6.10. Structure of enamel in wild boar teeth. (a) Fractured specimen showing the great thickness of enamel ("en") and dentine ("de"); (b) Prismatic structure of enamel, showing the change in prism orientations from one layer to the next; (c) Parallel prisms within an enamel layer; (d) Crystallites within a prism (SEM image); (e) Crystallites within a prism (AFM phase image); (f) Detail of the crystallites (AFM amplitude image). a, d, f: Dauphin and Williams (2007); b, c: Dauphin *et al.* (2007).

by mastication (chewing). The basic structure is similar to that of reptiles, being composed of small elongate crystallites. In the case of the wild boar (Fig. 6.10), the diameter of these crystallites is about 50 nm. In human enamel, the median length of crystallites exceeds 160 nm and the median width, 20 nm (Sakae *et al.* 1997). These crystallites are grouped in well-defined prismatic packets, whose diameter approximates 4 μm in the boar. This enamel is referred to as prismatic (Fig. 6.10c). Additionally, prisms are grouped into zones within which they are parallel, but with their orientation changing between adjacent zones. This property mechanically slows the propagation of cracks (Rensberger 1997).

Motifs for the arrangement of prisms are highly varied and characteristic for each taxon (Fig. 6.11). However, they are not always visible on unworn teeth because the most external layer is habitually nonprismatic. Marsupials have a characteristic structure: their enamel possesses long tubules whose origin is still controversial (Gilkeson 1997).

An attempt to unify the definitions of descriptive terminology for the enamel of both reptiles and mammals has been made by von Koenigswald and Sander (1989), but "cultural" differences between dentists, paleontologists, paleoanthropologists, etc., continue to exist.

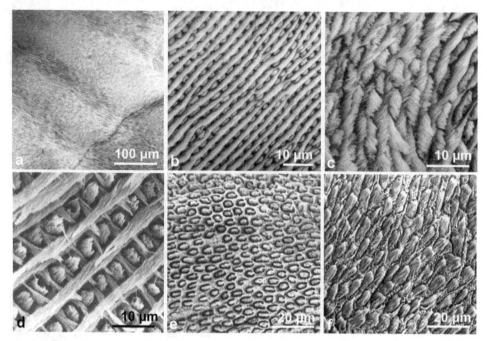

Fig. 6.11. Diversity of microstructural motifs of mammalian prismatic enamel. (a) Hunter-Schreger bands due to alternating orientations in enamel prisms, here of a deer incisor; (b–c) Incisor of the horse: a detail (c) of the prismatic structure in the horse; (d) Alternation of orientations of the prisms in the enamel bands of the jerboa (rodent); (e) Enamel of a mole (*Talpa*); (f) Marsupial tooth (*Macropus billardei*). d: Dauphin and Williams (2007); e: Dauphin (1989).

Bones and teeth: generalizations

Beyond the more or less conflicting definitions and classification of the microstructural organizations of bones and teeth, the essential difference between these highly diverse phosphatic materials produced by members of the Phylum Vertebrata, and the even more diverse calcareous structures, clearly appears when studying the process of crystal formation.

The structural observations at various scales applied to Ca-carbonate crystalline units have resulted in concluding that these Ca-carbonate crystal-like units are formed through a series of post-depositional crystallization changes, each of them involving a mineralizing layer a few micrometers thick. The size and crystallographic consistency of each microstructural unit is determined by the efficiency of the mechanism that maintains the relationship between superposed growth layers. When this mechanism is efficient, the layered growth mode can result in long polycyclic units that have a monocrystalline behavior, for example, calcite prisms or aragonite fibers. However, the overview presented in Chapter 4 has shown that such crystallographic consistency between the superposed mineralized layers is not the rule. Numerous organisms produce calcareous skeletons in which no visible microstructural organization appears at an optical scale (in polarized light).

This indicates that the superposed mineralizing growth layers can remain independent of one another with respect to their crystallography. In such cases, the spatial arrangements of the small calcareous units are controlled at the level of micrometer-thick layers, their orientation depending on the morphological configuration of mineralizing epithelium. The highly controlled barnacle microstructures, of which the taxonomic value has long been recognized, but in which no long, polycyclic monocrystalline units are formed, are examples of this. It is worth remembering that, whether or not polycyclic microstructural units are developed, the nanometer-sized units that form mineralized layers are always present as the first steps in the mineralization process.

Nothing equivalent has been observed in bone. The dimensions of the tiny platelet-shaped crystals stacked within the collagen fibrils are far smaller than the apparent diameter of the nodular roundish units that form the reticulate calcareous crystals. Research addressing the crystallization process in bone focuses on the organization of the unit cell, in the crystallographic sense, a parameter of major importance with respect to the physical properties of bones. As pointed out by Pasteris *et al.* (2007), the bone mineral component should be more correctly referred to as "carbonate-apatite," due to the presence of 6–7% carbonate. Experimental approaches indicate that the concentration of carbonate ions affects the grain size, the crystalline order, and the degree of hydroxylation of apatite, variations that can result in the modification of bone's ability to exist. Transfer of data derived from these in vitro investigations into the dissolution–deposition process of bone formation is a major goal of bone biomineralization studies at present. Formation of the variously shaped "prisms" of mammalian enamel and their three-dimensional arrangement also appear as direct results of a crystallization process in which concentrations of carbonate and hydroxyl ions are biologically imposed, and result in distinct physical properties of enamel crystals (Pasteris *et al.* 2004). The carbonate ion concentration in enamel, for example, does not exceed 3–3.5 wt%, instead of the 6–7 wt% typical of bones. Other experiments (Ouizat *et al.* 1999) have shown that OH^- concentration is likewise an important factor in the quality of crystallization, the OH^- acting as a bonding agent between apatite and collagen surfaces. The degree of crystallization is a critical point with regard to the durability of enamel, as this is a nonrenewable material that is submitted to hard physical and chemical usage.

Even more than mineralization of bone, the spectacular and species-specific arrangement of prismatic enamel in mammals (Fig. 6.11) exemplifies the major differences between Ca-carbonate and apatite modes of crystallization. Both structures result from the activity of a mineralizing epithelium producing layered structures that do not reflect any arrangements of the producing cells (this is not the case in bones). However, each epithelium also produces mineral units with well-defined morphologies. But, in the calcareous prisms of invertebrates, morphology is determined by formation of envelopes that laterally limit crystallization within the superposed growth layers. Thus, two distinct synchronous but independent mechanisms are involved in the formation of a calcareous crystal-like unit. In contrast, the enamel-producing epithelium controls crystallization of nanometer-scale crystalline units through direct bonding of the mineral component into the various arrangements of the extracellular organic framework. Crystalline units in calcareous and apatite prisms are therefore completely different with respect to their origin and mode of growth.

6.2 Silica-based structures: highly controlled morphologies formed of amorphous mineral material

Silica deposition occurs in a number of biological groups, of which three are the best known: sponges, because they build macroscopic skeletons, diatoms, the most geologically important group concentrating silica at the present time, and the Radiolaria, protozoans well known for their complex morphologies as illustrated by the drawings of Haeckel (1862) and, later, in his report on specimens collected by HMS Challenger (1887). In addition to these, several other modern groups of the Protista (e.g., Silicoflagellates), and many of the families of higher plants, such as the horsetails (Equisetaceae) and grasses (Poaceae), also concentrate silica (see the extensive review in Voronkov *et al.* 1975). In all cases, silica tetrahedrons are linked together by intermediate water molecules, creating an amorphous mineral known as opal or silica gel (Kaufman *et al.* 1981). From various viewpoints, the importance of this silica-concentration process by plants must not be underestimated, either quantitatively, as in considering *Equisetum*, with 20% silica (dry weight), and the 5 to 20% in rice or other Graminacea, but also qualitatively. In sugarcane, for example, internodal cells heavily charged with silica gel have been shown to play a role in transmitting light to the photosynthetic mesophyll tissues (Kaufman *et al.* 1981, Fig. 15.27). In general, silica exhibits specific morphological patterns in the intracellular vesicles where concentrations occur, demonstrating a well-controlled process of deposition.

6.2.1 The silica in sponges (Porifera)

From the point of view of their biomineralization, sponges certainly are one of the most original phyla, owing to the structural and chemical diversity of the skeleton they build. All sponges need a supporting skeleton because of their method for capturing nutrients, based on the presence of specialized chambers in which water is moved by flagellate cells (the choanocytes). Sponges have developed various architectures to support this nutritional system and utilize various combinations of mineral and organic components. Those with calcareous exoskeletons are poorly represented in modern oceans, but have considerable importance as modern analogs of fossil sponges; these have been previously discussed in some detail in Chapter 4 (Figs. 4.8 to 4.18). One should remember that among sponges, the calcifying Demospongiae are unique in that they are able to create both calcareous and siliceous structures through the simultaneous use of two completely distinct biomineralization mechanisms. Quite the opposite, a few species (e.g., those belonging to the widely utilized sponge genera *Hippospongia* and *Spongia*) build an entirely organic skeleton based on protein molecules with compositions close to that of collagen (Fig. 6.12a). An organized proteinaceous scaffold is formed, ranging from a main framework that defines overall morphology to delicate ramifications that support the choanocyte chambers (Figs. 6.12b–c). However, most

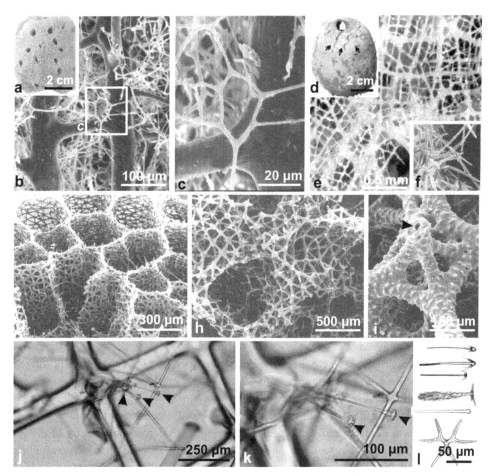

Fig. 6.12. Collagen and silica-based skeletons in sponges. (a–c) A soft sponge, with an entirely organic skeleton. Major exhalant oscula are easily visible (a), whereas close-up views of the organic framework show hierarchically organized rods (b) down to finer units that support individual choanocyte chambers (c); (d–e) Typical demosponge (d) with an internal reticulate skeleton built of individual spicules cemented by collagen (e–f); (g–i) Examples of three-dimensional frameworks formed of rigid spicules. In some species, skeletal rigidity is reinforced by bracing nodes (i). The arrow in (i) indicates the axial organic filament; (j–l) Distributed in the living tissues, microscleres (j–k: arrows) are generally dispersed after the death of the animal. Image (l) provides examples of the morphological diversity of microscleres.

sponges use silica or a combination of silica and collagenous proteins to build their architectural framework. Thus, these sponges are essentially made of two intricately interwoven networks composed of (1) the nutritional chamber system in which canals direct seawater in and out of the chambers and these canals coalesce in a major osculum by which seawater is evacuated; accompanied by (2), the skeletal support framework built by the resistant protein or protein/siliceous assemblage (Figs. 6.12d–e).

The mineral phase is organized into cylindrical rods or spicules, most frequently associated in three-dimensional arrangements with considerable diversity (Figs. 6.12f–g). In addition to this principal skeleton that has a complexity that requires the collaboration of multiple cells, smaller mineral units, the microscleres, are each built by a single cell, and both types are associated to form the framework scaffold (Figs. 6.12h–i: arrows). The very small microscleres also exhibit great diversity, as each sponge species produces a definitive assemblage of distinct morphologies. From a taxonomic viewpoint, the characterization of each species of a given sponge genus includes the description of its microsclere assemblage.

The structure of sponge spicules has been studied in considerable detail (Bütschli, 1901; Minchin 1909; Jones 1979), and has established the presence of an axial filament surrounded by thin silica layers. The shape of transverse sections of spicules reflects differences in the morphology of the axial filament, long described as "triangular" (Reiswig 1971). Numerous investigators have described it as hexagonal, and at times, both patterns can appear in a given species (Weissenfels and Landschoff 1977).

Scanning electron microscopy allows analysis of their detailed morphology in greater detail, also drawing attention to the consistent control of the spatial arrangement of spicules. Specimens of the genus *Euplectella* are widely used to illustrate the rules of construction maintaining the orientation of each distinct series of spicules that can range up to more than thirty centimeters in length (Figs. 6.13a–d). The resulting mechanical properties have recently been analyzed (Weaver *et al.* 2007). Even longer spicules are found in the deep-sea sponge *Monorhaphis*, where the sponge body is suspended above the sediment surface on a huge spicule, with length approximately 2 to 3 meters in length, with diameters of 1 cm (Figs. 6.13e–h).

However, information about the fine structure of axial filaments and the formation of silica layers was delayed until the recent development of microscopes that could permit studying them at the nanometer scale. The specific structure of the axial filament was studied by transmission electron microscopy (Garrone 1969). It has been known for decades that a simple linear spicule is built by a single pair of cells, whereas the formation of a complex multiple branching spicule requires the association of several pairs of cells, each of them producing its silica rod in a precisely controlled orientation. Before the start of silica deposition, the sclerocytes (single or associated) create a linear, highly structured organic filament. TEM observations have shown that a nanometer-scale reticular organization is visible in these organic filaments; the spatial arrangement of the subunits is very regular, with the result that the margins of the filament show planar growth faces. In *Tethya*, the axial filament is composed of proteins (up to 98%) that exhibit a regular periodicity of 17.2 nm (Shimizu *et al.* 1998) (Fig. 6.14). Recent synchrotron-based studies have drawn attention to differences in the nanostructural patterns of filaments between species (Croce *et al.* 2003), for example, the filament periodicity is 5.8 nm in *Geodia cydonium* and 8.4 nm in *Scolynastrea joubini*.

The filament forming protein has been identified as silicatein. In *Lubomirskia baicalensis*, the freshwater sponge from which silicateins were first sequenced, these are present in slightly different forms, with sizes of 23, 24 and 26 kDa (Müller *et al.* 2007a). These

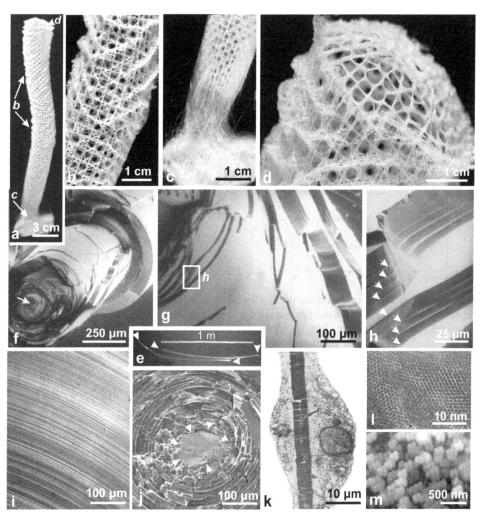

Fig. 6.13. From macroscopic organization to AFM characterization of sponge spicules in *Euplectella* (a–d) and *Monorhaphis* (e–m). (a–d) *Euplectella*, the most commonly illustrated sponge, due to the readily apparent arrangement of its long, translucent spicules; (e–h) The largest spicules, up to two meters long (e), produced by the deep-sea sponge *Monorhaphis*. The layered structure (g–h) explains their flexibility and efficient mechanical properties; (i) Three-millimeter-thin axial section of a *Monorhaphis* spicule, observed under crossed nicols. The distinct colors of the superposed layers suggest that some ordered material is present in the mixture of amorphous silica and organic components controlling silica deposition; (j) Axial view of the central portion of the spicule. Arrows indicate the limit of the axial area, with layered homogeneous silica surrounding the axial filament; (k–l) The axial filament in a young sclerocyte cell (k) and close-up view (l) of its reticular structure; (m) The nanometer-sized silica particles. b, f: Dauphin (2005); k, l: Garrone (1969).

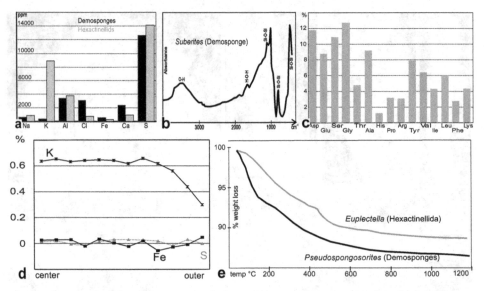

Fig. 6.14. Overall chemical characteristics of siliceous sponge spicules. (a) The diverse minor element concentrations between the Demospongiae and hexactinellid glass sponges; (b) Infrared absorption of hydrated silica from sponge spicules; (c) Amino acid proportions in spicule organic matrix; (d) Evolution of K, Fe and S concentrations during spicule formation; (e) Weight loss in spicules (*Euplectella* and *Pseudospongosorites*) when undergoing ATG. b: redrawn from Sandford (2003); e: data from Shimizu *et al.* (1998).

molecules exhibit enzymatic activity that allows the polymerization of silica. In spite of numerous studies (due to the great potential interest in determining the spatial arrangement of silica at the molecular scale), the relationship between the crystal-like arrangement of silicateins and surrounding amorphous silica was not definitely established at the molecular scale until recently. The presence of smaller molecules (galectins) was recognized in association with silicateins (Wiens *et al.* 2003; Eckert *et al.* 2006), and in addition, the presence of collagen has also been reported here (Erlich *et al.* 2005).

Owing to their layered structure and organomineral composition, sponge spicules demonstrate remarkable properties. The resistance and flexibility of the *Monorhaphis* giant spicules largely exceed comparable values measured for any kind of modern glass (Lévi *et al.* 1989). The optical properties of the filamentous spicules of *Euplectella* have been reported by Sundar *et al.* (2003) and by Aizenberg *et al.* (2005).

Until recently, there was no information available regarding the fine-scale organization of this layered amorphous material. AFM observation has shown that these layers, several micrometers thick and visible in the SEM (Fig. 6.13h), are composed of spheroids of condensed silica only 10–20 nanometers in diameter. These nodules are arranged in concentric elemental layers, with some suggestion that the silica observed could actually be deposited by superposed solitary layers of nodules (Weaver and Morse 2003; Weaver *et al.* 2003, Fig. 5). This nanometer-scale arrangement of the spheroidal opaline units represents the ultimate level of "internal stratification" of the siliceous spicules that was previously observed at lesser magnifications by Schwab and Shore (1971).

Overall, chemical and physical descriptions of these spicules indicate that, in addition to water (mean content 12.3%, established by Sandford 2003 and Croce *et al.* 2004), a complex organic phase is also present. In the giant spicules built by the deep-sea sponge *Monorhaphis chuni*, Müller *et al.* (2007b) identified two major proteins (150 and 35 kDa, respectively) with several others of sizes ranging from 50 to 250 kDa. Deglycosylation causes a variable reduction of the molecular weights. One of them (24 kDa) shows a clear silicatein affinity.

As was pointed out by Weaver and Morse (2003), if the properties of silicatein provide the explanation for the early fixation of silica on the axial filament surface, silica deposition obviously continues in a layered and well-controlled form, leading to specific morphologies of spicules and axons (rods) of the skeletal framework. Obviously, the layered deposition of the material is also under biological control of a cyclical nature. In practice, thick transverse sections of the spicules, observed in polarized light, allow for direct observation of the layering (Fig. 6.13i). Polarized light varies from layer to layer, suggesting that, although attenuated compared to the axial filament, some ordered structure exists in the organic material and that this controls post-axial silica deposition. Indeed, every new study adds to the complexity and taxonomic diversity of the organic compounds within spicules, both in the axial filament and in the layers of silica.

A detailed interpretation of the functions of these organic compounds in the process of mineralization has been proposed only recently (Schröder *et al.* 2007), and now explains the post-initial (i.e., intrasclerocyte) growth of spicules. Observation of silica-rich and silica-poor vesicles in sclerocytes surrounding growing spicules suggests that they play the key role of transporting silica to the locus of silica deposition. Additionally, the presence of layers of silicatein between silica layers has been proven. Regulation of the deposition process is insured by the previous emplacement of galectin molecules, themselves "morphogenetically guided by collagen" (Eckert *et al.* 2006; Schröder *et al.* 2007). According to this model, where the previously identified organic components play complementary roles in the silica deposition, it clearly seems to be a matrix-guided process that uses the spherical nodules, each a few tens of nanometers in diameter, as mineral units.

In terms of the relationship between mineral and organic components, sponge spicules provide a contrast to equivalent processes in the deposition of calcium carbonate and phosphate biominerals. In contrast to Ca-carbonate mineralization and, to a lesser extent, to the complexity of bone apatite, the biomineralization models suggested to date for siliceous minerals are consistent with respect to the condition of the mineral phase. The sizes of the nanoparticles of hydrated silica seem to agree with the morphological variations linked to pH during in vitro polymerization experiments, as shown by Iler (1979). From the capture of free, dissolved, silicic acid molecules in seawater to the formation of polymerized and hydrated opal nanospheres that are deposited in regular layers, the silica appears to be a stable mineral phase.

Meanwhile, the considerable progress that has been made recently implies that an unexplored coordination mechanism that produces the siliceous layers continues over broad surfaces. In addition to synchronism of secretion, localized modulation of the secretion mechanism is necessary to produce the various types of spicule morphology. From this viewpoint, coordination of cell functions during silica deposition, although carried out by cells not organized into distinct tissues, is quite comparable to the coordination demonstrated in forming layered growth units by the mantle epithelium in molluscs or the basal ectoderm of corals. In all cases, overall regulation must control the mineralizing system; this is not surprising for complex metazoans, such as molluscs or echinoderms, and we have also observed that such control exists in corals. The layered organization of spicule formation provides evidence of the same type of broad control in sponges as well.

In contrast to this similarity in the overall secretion mechanism, the behavior of the mineral phase differs markedly. Recent evidence of the "matrix-guided" deposition of layers of silica-gel granules in sponge spicules also suggests comparison to the layered accumulation of calcareous mineral precursors in shell growth layers. The main difference between them occurs in the post-depositional stage. In sponge spicules, the spheroidal nanometer-size grains of amorphous hydrated silica remain stable after being deposited on the organic substrate. In a calcareous growth layer, the analogous aspects result from very similar and equivalent transport – captured by mineralizing cells from internal body fluids, transiting through the cell in a stabilized amorphous state and exocytosed on the mineralizing side – are then submitted to a somewhat poorly understood crystallization process.

6.2.2 Diatoms

In the nature of their biominerals and their organization at the supramolecular scale (their nanostructure), diatoms appear to be a very homogeneous group. Without any known exceptions, these unicellular algae, with species numbering tens to hundreds of thousands (Davis and Hildenbrand 2008), are all built by a three-dimensional arrangement of particles of silica-gel in the size range of 40–100 nm in diameter (Sumper *et al.* 2005). Based on synchrotron-generated X-ray scattering, the silica has a completely amorphous nature, as demonstrated by Vrieling *et al.* (2000). Small-angle X-ray scattering allows very high resolution for study of silica arrangement, and specifically yields data concerning the nanoscale porous structure built by these particles. Two distinct pore sizes were found, the smallest less than 10 nm in diameter, the largest reaching 65 nm. All species studied show similar patterns, in spite of the species-specific diversity in terms of their size, shape and detailed morphology. These porosity patterns at the nanoscale apparently agree with results obtained by in vitro, chemically driven polymerization of silica, with size most dependent on pH. They reflect the ultimate small-scale porosity types in the silica structure covering the diatom cell.

At lower magnifications, an imbricate porosity appears to provide the constructional system for diatoms but, unlike the chemically produced nanoscale porosity, the micrometer-scale porosity is remarkably species-specific (Figs. 6.15a–c), providing a valuable basis for the taxonomy of this prolific and highly diversified group.

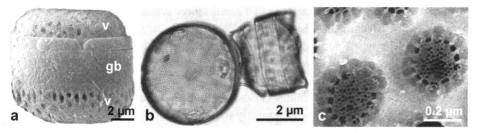

Fig. 6.15. General morphology of a diatom frustule of the common central type. These three images aim at emphasizing the precision of control of pore placement at the micrometer scale. Beneath the micrometer-sized pores (b) that have long been used as a test of optical microscope resolution, the SEM permits observation of a superfine silica grid (c).

Diatoms as a biological group appeared relatively recently, arising during the Mesosoic Era, but reaching true geological importance only during later geologic periods (the Late Cretaceous and Cenozoic). Due to their considerable ability to absorb dissolved silica, they exert an extremely large influence on the composition of the world's oceans, reducing an estimated oceanic average silicic acid concentration of $< 70\,\mu M$ to less than $10\,\mu M$ in the surface waters in which they live (Trequer *et al.* 1995). Initially marine organisms, they are now also present in freshwater environments, in most part since the early to middle Cenozoic Era. The purity of the mineral materials concentrated by these organisms has led to a broad spectrum of industrial applications. Their fascinating ability to create species-specific three-dimensional architectures using silica tetrahedra as basic mineral units in addition drives numerous investigations regarding the precisely controlled deposition mechanism. However, as also was the case with sponge spicules, a precise understanding of the transportation of silica from seawater to the mineralization sites in the diatoms had to wait for the availability of high-resolution methods of observation and new techniques of molecular biology. The same is true for understanding the mechanisms responsible for their perfect replication of patterns of micrometer and submicrometer morphology.

Silica deposition at the submicrometer range

Among the characteristics that led to the success of the diatoms as a group is their capability for developing vegetative multiplication, making possible the formation of huge populations in a short time as a rapid response to favorable conditions. In this process, the formation of new siliceous valves is critical, as separation of two daughter cells occurs only when each of them is covered by newly formed valves. Diatoms are primarily single-celled organisms, although they commonly occur as long series of nonseparated individuals. Cells are covered by a theca (specifically named a "frustule"), a three-component silica structure consisting of two symmetrical valves (Fig. 6.15a: "v") and girdle bands (Fig. 6.15a: "gb"), the number of which varies depending on growth conditions.

Cell division occurs within the mineral theca, as well as the formation of new theca for the daughter cells. Accordingly, separation of the daughter cells cannot occur before completion of the two resulting frustules. Each part of the initial theca is reused and two new valves are formed, each of them using the preexisting valve as the larger of the developing new frustules. Of the newly formed frustules, the one that uses the larger valve is an exact equivalent of the initial cell. However, the one using the smaller valve (the base of the initial frustule) must create a new valve that is still smaller. As this process is repeated at each cell division, the mean size of the resulting clones is statistically regressing. Sexual reproduction allows commencing a new clone of normally sized individuals.

Investigators have long reported the perfect reproduction of ornamentation and porosity in the silica deposited to form the new valves, a process obviously under effective genetic control. This process occurs within a cellular organelle, the silica deposition vesicle (SDV), in which the new half-frustule is developed to complete the theca of the newly divided cells.

The first step in the silicification process is the binding of silicic acid from water. The cell membranes of diatoms contain specialized sites whose genomic sequence has been deciphered (Hildebrand *et al.* 1997, 1998). The extension of this to various other species from both marine and fresh waters (Grachev *et al.* 2002) reveals a family of proteins that play this role; they exhibit comparable sizes (550 residues) and topologies (Thamatrakoln and Hildebrand 2005).

The methodology for the next step, the intracellular transportation of bound silica to the depositional vesicle (SDV) is not clearly established. The suggestion has been made (Azam *et al.* 1974) that silica is associated with an organic stabilizing compound, possibly within specialized vesicles (Schmid and Schultz 1979). Noticeably, fusion of vesicles has been observed during expansion of the silica deposition vesicle in which frustules are growing. In terms of the kinetics of this process, the complete extension of the newly silicified valve occurs within a 15-minute period (Hazelaar *et al.* 2005). Present mineralization models suggest that the formation and extension of the SDV membrane provide the substrate for controlled deposition of the silica, the species-specific pattern resulting from association of at least three distinct groups of biochemical compounds.

A general feature of silica-forming diatom frustules is the presence of long-chain polyamines (Kröger *et al.* 1999, 2000) that exhibit diverse taxonomy-linked structural characteristics. Additionally, frustules also contain modified protein compounds (mostly by phosphorylation). More recently, a new class of peptides, named silacidins (Wenzl *et al.* 2008), has been found here. They have been shown to contribute to the formation of the silica guide framework through their high degree of phosphorylation. This new group of phosphorylated peptides seems to control the size of the silica nanospheres, the building blocks of the frustules. It has been repeatedly shown (e.g., Crawford *et al.* 2001) that frustules are built by the coalescence of silica nanospheres. The reactive biochemical compounds that have been discovered to date, some of them taxonomically specific, suggest that the interaction between them creates the macromolecular assemblages that act as guides for silica deposition. Once the two valves are completed within the daughter cells (and still inside the initial frustule), they are exocytosed and the two daughter cells become fully independent. It is also noteworthy that, throughout the life of the cell, the external surface of the frustule is protected from direct contact with water by an organic membrane, thus preventing dissolution of hydrated silica.

More study is still necessary to determine precisely the three-dimensional arrangement of the various molecular components that are associated to form the SDV, and to understand the process driving silica-to-silica patterns. Meanwhile, the rapidity of recent progress in determining the overall mechanisms of silica deposition led Sumper and Brunner (2008) to emphasize similarities that are progressively being recognized between mineralization processes in diatoms and sponges. Although differences in their initial steps have been emphasized by Weaver and Morse (2003) as "catalytic in the demosponge axial filament and stoichiometric in polyamine molecules of diatoms," more detailed biochemical information now allows the recognition that comparable types of molecules serve to provide the essential functions in the silicification mechanism of both. The supposed difference between these two very distant groups is now considerably lessened.

6.2.3 Radiolaria

Data regarding biomineralization in the third major silica-producing group, the Radiolaria, are scarcer by far. Among the reasons contributing to the lack of research on mineralization in the Radiolaria, the principal one is perhaps the difficulty of maintaining long-term cultivation of these protozoans in the laboratory. It was reported by Pappenhöfer and Harris (1979) that no successful method of culturing them was known at that time.

The skeleton of Radiolaria is also more complex than sponge spicules or diatom frustules. It comprises a central spherical capsule (calymma), in which organelles are concentrated, thus insuring the animal's essential biological functions. Additionally, the Radiolaria, a subclass of the Actinopodia, are a complex group, even after separating the Acantharia (whose skeleton is built of strontium sulfate). The Radiolaria comprises two orders with a marked difference in the structure of their opal skeletons: the Tripylea and the Polycystina. The former has its skeleton built of hollow rods of silica, whereas the rods are compact in the latter. Detailed optical microscopic studies were carried out by Hollande and Enjumet (1960) and by Cachon and Cachon (1972), but access to the submicrometer-scale features bearing on silica deposition was first obtained by Anderson (1981) with a series of TEM observations. The growth edges at the surface of the spherical skeleton of a *Collosphaera* were observed growing by the accretion of silica granules in the size range of 100 nm. When they are distant from the more massive parts of the skeleton, the granules appear as distinct features, but they are gathered into "less-dense matrix" to form the skeleton (Anderson 1981, Fig. 13.14, p. 364). Although building a massive structure, the silica granules do not fuse into a compact silica structure. Scanning electron microscopic observation of cleaned skeletons allows *in situ* observation of these densely packed spheroidal mineral units (Anderson 1981, Fig. 13.16, p. 365).

It has been concluded that, beyond the biochemical differences and diversity of the cellular organelles involved in the process of silica biomineralization in these three major groups, an important similarity is present between them in terms of the state of the mineral phase involved. The ability of silicic acid to form polymers has led these phylogenetically distant organisms to develop biochemically distinct but quite comparable depositional processes. Each observation of these structures of silica at the relevant scale (approximately 100 nm) shows a similar particulate morphology on the silica surface (e.g., some examples in Bovee 1981 regarding the dinoflagellate *Actiniscus*, or more recently in the Testate amoeban *Netzellia*, Anderson 2007). The suggestion is that this mode of mineralization can be extended to most organisms using silica for their skeleton. Variability in polymerization can result from the diversity of taxonomy-linked compositions of organic compounds. Further research will identify precisely the mechanisms driving depositional processes in different organisms, allowing for the better evaluation of similarities and differences between the resulting biochemical configurations. However, with respect to the mineral phase itself, it appears that the basic point is essentially that these spheroidal units of polymerized silica can form stable structures when linked to the driving biochemical scaffolding. Even when a "fusion" phase seems to occurs, as in the axial part of hexactinellid spicules (Müller *et al.* 2009), the process leading to formation of a massive core of silica surrounding the axial filament (Fig. 6.13j) does not generate a specific microstructure. From this viewpoint and to the degree that we can compare these three major mineralizing systems at present, polymerized-silica apparently allows very precise control by an organic biochemical framework. Collagen and associated matrix materials permit additional accretion to occur after the well-controlled initial phase.

Undoubtedly, the post-depositional formation of highly diverse Ca-carbonate microstructures is still the most intriguing question concerning the roles and mode of action of glycoprotein assemblages in the invertebrates, and this differentiates the biomineralization processes in carbonate producing groups from those depositing silica.

6.2.4 Comparisons

The present state of biomimetic approaches concerning these three materials provides major examples of the considerable differences between the amorphous or nanocrystallized materials on the one hand, and on the other, the large crystal structures resulting from calcareous biomineralization. The "direct-ink" method now allows silicification of in vitro-formed organic assemblages (Xu *et al.* 2006) and parallel bone-compatible biomimetic materials that are beginning to be used as bone repair materials (Liao *et al.* 2005). To date, no equivalent biomimetic calcareous structure has been obtained. Attempts to use the "sizing and shaping" method have never succeeded in forming any sort of microstructural unit from calcareous crystals growing in a saturated solution. No handmade nacreous tablet has ever been formed; to obtain morphologically mimetic objects, researchers are presently obliged to utilize natural envelopes previously prepared during the complete calcification of nacre. The insoluble matrix framework then can be infilled by in vitro crystallization, resulting in "nacreous-like" units made of calcite (Cölfen 2007, Fig. 4). Of these two examples, the former is a "crystal growth-based" approach, and the latter focuses on using envelopes; both produce unsatisfactory results for comparison with naturally occurring materials.

The preceding survey of calcareous structures has shown that calcifying organisms use a unique method to produce their calcareous hard parts, simultaneously controlling size, shape, mineralogy, and crystallinity within each superposed micrometer-thick layer. The two-step process that produces a micrometer-thick growth unit in calcareous structures is apparently considerably more complex than silica deposition on an organic framework, as it forms the final step of the crystallization process. Meanwhile, the success of methods utilized in research dealing with materials of amorphous silica indicates that very similar and parallel control by matrices to generate the mineral phase, as well as the surrounding envelopes, provides us with an effective way of increasing our understanding of calcareous microstructures.

7

Collecting better data from the fossil record through the critical analysis of fossilized biominerals

Case studies ranging from the interpretation of individual samples to the distribution of fossils through time

Any improvement in our knowledge of the mineral structures formed by living organisms contributes to a more accurate analysis of their fossil remains. Such a simple remark is important, because selection of the methods used for assessing the preservation of any fossil, as well as interpretation of numerical values resulting from measurements carried out on fossil material, depends heavily on concepts regarding its original state and mode of growth. In corals, for instance, the amount of confidence in the reliability of isotopic or chemical measurements has long been based on a simple X-ray diffraction diagram, owing to postulation of a purely mineral composition of these "physiochemically" crystallized materials.

As defined by Berner (1980) the term "diagenesis," as applied to any sedimentary object, refers to "the sum total of processes that produce changes – mineralogical, chemical and physical – from the time of deposition." Such an extensive definition (see also Bates and Jackson 1980) obviously includes fossilization, the term we use when sedimentary processes are modifying materials that have been formed by living organisms. From this standpoint, the methods by which the three major biomineralization mechanisms control the deposition of their mineralized structures allow us to assume that diagenesis of the resulting materials will follow very different and specific pathways. A major difference from chemically precipitated crystals is that biominerals exhibit very distinct structural parameters at the micrometer and submicrometer scales, even if the chemical compositions of their mineral parts do not greatly differ from purely chemical equivalents. As a result, their sensitivity to environmental, burial, and sedimentary conditions produces a variety of fossilization pathways within any given environment. In addition, the organic components present at all structural scales create a major factor differentiating them. In their structural and general compositional patterns, biominerals differ from their chemical equivalents, and the probability of preserving them in the fossil record cannot be evaluated only through the physical properties of the mineral material.

From establishing the character of an individual sample to the most general syntheses, a few examples suffice to make it obvious that specific approaches must be developed in order to understand the multiple changes that mineralized skeletons undergo during the post-depositional portion of their history. Biominerals are seen to differ in this among themselves as much as between them and their chemical counterparts. Owing to the molecular

349

organization of polymerized silica, fossilization of siliceous biominerals can usually be understood by taking into account their sensitivity to pH conditions in the local environment. In contrast, the fossilization of bone primarily depends on the possibility in a given environment of avoiding the rapid destruction of collagen fibers; here, microbial invasion, hydrolysis, and oxidation are leading factors. Between these two extremes of siliceous and organic materials, calcareous structures provide very different case histories, frequently with unexpected results. They exhibit such differences in the organic-mineral ratio and variety of microstructural patterns that assessing the status of a given sample first requires a careful examination of its microstructure and composition. Comparison to equivalent materials known to be in a pristine state must be carried out at relevant scales. The diversity of the biominerals produced within given skeletons and also, the variability of their resulting behavior when subjected to fossilization processes, makes extreme care necessary during their study, and commonly results in producing examples of unanticipated preservation that is independent of geological age. Sometimes, this approach may also result in considerable modification of our understanding of the evolutionary history of a major taxon. The corals are a good example of such a re-examination.

Whether at the scale of a single fossiliferous locality or a regional synthesis, a compilation of the preservational conditions of all of the various taxa present can provide significant information on the overall diagenetic environment. In a more general synthesis, examination of these relationships according to their fossilization through geologic time, opens a new research area in which biomineralization and fossilization are linked to overall longer-term environmental oscillations, as suggested by geochemical data. Data presently available concerning the historical development of life during the Phanerozoic portion of geologic time is certainly pertinent to the study of an individual fossil, and just as true for the huge amount of data that has been gathered by paleontologists during the past two centuries. In contrast to the predominance of morphological characteristics that have been used most commonly to classify fossils, biomineralization-based criteria now lead to the in-depth reexamination of the evolutionary history of broad groups, such as the sponges or corals. In addition, when longer time periods are taken into account, researchers face additional phenomena connected to changes in global environmental conditions. Far from relying on the strict application of uniformitarianism, the main basis of geological thinking since its formulation by Lyell (1830–1833), the modern paradigm of a changing world has incorporated, during the last decades, study of the long-term modification of physical and chemical characteristics over the whole surface of the Earth.

The resulting views of Earth history now offer fascinating new perspectives for paleontologic investigation. Recent descriptions of the sensitivity of calcification processes to environmental conditions can now be analyzed accurately, utilizing experimental approaches developed for each biological group. Through the multiple structural and biochemical methods for characterization that are now possible, the specific reactions of given groups to carefully controlled environmental change can be precisely evaluated. Additionally, paleontologists should also consider the extent to which these environmental changes have either favored or retarded fossilization processes. Clearly, in-depth analysis of the consequences of environmental change on synchronous processes of biomineralization and fossilization will result in substantial modification in our interpretation of life history as reflected in the fossil record. With respect to such restudy, the Permo-Triassic provides particularly appropriate examples because the nearly complete extinction of Paleozoic faunas and the Mesozoic faunal recovery has led to the general view that most modern lineages arose as a consequence of these major biological events.

An interesting parallel suggested by the rapid changes observed in environmental conditions at present provides further striking illustration of the importance of new paleontological thinking and use of improved analyses. Our ability to model and accurately predict the near future of our planet's biota directly depends on the adequacy of concepts utilized in describing the closely related processes of biomineralization and fossilization as they concern living organisms. Interpreting data from the last millennium is just as dependent on a reliable paleontological approach as is our understanding of the records of the deeper past throughout geological time.

7.1 Fossilization of siliceous biominerals

Siliceous structures formed by the three major groups that contribute to the geological cycle of silica are, throughout the life of the producing organism, protected from direct contact with seawater. Radiolaria and sponge spicules (except perhaps for the distal parts of giant spicules in *Monorhaphis*) are produced intracellularly, and diatom frustules, likewise, remain covered by an organic membrane, even after exocytosis of newly formed half-frustules and separation of the two daughter cells. The mineral material comes into contact with seawater only after decay of this protective layer. At that time, owing to the slightly alkaline pH of seawater, the opaline silica produced by these marine organisms is subject to conditions inimical to its preservation.

Solubility curves for opaline silica and Ca-carbonate (Fig. 7.1a) clearly illustrate the opposing stability conditions for these two main sources of biominerals. This contrast is best illustrated by the siliceous spicules of calcifying demosponges, as in the ceratoporellids, where they are included within surrounding calcareous (aragonite) fibers (Figs. 7.1b–d). Since the living tissues of these sponges are present in the upper part of the skeleton only, seawater infiltrates the basal fibrous structures. The resulting high pH solution causes the rapid dissolution of the siliceous material. It is a common observation that a few millimeters below the growing surface of a living ceratoporellid sponge, the positions of siliceous spicules are already empty voids (Fig. 7.1e). Depending on subsequent burial conditions, coarse euhedral calcite may be deposited in place of the initially siliceous spicules. Occupying the void spaces, the crystallized calcite only roughly reflects the overall morphology of the dissolved spicules. However, their presence affects our correctly placing the calcareous materials within these demosponges (see also Section 4.2).

Fossilization of siliceous structures buried within oceanic sediments depends primarily on dissolution that occurs within the water column prior to deposition. The mass balance between production and deposition of biogenic silica is still poorly quantified: in 2007,

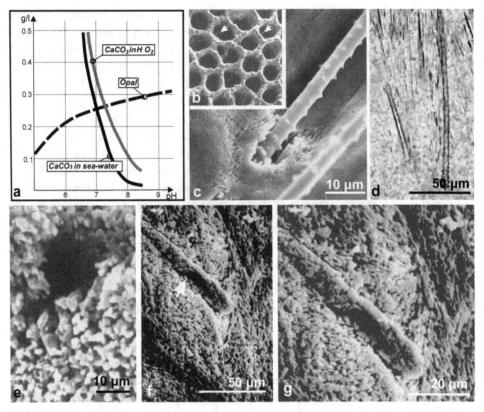

Fig. 7.1. *In situ* dissolution of siliceous spicules in the fibrous aragonite skeleton of a calcifying demosponge and diagenetic deposition of calcite pseudomorphs. (a) Opposing solubility curves for opaline silica and Ca-carbonate; (b–d) Siliceous spicules of the demosponge *Ceratoporella*, inserted in its fibrous skeleton (aragonite); (e) Closer view of a spicule-shaped void space; (f–g) Scanning electron microscope view of fossil ceratoporellid showing calcite pseudomorph of an initially siliceous spicule. The coarsely crystallized calcite does not allow the exact determination of spicule morphology (Triassic).

only 57 complete profiles were known that chart the amount of silica dissolution between the seawater surface and the sediment surface (Fripiat *et al.* 2009). Due to this, the long-reported difference between maps of production and maps of deposition of biogenic silica in the world oceans (e.g., Lisitzin 1972) is yet to be fully explained. Because neutral pH forms the crossover point for the dissolution curves of both silica and Ca-carbonate, lowering of pH in sediments can increase the amount of silica in them.

The preservation of biogenic silica after deposition and the ratio between surface production of silica and its dissolution during transit through the water column, is complicated by the presence of concomitant production of calcareous biominerals, such as coccoliths and/or pelagic foraminifera. An additional component has to be considered: most biogenic mineral particles do not move freely down through the water column, but are included as components of "marine snow," the agglutinated organic material resulting mostly from fecal pellets of plankton. Therefore, the composition of this organomineral melange that settles on the ocean floor differs greatly from place to place, and has varied much more through geological time, as exemplified by the occurrence of chalk.

Chalk covered broad surfaces of the European marine platform during the upper part of the Cretaceous Period and the lowermost part (Danian) of the Cenozoic. This micrograined sediment is mostly composed of pelagic biogenic components (apart from very littoral zones). Chalk is also of great interest because it exemplifies at a regional scale, a characteristic segregation between its siliceous and calcareous components during diagenetic evolution. Chalk deposition occurred at depths of 300 to 400 meters (the mean depth of epicontinental seas). Owing to this rather shallow depth, the major part of the abundant calcareous production of coccoliths was deposited in bottom sediments, resulting in carbonate mud with a high pH, where siliceous components of the sediment had a very low probability of being preserved morphologically. Because of this, dissolved silica was present and moving through the uncompacted sediments, encountering physical conditions that caused its deposition. Layers of irregular nodules of flint are commonly found in numerous places within the chalk (Figs. 7.2a–b), demonstrating that in these layers, which remain strikingly parallel over considerable distances, free silica molecules that resulted from the solution of opaline biominerals were able to cluster together to form solid nodules that are exclusively silica. The parallel nature of flint layers at kilometer scales makes it clear that the sediment composition at the seafloor was modified cyclically over broad surfaces by a factor that changed the overall pH of the sediment. A number of researchers, beginning with White (1842), studying the chalk from Suffolk (Eastern England), to Deflandres (1936), reporting on it from Northern France and Belgium, have pointed out the frequent presence of peridinian cysts (Figs. 7.2c–g) in the flint nodules. Preservation of these cellular cysts is so excellent that it is still possible to stain them using biological dyes.

It is understood that formation of peridinian cysts is a part of the biological cycle of the peridinian dinoflagellates that occurs at the seafloor. Such an occurrence of silica in layered concentrations over broad areas suggests that periods of proliferation (blooms) of peridinians may have created parallel layers enriched in organic materials that lowered the overall sediment pH sufficiently to provide sites for deposition of dissolved silica molecules. This silica concentration included some resistant organic remains, such as the peridinian cysts themselves, and in addition, invertebrate skeletons, such as the tests of sea urchins common in some localities, and at times, even sponge skeletons that have been protected from dissolution by local organic concentrations (Figs. 7.2i–j). The sponge body itself can create localized conditions that lead to the formation of pyrite, thus reflecting the presence of H_2S, and forming replicas of the framework morphology of siliceous sponges (Fig. 7.2k).

At a much finer scale, an equivalent situation is observed in the partial silicification of corals, where microprobe mapping (Figs. 7.2l–m) shows that the silica is in most part located in the organic-rich medial zones of septa (the line of distal mineralization) and

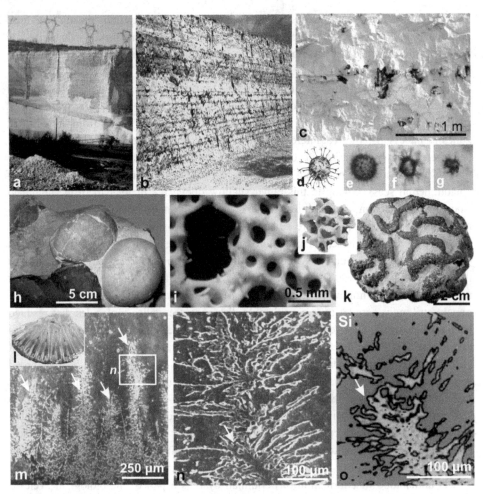

Fig. 7.2. Silica concentration in a calcareous environment – fossils from chalk; diagenetic silica deposition in coral. (a–c) Layered concentrations of flint: (a) quarry in chalk (U. Cretaceous, Maastricht), (b) parallel flint layers in chalk at Mantes (Paris Basin), and (c) irregularity of individual flint nodules within layer; (d–g) Peridinian cysts (commonly known as hystrichospheres) observed in chalk flint. (d) One of the remarkable drawings by White (1842); (e–g) Specimens from the Deflandres collection (MNHN, Paris); (h–k) Fossils from organic rich layers in the chalk: (h) sea urchins *Ananchytes* sp., (i–j) the sponge *Plocoscyphia* from Cap Blanc Nez, and (k) pyritized sponge cf. *Plocoscyphia* from Cap Blanc Nez; (l–o) Silica deposition within a coral skeleton controlled by a concentration of organic material in the distal mineralization zone of septa (arrows) and at a greater enlargement, by fiber envelopes. Silica is visible as a result of etching (m) or microprobe mapping. (n–o) *Coelosmilia*, late Cretaceous from Piotrawin, Poland (collection J. Stolarski).

also in locations of fiber envelopes (Gautret *et al.* 2000). The local development of organic concentrations explains the preservation of siliceous structures such as the sponge networks, but silica molecules in solution have also been found to fossilize nonsiliceous materials such as shells or plant cells.

Silicification of biogenic structures: a highly destructive process

It is commonly stated that silicification is a fossilization process providing excellent fossil material. The silicified carapaces of trilobites exemplify the ability of silicified materials to reveal minute morphological details of their surfaces (e.g., Whittington and Evitt 1954; Klug *et al.* 2009). Silicified plant cells frequently provide possibilities for histological studies truly equivalent to those carried out on modern plants (Figs. 7.3a–c). However, silicification is also, in almost all instances, a severely destructive process in terms of preservation of skeletal microstructure. As pointed out by Maliva and Siever (1988a), little research has studied the process through which silica is substituted for Ca-carbonate in biogenic materials. Their study, based on several hundred thin sections from Europe and North America and of different geologic ages, has shown that in silicified fossils, the silica is now quartz and was directly precipitated as such, although with different habits (chalcedony, euhedral, or microcrystalline). Previous studies (Meyers 1977; Jones and Knauth 1979) suggested that initial deposition of silica might have occurred as opal-CT, a form of silica that consists of packed microscopic (150–300 nm) spheres made up of microcrystalline blades of cristobalite and tridymite. Studies of Cenozoic and modern deep-sea chert have shown that this process does exist, and some supporting evidence has also been found in fossils, specifically in silicified wood (Stein 1982). Normally, this transient form of silica recrystallizes as quartz. Multiple examples of the progressive silicification of fossils, commencing in places where organic components are normally concentrated within shell materials (owing to species-specific biomineralization processes), suggest that the initial step in the silicification process may involve bonding between hydroxyl groups and silicic or polysilicic acid (Leo 1975).

This conclusion was proposed at a time when the nanoscale interplay between organic components and mineral materials had not yet been explored at high resolution. This interpretation is even more valid now, and allows us to understand more about the process of silicification, not only guided by external envelopes of skeletal units, as in coral fibers (Figs. 7.2l–o), but as a process continuing until there is complete removal of calcareous components of the shells. However, AFM examination of silicified biogenic carbonates, such as a belemnite (Figs. 7.3e–k), indicates that, in spite of the massive appearance of these silica deposits and the presence of euhedral quartz crystals (Figs. 7.3h–i), a markedly disordered arrangement of silica units occurs here, and this contain numerous inclusions, among which organic remains can be recognized by phase-contrast imaging (Fig. 7.3j).

The distribution of silicified fossils and bedded chert throughout the geological column has been investigated by Kidder and Erwin (2001), to evaluate the respective roles of faunal modification and climate change in the long-term compositional evolution of the world oceans. They agree with the previous studies of Maliva *et al.* (1989) and others on the leading role of silica-concentrating organisms as the source of the silicifying fluids that cause the formation of bedded chert and silicified fossils. Surprisingly, silicification of calcareous or organic fossils was more common in the Paleozoic ("just over 20%," Kidder and Erwin 2001, p. 511) than in younger strata, which have a proportion of

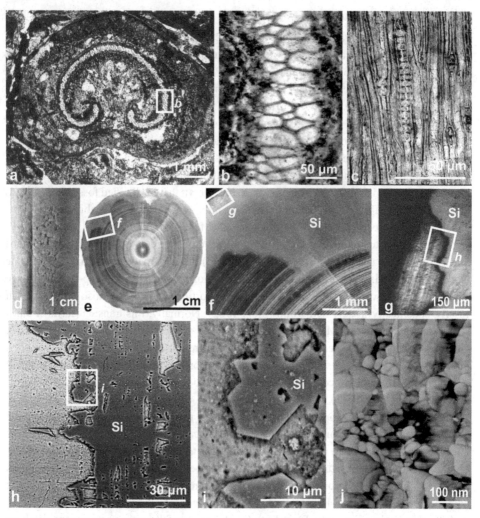

Fig. 7.3. Silicification. (a–b) Example of a histological thin section in a silicified fern (Fam. Osmundacea); (c) Silicified wood (cf. *Taxodium*, Stampian, Paris Basin); (d–e) External view (d) and section (e–f) of a partially silicified belemnite: abrupt transition at a macroscopic scale between the layered carbonate and the homogeneous silicified portion; (f) Closer view of silica deposition. Numerous remains of calcareous material are visible within the silicified part, indicating that the layered microstructure acted as a guide for the passage of silica-rich solutions (g); (h–i) Euhedral quartz. The planar growth surfaces contrast with the disordered arrangement of the majority of the silica nodules and the mixed composition of the silicified areas, including numerous remains that are probably organic, as indicated by the AFM phase-contrast image (j).

approximately 4% silicified fossils. This contrasts with the marked increase in biogenic silica concentration resulting from the development of the diatoms. During the Paleozoic Era, sponges and radiolarians (with the latter having a markedly greater influence) were the main silica providers. Owing to the high proportion of silicified fossils, the two first major faunal extinctions are reflected in the occurrence of silicification. The Upper Carboniferous is characterized by a decrease in it, followed by a new high level in the Permian Period. By the end of Permian, the role of the two main Paleozoic silica providers had declined rapidly (sponges) or progressively (radiolarians), whereas the role of diatoms increased rapidly during the Jurassic, reaching present-day levels that are unprecedented in geologic time, according to the results of Kidder and Erwin (2001, Fig. 4). However, the number of silicified fossils remains very low compared with the large number of basinal bedded chert units, and this is also true for shelf bedded cherts compared with basinal cherts (Schubert *et al.* 1997). Rapid recycling of silica by the Diatomacea themselves in shelf seas might also explain the contrasting decrease in formation of silicified fossils during the same period.

7.2 Fossil bone, dentine, and enamel: multiple causes of analytical obstacles

As the configuration and arrangement of mineral building units in modern bone materials are still debated, it is understandable that no comparative studies exist at relevant scales that focus on the structural or compositional modifications that occur in apatite crystals at the nanoscale. Owing to their anatomical organization, bones (even so-called "compact" bones) are highly porous structures. From the Haversian canals, several tens of micrometers in diameter, to the filamentous expansions of mineralizing cells (see Figs. 6.1 and 6.2), bone is in large part open to the circulation of diagenetic fluids. In addition, bone contains approximately 20 to 25% by weight organic compounds, a proportion that is much higher when it is expressed as volume. The natural decay of this organic material greatly increases the initial porosity of bone material. Porosity here is estimated to reach as high as 50% of the total volume of dead bone, simply due to the natural decay of organic material (Hedges 2002). An even more important factor that results from the breakdown of the original bone composition, the internal organic components provide nutrients for unicellular organisms (bacteria) and/or filamentous endolithic borers (Fig. 7.4). As a consequence, additional porosity is added to original structural and biochemical porosity.

Fossilization of bone is improbable in surficial terrestrial environments, due to a number of factors, the main one of these following from vertebrates themselves being predators. Fossilization of bone fragments, and exceptionally, entire skeletons, only occurs in protected environments that featured rapid burial, thus limiting bacterial and endobiont activity. Regardless of these unfavorable conditions for the study of microstructural and compositional properties, various attempts have been made to derive biological information from fossil bones and teeth. Two outstanding causes have hindered this, at times such investigations are relatively unsuccessful.

Observations at the micrometer and submicrometer scales reveal the variability resulting from various combinations of biological and sedimentary mechanisms that modify bone structure. Collins *et al.* (2002), reviewing the structural and chemical mechanisms that enable preservation of organic compounds in bone (for the most part, collagen and osteocalcin), compared the main mechanisms potentially acting on (i) the organic fraction of

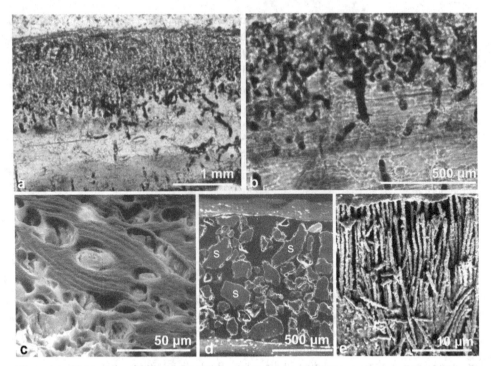

Fig. 7.4. Endobiont microboring, sedimentary infilling, and dissolution in bone and dentine. (a–b) Intense destruction of bone structure by filamentous borers (Lower Oligocene, Quercy, France); (c) Haversian systems partially preserved in a 200-year-old horse bone (Recent, France); (d) Bone of a fossil rodent: large dissolution cavities are now filled by sediment ("s") (Plio–Pleistocene, Tighenif, Algeria); (e) Structural inversion in dentine: around tubules infilled by sediment, dentine has been removed by dissolution.

bone, (ii) the mineral part itself, and (iii) the influence of biodegradation. The complexity of the relationships between the main causes is illustrated by the diversity of current interpretations regarding the role of each major bone component. The presence of organic molecules in contact with apatite crystals can cause the apatite crystal surfaces to become passive (Putnis *et al.* 1995), slowing dissolution of the mineral phase. Conversely, Collins *et al.* (1995a) suggested that the presence of mineral materials acting as "pH buffers" prevents the occurrence of extreme pH's, to which collagen is particularly sensitive. The conclusion that microbial deterioration is likely to be the most common cause of bone destruction (Child 1995) emphasizes that the presence of neutral pH is important for the development of microbial populations. On the other hand, Carpenter (2007) has pointed out numerous experiments that established the role of precipitated mineral materials occurring within biofilms produced by microbial proliferations (Briggs and Kear 1994; Martill and Wilby 1994; Briggs 2003) that promote the fossilization of bones and at times even the preservation of cells within them.

Within this negative context regarding fossilization and the utilization of chemical data from fossil bones, the recent discovery of soft tissues in bone samples as old as Cretaceous

Table 7.1 *Categories of predators and modifications caused to fossil micromammal material (from Andrews 1990 and Fernández-Jalvo and Andrews 1992)*

Categories	Predators	Alterations
1	Barn owls, snowy owls, great gray owls	Very little modification; absent or light digestion
2	Long-eared owls, African eagle owls	Little modification; light degree of digestion; enamel removed from tips of incisors
3	Tawny owls, European eagle owls	Greater destruction; moderate/heavy digestion over enamel
4	Little owls, kestrels and peregrines, viverids and mustelids	Extreme enamel and dentine corrosion; mustelids produce extreme modification, but digested elements appear in low %
5	Buzzards and kites, canids and felids	The most destructive effects; mammalian carnivores produce rounded edges of skeletal elements; gnaw marks rare

(Schweitzer *et al.* 2007) emphasizes the possibility of "organic preservation" noted by Butterfield (1990) and numerous others.

One particular aspect of bone diagenesis, the action of the digestive process, has been investigated in bony material by Raczynski and Ruprecht (1974), Dodson and Wexlar (1979), and more recently by Andrews and Nesbit-Evans (1983). This approach has been of particular use in the study of small mammals that are the usual prey of birds such as owls, eagles, and buzzards, and in addition, mammalian predators such as the Viverridae (genets, civits) and Mustelidae (weasels). Andrews (1990) and Denys *et al.* (1992) have established empirical scales of degradation based on the respective dissolution of bone and enamel (Table 7.1). As climate-change studies for continental areas now frequently rely on statistics of population changes in small rodents (mice and gerbils), an important bias can be avoided by taking into account the role of predators as being at least partly responsible for causing artifacts in data concerning accumulations of selected species (see below).

Dentine and enamel

In spite of the presence of parallel canals, dentine is generally less porous than bone. The compact mineralized material frequently provides examples of good preservation (Fig. 7.5). Dentine canals may remain empty (Figs. 7.5a–d). In such cases (e.g., *Kubanochoerus*, Miocene, Libya) the nanostructure of the dentine can be well preserved. However, both overall composition and nanostructure are taxonomy-linked parameters, making precise evaluation of low-level changes difficult in the absence of living representatives. The incorporation or loss of chemical elements has been documented for dinosaurs belonging to distinct taxa and in differing sedimentary environments (Dauphin 1989b, 1991). To date,

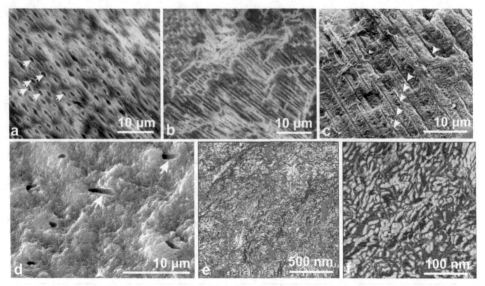

Fig. 7.5. Dentine. (a) Well-preserved dentine of a rodent incisor containing empty tubules (arrows), collected from a modern regurgitation pellet; (b) Empty tubules of dentine of a large fossil mammal tooth from Malawi (Plio–Pleistocene); (c) Parallel tubules (arrows) in dentine of an incisor from a fossil rodent from Tighenif (Plio–Pleistocene, Algeria); (d) Empty tubules with peri-tubular dentine still distinct (arrows), from a *Kubanochoerus* specimen (Miocene, Libya); (e–f) Nanostructure of the dentine of *Kubanochoerus*. c: Dauphin (1999b); d–e: Dauphin *et al.* (2007).

the molecular mechanisms leading to these distinct fossil facies have not been investigated. In contrast to easily recognizable dentine materials, negative casts can also occur when, after infilling of tubules, complete dissolution of the dentine material has occurred (Fig. 7.4e).

Enamel, well known for its hardness and low organic content, frequently shows very recognizable microstructural patterns (Figs. 7.6a–d); however, it does not escape nano-structural (Figs. 7.6e and f) and chemical change (see Fig. 7.11). Statistical studies of chemical composition between a number of fossiliferous sites, and between successive layers within given sites, indicates some overall trends of fossilization, and these emphasize the specificity of each site (see Fig. 7.11).

It seems a paradox that excellent preservation of structural patterns, as demonstrated by these figures, is not sufficient to insure an equivalent quality of chemical information retrieved from bones and teeth. The mode of fossilization is probably the leading factor in providing for the type of preservation where chemical data does persist, as was indicated by the research of Lee-Thorp and Sponheimer (2003). They showed in addition that major effects can be due to methods of sample preparation.

At present, recognition of the diagenetic state of fossil bones and teeth is obtained by X-ray diffraction or by infrared absorption measurement of the "splitting factor" based on two PO_4 absorption bands at 560 and 605 cm^{-1} (IR-SF). Increases in crystallinity and/or loss

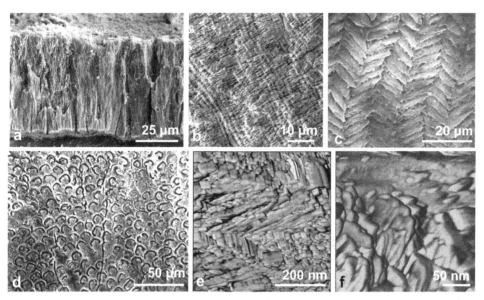

Fig. 7.6. Fossil enamel. (a) Prismatic structure preserved in a tooth of the dinosaur *Carcharodontosaurus* (L. Cretaceous, Albian, Tunisia); (b) Preserved growth lines in dinosaurian enamel; (c) External pattern on enamel of a fossil rodent molar (Plio–Pleistocene, Tighenif, Algeria); (d) "Keyhole" pattern of prisms on the surface of a tooth of *Numidotherium* (Eocene, NW Africa); (e–f) AFM views of the enamel nanostructure of *Numidotherium*. a: Dauphin (2005); d: Dauphin and Williams (2004); j: Dauphin *et al.* (2007).

of organic content are the two basic features, more or less correlated, utilized to characterize diagenetic condition. As emphasized by Lee-Thorp and Sponheimer, the pathway of fossilization apparently is more important than the absolute values of these two specific character indices. It has also been found that sub-modern (archeological) bones commonly make inadequate objects for chemical measurement (Wright and Schwarcz 1996). However, the "slow removal of collagen together with a relatively rapid increase in crystallinity," seems to be the best condition for preservation of chemical data over longer periods of time (Lee-Thorp and Sponheimer 2003).

In terms of their application, the following examples of research based on chemical or isotopic properties of bones and teeth reveal, simultaneously, the potential of these materials, and also that "we are still a long way from being able to regularly use the stable isotopes in bone minerals as a window in the past" (Lee-Thorp and Sponheimer 2003).

Biological or environmental applications based on chemical or
isotopic properties of fossil bone, dentine, or enamel

As past climate data are rare regarding continental areas, it is of particular interest to obtain reliable information from fossil terrestrial organisms, allowing correlations to be made with contemporaneous marine faunas. Since the pioneering studies by Longinelli (e.g., Longinelli 1984), numerous attempts have been made to use phosphatic biominerals as alternative sources to Ca-carbonates for environmental data. As was noted by Sharp *et al.* (2000), "phosphates were thought to be the panacea ... until recently it was believed that once formed the oxygen isotope composition of phosphate would not change." More specifically, inference of the dietary regime of ancient vertebrates from isotopic signals preserved in bones and teeth has been widely explored for about 30 years since the seminal paper by Sullivan and Krueger (1981). An immediate and opposite example (Schoeninger and De Niro 1982) triggered controversy regarding the significance and reliability of the data collected, and resulted in new research to establish precise relationships between diet and the chemical composition of whole bone, or the collagen or enamel.

Modification caused by microbial enzymatic activity has been found, thus, the influence of diagenesis on fossil bone composition was studied by De Niro and Weiner (1988). They also focused on determining what precautions must be taken when using physical–chemical parameters as the basis for paleoenvironmental reconstruction or for dating. In a series of three papers, they compared the compositions of the insoluble organic material, or "collagen," extracted from bone at multiple localities, and noted the great variability of the results they obtained. In comparing $\delta^{13}C$ values obtained from "collagen" and from organic aggregates from the same samples, the differences reflected in analyses of their amino acid compositions confirm their extreme alterations. Therefore, they recognized that isotopic values obtained from these samples were so greatly modified that they were useless. The term collagen, commonly used to refer to the insoluble organic matrix of bone, has limited applicability, as noted earlier in the chapter concerning modern bone. This is especially true of the insoluble organic compounds extracted from fossil bone: these components certainly do lessen the accuracy of isotopic analyses. In the same way, study of the development of crystallinity in bones and dentine by Koch, Tuross, and Fogel (1997) has shown that scattering of measured values was lower in enamel than in either bone or dentine. As microbial activity appears to be the major factor in early diagenesis, this result is consistent with the simple fact that the concentration of organic materials in enamel is low to begin with.

Parallel with this, the significance and reliability of ratios of elements such as Sr/Ca and Ba/Ca have also been explored, assuming that the variation seen between different vertebrate groups was related to their respective trophic behaviors (from Toots and Voorhies 1965 to Dos Santos *et al.* 2006). The Ba/Sr ratio, for instance, allows a clear-cut separation of herbivorous and carnivorous animals. Again, various opinions have been expressed, specifically concerning the controversial question of the diet of ancient hominids. The Ba/Ca or Sr/Ba ratio in enamel has indeed made it possible to differentiate between the diet of paleohominids and that of other vertebrates (grazers, browsers, and carnivores) in the Sterkfontein and Swartkrans environments (Sponheimer and Lee-Thorp 2006). However, as stressed by these authors, locally significant ecological patterns cannot be transferred to similar groups in other environments.

In practice, despite the occasional statement that no diagenesis has occurred (e.g., regarding a dinosaur skeleton, Barrick and Showers 1995), fossil bone as a whole is always a mixture of variously modified original structures that are penetrated by allochthonous materials. Depending on details of the burial and fossilization processes, distinct isotopic values vary over very short distances within a single tooth (Sharp *et al.* 2000). Only microsampling techniques associated with careful study of localized recrystallization offer opportunities for an improved, more reliable use of phosphatic biominerals. Still more necessary are baselines (Sponheimer and Lee-Thorp 2006) that deal with both elemental distribution in modern ecosystems and diagenetic modifications. According to these authors, improving the former is most necessary, as its improvement is the *sine qua non* for progress in the latter.

Bone histology as phylogenetic tool or environmental tracer

The histology of the bones of fossil vertebrates has long been studied (Gros 1934; Enlow 1963), but among the wide diversity of structures described it is still difficult to definitely correlate a given pattern to one of the four possible factors that are determinants for microstructural organization: (i) phylogeny, (ii) ontogeny, (iii) biomechanics, and (iv) environment (de Ricqlès 1975). Comparative analysis restricted to the limited number of case studies can allow the proposal of correlations. Botha and Chinsamy (2000), for instance, studied two contemporaneous cynodonts with readily distinguishable bone growth patterns and were able to propose that the cyclical growth mode of the observed herbivore/omnivorous species was linked to a seasonality of local vegetation and other nutritional resources, whereas, in contrast, the carnivorous species is characterized by continuously sustained bone growth.

The basic biological processes governing bone microstructure are still debatable in large part, and there are divergent opinions regarding the interpretation of even the major histological patterns, such as the Haversian canals. These easily recognizable structures are commonly tied to high metabolism, but they have also been found in typically "low metabolism" animals such as crocodiles and turtles. Conversely, they can be absent in the skeletons of rodents or birds, undoubtedly high-metabolic animals. The adaptability of the bony structure that makes interpretation of fossil samples so difficult would appear directly linked to the fact that, unlike calcareous structures, bones have true cellular architectures, subject to many deep, diversified, and still poorly understood biological influences. It is disappointing that the interpretations of the many spectacular and decipherable histologies of the bone secreting cells are yet so uncertain.

Bone and tooth composition through statistical studies

If interpretation of characteristic points (whether chemical or histological) is still subject to multiple uncertainties, the statistical treatment of measurements made on fossil bones and teeth appears to be a useful tool when applied to large collections of material from a number of separate localities, or from various levels at a given locality. This type of analysis is particularly pertinent to the study of fossil bones of micromammals, which are generally numerous and diversified in some specific sites (often caves). In recent years, particular attention has been given to European or African deposits of this type. As these accumulations of micromammals are generally related to predators (birds or other mammals), digestive effects have to be taken into account as an important diagenetic factor

(Andrews 1990). Research by Raczynski and Ruprecht (1974) has shown the differing results of ingestion by various owls, results confirmed by Dodson and Wexlar (1979), and then by Andrews and Nesbit-Evans (1983). Proceeding from the analysis of a large number of samples, Andrews (1990) was able to define five categories of predators based on the number, fragmentation, and state of bone surfaces and teeth found in gastric balls of raptors and in feces of carnivorous mammals (Table 7.1). Accordingly, numerical indices expressing the state of preservation of bones have also been defined (Denys *et al.* 1992).

These analyses indicate that predation and digestion of micromammals are responsible for important biases in statistical analyses of bone abundance and fragmentation. As interpretations of climate change in continental areas are frequently based on variation in the proportions of micromammal species, the detection of predation and evaluation of its effects on the composition of given assemblages must be carefully studied. Thus, in the case of *Bubo bubo* (the horned owl), skulls of the Dipodidae, the Gerbillidae and Crociduridae better resist digestion than do those of the Murinae, the old world mice (Denys *et al.* 1996). In the case of a preservation differential, the greater abundance of gerbils, observed in the modern owl gastric hair balls from Algeria, does not necessarily indicate their greater initial abundance in the fauna. This type of observational bias indicates that we need to be prudent in using variation of abundance data for taxa.

Not only are the respective abundances of different bones or skulls modified by predator activity, but the composition of the remaining bones is also differentially modified. Infrared spectra of bones from modern gastric balls collected at Olduvai (Tanzania) show that the amide bands I and II (organic matrices) are less intense than in those of fresh bone, while the v3 band indicative of PO_4 is more intense in bony material found in the balls (Dauphin *et al.* 2003d). In these samples, no enrichment in fluorine was recognized, and the level of bone crystallinity was found to be high in the regurgitated balls (Fig. 7.7). At the same time, there is no evidence to suggest whether such augmentation results from recrystallization of amorphous bone, or the opposite, from its dissolution.

Variations depend on the prey, but equally as much on the predator, resulting in complex data. Therefore, statistical analyses carried out on the elemental chemical composition of modern rodent bone, originating from different predator neighbors, show differences both in their content of major elements (P and Ca), and in minor elements (Dauphin *et al.* 1989, 1997).

Amino acid analyses of bones from *Bubo* (owl) gastric balls (Olduvai, Tanzania) also indicate that organic matrices have been differentially modified (Fig. 7.8) (Dauphin *et al.* 2003).

To these observed modifications at the "gastric-ball" stage, changes that occur during and after burial need to be added. This diagenetic modification differs from one fossiliferous locality to the other, and also varies in successive levels at each given site, even if the amount of time represented is very short on the geological scale. Thus, in the Pleistocene locality at Tighenif (Algeria), quantitative relationships of the organic and mineral phases, crystallinity, etc., are unequally modified in rodent bones originating in three different levels although the included sediments do not fundamentally differ (Fig. 7.9) (Dauphin *et al.* 1994).

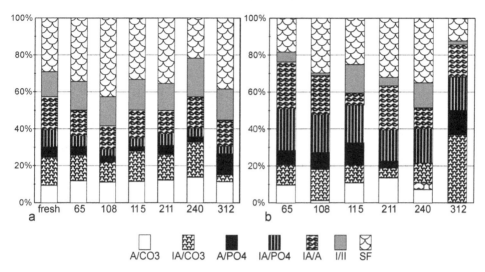

Fig. 7.7. Modification of composition and crystallinity of bone collected in different modern regurgitated gastric balls from Olduvai, Tanzania, according to infrared spectrographic data. (a) A/CO$_3$, IA/CO$_3$, A/PO$_4$, and IA/PO$_4$: modifications of the relationship between organic matrix and mineral component; (b) IA/A, I/II: modifications of the composition of organic material, based on amide bands; SF: crystallinity factor.

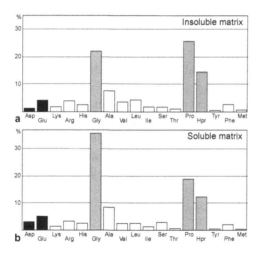

Fig. 7.8. Amino acid composition of soluble and insoluble matrices of bones extracted from modern regurgitated gastric balls from Olduvai, Tanzania.

Not only is bone not a "closed system," as long claimed by some geochemists, but dentine and even enamel undergo chemical changes that vary considerably between different regions, as indicated by the tooth composition seen within five dinosaur genera (Fig. 7.10).

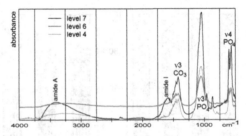

Fig. 7.9. Infrared spectra showing the marked difference in composition of rodent bones recovered from three levels within Pleistocene strata at Tighenif (Algeria). Dauphin *et al.* (1999).

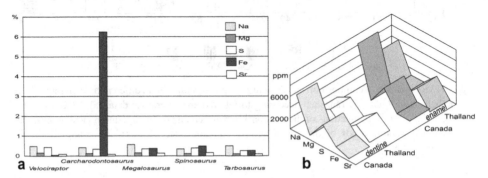

Fig. 7.10. Variation of element concentrations in enamel and dentine of five dinosaur genera from different areas.

At present, much of the chemical data at hand regarding bone and tooth composition lacks precise information concerning conditions of chemical exchanges between vertebrate remains and their diagenetic environment. In contrast to the earlier prevailing opinion regarding the chemical stability of isotopic ratios, it is now commonly acknowledged that bone, being a complex and fragile tissue, is almost always poorly preserved, and of little use for reconstruction of ancient environments and alimentary regimes. These statements are based on analyses carried out with generalized sampling methods, thus including both bone and secondary deposits, making evaluation of the bony tissue itself impossible. Identification of the preservation state must be made at relevant detailed scales (Dauphin *et al.* 2007b).

When chemical measurements are made with bone structure taken into account, chemical trends during fossilization can be evaluated by the statistical analysis of chemical data (Fig. 7.11).

Relevant comparisons can be carried out between comparable materials from distant areas (Figs. 7.11a–c), or between successive layers at one site (Figs. 7.11d–f). This approach provides acceptable case studies through re-examination of long-known sites, such as the caves at Quercy, France – a site that has been explored for decades – as well as their comparison with materials from Olduvai (Kenya) or Tighenif (Algeria).

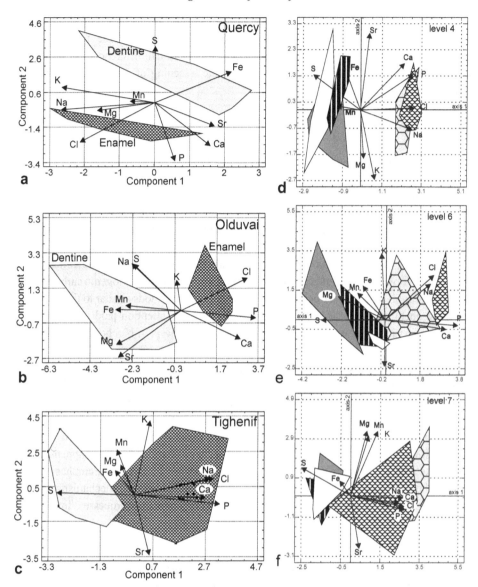

Fig. 7.11. Principal Component Analysis of chemical measurements carried out on fossils from three localities (a–c) and from three levels of the Tighenif locality (d–f). (a) Enamel and dentine are very distinct in the Oligocene of Quercy, France; (b) Enamel and dentine are distinct at Olduvai, Kenya (Lower Pleistocene), but the main chemical elements involved differ from those at Quercy; (c) Enamel and dentine are not very distinct in teeth from Tighenif, Algeria (Plio–Pleistocene); (d–f) Chemical alteration differs in three levels studied at Tighenif. a–c: Dauphin and Williams (2007); d–f: Dauphin *et al.* (1999).

Principal Component Analysis of microprobe data on the major chemicals involved in the fossilization process permits precise evaluation of diagenetic trends in each locality; two of them (Figs. 7.11a–b) accentuate compositional differences between fossil remains, whereas the third (Fig. 7.11c) leads to the lessening of differences in chemical composition.

7.3 Fossilization of calcareous skeletons

Owing to their importance as major factors in equilibria between sediments, seawater and atmosphere, and chemical properties, the behavior of Ca-carbonates have been thoroughly investigated by geologists for more than a half-century (Bathurst 1971; Morse and Mackenzie 1990). Among these, biogenic Ca-carbonates occupy a special position due to their role in recording environmental data. Theoretical and field or experimental approaches have been continuously developed and modified during recent decades in order to improve understanding of how isotopic or chemical signals form. However, in geochemical investigations involving environmental and paleoclimate reconstruction, biogenic carbonates are usually recognized and described by physical or chemical methods similar to those used for the pure chemically produced polymorphs. Only rarely is attention paid to their specific method of crystallization or to the influence during diagenesis of biochemical compounds included within them.

The goal of this section is to illustrate the potential of reinterpretation associated with in-depth analyses of the fossil record. Examples focus in most part on the diversity of possible applications and possible consequences. At times, observations simply fail to explain an unexpected action during fossilization, or the effectiveness of different methods of gathering reliable chemical data in biogenic structures that are partly altered or contaminated. However, by gathering apparently independent observations, it also happens that we can make substantial contributions to long-standing issues. In each case, emphasis here is put on the continuous interplay between biological and biochemical mechanisms as clues to the improved understanding of individual fossils and the processes involved in preserving them.

7.3.1 *Converging processes leading to the complete degradation of skeletal components, with some unexpected means of preservation*

Most biologically produced Ca-carbonate objects are submitted, postmortem, to a series of physical, biological, and chemical factors that each contribute to the rapid and complete recycling of their organic and mineral components. When there is a complete disappearance of biogenic carbonates, the main cause is generally not a physical force such as wave action, but rather, bioerosion by varied organisms that create a hierarchical series of cavities inside the mineralized structures themselves. Some of these organisms create dwelling cavities within shells, such as worms (Figs. 7.12a–b), sponges (Figs. 7.12c–d), or bryozoa (Figs. 7.12e–f).

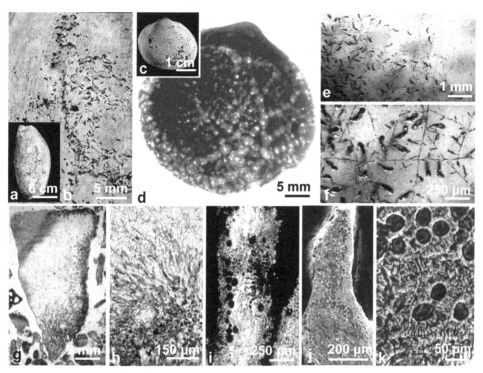

Fig. 7.12. Examples of common endolithic borers. (a–f) Macroborers residing in shells: worms (a–b), sponges (c–d), Bryozoa (e–f); (g–k) Borers using shell matrix material as nutrients: algae and fungi in a mollusc shell fragment (g–h), and in a coral skeleton (j–k).

Other organisms, the highly diversified filamentous borers, both algae and fungi, that bore at fine scales (but with extreme efficiency), continue the degradation of material (Figs. 7.12g–k) to micrometer scales. Calcareous shells are commonly reduced to fragments whose mean grain size becomes progressively finer, increasing their surface area subject to dissolution. Therefore, these multiscale biologic agents play a major geochemical role by accelerating dissolution and recycling of Ca-carbonate.

This biological process is not linked to the mineralogy of the shells, the property that is very commonly considered as the major factor affecting diagenetic processes. The development of endolithic boring is essentially driven by the organic content of the biologically produced Ca-carbonates. Thus, within identical faunal assemblages submitted to identical depositional conditions, the probability for fossilization of different taxonomic components does not depend primarily on the mineralogy of their skeleton. The proportion of organic content, which varies greatly between types of microstructures (and even within different layers of a single shell), causes considerable bias in post-depositional change, different from that assumed on the basis of their mineralogy. For example, low Mg concentrations in the prisms of *Pinna*, for instance, should make this carbonate more resistant to dissolution than that of the echinoderm skeleton formed of high-magnesium calcite, but observation disproves this. The calcareous plates and spines that form the tests of echinoderms are formed by mesoderm, and throughout the life of the animal they remain covered and are protected from contact with seawater, and also from the action of endolithic borers. Additionally, the level of their content of organic material is one of the lowest among biominerals; so that they are not attractive to filamentous borers as a source of nutrients. This could be one reason why the contribution of echinoderm debris to biogenic sediment is among the greatest in the geological column, especially during the Paleozoic. In the Carboniferous Period, the crinoids, then diverse and broadly distributed in shallow waters, produced huge accumulations of debris over thousands of square kilometers of North America and northern Europe. Actually, the lack of destruction of their skeletal segments by boring microorganisms is one of the characteristic aspects of the fossilization of echinoderms. Their unique mode of biocrystallization has been maintained without change since the appearance of the early echinoderms, thus diagenesis of their skeleton results in typical features (see below, subsection 7.3.2).

For a similar reason, the aragonite of coral skeletons is more sensitive to microboring activity than the cross-lamellar structures of molluscs (almost always aragonitic), as the latter generally contain a lower content of organic material. Even in the earliest changes affecting skeletal materials, the consideration of its mineralogy alone does not provide a reliable indication of the probability of its preservation based only on the early – but very important or crucial – steps to its becoming part of the fossil record. Nevertheless, a clear framework for re-examining the rules for diagenesis can be provided by taking into account both the approach, at various scales, of microstructural patterns (microstructure, growth layering, and nanoscale organization), and also, evaluating the quantitative and qualitative evolution of the contents and positioning of organic components.

Biological processes are not always destructive; they can also cause fossilization as shown by the calcareous structures built by the green algae, especially shown here by *Halimeda* (Udoteaceae, Fig. 4.1) or the Dasycladales (Fig. 4.2). In these algae, their mineral parts consist of irregularly dispersed or weakly associated aragonite needles formed within an organic gel located between ramifications of the cellular wall. This organization makes their fossilization unlikely and, at present, these organisms are the main providers of fine calcareous mud in shallow tropical seas (see Fig. 4.1). But, fossils belonging to the Dasycladales are extremely abundant in the shallow, warm water deposits of certain geological periods, such as the Upper Triassic or Cretaceous. In addition, they can be very precisely identified, based on their tube ramifications and positions of conceptacles (the reproductive cavities), so that, in spite of apparently unfavorable conditions for fossilization, the dasyclads are widely used as chronologic indicators. Usually, microscopic observation indicates that their mineralized material, located between vegetative tubes, has been replaced by coarsely crystallized calcite. Some specific process must be involved to explain this preservation of the overall morphology of the ramified cylinders (Figs. 7.13a–c). A spectacular specimen from the Eocene of the Paris Basin adds to our understanding of the process (Figs. 7.13d–l).

This sample has no such coarsely crystallized calcite developed between the ramified tubes; instead, they are casts, seen to have a branched central cylinder (Figs. 7.13d–i). This

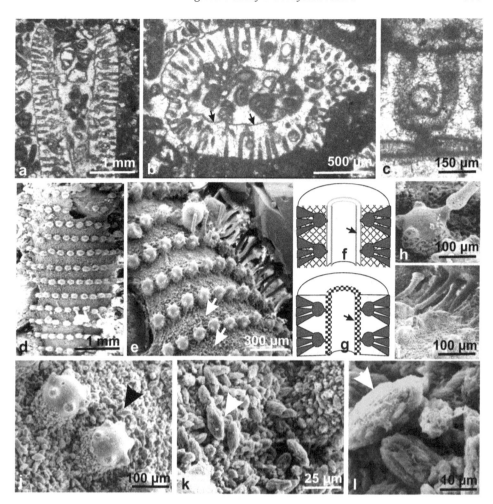

Fig. 7.13. Fossilization of green algae (Dasycladales). (a–c) The most common modes of fossilization: fossilization of the cell wall (a, b: arrows), infilling of ramifications and the secondary formation of euhedral calcite between them (c); (d–l) A mixed fossilization process: fossil of an Eocene dasyclad from the Paris Basin; (d–e) Ramified tubes are attached to basal heads arranged in regular verticils (cycles of simultaneously produced ramifications). Each head (first rank ramification of the central cylinder) typically has six secondary tubule prolongations; (f–i) Initially formed as elongate diverticuli of the cellular membrane, the primary and secondary tubules have here been filled in by sediment before dissolution of the aragonite needles (f–g). The tubules appear as positive casts of the cellular ramifications (h–i); (j–l) Microstructure of the membrane of the axial cylinder. The cellular wall forming the central cylinder is now built by small calcite granules that are calcified bacteria. Closer observation shows the typical "weizenkorn" (wheat kernel) appearance of the calcareous deposits on the bacterial coating (arrow).

gives us a direct view of the ramification as it was when the algae were alive. It is likely that the organic tubes underwent early and rapid decay and subsequently, fine sediment filled in voids prior to dissolution of the aragonite needles, a rather unusual sequence. Closer observation of the central cylinder indicates that the rounded axial cell wall (normally nonmineralized) has been replaced by granular carbonate that shows very specific morpho-logic types (Figs. 7.13j–l). These grains apparently are calcified bacteria, exhibiting the typical "weizenkorn" (wheat kernel) grained pattern, comparable to those often noted in beach rock environments. Krumbein (1974) showed that this crystallization pattern can be obtained by in vitro bacterial cultivation.

Evidence of a bacterial contribution to fossilization explains the common fossilization of the organic, cellular wall of green algae in the shallow water of carbonate platforms, places where well-oxidized water does not generally allow fossilization of organic compounds. The thin and continuous black lines visible in optical thin sections precisely delineate the contours of the central cylinder wall and its ramifications (Fig. 7.13b: arrows). Based on the previous example, we can hypothesize that the organic walls were utilized as nutrients by calcifying bacteria. As a result, nothing exists of the dasyclad itself: the cell membrane is now a thin mineral layer composed of calcified bacteria. Within the space outlined by this mineral film, aragonite needles have been dissolved and replaced by coarsely crystallized calcite. However, the mechanism is so precise and widely distributed that it provides easily identifiable specimens useful for stratigraphic dating of shallow-water tropical platforms throughout the later Paleozoic, the Mesozoic, and early Cenozoic Eras.

7.3.2 State of mineral material in some older calcareous biominerals

The case of the Dasycladales is a particular one; most calcareous biogenic materials appear as solid and sometimes massive structures to which the rules of "respective solubility" seem to be reasonably applicable. A sizable literature has been dedicated to the study of the mineralogical transformations that may affect biogenic carbonates. Reconstructing the evolution of paleoenvironments reflected by sedimentary sequences through focusing on recrystallization patterns of fossilized organisms has been a typical application in geological research. Modern marine areas where carbonate sedimentation dominates, for instance, the Bahamian Platform or the east coast of Arabia, became the location of numerous and detailed studies that have produced comparative models applicable to ancient carbonate rocks. Studies dealing with reefs and their diversity based on the types of frame-builders benefited from this sedimentological approach. From the studies of Dunham (1970), Bathurst (1971), Heckel (1974), Wilson (1975), James (1983), and Schröder and Purser (1986), has come a continuous series of publications that illustrate the development of pioneering investigations in this area, begun by Walther in 1888. Generally speaking, more attention is paid to the development of the diverse neomorphic cements occurring there than to changes that may have occurred in the skeletal materials. Interpretations are primarily based on petrographic studies employing the polarizing microscope, and X-ray diffraction methods have been used to assess the overall status of the materials analyzed.

Recently, studies to investigate the extent to which chemical interpretations are affected by the ever-increasing proportion of recrystallized aragonite in corals (e.g., McGregor and Gagan 2003) suggest that such empirical approaches can still be useful.

In the area of geochemical research, rapidly developing microsampling methods enable an ever more precise approach to the effect of diagenesis on biogenic structures to be followed. Of course, spatially restricted measurements require better selection of places to be measured. It was noted above that, at the dimensional range of 10 micrometers, coral skeletons contain distinct areas of mineralization, each of them producing specific signals. Avoiding the mixing of these signals by the use of microsampling methods may lead to a significant reduction in numerical scatter (see Fig. 2.17).

Equivalent improvement of data can be obtained when chemical measurements need to be carried out on older material affected by early diagenetic processes that have resulted in deposition of additional minerals. The skeleton of the scleractinians, largely open to aqueous solutions, provides typical examples of deposition of these diverse neomorphic carbonates in the early stages of cementation (see, for example, Schröder and Purser 1986). Such cementation can differ considerably within samples (Figs. 7.14a–c), and these neomorphic crystals have variable chemical and/or isotopic ratios that can greatly reduce the accuracy of any overall measurement. However, closer examination by serial polished sections can reveal suitable surfaces showing the fibrous skeletal carbonate just below its cortex of neomorphic crystals (Figs. 7.14d–f). Although some deposition of nonbiogenic carbonate between fibers cannot be excluded, the possibility exists here for obtaining significant chemical measurements.

By contrast, Fig. 7.15a illustrates an example of the complete dissolution of a coral skeleton where the resulting void spaces have been filled in by blocky calcite. Between these two extremes of coral diagenesis, Figs. 7.15b–g provide examples of progressive recrystallization in which particular patterns of skeletal microstructure influence early diagenesis. Early mineral change in the "centers of calcification," characterized by their high concentration of organic components and micrometer-sized aragonite crystals, begins by creating loci for preferential dissolution. In typical trabecular structures (like the microstructure pictured in Figs. 7.15b–g), it has been determined that, beginning at this central area of dissolution, the aragonite–calcite inversion progresses centrifugally. Mapping of strontium distribution using a secondary ion mass spectrometer shows a clear differentiation between original coral aragonite at the periphery of fibers, and more central parts that are only partially recrystallized (Fig. 7.15g). It is interesting to note here that, in the transitional area between the aragonite and calcite, remnants of aragonitic units occur within fibers that are mostly recrystallized (Fig. 7.15g: arrows).

It can be seen at this early stage of diagenetic change that the arrangement of recrystallized units in the central parts of trabeculae (Fig. 7.15g) may reflect the microstructure in its initial state. However, the loss of strontium and development of internal porosity (white arrows in Figs. 7.15e–g) are characteristic of this early step in recrystallization. Atomic force microscopy shows that some heterogeneity still exists within the newly formed mineral material (Figs. 7.15h–i). Isolated globules producing strong phase-contrast signals indicate the possibility of them being remnant organic matter, probably composed in most part of insoluble matrix.

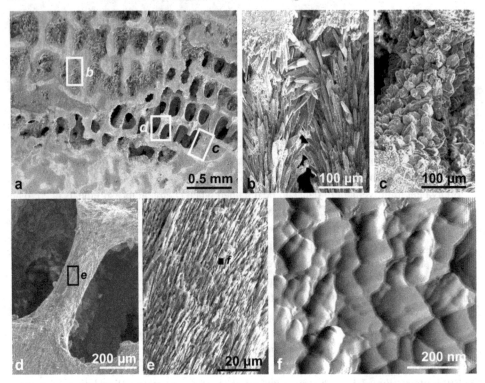

Fig. 7.14. Surface condition of coral skeleton from a core in a modern reef, and the microstructural state of basic skeletal fibers. (a–c) Distinct neomorphic cements developed on the coral skeleton at a 1 mm scale. In most isotopic measurements, these sedimentary cements would be included in the chemical measurement of coral skeletal carbonate. (d–f) Skeletal axes of the coral skeleton (d) exhibit good microstructural preservation at the micrometer (e) and at a nanometer scale (f), showing the possibility of obtaining accurate chemical data, even on biogenic structures modified by early diagenesis.

Further changes can occur, depending on localized post-depositional history, but this early diagenetic status can also remain quite stable, as clearly shown by the specimen illustrated in Fig. 7.15. This coral belongs to the Triassic genus *Rhopalophyllia*, and was collected in lower Norian units in the Lycian Taurus (South Anatolia, Turkey). It demonstrates how stable the coral structure can remain over time, this conclusion reinforced by biochemical data from mineralizing matrix remaining within it (see Figs. 7.25 and 7.26).

Echinoderm skeleton

The Phylum Echinodermata is characterized by the diverse modes of life and ecological adaptations of its numerous classes. In addition to the five classes living at present, 15 additional classes are found as fossil forms seen in Paleozoic strata (Ubaghs 1967), and up to 23 according to Paul (1979). Phylum diversity was at its peak during the Ordovician Period,

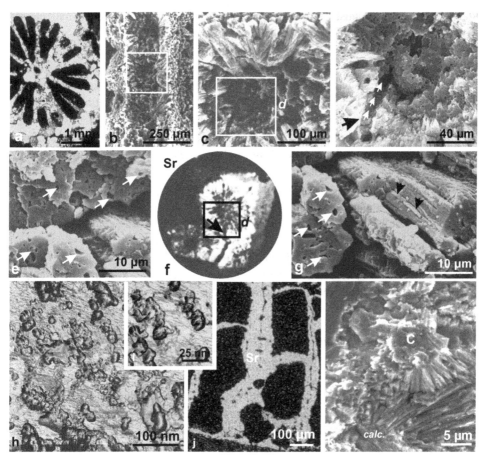

Fig. 7.15. Partially recrystallized coral skeletons. (a) Tranverse section of a corallite showing complete replacement of the skeleton by coarsely crystallized calcite: typical example of dissolution leaving no traces of the original material; (b–e) Broken tips of trabeculae at growing edge of a septum. Radiating fibers are still visible at the periphery of each trabecula (b–c), whereas the centers show irregular crystallization developed in the zone of distal mineralization and the more central parts of fibers (d–e); (f) Distribution of Sr showing aragonite recrystallization to calcite: the axial zone of trabeculae and internal portions of fibers have lost their high Sr content, whereas external portions of fibers are still preserved (Image: Slodzian, LPS, Orsay). Note the path of an endolithic borer towards the formerly organic-rich distal mineralization center (arrows); (g) Presence of aragonite remnants (arrows) within recrystallized fibers; (h–i) AFM views of recrystallized fibrous material. Mineral granules no longer exist, but phase-contrast imaging suggests that some patches of organic-rich materials may be present within newly formed calcite (i); (j–k) Microprobe mapping of the Sr distribution in partly recrystallized septum of a Triassic *Retiophyllia* from the Richthoffen Reef (Dolomites) (j) and corresponding structures (k) of the calcium carbonate (J. E. Sorauf original data and specimen). Owing to the spatial resolution of microsampling methods, these images indicate that it is possible to collect a valid signal from fossil corals, providing that previous microstructural study has been done. b–c, f: Cuif (1977).

then declined continuously throughout the Paleozoic, the last of these fossil classes (Blastoidea) becoming extinct during the Late Permian.

Throughout this long history, the mechanism of biomineralization in the phylum remained very constant in these abundant and diverse forms. There is no known exception to their mesodermal network of high-magnesian calcite, where each distinct skeletal unit exhibits monocrystalline behavior (see Fig. 4.48), except for echinoid teeth, each of which are polycrystalline, resulting from fusion of several skeletal units.

In accordance with this, fossilization of echinoderm skeletons seemingly is uniform throughout the entire phylum. As each mineralized skeletal component shows a mono-crystalline behavior in its pristine state, diagenetic processes transform these porous monocrystalline units (Figs. 7.16a–b) into a massive monocrystal, preserving the external shape of each skeletal unit. When broken, any fragment of an echinoderm skeleton, even if unidentifiable at the class level, can be definitely attributed to the phylum by observing the rhombohedral cleavage faces on the broken surfaces (Fig. 7.16c). This fossilization process, unique to the Echinodermata, has remained identical from the first-known representatives of the phylum, as illustrated in Figs. 7.16l–m by a specimen belonging to the carpoids, one of the most distant from modern Echinoderms, both in time and phylogeny.

The ultrastructure of the shells of some Paleozoic brachiopods

As the main group of bivalved shells throughout the Paleozoic Era, the brachiopods usually play a major role as environmental indicators. Parkinson *et al.* (2005) and Cusack *et al.* (2008d) have shown that reliable environmental information can be obtained from these shells, generally expected to be somewhat nonsensitive to diagenetic change due to their uniform calcite mineralogy. Some examples of Devonian brachiopods, not selected partic-ularly because of their exceptional preservation status, illustrate what can be considered the normal and expected preservation of brachiopod shell fibers (Fig. 7.17).

The fibers appear well preserved at the morphological scale (i.e., in the 10 µm range) and atomic force microscopy likewise fails to reveal obvious patterns of recrystallization in them. Carbonate granules here still show their usual disordered arrangement, with no traces of fusion into higher-level clusters.

Such good nanoscale structural patterns explain the quality of isotopic signals collected by Popp *et al.* (1986) in their extensive investigation of Paleozoic cements and brachiopod shells. After analyzing approximately 200 shells of brachiopods tested by cathodolumines-cence (and discarding samples whose luminescence revealed diagenetic alteration reflecting enrichment in secondary calcite with high Mn and Fe contents), their conclusion was that shells of brachiopods would be the "best choice" to establish reference standards for Paleozoic carbonates.

From the perspective of their use as standards for environmental analysis, taking advantage of the recently developed secondary ion mass spectrometers that allow localized measure-ments in the micrometer range, it is important to be reminded that brachiopod shells record somewhat differentiated signals between distinct shell layers, as recently demonstrated

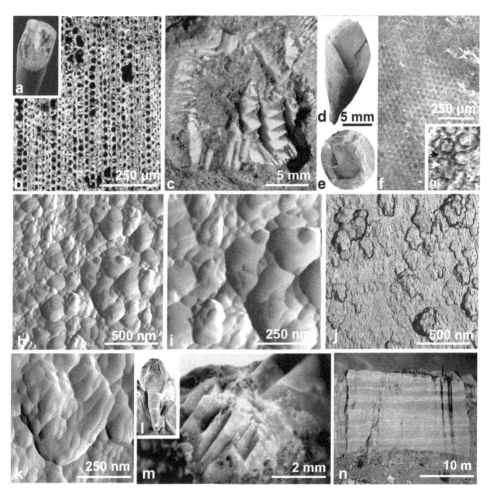

Fig. 7.16. Diagenetic behavior of the Echinoderm skeleton and persistence of its calcitic nature and granular nanostructure after infilling of the original porosity. (a–b) Recent spine of *Cidaris*, fracture surface (a) and longitudinal thin section (cross-polarized light); (c–e) Fracture surfaces of fossilized echinoid test (c) and spines (d–e). Note the perfect rhombohedral fractures contrasting to the conchoidal fracture surfaces observed on modern specimens (Figs. 4.48k–l); (f–g) Aspect of a polished surface in fossilized material, showing that the original organic network is still visible; (h–i) AFM views of fossilized biogenic calcite (g: black arrow). To be compared to Fig. 4.50; (j–k) AFM views of calcitic infilling of porous network (g: white arrow); (l–m) Rhombohedral fracture of the skeleton of an ancient echinoderm (*Callocystis*, Ordovician), with typical fracture; (n) Example of a sedimentary unit formed by the accumulation of echinoderm skeletal fragments (mostly echinoids and crinoids; Oxfordian, Pouillenay, France).

(Cusack *et al.* 2008b). Practically, the ability to measure remnant isotopic differences between layers of a single shell might also serve as a test for reliability of the chemical data.

From a biological viewpoint, images such as these invite the development of more extensive investigations at the nanoscale, in order to note precisely the similarities and

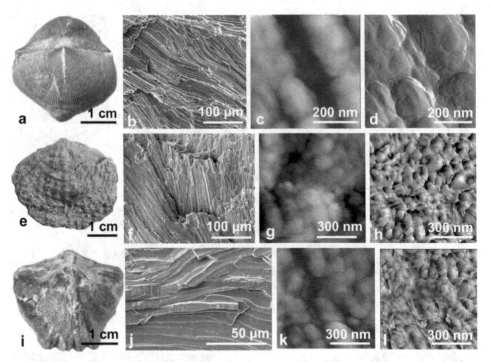

Fig. 7.17. Preservation of fibers in three Upper Devonian brachiopods: (a–d) *Theodossia hungerfordi* (spiriferid); (e–h) *Spinatrypa rockfordensis* (atrypid); (i–l) *Calvinaria* sp. (rynchonellid). These three parallel series of enlargements are examples of the preservation of brachiopod shells in representative specimens from the Upper Devonian Lime Creek Formation at Rockford, Iowa, USA, a locality noted for its great abundance of well-preserved brachiopods. The granular patterns in fibers are still clearly visible, and comparable to those illustrated in Figs. 4.37 to 4.41 (JES collection, Paleontological Research Institution, Ithaca, New York, USA).

differences between the fiber secretion processes in the main branches of this phylum that predominated during the Paleozoic Era.

Since the atomic force microscope does not require coating of samples for observation, the sample observed can subsequently be used as the source of desired chemical data, avoiding any bias between the areas observed and the points sampled for data. This is of particular importance with respect to possible diagenetic variation within a sample, where distinct parts were subjected to different, localized taphonomic conditions.

7.3.3 Fossilization of organic components of calcareous biominerals: possible reevaluation of their influence on diagenesis and their role as indicators of fossil condition

There is a striking contrast between the extensive literature dealing with diagenetic change in the mineral portion of biominerals and the small amount of attention paid to their organic content. Owing to the submicrometer interaction of the glycoprotein assemblage with the mineral phase, it is to be expected that the identification of the nature of their remains in fossil biominerals would be a useful indicator of the level or intensity of diagenetic change. In addition, the chemical composition of this organic component of shell material and other calcareous structures is so exact, and their positioning within the mineral phase is so intimate, that we can assume they play a direct role in the fossilization process. Establishing the particular influence of the organic content within biominerals on their diagenetic pathways appears to be a potential clue to a novel and realistic approach to the fossilization process. The analytical methods available at present make this type of study feasible. Even more important, the presence and recognized properties of matrices in fossils provide direct evidence to support biological interpretations that cannot be made with only an examination of the mineral component. Of course, it is primarily during the early stages of diagenesis that the specific influence of organic components can be evaluated.

Evidence of the decay of shell organic components and its consequences

Theoretically, mollusc shells are protected from direct contact with seawater by the external organic membrane, the periostracum, whose main role is to seal the mineralizing edge of the shell on its external side (see Fig. 3.21). Having played this essential role during early mineralization stages of the shell, the periostracum remains in place simply as a membrane covering its outer surface. However, examination of freshly collected shells of living *Pinna* or *Pinctada* reveals that both the periostracum and the underlying insoluble envelopes surrounding the prisms are already much degraded (Fig. 7.18).

Alteration can develop to such an extent that the earliest processes of fossilization properly said (conventionally commencing at the organism's death) will act on materials that have already been greatly modified. Microboring by algae or fungi (Fig. 7.18a: arrows) creates pores that provide access to the calcareous shell material. Envelopes in the prismatic layer are the first material degraded by bacterial action, to such an extent that prisms are dissociated as a result. This is clearly seen in the red-colored shells of *Pinna nobilis*, as, in this species, the red pigment (a canthaxanthine molecule close to carotene) is located in the insoluble organic envelopes. Decay of prism envelopes results in forming white spots or white surface layers (Figs. 7.18c–d), continuing until there is a complete dissociation of the external portions of the prisms.

In addition, due to their high content of mineralizing matrix material (see Fig. 3.24), the calcite prisms are highly vulnerable to biological activity by endolithic borers (Figs. 7.18c–h). The submicrometer-size particles held together by organic framework (e.g., Figs. 2.56 and 3.5) can also be dissociated by enzymatic activity. This explains the field observation that in some occurrences the nacreous layer of *Pinna* is more frequently preserved than the calcite layer, truly a mineralogic paradox (Fig. 7.18i).

This exemplifies the diversity of behavior of various microstructures with respect to the earliest stages of their diagenetic evolution, and illustrates how fossilization of biogenic carbonates can lead to paradoxical results. Analyzing the diagenetic evolution of calcareous biominerals based only on mineralogy must be regarded as an oversimplification. The diverse nanostructures of biocrystals, along with the varied behavior of their organic

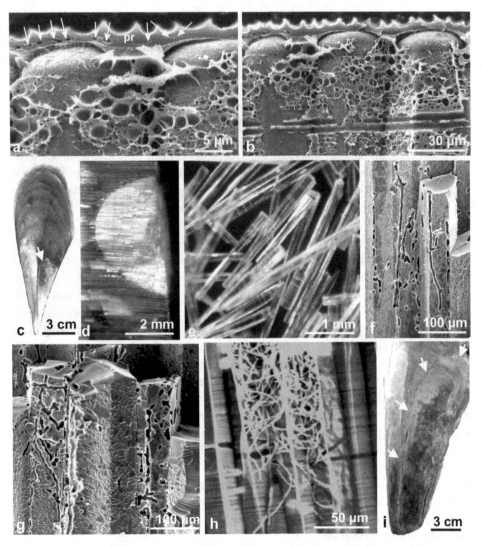

Fig. 7.18. Decay of periostracum and organic envelopes surrounding calcite prisms in *Pinctada* and *Pinna*. (a–b) Early decay of prismatic envelopes at the outer surface of the prismatic layer of a *Pinctada* shell. Multiple galleries and boreholes ("pr": arrows) allow access to the prism envelopes that are already highly degraded; (c–i) A similar process commonly occurs in *Pinna* shells (c), where destruction of the red-colored prism envelopes creates white areas visible in reflected light (d). Prisms are then directly exposed to seawater and nearly dissociated in older portions of the shell (e). The surfaces of prisms (f–g) and subsequently their interior (h) are frequently invaded by borers, actively destroying the organic compartments. As a result, in (i), in contrast to the mineralogy-based view, the nacreous part of the shell (c: arrow) is more commonly fossilized (generally after diagenetic calcitization) than are the apparently stronger prismatic (and primarily calcitic) layers. Fossils of the *Pinna* shells are generally limited to the apical part (formerly aragonitic) of the shell (arrows).

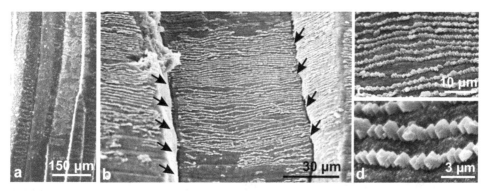

Fig. 7.19. An early diagenetic process linked to organic layering within *Pinna* prisms. (a–b) Overall (a) and enlarged (b) side views of *Pinna* prisms in which decay of organic envelopes (b: arrows) provided access to seawater, resulting in localization of the dissolution and recrystallization process; (c–d) Preferentially located in organic-rich layers, small euhedral inorganic crystals have replaced biogenic calcite in prisms.

components, are essential factors in this process. In addition, the starting point of their diagenetic history as fixed by the Berner statement ("the time of deposition") can be shifted for biominerals to the time of biocrystallization. As soon as the crystallization growth layer is formed, it becomes isolated from the biological influence of the organism that produced it. The complex compositional assemblage then commences changing simply because of its composition and, depending on the local environment, is rapidly exposed to multiple factors that affect its transformation. *Diagenesis of the growth layer begins immediately after its formation.*

Changes in skeletal mineral components are also deserving of reexamination, taking into account modern data concerning the scale at which the mineral–organic interaction takes place. Sometimes, the relationship between removal of organic material and development of recrystallization is obvious, recognizable simply by the location of newly formed material. The formation of layers of small, euhedral calcite crystals within the prisms of *Pinna* (Fig. 7.19) is demonstrative of this. In this case, the "insoluble" prism envelopes were removed (Fig. 7.19b: arrows), providing seawater an access to the calcite prisms. The small crystals have been formed by layers of truly recrystallized calcite (Figs. 7.19c–d). The parallel arrangement and distance between these layers suggests that they result from local dissolution of biogenic calcite and the almost immediate precipitation of the small, euhedral calcite crystals. Here, use of the term "recrystallization" is justified as it describes the process of transformation of biogenic calcite to nonbiogenic calcite. Comparison of the layering of these small euhedral crystals to the general distribution of the organic-rich sheets that separate superposed mineralization layers in *Pinna* prisms (see Figs. 1.7j, 2.50, and 3.24a–b) indicates that the dissolution process preferentially removed layers with higher concentrations of soluble matrix by penetrating into the internal structure of prisms.

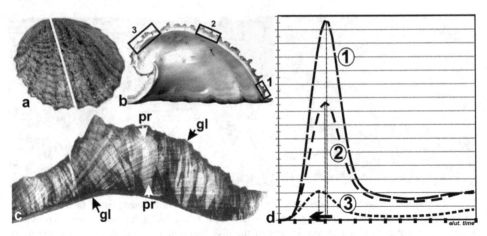

Fig. 7.20. Gel-filtration chromatography of soluble matrices from three distinct areas of a living *Concholepas* shell. (a–b) Shell of *Concholepas*: external (a) and cut radially (b); (c) Optical view of microstructure: the main part of the shell is roughly prismatic, with irregular elongate elements (white arrows: "pr") formed of highly oblique growth layers (black arrows: "gl"); (d) In this species, microstructure remains unchanged from the apex to the growth edge. Using an identical amount of shell from the regions "1" to "3," the clear difference in concentrations of soluble matrix through time can be seen. Additionally, a small shift in molecular weight is observed (black arrow).

In addition to morphological degradation of the shell resulting from aging of the animal, chemical alteration of mineralizing matrices prior to actual fossilization (in the strict sense), can be shown by chromatographic measurement of the amount of soluble matrices extracted from various parts of the shell of a living *Concholepas* (Fig. 7.20). Between recently deposited growth layers at the distal growing portion of the shell, and shell that is 3 to 4 years old near the shell apex, the layered microstructure is visibly unchanged, but chromatographic elution of soluble matrix material extracted from identical amounts of shell indicates a loss through time of the total amount of soluble organic phase. This provides more evidence of pre-depositional diagenesis.

A comparable process is illustrated by study of a core of *Porites* from the New Caledonia lagoon (Fig. 7.21a). This was obtained by submarine drilling of a colony living at 9 meters (thus never exposed to subaerial diagenesis). Samples were taken in the upper 30 cm, and these powders used for parallel $\delta^{18}O$ measurements and chromatographic study of soluble organic contents for each sample. Based on the assumption that mineralizing matrix was produced in equal amounts during growth, the series of chromatographic elution profiles provides a direct expression of the rapid loss of soluble organic contents. These data can then be interpreted as an example of the heterogeneity of diagenetic processes within a single specimen, focusing our attention on the instability of these skeletal organic compounds, as well as the ease with which they can be removed from skeletal carbonate. Using ordinary methods of determining skeletal status, generally limited to X-ray diffraction, researchers

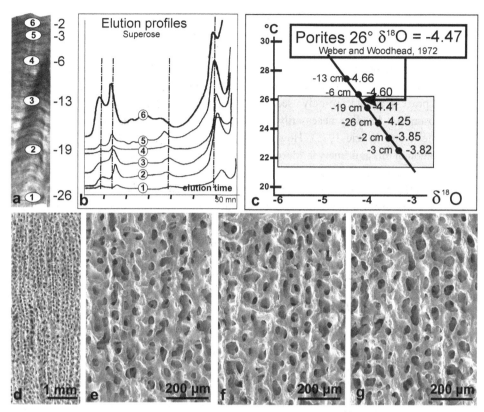

Fig. 7.21. Parallel study of samples from the upper part of a core of modern *Porites* from the Noumea Lagoon (New Caledonia). (a) Sampling location; (b) Elution curves of soluble matrices showing the overall decrease in concentration and the slight shift towards the right side of the chart, indicating lighter molecular weights; (c) Values of $\delta^{18}O$ (double measurements). Two samples provide calculated temperatures outside the upper limit of the continuously recorded temperatures at the colony site; (d–g) Overview and enlarged views of *Porites* skeleton at –13 cm (e), –19 cm (f), and –26 cm (g) below the surface of the colony. b: Cuif *et al.* (1997).

would be unaware of this removal of mineralizing matrix compounds and its probable consequences on the mineral component.

Measuring oxygen isotope fractionation in this *Porites* colony, no reliable explanation can be provided for the two samples (–6 and –13 cm) that provide improbable temperatures, outside the range of seawater temperatures that were continuously recorded at this site (Noumea Lagoon, New Caledonia). It is usual to eliminate this type of "outlier" from consideration in a series of measurements, as supposedly they result from technical errors. But here, these outliers effectively draw our attention to an insufficient understanding of diagenesis in biogenic carbonates. The suggestion can be made that a correlation does exist between the compositional alteration of organic matrices and the presence of unexpected $\delta^{18}O$ values at these levels. Because the coral skeleton has been considered as the result of chemical crystallization, no investigation has yet been carried out to check this possibility.

The pattern of banding visible in this 30 cm long core (Fig. 7.21a) also illustrates the limit of present day interpretations based on an inadequate understanding of the biomineralization process. The banding is generally regarded as reflecting annual variations in the biomineralization process, and accordingly used as a "paleontological clock" (Runcorn 1966). Such X-ray absorption banding necessarily relies on variation of calcification during growth (Buddemeier and Kinzie 1975). However, as pointed out by these authors (p. 143), an interpretation of this as primary is "deceptive," because close observation of the morphology of the vertical trabeculae, the basic skeletal unit of *Porites* (see Figs. 7.21d–g), indicates that no visible difference exists in their morphology (e.g., diameter) that can be correlated to the variations in X-ray absorption. Such visible banding, in the absence of increased trabecular diameters, indicates a periodically increased density of the skeleton, possibly correlated to variations in the rate of skeletal growth. This example draws our attention to the important gaps in our understanding of the biomineralization process, and how these deficiencies preclude making reliable interpretations of very basic measurements.

Long-term preservation of skeletal matrix

Examples of opposite conditions also exist that demonstrate the ability of soluble skeletal matrices to maintain the characteristics through geologic time that significantly relate to their mineralizing activity.

The presence of organic compounds within fossil mineralized structures was first established by the research of Abelson in the early 1950s during seminal work in paleontological research (e.g., Abelson 1954). The presence of amino acids was demonstrated in Miocene mollusc shells, from dinosaur eggs, from trilobite carapaces, and so forth. In practical terms, the methods used to investigate recent biominerals were rapidly applied to studying fossil material. The comparison of proteinaceous materials extracted from oyster shells having different origins and ages from Pliocene to Cretaceous, or from cephalopods as old as Devonian, have each established that the conditions existing during early fossilization determine the possibility of preservation of organic materials within biominerals (Matter *et al.* 1969). In contrast to a common assumption, there is no correlation between the preservation of biominerals and the absolute age of fossils. It was in 1972 that Wyckoff presented his first synthesis regarding his analyses of several dozen organisms belonging to major taxa. His results confirmed that, generally, after an early phase of quantitative diminution of the organic matter in biominerals, accompanied by some modification of the proportions of their constituents, the samples showed a surprising stability through time.

Once established, molecular paleontology took advantage of the astonishing technical and methodological progress that has occurred during the last several decades. Paper chromatography was used by Abelson, a method that requires a relatively large quantity of fossil material. To carry out research on amino acids, Matter, Davidson and Wyckoff needed to use 50 grams of oyster shell, without separating distinct shell layers. A few years later, methodology based on chromatographic columns completely transformed the necessary amount of sample needed for this research, which depends on the ability to identify the characters of small (and often unique) samples. This is even more important than in the study of modern materials. At the present, ion-beam microsampling methods associated with time-of-flight methods make biochemical characterization possible at the scale of several square micrometers (see Figs. 2.47–2.49).

Causes of contamination and methods of control

It is important to pay attention to a critical aspect of determining the organic content of fossil carbonates. If there is commonly the removal of degraded glycoprotein gels from mineralized structures, the reverse can also occur, that is, the fixation of free organic molecules from seawater on biogenic calcareous structures during their diagenesis. A variety of experiments have shown that organic molecules in solution can interact with Ca-carbonate crystals, depending primarily on their isoelectric point and on the pH of the water they are dissolved in. In natural seawater, one clear example of capture of free organic compounds by growing chemical crystals is provided by the nodules of botryoidal aragonite frequently developed in reef carbonates. These nodules, with dimensions in the centimeter to decimeter range (Figs. 7.22a–b), are composed of elongate radiating fibers having transverse sections with growth faces and twinned assemblages (Fig. 7.22c).

Infrared absorption diagrams of powders quickly indicate the presence of organic materials. Results of amino acid characterization using standard preparative processes establish that complex organic materials can be included within these chemically precipitated carbonates. Clearly, there is also a question regarding any possible contamination of fossil biogenic carbonates by the free organic components in seawater, so that it is clear that additional determinations of these should be carried out in order to insure the accurate identification of organic remains extracted from fossil biominerals.

In the process of assessing the authenticity of these intraskeletal organic materials, special care must be taken to make sure that, if original mineralizing skeletal components did not remain stable, neither should it be expected that fossil molecules will exhibit biochemical characteristics identical to their modern equivalents. In contrast to the common expression, "perfectly preserved fossils," such do not exist.

Aging itself is established as one possible cause of modification, the result of the thermodynamic instability of its basic components, such as in the amino acids; this is known as racemization. Mineralizing matrix materials, being complex glycoprotein hydrated materials, can undergo spontaneous hydrolysis that is potentially accelerated by a thermal rise. Some experiments have been carried out by artificially accelerating the aging of these organic compounds, in most part focusing on the proteins (Grégoire 1972; Voss-Foucart *et al.* 1974). Initially, these experiments relied on an oversimplified approach to the compositions of mineralizing matrix. Results showing their biochemical complexity make it clear that no significant results have yet been attained that allow formulating a reliable model of biochemical diagenesis. Now, just as it was at the beginning of studies on organic materials in fossils, any assessment of the authenticity of organic compounds extracted from fossil Ca-carbonates has to be approached on a case-to-case basis.

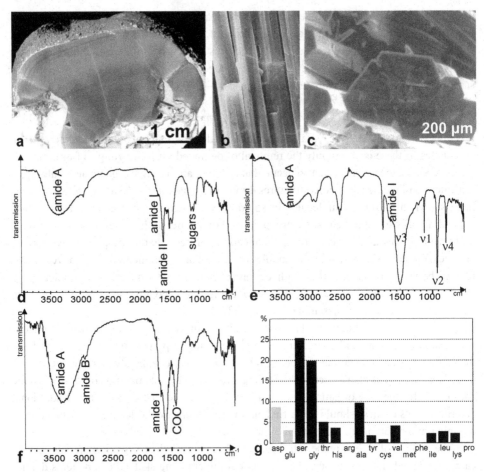

Fig. 7.22. Organic contents of fibrous aragonite nodules from reefs. (a–c) Botryoidal aragonite from Red Sea reef carbonates; elongate parallel fibers (a–b) with typical twinned crystal sections (c); (d–f) FTIR absorption spectra: total organic material (d), soluble component (e), and insoluble component (f); (g) Amino acid concentrations in the soluble organic content of the botryoidal aragonite. c: Dauphin and Perrin (1992).

Racemization of amino acids

Amino acids are characterized by the attachment of four different groups on one atom: the carboxylic acid group (COOH), an amine group (NH_2), a hydrogen atom, and a radical whose simplest form is a hydrogen atom (in the case of glycine). For each amino acid, there exist two optical isomers called "enantiomorphs" (meaning, symmetrical with respect to a plane). At times there exist as many as four of these, when the radical itself carries an asymmetrical carbon (isoleucine and threonine). These amino acids are denoted D and L (the terminal atom of hydrogen being placed in close proximity to the plane of symmetry). Thermodynamically, D and L forms appear in equal proportions of the enantiomorphic amino acids. A 50:50 mixture of them is called racemic. In living organisms, selection does exist; there is an absolute specificity in the amino acids employed, as all of these belong to the L series. This rule has only rare exceptions, principally in the case of the bacteria (Corrigan 1969). At the organism's death, proteins are decomposed, and the amino acids freed by hydrolysis are progressively re-equilibrated until they reach the normal statistical condition of 50% L, 50% D. This is the process of racemization, the return to a racemic mixture. The theoretical speed of various reactions has been calculated as a function of temperature, as in the following examples:

Aspartic acid: 4.6×10^6 years at 0 °C, 518×10^3 years at 110 °C;
Isoleucine: 47.6×10^6 years at 0 °C, 5.4×10^6 years at 100 °C.

This opens two possibilities of application:

– to determine the age of a sample as a function of the degree of racemization of its amino acids. Thus, Masters-Helfman and Bada (1976) detected the racemization of amino acids in proteins making up the mineralizing matrix of tooth enamel;
– or, if its age is known, to determine the conditions under which it was fossilized as a function of its relation to the theoretical degree of racemization. Different amino acids each have their own coefficient of racemization, and a small number among them are truly useful in research on fossil material (alanine, valine, leucine, isoleucine, aspartic acid, phenylalanine). Among these, isoleucine is often favored, as its racemization (called epimerization, a term applied to amino acids carrying two asymmetrical carbons) yields D-alloisoleucine, easily separated from L-isoleucine.

However, it soon became apparent that the process proceeding in natural environments gives perceptibly different data than that acquired during experiments with artificial racemization. The amino acids of proteins are racemized more slowly than free amino acids (Wehmiller and Hare 1971).

In a critical analysis of organic material extracted from fossils, racemization of the amino acids may also provide useful data, that which can establish contamination of an organic residue. In any fossil with a few million years of antiquity, the racemic state of amino acids must have been reached in the original mineralizing matrix. If some chirality (left or right prevalence – in this case, left as we deal with eucaryotic cells) is still observed as the result of the presence of some molecules that are not re-equilibrated, this is counter to these organic remains being indigenous, or at least indicates the contamination of them by fluids containing more recent organic compounds.

Evidence of high molecular weight of fossil skeletal matrix materials

Establishment of relatively elevated molecular mass for remnant organic material is a primary index, useful in assessing the authenticity of fossil matrix. The techniques used by Weiner, Lowenstam and Hood (1976) were relevant for the comparison of organic compounds extracted from shells of trigonid bivalves from the Upper Cretaceous of Tennessee with those of modern trigonids. These authors found that nearly all of the insoluble fraction had disappeared from the fossil specimens, indicating that the determinants for preservation of organic mineralizing matrix actually depend on their relation with the mineral phase, and not due to their initial insoluble state (as envelopes of microstructural units). In their fossil specimens, the residual organic matter only represents 1/280 of the amount of organic matrix compounds in the modern specimens. In this fragmentary sample, at the molecular level, the proportion of three amino acids, aspartic, serine, and glycine, seems to be similar in both modern and fossil specimens.

Preservation of functional properties: persistence of high acidity levels of mineralizing organic matter and its ability to fix calcium ions

The capacity for organic matter from fossils to fix calcium has frequently been used as a criterion for establishing their authenticity in earlier research on this topic. For instance, Krampitz *et al.* (1977), in isolating organic matter from dinosaur egg shells, and Samata *et al.* (1980), studying the same from Miocene oysters from Languedoc (France), obtained Ca-fixing molecules with masses of about 10 kDa (much lower, however, than masses of these mineralizing compounds in their original state). In addition to this property of fixing Ca, fossil soluble matrix compounds show overall compositions of amino acids comparable to those of living forms.

Fossils from the Eocene of the Paris Basin, as an example, are known for their excellent preservation, and aragonite is common in both corals and molluscs. Ion exchange chromatography of soluble matrices has been carried out on three different coral species, and these compared to soluble matrix from modern corals obtained by identical methods of preparation (calibrated pulverizing of cleaned skeleton, previously checked for the absence of endobionts by scanning electron microscopy (SEM), acetic acid demineralization at a continuously controlled pH, centrifugation and rinsing with MilliQ water and electrically controlled desalting, followed by concentration). Elution of the modern and fossil matrix materials subjected to increasing salt concentration indicates that comparable profiles result (Fig. 7.23a), in spite of a slight shift towards decreasing acidity observable in profiles for the three fossil specimens.

These results suggest that, as long as the mineralizing components remain attached to the mineral structure, they preserve at least part of their specific biochemical characteristics. SEM images (Figs. 7.23b–d) confirm that aragonite skeletons still resemble original appearances. Only AFM amplitude and phase-contrast images indicate that some change has occurred at the ultrastructural level. Fibers have a more compact appearance, but in the closely packed subunits, mineralized grains are still visible (Fig. 7.23g). This example reinforces the need for research focusing on diagenetic processes at a relevant scale, that of the biomineralization process itself. There does not exist, at present, any relevant database to allow consistent correlations to be made between the diversity of micro- and nanostructural features in coral families and the diagenetic conditions that produced the progressive elimination of specific taxonomy-linked features.

Immunological characterization

A valuable indication of the authenticity of vestigial organic phases lies with the possibility of obtaining immunological reactions using antibodies prepared from organic compounds

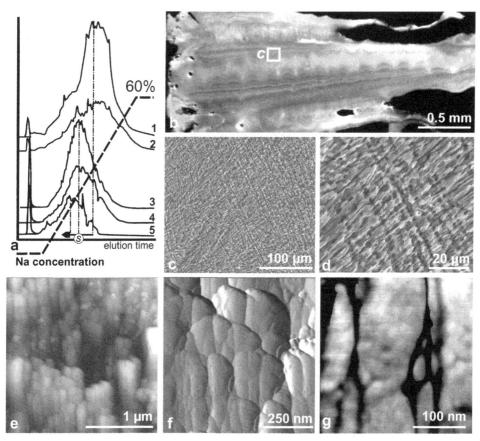

Fig. 7.23. Soluble matrices and fibrous skeleton in Cenozoic corals from the Paris Basin. (a) Ion exchange profiles for soluble matrix from two modern corals (1: *Acropora* sp.; 2: *Pocillopora* sp.) and three Cenozoic specimens from the Paris Basin (3: *Trochosmilia*; 4: *Eupsammia*; 5: *Dendrophyllia*); (b–d) Surface (b) and SEM views of fibrous portion of septum in *Trochosmilia* sp.; (e–g) AFM view of fibers. Even at much higher enlargement, no recrystallization patterns can be seen here: height image (e), amplitude (f). High-resolution phase imaging (g) allows the nanogranular structure to be clearly pictured. a: Cuif *et al.* (1992).

originating in taxonomically related modern shells. Such an experiment was carried out by de Jong *et al.* (1974), utilizing organic materials extracted from recent cephalopod shells (*Sepia* and *Nautilus*) and the rostrum of a Maastrichtian belemnite (*Belemnitella*). Each of these extracts was injected into an animal (rabbit) and resulted in serums containing species-specific antibodies. A series of cross-reactions explored the reactivity of these different serums (notably, verification of nonprecipitation between serums before injection). To summarize, the experiment was conclusive: in spite of biochemical modifications resulting from fossilization, distinct reactivity was observed when antibodies were placed in contact with soluble matrix materials from shells of the two modern cephalopods.

Understanding the preservation of fragments containing organic molecules in fossils can also be based on broad sampling, in order to derive an evaluation of immunological distances between different fossil species. Collins *et al.* (1991) carried out research on serotaxonomy, beginning with organic matrix originating in fibrous brachiopod shells, for use in resolving phylogenetic questions. However, in spite of the attractions of the method and its strictly defined operating conditions, the interpretation of the results of cross-reactions is very difficult. The production of antibodies is a complex process and more than one antibody can be produced by different antigen portions of the same molecule (typically a sequence of 4 to 8 amino acids in one protein can constitute a determinant antigen).

7.3.4 Validation of biological information by taking advantage of convergent methods

Corals from Triassic Lower Norian strata in the Lycian Taurus of southern Anatolia provide an interesting case study of the application of multiple characterization methods. The organisms that built small patch reefs here (early scleractinian corals, calcifying demo-sponges, and solenopore algae) all exhibit excellent preservation. In the corals, for instance (Fig. 7.24), microprobe mapping indicates the high strontium contents of the septa (0.7%, Figs. 7.24a–c), with interseptal infilling by iron-rich euhedral calcite (0.4% Fe, Fig. 7.24d) and with sulfur content of septa as high as 0.25%. XANES measurements and mapping indicate that this sulfur is still in the typical coordination state of sulfated polysaccharides (Fig. 7.24f).

The multiple closed cavities of the coral skeleton between septa and dissepiments have been entirely filled by crystallization of ferroan blocky calcite (they never contain particulate sediments) (Fig. 7.24g; see also Fig. 2.1c). Transit of calcifying solutions through the fibrous structures has had no visible effect on the aragonite fibers. The only part of the skeleton that bears traces of diagenetic change is the zone of distal mineralization, which has been largely dissolved, with the resulting void spaces now occupied by calcite (Figs. 7.24c–e: arrows). Synchrotron-based mapping shows that the distribution of strontium and that of sulfated polysaccharides are closely associated within the skeletal microstructure.

The contribution of symbiotic algae to the overall metabolism of coral polyps has long been investigated (from Yonge 1931 to Muscatine and Porter 1977, etc.). It was noted early on that, from a purely microstructural viewpoint, the skeleton found in nonzooxanthellate corals quite closely resembles that of reef corals (Sorauf and Podoff 1977). The incremental growth of fiber bundles, and the presence of a medial septal line with specific microstructural features, all indicate that the biomineralization process is essentially the same in both zooxanthellate and nonzooxanthellate corals. The result is that it is not possible to characterize those corals with and those without symbionts by only describing their skeleton microstructure.

However, from a geochemical viewpoint, isotopic differences in the isotopic fractionation in skeletons of zooxanthellate and nonzooxanthellate corals have long been noted (review by Swart 1983; see also Fig. 2.13). These are generally explained as resulting from the presence of symbiotic algae that influence the composition of a "sub-ectodermal liquid

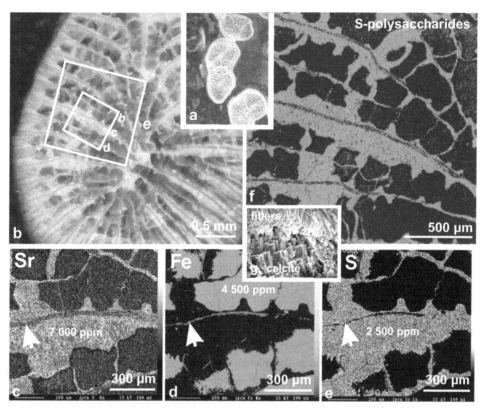

Fig. 7.24. Sulfated polysaccharides in Triassic corals (*Retiophyllia*). (a–b) Section of the corallite perpendicular to growth axis; (c–e) Distribution of Sr (c), Fe (d), and S (mapping by C. T. Williams, Mineralogical Dept., NHM, London; CAMECA SX50 Microprobe) (c–e: arrows indicate the medioseptal line – the zone of distal mineralization – strongly altered by diagenesis); (f) Distribution of sulfated polysaccharides (mapping by M. Salomé, ID 21 ESRF, Grenoble); (g) Contact between septal fibers and euhedral calcite. c–f: Cuif *et al.* (2008c).

layer" (McConnaughey 1989a, b). No explanation has yet been proposed to explain the biochemical pathway by which symbiotic algae located within endodermal cells can modify CO_2 concentrations within this external hypothetical "liquid layer" by passing through the mesogloea and ectodermal layers.

On the other hand, a biochemical difference between the mineralizing matrices of the two types of corals has been shown. Based on 24 species (13 zooxanthellate and 11 nonzooxanthellate), comparative study of mineralizing matrices extracted from these skeletons indicates (Cuif *et al.* 1999a) that a compositional difference exists in the sugar and amino acid compositions in mineralizing matrix from the two groups of species. Confirmation was obtained by use of Principal Component Analysis of the biochemical data (Fig. 7.25).

This statistical approach reveals that concentrations of sugars and amino acids are significantly distinct between the two types of metabolism. Aspartic acid, for instance,

Biominerals and Fossils Through Time

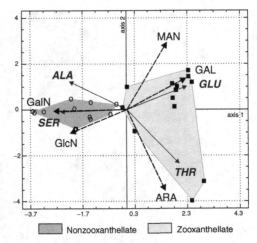

Fig. 7.25. Principal Component Analysis of sugar and amino acid concentrations in mineralizing matrices of zooxanthellate and nonzooxanthellate corals. Cuif *et al.* (1999).

although dominant in both cases, is not discriminant, whereas glutamic acid shows a discriminant high concentration in zooxanthellate corals. This result suggests that, in addition to an overall metabolic benefit brought by the peridinian symbionts to the coral polyp (using poorly understood mechanisms), a transfer of metabolic products does exist from endodermal cells of the coral (where peridinian algae are hosted) to calicoblastic ectoderm cells responsible for skeletal mineralization. It is noteworthy that in *Aiptasia pulchella* (a marine anemone) it has recently been shown that transport of labeled glutamate from zooxanthellae to host cells can be detected after 5 minutes (Swanson and Hoegh-Guldberg 1998). In the still controversial question of the method of mineralization for coral skeleton, establishing a link between the biochemical composition of the mineralizing matrix and the two distinct modes of life observed in corals provides additional weight to the hypothesis of biochemically driven crystallization of the coral skeleton.

Evidence of a biochemical exchange between zooxanthellae and the organic mineralizing matrix provides an alternative possibility for explaining the difference in the two types of isotopic ratio. It is now clear that hydrogels that form the crystallization medium for mineralizing growth layers have distinct and differing compositions in symbiotic and non-symbiotic corals, as indicated by the statistical analysis above. These differing hydrogels form distinct crystallization environments, and can have an influence on crystallization kinetics, and consequently, are likely responsible for the long-known differences in fractionation (see Fig. 2.13).

The recent measurement of isotopic ratio in mineralizing matrix materials of zooxanthellate and nonzooxanthellate corals now brings more direct evidence of symbiotic metabolism. By studying organic matrices of a series of coral skeletons produced by symbiotic and nonsymbiotic living species (17 distinct species), Muscatine *et al.* (2005) showed that $\delta^{13}C$ and $\delta^{15}N$ ratios differ significantly between corals with zooxanthellae and corals lacking

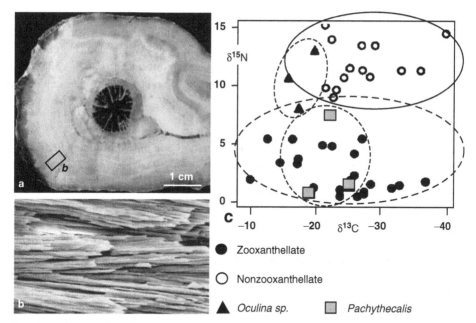

Fig. 7.26. Evidence of symbiotic metabolism in a Triassic coral (*Pachythecalis*) by measurement of the δ^{13}C and δ^{15}N in soluble organic matrix from wall fibers. (a–b) Section of the *Pachythecalis* corallite (a) and view of its radially oriented fibrous bundles of aragonite (b); (c) Distribution of δ^{13}C and δ^{15}N values for skeletal matrices extracted from zooxanthellate and nonzooxanthellate corals. c: redrawn from Muscatine *et al.* (2005).

them. Their data indicate that symbiotic algae have a significant effect on the δ^{15}N content in organic matrix of the coral skeleton (Fig. 7.26), a result that is consistent with the statistical discrimination of typical amino acid and sugar concentrations in their soluble matrices. Taking advantage of the compact wall structure of *Pachythecalis*, the type genus of the Triassic family Pachythecalidae (Fig. 7.26a) from the same outcrops and equivalent preservation state as shown in Fig. 7.24, it was possible to isolate mineralizing matrix material still present within the closely packed fibrous bundles. The positions of three measurements made on organic matrix from *Pachythecalis* on the δ^{13}C/δ^{15}N graph of Muscatine *et al.* (2005; Fig. 7.26c: gray squares) indicate that there was symbiotic metabolism in this species.

Thus, these multiple methodologies indicate just how informative calcareous fossils can be. This biochemical assessment of symbiotic metabolism in a Triassic coral has an increased significance when considering the microstructural and architectural features of this group of corals in which symbiosis has been demonstrated.

A summary of presently available data on the general evolution of corals illustrates the opportunities provided by convergent studies that examine fossil skeletons from various viewpoints; from nanoscale structure related to mineralogy, to changes in skeletal microstructures due to localized modifications in ectodermal growth control of fibers, all of which are linked to the biomineralization process. By combining results obtained from modern corals, such as molecular data and/or indicators of symbiotic metabolism, and corresponding observations of fossil specimens, consistent information can be collected to contribute to the better understanding of evolutionary relationships between extant and fossil corals.

7.4 Microstructural analysis applied to reconstruction of evolutionary history as exemplified by recent data on fossil corals

Since the pioneering investigation by Ogilvie (1896), it has been realized that the diverse spatial arrangements of fibers in coral skeletons are taxonomically significant. What has been shown in previous chapters makes it clear how this control is exerted. Polypal control of the characteristics of micrometer-thick growth layers throughout corallite ontogeny results in the development of species-specific morphology and septal architecture (see Fig. 1.31). However, as pointed out by Wang (1950), who carried out a detailed study of skeletal microstructures among Paleozoic corals, although authors of general syntheses usually recognize the importance of the three-dimensional organization of fibers as characters, little use is made of them, and little relevant information is provided in taxonomic descriptions. Hill and Wells (1956), after having dedicated some figures to microstructural information for Paleozoic and modern corals, respectively, illustrated this use of morphological features as black and white drawings. Although Hill (1956) presented photographs in the Treatise revision of the Paleozoic coral's microstructural organization as a character of great importance it has yet to be accepted in practice.

7.4.1 Converging results of microstructural analysis and molecular biology

During recent decades, the development of techniques of molecular biology has introduced a new approach to taxonomy (McMillan and Miller 1990; Chen *et al.* 1995; Romano and Palumbi 1997; Medina *et al.* 1999; and numerous others). Use of this method in coral studies is of particular importance because, for this group, classification has been primarily based on skeletal characters since the early nineteenth century. A purely genome-based classification, using only polyp tissues as the source of data, could result in complete disregard for information from fossil corals, the record of 500 million years of evolutionary history.

In order to test whether molecular taxonomy and skeleton-based classifications are compatible, a comparative study involving 40 species of modern corals was carried out. The objective was first to evaluate the families proposed in the most widely accepted classification of the Scleractinia, by J. W. Wells (1956), and in parallel, to examine whether the results of molecular comparisons using polyp tissue are compatible with skeletal features, with an emphasis on microstructural characters. This study focused on the type species of genera (excepting a small number of individuals with an importance based on historical taxonomic problems), and as much as possible, belonging to families important to classification of the Scleractinia (e.g., *Favia fragum*, the type genus of the Family Faviidae and the Suborder Faviida). Twenty-five genera representing 12 families thus were represented in this comparison. The specimens utilized were collected alive in their original biogeographic areas (Caribbean, Polynesia, New Caledonia). The 5th branch of segment 26S of ribosomal DNA was selected as the indicator of long-term evolution. Additionally, the method used for extraction of DNA permitted obtaining very long segments (more than 700 base pairs), providing significant comparative data (Cuif *et al.* 2003b).

The results were as follows: comparisons of molecular data (MUST software was used) led to the conclusion that the 12 families involved in the study belong to a monophyletic group. An important additional conclusion was that several major taxa in the present classification system should be divided; the main one of them is the Suborder Faviicae, a major component of coral evolution through time (see the classic figure by Wells, 1956, Fig. 259). The work of Romano and Palumbi (1996) using the 16S mitochondrial gene, continuing on into that of Fukami *et al.* (2008) establishing the molecular positions of 128 species by the use of mitochondrial and nuclear genes simultaneously, have resulted in diverse propositions being made regarding major subdivisions of the Scleractinia. In contrast, all agreed to dismantle some of the most important groups in presently used classifications. As a result of the Fukami *et al.* analysis, where molecular positions of 128 species were compared, 11 of 16 scleractinian families were shown to be polyphyletic, but monophyly of the complete group of the Scleractinia itself was demonstrated.

In the comparative study, involving both molecular taxonomy and microstructural analysis, families that were confirmed are those that exhibit such peculiarities in skeletal architectures and microstructural patterns that molecular validation of their taxonomic value is not surprising. The Dendrophyllids exemplify such a case, with a porous skeletal architecture associated with typical clusters of tiny "centers of calcification." It also appears that, among new groups resulting from dismantling of major taxa in the present taxonomic system, the new clusters in molecular classifications have comparable microstructures (see Cuif *et al.* 2003a, Fig. 7). The suggestion can reasonably be made that close control of the spatial organization of fibers through the biomineralization mechanism provides us with a way to correlate molecular- and skeleton-based taxonomies.

7.4.2 Microstructures of fossil corals as tracers of their evolutionary history

Although somewhat fragmentary, these results are of considerable interest in validating study of fiber growth controls and their spatial arrangement in order to create a new skeleton-based taxonomy of fossil corals. In recent years, efforts have been made on a limited scale to re-examine the existing taxonomic framework. Thus, Stolarski and Roniewicz (2001) and Stolarski and Russo (2001) have carried out extensive studies dedicated to the amphiastreid corals and the post-Triassic pachythecalid corals, respectively.

Beyond their specialized conclusions, results from this research converge to suggest entirely new taxonomic assemblages, extending from Middle-Triassic to Late Cretaceous time (see Fig. 6 in Stolarski and Russo 2001). From an overall viewpoint, the definite discrepancy between these proposals and the classical taxonomic framework agrees with the main results of molecular taxonomy. Furthermore, the conclusions of these two studies based on fossils, led to a more important hypothesis, that a phylogenetic link exists between corals from the lower Mesozoic and some of the Late Permian (Stolarski and Roniewicz 2001). This would definitely bring into question the century-old diphyletic theory of Haeckel (1896), who postulated a complete extinction of corals at the end of the Paleozoic, with a discrete and distinct origin of the Scleractinia during the Triassic Period.

Dealing with a more restricted time interval (the Upper Triassic only), a similar attempt had been made (Cuif 1980) after extensive investigation of the coral faunas in the classical fossil localities of the Salzkammergut (Austria) and Dolomites (formerly the Süd-Tirol), and in addition, including newly discovered reef faunas from the Lycian Taurus (Turkey). Synthesis of these microstructural results also led to the conclusion that the Wells phylogenetic tree (1956) indeed does not reflect actual relationships among Triassic corals. Taking advantage of the exceptional preservation of these Triassic faunas, some examples will show how precise microstructural analysis can provide a more representative view of their mineralization and the controls exerted by basal ectoderm in these corals (from approximately 220 Ma).

Among the abundant and diversified fauna from the Middle and Upper Triassic, a group with unusual morphological and microstructural characteristics is easily recognizable. In Turkey, representatives of this group may reach large sizes, and several species with two main morphotypes (solitary and branching forms) are present (Cuif 1980). They are characterized by a very thick wall (see representative transverse and longitudinal sections in Fig.7.26). In the Dolomites, where the first representatives of this group were found, the Ladinian and Carnian coral *Zardinophyllum* builds small cylindrical corallites with a diameter of a few millimeters and weakly developed septal systems. Apparently, it was for this reason that the specimens escaped the attention of numerous investigators during more than a century and a half following 1841, when the first description of the Triassic corals was published by Münster.

When looking at these corals as products of biomineralization, it is clear that their originality does not rely on morphology alone (Fig. 7.27).

Members of the family Pachythecalidae (e.g., *Pachythecalis*, Figs. 7.27a–h, and *Pachysolenia*, Figs. 7.27j–l) are characterized by a specific wall microstructure. The wall is built by inward growth of radially oriented and densely packed fibrous bundles. This results in forming a thick and massive wall (Figs. 7.27a, d–f, j). Each fiber bundle is developed from an initial mineralizing zone, a radial axis located at the periphery of the wall (Fig. 7.27k: arrows). Synchrotron-based characterization of polysaccharides shows that these wall bundle axes have been diagenetically altered (Fig. 7.27l; compare to the distal mineralization zone in septa of Fig. 7.24). In contrast to the common centrifugal construction of the external wall (epithecal) positioned on the outer margin of the septa, an enlarged view of the contact between the wall and septa (Figs. 7.27f–h) shows that in pachythecalid corals, walls are fully formed prior to development of the septa. Illustration of fiber orientations at the contact between the internal surface of the wall and the outer part of

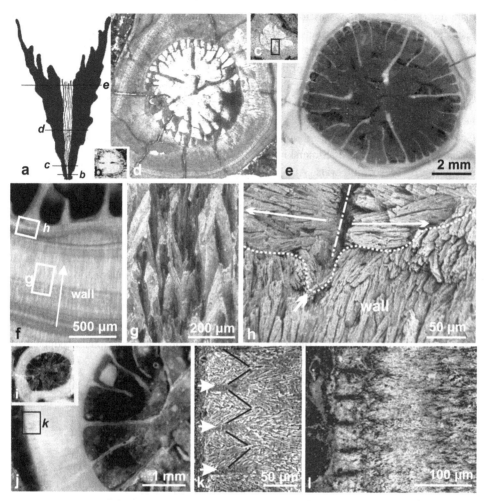

Fig. 7.27. Microstructure as an indicator of the wall/septa relationships in pachythecalid corals. (a) Drawing of a longitudinal section of *Pachythecalis*; (b–e) Serial sections through a *Pachythecalis* corallite (see microstructure of (c) in Fig. 7.28); (f–h) Fiber orientations in the wall and septa of *Pachythecalis*. Note the pattern resulting from inwardly divergent fiber bundles (g) and the right-angle relations of fibers between the internal surface of the wall and the outer part of the septa (h); (i–j) *Pachysolenia*: corallite section (i) and enlarged view of the wall and septa (j); (k) Microstructure of inward-growing fiber bundles in the wall of *Pachysolenia*. Arrows indicate the zone of distal mineralization for each of them; (l) Synchrotron-based XANES mapping of the sulfated polysaccharides in the outer part of the wall. The orientation of fibers is still recognizable, but the mineralization axes have been altered by diagenesis (compare to Fig. 7.24). e, g: Cuif *et al.* (2008c).

the septum (Fig. 7.27h) clearly indicates that radial growth of wall fibers (lower part of the picture) is complete when septal fibers are formed at a right angle to fibers of the wall.

Chemical mapping confirms the exceptional preservation of the skeletal structure in the Turkish fossil-bearing localities of the Lycian Taurus; high-resolution XANES-based mapping (Fig. 7.27l) allows us to pinpoint the location of sulfated polysaccharides with respect to the fibrous microstructure. The latter image also provides us with an additional example of an *in situ* biochemical identification, a further indication of the utility of skeletal organic material to act as a tracer of both biomineralization and diagenesis.

It is important to note that this mode of corallite construction, unknown among the other Scleractinia, confers to the pachythecalids an interesting position in coral history. Authors have recognized their "primitive" character (this term was first used by Montanaro-Gallitelli, who described *Zardinophyllum* from the Dolomites), whereas Roniewicz and Stolarski (1999), as well as Stolarski and Russo (2001), placed this family at the base of one group of Mesozoic corals. Also, they have been reliably identified as having been zooxanthellate (see Fig. 7.26). Thus, from the aspect of their biomineralization, *Pachythecalis* and its allied species present a paradox: their skeletal organization is unique among the Scleractinia, but algal symbiosis was already active. As assessed by isotopic fractionation of their skeletal matrices, their biomineralization process was probably comparable to that of living zooxanthellate corals. It is noteworthy also that these primitive Triassic corals are composed of aragonite, and, based on isotopic fractionation of oxygen and carbon, Stanley and Swart (1995) have also detected symbiotic mineralization in Triassic corals.

By focusing on the evolution of microstructures and involving careful comparison of fiber arrangements within this Triassic fauna one can establish an evolutionary sequence of them on the basis of their microstructures. This approach is illustrated by Fig. 7.28, showing that, from typical pachythecalid skeletons (Figs. 7.28a–c), the evolutionary process created new skeletal organizations reflecting the change in timing of the development of wall and septa, respectively.

Assessing the main evolutionary trends in Triassic corals

From these rare and surprising corals of the Family Pachythecalidae it is possible to produce a consistent outline of the microstructural modifications that occurred during their radiation, which resulted in the large and diversified coral fauna known from Upper Triassic strata. From the *Volzeia* microstructural type that exhibits many common features with the pachythecalids, the first trend identified – the prevalence of the septal system and the accompanying suppression of the pachythecalid type of wall – is accelerated, leading to the formation of the different types of walls present in modern corals. Additionally, the microstructural arrangement of fibers within the septa undergoes considerable change. The straight and continuous line forming the distal mineralization zone in *Volzeia* changes into a zigzag line along the growing edge, well illustrated in *Distichophyllia* (Figs. 7.29a–c). This major modification in the profile of the septal growth edge is accompanied by the development of a new arrangement of the fibrous material, namely the first occurrence of mineralization axes that diverge to the sides of the crests of the laterally offset margin. These lateral axes produce the granules that are seen on lateral faces of the septa (Fig. 7.29d), granules that were long described as only being ornamental. In fact, they are an expression of fundamental changes in the control of spatial relationships of fibers that result in a great

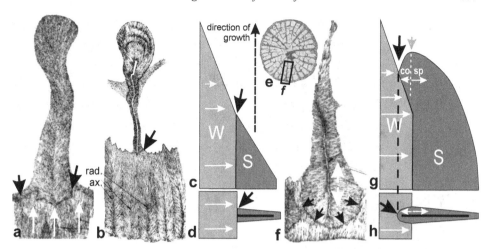

Fig. 7.28. First step in the transition from *Pachythecalis* wall/septum structure to modern corals: initial indication of the accelerated development of the septal system in the coral Family Volzeidae. (a–d) Wall/septa contact in *Pachythecalis* (a: young septum from Fig. 7.27c) and *Pachydendron* (b) summarized by drawings in (c) and (d). Wall fibers are oriented axially (white arrows) and wall structure ("W") is completed prior to the appearance of septa ("S"). Contact between them is essentially flat (black arrow); (e–f) In a section of *Volzeia* (e), orientation of the fibers in wall and septum (f) reveals a clear difference in the timing of the formation of wall and septum. Wall fibers are still oriented axially, but, in the outer part of septa, fibers are now growing outwards (see the two opposing white arrows in g). The contact between wall and septa is a curved surface (black arrows); (g–h) The scheme of the wall/septum relationships in lateral view (g) and transverse section (h). Orientation of fibers as shown in (f) indicates that, at the growing edge of the corallite, the septa now are composed of a costal portion (growing outwards) and a septal portion (see the opposite white arrows "co" and "sp"). a, b: Cuif (1975c); c, d, g, h: Cuif (1980); f: Cuif (1974).

diversity of corallite microstructures. As long as these axes remain laterally directed, they produce various arrangements on septal flanks. However, a new and important evolutionary step occurs when these axes are formed at the septal growth edge itself. The axes then become vertical and their growth is no longer limited (Figs. 7.29j–k). This is the development of the "trabeculae" in the traditional sense. The next evolutionary stage is reached when, instead of being closely packed longitudinally, regularly occurring spaces between these vertical axes lead to the formation of porous septa (Figs. 7.29l–o). The genus *Araiophyllum* best illustrates this evolutionary stage. It also is apparent that this microarchitectural evolution of septa results from the diversification of mineralizing epithelium. This correlates to the continuing trend of the septal system becoming dominant over the development of the outer wall. Thus, in the genus *Araiophyllum*, septal axes are vertical in the axial portion of the corallites and bend outwards during growth, resulting in the upper surface of the corallites becoming hemispherical.

These corallites, judged to be highly evolved, based on their microstructure, were present within the same small Norian patch reefs where pachythecalids were also living. This suggests that the evolutionary radiation of the Triassic coral fauna was very rapid, allowing

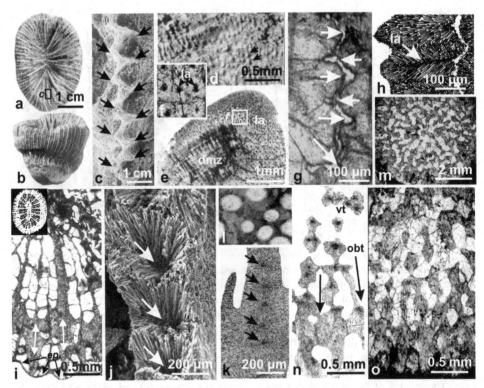

Fig. 7.29. Major microstructural types among the Triassic coral fauna, from the first appearance of fibrous axes to fully perforate septal architectures. (a–c) *Distichophyllia norica*, a solitary coral whose septa are built by a continuous median plan showing a zigzag profile, easily seen at their growing edges (c); (d–f) On both sides of the septa, fibrous axes are inserted onto the crests of the median plan and thus regularly distributed. These lateral axes form granules on the septal surfaces (d) and their sections are clearly visible on slides oriented parallel to the median plan of the septum (e–f); (g–h) Closer view of a thin section perpendicular to the corallite growth axis. Here lateral axes are cut longitudinally. Note the empty zigzag line corresponding to the dissolved median plane of the septum; (i) *Retiophyllia*, one of the most abundant branching corals among the Triassic fauna, shows a developed costal part (white arrows) that confirms the overall trend of septal prevalence in architecture of the modern corals. Here, the wall is now a true epithecal lamina deposited onto the costal edge (black arrows); (j–k) Two examples of true trabeculae: the fibers radially diverge from a vertical axial part (see additional explanation in Fig. 1.27); (l–o) *Araiophyllum*: in this Norian genus, evolution to a porous septal structure is already accomplished. Trabecular axes are vertically growing in the axial part of the corallite and bending towards the exterior during their growth. a–i: Cuif (1974); j: Cuif (1976); k: Cuif (1975b).

the concomitant existence of microstructurally "primitive" and quite highly evolved corals. It is also important to note the existence of comparable taxonomic assemblages in Triassic strata of the Pamir Mountains (Melnikova 2001).

These results also support the use of the distal mineralization zone configuration as a way to trace the course of scleractinian evolution. Each specific arrangement of the "centers of calcification" at the septal growth edge is the result of an evolutionary branching, as the control of fiber growth and organization emphasizes the diversity in geometrical arrangements of the distal mineralizing zone. The evolution of the Scleractinia has been superimposed on the post-larval, first-stage development of a basal microgranular plate and linear microgranular septa, first shown by Vandermeulen and Watabe (1973) in their study of early post-larval biomineralization in *Pocillopora damicornis* immediately after settlement (see subsection 1.5.7). The fragmentation of this continous central zone of distal mineralization into short and variously arranged segments, and then eventually to restricted point-like areas of distal mineralization, resulted in numerous and varying septal microstructures. On a broader temporal and geographic scale, the microstructural types that can be identified within the Triassic coral fauna reflect this process very closely.

The "primitive" microstructural patterns of the Pachythecalidae compared to a Permian polycoelid coral

Interpretations regarding pachythecalids as primitive corals find additional support in the comparison of their wall and septal structures to equivalent parts of some Permian corals. The calicinal morphology of one of the solitary corals belonging to the Family Polycoelidae is shown in Figs. 7.30a–e. This oblique view focuses on the appearance of the septa, showing their initial formation on the interior of the wall. In thin sections, examination of the contact between the wall's internal surface and the outer margin of the septa (Figs. 7.30b–c) confirms that the thick fibrous wall is fully completed before appearance of the septa. The contact line, although not perfectly flat, can be compared to the corresponding area in *Pachythecalis* (Fig. 7.27h). The wall itself is made of fibers that grow inwards. Fibers are produced around short axes, obliquely cut on sections perpendicular to the corallite growth axis (Figs. 7.30d–e: arrows). These units of wall formation are in close lateral contact, producing a dense compact wall.

In these two well-preserved samples, the nano-reticulate aspect of both calcite and aragonite fibers is still visible (Figs. 7.30k–n). Groupings of grains into larger clusters are sometimes visible, however (see Fig. 7 in Cuif *et al.* 2008), suggesting very early stages of recrystallization that do not change the microstructure seen at the micrometer level.

In the three best known genera of the pachythecalid family (*Pachythecalis, Pachysolenia*, and *Pachydendron*) a similar "bauplan" is observed that, in fact, contains elements of similarity to these Permian corals. This result strongly reinforces the concept of "primitive corals" suggested by different authors for this Triassic family, and is reviewed in a broader examination of what happened to corals in the final part of the Permian Period (see Section 7.5).

Additionally, it has to be noted here that the hypothesized series of microstructural types shown in Figs. 7.28 and 7.29 do not represent the total complexity of the Middle Triassic coral fauna. Besides many corals of modern-appearing aspect, there are also numerous other different Triassic corals, the appearances of which have suggested various interpretations by various authors. *Protoheterastrea* and *Gigantostylis* exemplify this faunal heterogeneity and the resulting evolutionary questions still being debated.

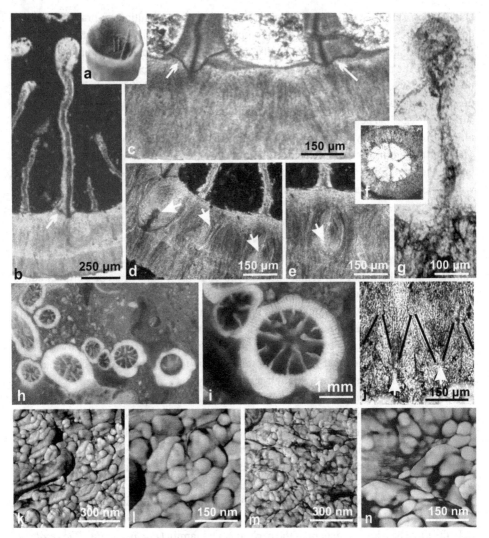

Fig. 7.30. Comparison between a Permian polycoelid coral and the Triassic Pachythecalid *Pachydendron*. (a) Calicinal view of *Calophyllum angustum* (collections of Münster University, Geologisch-Paläontologisches Institut, Germany); (b–e) Wall/septum contacts and wall microstructure in *Calophyllum angustum* and *C. parvum* (MNHN collection, Paris). The short axes of fiber bundles are shown here (arrows), although cut somewhat obliquely; (f–i) *Pachydendron*: thin sections (f–g) and polished surfaces (h–i) showing wall and typical septal arrangement; (j) Orientation of fibers and fibrous units forming the wall: fibers diverge from short axes and grow inwards. Contacts between fibers from neighboring wall-building units produce angular patterns that are biogenic (whereas recrystallization can produce angular patterns, but they are oriented differently with respect to the original microstructure and not specific to the wall organization); (k–n) The granular patterns in fibers of both Permian calcite (k–l) and Triassic aragonite (m–n). a–c: Cuif (1972).

Protoheterastrea (formerly *Hexastrea*) forms branching corallites, with sections that are morphologically distinguished by their thick wall, and the regular subdivision of the internal tube-like corallite by five major septa (Figs. 7.31a, b). Both the wall and septa exhibit distinctive features with respect to the common organization of the coral species occurring with them.

Their major septa are complexly organized. The fiber arrangements indicate that their radial growth was symmetrical. From a medial vertical line, fiber bundles are oriented toward the corallite axis and toward the corallite wall (Fig. 7.31d: arrows). External fiber bundles are not inserted directly into the internal side of the wall, but, instead, an intermediate zone is present between them with a specific microstructure (Fig. 7.31d: black and white arrows; and Figs. 7.31c, e). In this area, septal fibers are oriented perpendicular to the median plane of the septa.

The wall here is also extremely thick. Its external part is built by thick, rounded fiber bundles, formed around short axes (Figs. 7.31g–h: arrows), a type of microstructure that recalls the pachythecalid wall structure. It is difficult to find a better example than this of precise control of fiber growth during corallite ontogenesis. It is also significant that this genus was created by W. Volz (1896), who was the first to apply microstructural analysis to fossil faunas, taking advantage of the excellent preservation of corals from the Dolomites (at that time Süd-Tirol). Of additional interest is the work of M. Ogilvie, whose pioneering research on coral microstructures and taxonomy was published the same year. This was in great part due to her extensive contacts with paleontologists from Munich and Vienna, where she was involved in the geological description of the Dolomites, the source of Volz's material.

Six years prior to this, Frech (who was Volz's research director) had published his classical study of Triassic corals from the Zlambach Beds. He noted a small solitary coral that is very unusual among this Rhaetian coral fauna from the area near Salzburg, Austria (the Salzkammergut), and created for it the genus *Gigantostylis* (Fig. 7.13i–m). A microstructural analysis of the thick wall in this genus reinforces the singularity of its morphology. This results from its having a massive axial columella and a reduced system of septa, the latter suggesting to him that the biological affinities of this species should be looked for in the past. At that time (1890), Haeckel's theory was not formally established, thus for Frech it was no problem to suggest a Paleozoic origin for this surprising species.

Frech's research emphasizes one aspect of paleontology that is pertinent to the study of fossil corals of various geologic ages. As we have learned more regarding biomineralization in both modern and fossil scleractinian corals, it has become clear that the overall mode of skeletal formation and matrix controls of the Scleractinia are very close to (or even identical to) those of the older coral groups of the Paleozoic. Therefore, the use of analogy based on the biology of fossil and modern Scleractinia is certainly justified in interpreting the biology of the Paleozoic corals. It also is clear that the taxonomic differentiation between Paleozoic and younger groups of corals has become much less sharp as we have learned more about microstructure and biomineralization their. The nomenclatorial difference commonly separating the Paleozoic Rugosa and the

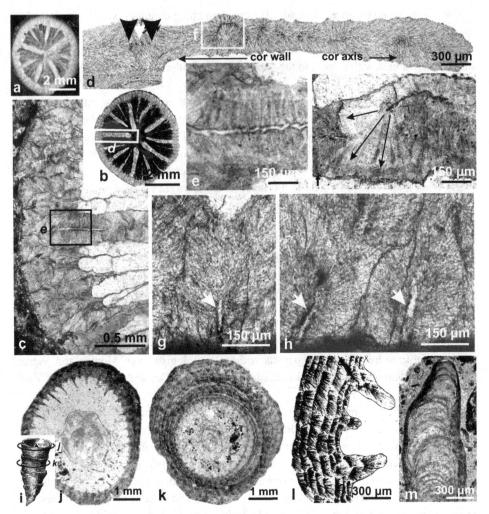

Fig. 7.31. *Protoheterastrea* and *Gigantostylis*. (a–b) Overview of sections of a *Protoheterastrea* corallite. Note the nonhexameral septal pattern; (c–d) Wall and septal microstructure: on (d), the two opposing arrows indicate growth direction of fibers ("cor axis" and "cor wall" = toward corallite axis and wall, respectively); (e–f) The intermediate area between the wall and the main part of the septa. Note the specific opposing orientation of fibers (e), as contrasted with the terminations of fiber bundles from the median septal part; (g–h) Examples of spheroidal fibrous bundles that have short axes in the zone of distal mineralization. These inward-growing fiber bundles appear as remains of the "primitive" architecture of the thick-walled corallites; (i–m) Morphology (i) and microstructural features of *Gigantostylis epigonus* Frech 1890. a, d: Cuif (1972); i–m: Cuif (1965).

Triassic to Recent Scleractinia cannot now be based solely on skeletal mineralogy or biomineralization mode, and certainly not on their microstructure. Other differences in the paleobiology of these corals are applicable to this problem, but these lie outside the scope of this book.

This suggests possible directions for more incisive study of other types of fossil materials that would be very productive, beginning with those rare specimens that have been protected from complete diagenetic change (or recrystallization) by ideal post-mortem conditions. Commencing with these materials, which have a highly variable frequency of occurrence according to geologic period, realistic classifications can be established, based on, and accompanied by, consistent evolutionary models.

7.5 Fossils and biomineralization in geological time

Examples from the previous chapters have drawn attention to the extraordinary properties of well-preserved Triassic fossil corals with aragonite fibers that have preserved intraskeletal biochemical compounds for a very long time; the specimens of *Pachythecalis* used to assess the symbiotic metabolism of this family (Fig. 7.26) have an age of about 220 Ma. Examples seen in the sponges and cephalopods of this fauna establish that highly favorable conditions for fossilization here resulted in outstanding preservation of specimens from these Triassic outcrops scattered from northern Italy to Turkey. Such specimens appear to be ideal illustrations of the relationship between biomineral formation and preservation on the one hand, and the "oscillation" of seawater composition as proposed by Sandberg (1983) on the other (Fig. 7.32). Based on a statistical study of Ca-carbonate cements in sedimentary rocks throughout the geological column, Sandberg established that the proportion of calcite and aragonite cements changed though time, thus producing an oscillating curve. Over the long term, variations in seawater composition have promoted the production either of calcite or aragonite as the dominant cements in sedimentary rocks.

These well preserved Triassic reef faunas lived within the period of the first "aragonite sea." What happened to organisms during the following time of "calcite seas" (and specifically to fossil corals) provides us with a demonstration of an opposite effect produced by the diagenetic environment at the later time. In the Early Jurassic, coral skeletons, and also those of the coralline sponges (common name for the calcifying Demospongia, see Sections 4.9 to 4.18), were suddenly either very poorly preserved, and are strongly recrystallized, or almost totally absent from formerly fossiliferous sediments after diagenesis. Such a drastic change was not understood prior to the study of Sandberg. This long-term "oscillation" between aragonite and calcite seas created a new paradigm for understanding variations in the fossil record throughout the geological column.

However, this oscillation based on sedimentary cements noticed by Sandberg can also be compared with a second type of research begun in the early 1960s, aimed at analyzing changes in overall taxonomic composition of fossil faunas throughout Phanerozoic time. The seminal paper "Revolution in the history of life" by Newell (1967) marked the first major publication in this line of investigation. Taking advantage of technical progress and widespread use of radiometric chronology, it became possible for the first time to associate actual time-based data regarding traditional stratigraphic time units (based on relative positions of fossiliferous sedimentary units), to produce an overall compilation of faunas and floras for each period of the geological time scale. Researchers were of course aware of multiple errors in fossil identifications, but nevertheless, in spite of this permanent "background noise," the resulting first overall treatments of life history to result from compilation of general taxonomic data in the fossil record differed substantially from traditional views of this history.

In addition to the two major historic boundaries identified by stratigraphers since the middle of the nineteenth century (Permian–Triassic and Cretaceous–Tertiary), four biological crises were identified by Newell. Later studies, based on a taxonomic compilation carried out at the family level (Raup and Sepkowski 1982), emphasized five major extinction events in the fossil record. In addition to the Permo-Triassic and Upper Cretaceous/ Paleocene events that were reinforced by this compilation, other major taxonomic revolutions were identified as occurring between the Ordovician and Silurian Periods (Ashgill), in an Upper Devonian event (post-Frasnian), and in an extinction event separating Upper Triassic from Early Jurassic times. These are the so-called "big five." In addition to these main "biological crises," taxonomic compilations also revealed that multiple oscillations in the diversity of marine faunas have occurred, although of a lesser amplitude. Beyond discussions bearing on which of these deserve to be regarded as "mass extinctions," the merit of these statistical approaches is to illustrate that, far from being a regularly expanding process, the evolution of life throughout Phanerozoic time has been subjected to long-term oscillations of highly variable amplitude.

Prior to the Sandberg theory of seawater composition changes through geological time, various hypotheses had been proposed to explain alternating phases of biological decline and recovery. Suggestions were made that depended on cosmic origins: variations in the terrestrial orbit generating variations in solar energy received at the surface of the Earth (Milankovitch), or solar energy that could have been reduced by large-scale dust clouds following the impact of major meteors on the surface of the Earth (McLaren 1970). The well-known Alvarez discovery of the Chicxulub meteorite impact at the end of the Cretaceous Period, with other evidence of additional impacts provided by large meteorite craters (Siljan, Sweden 52 km in diameter at 368 Ma; Manicouagan, Quebec at 212 Ma) supported this hypothesis.

However, the long-term decline of important groups, such as the ammonites or dinosaurs in the latter part of the Mesozoic Era (thus prior to the arrival of the Chicxulub meteorite), apparently requires recognition of additional causes. Thus, physical and chemical processes linked to an active Earth producing long-term influences were emphasized. Therefore, dust projected into the atmosphere during volcanic activity has been considered, as were changes in atmospheric composition by the emission of various gases. This was Sandberg's interpretation, based on an increase of CO_2 partial pressure related to the expansion of oceanic crust (Mackenzie and Pigott 1981).

A third line of general research that focused on the influence of marine chemical parameters on sedimentary materials (limestones and potash evaporites) resulted in bringing new data to beat on our understanding of the evolution of the global oceans. In terms of the formation of $CaCO_3$, relevant research, beginning with the pioneering work of Lippmann (1960) and Füchtbauer and Hardie (1976, 1980), indicates that the selection of the $CaCO_3$ polymorph being precipitated depends on the ratio of Mg and Ca ions. Low-magnesian

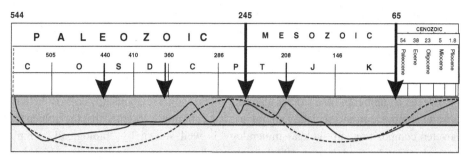

Fig. 7.32. Oscillation of seawater composition and chronological positions of major mass extinctions. Redrawn from Sandberg (1983) and Stanley and Hardie (1998).

calcites occur when the value of the Mg/Ca ratio is below two, whereas values above two lead to the formation of high-magnesian calcite or aragonite. Periods when the rate of production of oceanic crust was elevated (from Middle Cambrian to Middle Carboniferous, and later from Middle Jurassic to Paleogene) would thus have been times of precipitation of low-Mg calcite, separated by other periods of time that favored the formation of aragonite (Hardie 1996). This hypothesis is corroborated by study of evaporite mineralogies, which indicate that during the Phanerozoic an alternation of phases, the "aragonitic" periods, are rich in magnesian evaporates, just as the calcitic phases are dominated by potassic evaporites. In the two instances, a clear and unforeseen correlation exists between the oscillations detected by Sandberg, and the curve resulting from Hardie's calculations, which describe precisely the variations in geochemical parameters that directly influenced formation of nonbiological carbonates through time. Thus, seawater compositions apparently promote deposition of one of the two polymorphs depending on geological activity. As pointed out by Hardie (1996), the agreement between the predicted curve of types of evaporites deposited and Sandberg's synthesis of carbonate deposition is "extremely good."

Comparing the positions of the "big five" extinction events to Sandberg's oscillation or Hardie's curve does not indicate an obvious correlation. In terms of oceanic conditions, the Permo-Triassic and Cretaceous–Tertiary mass extinctions seem to have occurred in rather different seawater conditions. The Permo-Triassic crisis, the most devastating of all, occurred in "aragonite sea" conditions, whereas the Cretaceous–Tertiary event occurred during the passage from a calcite sea to an aragonite sea. Of the three other "big five" events: the Ordovician–Silurian occurred in the middle of long and continuous calcite seas, typical for the lower and middle Paleozoic; the Late Devonian (Frasnian–Famennian) took place during the beginning of the passage from a calcite to an aragonite sea; and the Triassic–Jurassic event occurred at a peak of the "aragonite sea condition," just prior to the rapid change to the calcite sea of the Cretaceous Period.

In contrast to the constancy demonstrated by sedimentological and chemical profiles throughout Phanerozoic time, the positions of the major discontinuities in the paleontological record do not lend themselves to simple, straightforward interpretation. It has previously been shown that living organisms exert a close control on the crystallization of their skeleton, and this could be considered as an acceptable reason for some apparent independence shown between the chemical evolution of seawater and faunal variation through time. In such an hypothesis, evolutionary mechanisms causing complex interactions between groups, possibly associated with "catastrophic" external events, could cause the long-term or accelerated variation observed in the fossil record.

However, a number of case studies also suggest that some relationships do exist between seawater composition and skeletal mineralogy. It is commonly accepted that the influence of seawater composition on skeletal mineralogy is well shown by "biologically simple" organisms (Stanley and Hardie 1998), such as algae, sponges, and corals, which have been the main reef builders throughout all of Phanerozoic time. Whether sponges and corals are "biologically simple" appears to be a somewhat questionable evaluation, as recent investigation indicates that all essential biological mechanisms were present in sponges, for instance (Müller 2003). As has been indicated previously, taking into account the level at which biochemical controls are exerted on their calcification produces a significant additional element in qualifying the sponges as fully differentiated organisms. With respect to the scale at which the interplay between biochemical and mineral compounds is detected, methods used by these groups to control their skeleton are comparable to those of all other calcifying invertebrates.

In addition to sponges and corals, some groups of "biologically complex" organisms also demonstrate mineralogic variations through geological time. The Rudists, an important group of bivalve molluscs, the cheilostome bryozoans, and coleoid cephalopods, are the most notable. The occurrence of aragonite among Rudists and bryozoa has long been recognized in these two groups, each of which generally produce calcitic skeleton. During recent years, extensive investigation has been carried out to securely establish the development of these innovative methods of mineralization and their relationship to the Sandberg–Hardie oscillation. Unfortunately, most of these mineralogical characterizations are not associated with microstructural descriptions that could have allowed a more precise analysis of the mode of deposition of these structures. The Coleoidea (a major subdivision of the Mollusca Cephalopoda) are the best documented in spite of their trend toward reducing skeletal mineralization. The following brief survey (in subsection 7.5.3 below) simply has the aim of suggesting that careful treatment of each case history is a prerequisite to avoiding oversimplification.

7.5.1 Changes in the mineralogy of reef-building organisms accompanying long-term variation in seawater composition

The history of the first organisms that could build solid structures able to resist wave and current action of water (a broad definition of "reef") commenced near the base of the Cambrian Period, but the group soon disappeared; only one genus of the Archaeocyathida persisted into the Middle Cambrian. Now accepted as an early form of calcifying sponge, the archaeocyathids exhibit a simple body plan (Fig. 7.33). They consist of two laminar cones, separated by two series of crossed sheets: one longitudinal, the other transverse with respect

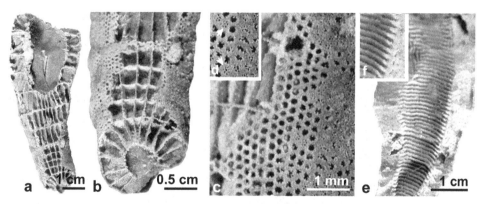

Fig. 7.33. Archaeocyatha. (a–b) Basic structure of an archaeocyathid, with its perforated outer layer; (c–d) The outer layer shows an elaborate two-layered structure, a layer with small pores lying outside the internal layer with large pores, thus providing a protective system to limit the size of particles entering the assumed food-filtration system located within the chambers; (e–f) An example of the deflecting lamellae on the outer surface of the internal cone. Water passing through the holes of the internal cone may have been deflected towards the upper part of the organism, thus increasing the flux of water.

to their growth direction (Figs. 7.33a–b). All of these structures are finely porous, a skeletal pattern that suggests internal circulation of seawater for filter feeding. Additional structures strongly reinforce this interpretation. Assuming that water circulation here followed the mode observed in sponges, water in archaeocyathids would have entered the mineralized skeleton through the complete external cone. After circulating within internal chambers, it would be ejected via the internal cone. Close observation of features visible on the outer side of both external and internal cones supports this interpretation. The major openings of the external lamina were covered by an additional thin calcareous layer, containing much-diminished openings, with diameters approximately one order of magnitude smaller (Fig. 7.33d). This additional layer could have prevented large-sized particles from entering the feeding system. On its outer flank, the internal cone is ornamented with concentric incurved lamina (Figs. 7.33e–f), which have orientations that indicate their role of deflecting water currents towards the upper part of the cone. Taxonomists of the archaeocyaths (Debrenne 1992; Zhuravlev 1989) have fully described the multiple variations of these water deflectors, and have concluded that they must have had a key role in developing greater efficiency in the feeding system, based on water circulation.

During their relatively short existence, the Archaeocyatha developed a great diversity of skeletal architectures, but seem to have been uniform with respect to microstructure and mineralogy. Among 180 genera identified thus far, skeletons were formed of undifferentiated microgranular calcite. Their biomineralization is also of interest, as the short history of the archaeocyathids corresponds to the part of Hardie's curve that descends from the highest level of Mg/Ca down towards that of the "calcite sea" that occupies the major portion of Paleozoic time. As yet, no research within this group has focused on possible structural or compositional variations at the submicrometer level.

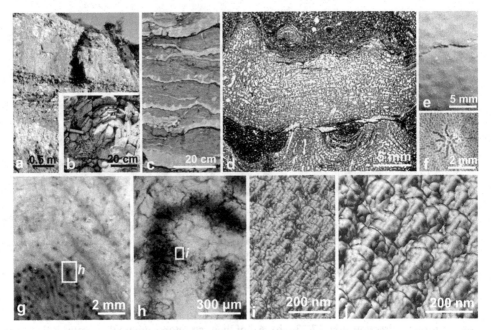

Fig. 7.34. Stromatoporida. (a–b) Silurian calcareous unit largely built by massive stromatoporoids (Gotland, Baltic); (c) Sawed surface of stromatopore-rich limestone (Upper Devonian, Tailfer, Belgium); (d) Typical layered architecture of stromatoporoids; (e–f) Dimpled growth surface of individual stromatoporoid, here marks the presence of radiating canals (astrorhizae); (g–h) Optical microscopic view (g–h) to nanostructure (i–j) of the fibers within the skeleton of a stromatoporoid. Example from Upper Devonian Frasnian strata at Boussu en Fagnes (Belgium). Atomic force microscopy, tapping mode, phase-contrast imaging.

The Stromatoporida: a group of sponges with a long history with respect to seawater change

Stromatoporoids were important calcifying organisms during the main part of the Paleozoic Era, and were major frame-builders contributing to the formation of massive buildups. Figures 7.34a–c illustrate two characteristic morphologies in this group: massive cylindrical forms (Figs. 7.34a–b, Silurian of Gotland) or flat elongate structures most commonly found in sites with turbid sedimentation. The basic organization of the stromatopores remained essentially unchanged, as their skeleton is formed of superposed calcareous laminae with their separation maintained by small pillars. Such architecture results in a series of layered void spaces. Sometimes, their functional interpretation is obvious, based on the presence of radiating canal systems (Figs. 7.34e–f) that strongly suggest a sponge-like feeding mechanism, with water entering choanocyte chambers through the entire surface and being ejected by specialized canals (astrorhizae).

They have been extensively studied for stratigraphic dating and paleogeographic reconstruction. Careful microstructural descriptions have furnished results that are optimal in what is accessible by optical microscopy. The studies carried out by M. Lecompte on the rich Devonian faunas of Belgium (1951, 1952) are most notable in this area, the culmination of a series of research by Carter (1879), Nicholson (1886), Waagen and Wetzel (1886), and numerous others.

With various interpretations, at times linked to taxonomy, authors have recognized the presence of "tiny granules" in the calcite fibers forming the stromatoporoid skeleton, where these readily visible dark spots (Figs. 7.34g–h) have long been interpreted as concentrations of organic material (Nicholson 1886). Authors have all been aware of the information potential of higher-magnification observations of skeletal microstructures for both taxonomy and studies of depositional environments. In the introductive part of his extensive investigation on the stromatoporoids from the Dinant Basin (Belgium), Lecompte wrote (1951, Vol. 1, p. 16): "Solving this question requires new observation techniques." Figures 7.34i–j reveal how farsighted this statement was. Atomic force microscopy provides surprisingly good images of these Paleozoic materials, indicating that the new scale of investigation that was hoped for by Lecompte is now available. In this Upper Devonian sample, calcite fibers are seen formed by well-defined, complex, rounded grains. It is of note that there is no trace of rhombohedral faces visible to suggest recrystallization. These phase-contrast images, in which the gray to black levels indicate the degree of interaction between the mobile tip and the substrate (i.e., the variation in concentration of the organic component, see Fig. 3.2), provides us with the type of information anticipated by Lecompte 60 years ago, needed to complete the evaluation of skeletal microstructures.

The main development of the stromatoporoids was during the Ordovician, Silurian, and Devonian Periods. The progressive decline of the group began during the Carboniferous, possibly due to the rise of sponges producing skeletons built of densely packed tubes (the chaetetid bauplan). It has been suggested that "true stromatoporoids, a presumably monophyletic group," were limited to Ordovician through Devonian time, the post-Devonian forms rejected as a "polyphyletic grouping of apparently unrelated taxa" (Stock 2001). Even assuming this is correct, note should be made that the Carboniferous and Permian organisms utilizing the stromatoporoid plan correspond in time to the rise of both Sandberg and Hardies curves, indicating the transition from calcite to aragonite seas.

Since acceptance that these organisms were calcifying Demosponges, attempts to classify them at the family level show a characteristic difficulty in integrating these calcareous structures into a taxonomic framework based on the biological properties of living sponges. Even when traces of spicules can be detected within the skeletons, these are pseudomorphs and cannot provide sufficient information to be significant. The prevailing weight formerly given to the overall geometry of skeletal architectures (e.g., laminar, tubular, reticulate) has now been largely replaced by analytical efforts to improve the status of our understanding of modes of biomineralization among these "coralline sponges." Fortunately, as pointed out by Reitner and Wörheide (2002) "their biomineralization processes are extremely conservative." Restudy of possible affinities between these long-living groups of formerly abundant and diverse organisms now can rely only on fine-scale characteristics of their skeletal components, primarily due to a loss of information due to diagenesis.

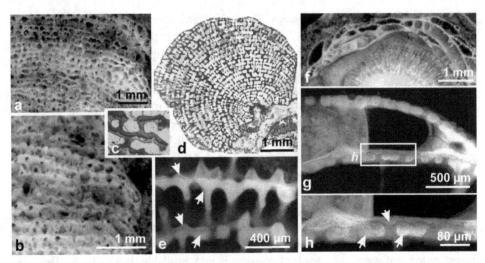

Fig. 7.35. Three sponges with a laminar bauplan and granular magnesian-calcite skeleton from Triassic (Norian) units in Turkey. (a–c) An unidentified stromatoporoid-like structure that shows primary laminae formed of two superposed layers; (d–e) *Cassianothalamia*: parallel growth steps also result in two-layered laminae; (f–g) *Uvanella*: a more sphinctozoan-like bauplan that also shows a stepping growth mode by superposition of irregular cupolae; (h) UV fluorescence allows recognition of the mineralization patterns within these granular calcite skeletons. a, b, c: Gautret (1991); g, h: Cuif *et al.* (1990).

It is not surprising that this approach commonly results in the hypothesis of far-reaching affinities, such as those resulting from the microstructural investigations of Stearn and Pickett (1994), who emphasized the similarities between the biomineralization modes in some "true" stromatoporoids (stromatoporellids and clathrodictyids) with those of modern "coralline sponges." In some of these, the authors concluded that skeleton is secreted in "modules that are homologous to the chambers of sphinctozoans." Interestingly, during the Triassic Period, among the highly diverse sponges mostly producing aragonite skeletons, some forms were present with skeletons formed by microgranular magnesian calcite. Figure 7.35 illustrates that, from the stromatoporoid-like bauplan to sphinctozoan architecture, a comparable mode of biomineralization can be observed in the laminae, here revealed by simple UV fluorescence.

Microstructural studies have shown how different the chambers of sphinctozoans can be, thus justifying rejection of the taxonomic use of this ancient term. The Stearn and Pickett suggestion that emphasized the resemblance between the mode of growth of some stromatoporoids and that of species formerly classified as coralline sponges provides a significant example of a present-day effort to evaluate their biologic affinities.

Study of the sedimentary dynamics of the Baltic platform in the Upper Silurian indicates that, at that time, the stromatoporoids lived in shallower water than did corals (Bassett, Kaljo and Teller 1989). They were also able to adapt to considerably more clastic material in their living environments, as shown by samples in the Tailfer Quarry, Belgium (Fig. 7.34c). The

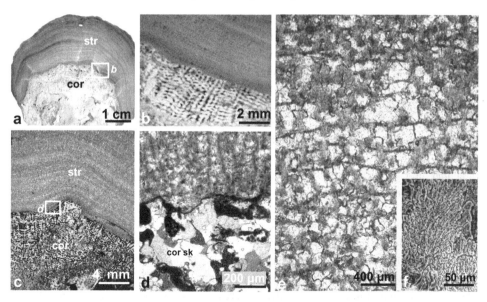

Fig. 7.36. Coral and stromatoporoid from Oxfordian strata (Jurassic, France). (a–b) Overall view of the sample. The coral skeleton (a: "cor") has been filled in by dark gray to black sediment, whereas the overlying stromatoporoid ("str") has been protected by its layered architecture. Void spaces were filled in by crystallization of calcite; (c–d) Identical sections observed in thin sections. The coarse calcite of the coral skeleton ("cor sk") clearly differs from the skeleton of the stromatoporoid; (e–f) Microstructure of the stromatoporoid skeleton: typical layer and pillar architecture (e), fibrous microstructure (f).

progressive decline of the group began during the Carboniferous Period, in apparent conformity to the Sandberg–Hardie oscillations. The stromatoporoids reappear in the Jurassic Period, again corresponding to the same oscillations. Figure 7.36 shows a specimen from the Oxfordian (Upper Jurassic) of the Jura Mountains (France) that has grown on a coral skeleton. This is of interest with respect to conditions in the "calcite sea," as simple optical microscopy indicates that the coral skeleton has been dissolved and replaced by coarse calcite (Figs. 7.36a–d), whereas the stromatoporoid skeleton still contains a readily deciphered fibrous microstructure (Figs. 7.36e–f).

By itself, this single sample summarizes two primary questions regarding the relationships between fossil organisms and the marine sedimentary environments that they inhabited. Dissolution of coral skeletons was common (practically the rule) during Jurassic time, and after the Triassic Period the microstructure of fossil coral materials is only rarely visible. It can be seen only where recrystallization to calcite has been moderated by geochemical activity that is poorly understood.

In contrast, the stromatoporoid exhibits the typical features of its group. This sample provides an excellent illustration of the sensitivity of the biomineralization process to environmental changes on a global scale. Supporting data for this statement can perhaps be found by an in-depth analysis of the skeletal features with modern tools such as AFM. If we assume that this conclusion is correct, it is still noteworthy that there are Carboniferous and Permian animals that reproduced the stromatoporid plan. This corresponds to the rise of both the Sandberg and Hardie curves, a time that was characterized by a transition from calcite to aragonite seas. During the Triassic Period, very few (if any) comparable fossils exist, but among the highly diverse sponges producing aragonitic skeletons, some rare forms do exist that produce closely spaced, parallel laminae constructed of microgranular magnesian calcite.

When "calcite sea" conditions reappeared in the Mesozoic, some organisms were able to again reproduce the stromatoporoid bauplan. At present, no phylogenic conclusions can be made to clarify possible relationships with the Carboniferous to Permian forms. We also note that these Mesozoic stromatoporoids, although living in a calcite sea, never developed carbonate mounds of a size or importance comparable to those created by their predecessors in the Paleozoic Era. The reason may have been that corals living in the Jurassic and Cretaceous seas had acquired symbiotic metabolism during the Triassic, and this resulted in a dramatic change in their position within Triassic marine ecosystems. Instead of living in the deeper parts of shallow-water areas, as they did throughout most of Paleozoic time, corals were now colonizing shallower water, producing dense and efficient populations, and above all, growing primarily in the vertical direction. With their mostly horizontal model of growth and mineralization, the stromatoporoids were reduced to a secondary role.

Another question arises regarding the form of the soft-bodied organisms that reproduced the stromatoporoid mode of mineralization in the Mesozoic calcite seas. The hypothesis can be reasonably advanced that they ceased mineralization when aragonite seas reappeared at the end of the Mesozoic, potentially leading to a possibility that they might still be living in modern seas. In order to check this, experiments on calcification might be a fascinating research area from a paleontological perspective. If so, perhaps highly valuable data on biomineralization in organisms that do not calcify under present seawater conditions might produce skeletons if placed in "calcite sea" conditions.

Other calcifying Demospongia: examples of permanency in their control of skeletal units and mode of biomineralization, reappearing after temporary suppression of mineralization

With respect to Sandberg's and Hardie's curves, no biological group offers a more spectacular illustration of recurring habit than do the demosponges. Here, the production of siliceous spicules was reduced, commonly until their complete suppression. The recent discovery of some long-known taxa (see Figs. 4.9 to 4.18) has led to considerable change in the way we interpret their fossil representatives. The Demospongiae were so abundant in the Triassic that (for the last time in the history of sponges) they were comparable to corals as contributors to formation of reef framework. Owing to favorable diagenetic conditions, fossils from Triassic units allow us to examine precisely their control of skeletal growth.

Astrosclera, the first of these recently rediscovered sponges, formed its skeleton through a very specific process, in which an intracellular fibrous nodule was first produced during the transit of a sclerocyte from the internal part of the sponge body to the upper portion of the growing skeleton. Figure 4.10 shows the general aspect of the resulting skeleton. Many

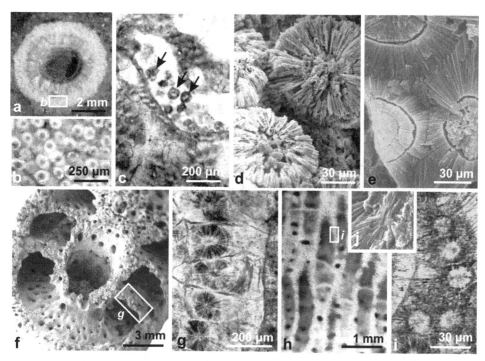

Fig. 7.37. *Astrosclera*-like microstructure in various architectural arrangements. (a–e) *Astrosclera*-like skeletal growth shown by fossils from Triassic outcrops in South Anatolia, Turkey. Note the evidence of a two-step process visible on polished sections (a–b) and in thin section (c). Natural surfaces of samples (d) or polished and etched surfaces (e) confirm the presence of this particular mode of skeletal formation in the Triassic; (f–i) The spherulitic mode of growth is seen in sponges with sphinctozoan architecture (f–g) and with a packed-tubular organization (h–j). In both cases the *Astrosclera*-like, two-stage cellular mode of growth is clearly shown. h: Cuif (1983).

Triassic sponges provide excellent examples of this specific growth mode of spherulitic skeletal units in a two-step biomineralization process. Polished samples allow recognition of this type of skeletal structure (Figs. 7.37a–b), and thin sections at times show the initial stage of spherulite growth preserved during development (Fig. 7.37c: arrows).

This specific manner of growth can be very precisely identified as occurring in various architectures in the Triassic. Although widely distributed in the Pacific and Indian Oceans, living *Astrosclera* are not differentiated architecturally. In the Triassic, however, this method of two-stage biomineralization can be identified in sponges built of superposed cupolae formerly classified as the Sphinctozoa (Figs. 7.37f–g), or occurring with tubular architecture (Fig. 7.37h–j) , and forming an abundance of yet undescribed species.

The Triassic sponge faunas contain two other major types of calcifying demosponges allowing accurate comparison with living species regarding their growth and methods of regulating microstructure. One of them shows remarkable similarities to the modern ceratoporellids.

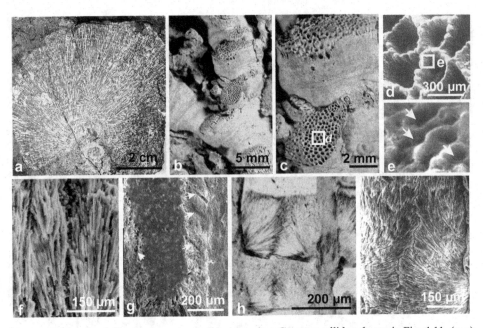

Fig. 7.38. Triassic tubular sponges comparable to modern Ceratoporellidae shown in Fig. 4.11. (a–e) Densely packed tubular chambers (a–c) with traces of inserted spicules (d–e: arrows); (f–g) Walls are formed of parallel vertical bundles of fibrous aragonite (f) including spicule molds (g); (h–i) Fibrous infilling progresses upwards, reducing the functional volume of chambers. d, e: Cuif (1973).

Close observation of individual tube morphology reveals multiple empty holes that correspond to places where siliceous spicules were apparently inserted, but have since been dissolved (Figs. 7.38a–e; compare to Fig. 4.11e). Tube walls are formed by elongate bundles of fibrous aragonite (Fig. 7.38f; compare to Fig. 4.11i). Calcite pseudomorphs of the spicules can be seen on polished and etched surfaces, although presenting no possibility of comparing morphologies. The formation of fibrous spheroidal nodules results in development of carbonate at the chamber base, thus restricting the volume of functional space to only the upper parts of tubes in *Ceratoporella*. In Triassic specimens, just as in living forms, these fibrous nodules develop symmetrically from the wall tube (Fig. 7.38h–i). In all visible details, the skeletal microstructure and growth mode are closely comparable.

An equivalent comparison can be made between modern *Vaceletia* (see Figs. 4.13 and 4.14) and some of the Triassic sponges formerly classified as Sphinctozoa (Fig. 7.39). These are formed of superposed perforated cupolae supported by pillars. Including the minute details of pore morphology (compare Figs. 4.13c and 7.39c), a close comparison is possible with the modern species. In both cases, skeletons of pillars and cupolae are formed by juxtaposition of irregular spheroidal nodules (see Figs. 4.13e–g for the modern specimen and Fig. 7.39d for the Triassic one). In fossil specimens, the undifferentiated granular aragonite skeleton is still preserved, in spite of its close proximity to surrounding blocky calcite (Fig. 7.39e–f).

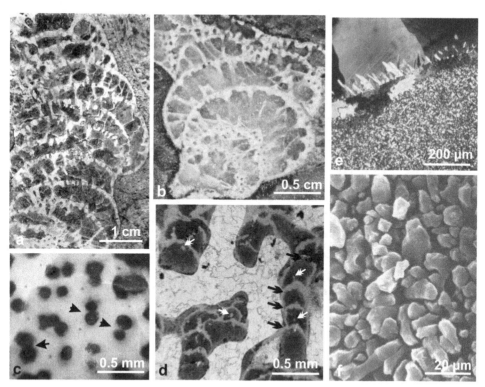

Fig. 7.39. A Triassic sponge (*Stylothalamia*), with skeletal features that are quite comparable to modern *Vaceletia*. (a–b) Skeletons built by regular spheroidal cupolae supported by pillars; (c) Details of porosity of the cupola (in tangential section) exactly corresponding with those of modern *Vaceletia*; (d) Typical vaceletid growth mode of the skeleton, formed by superposition of nodules with a distinct center; (e–f) Granular microstructure of Triassic specimens.

These three examples, in addition to the Triassic corals previously mentioned, suffice to illustrate the exceptional preservation state of these fossils. However, without a doubt, the main point to be made here is the close correlation between the Triassic fossil species and the three modern species that they resemble. It is doubtful that biologically independent organisms could have simultaneously been able to reproduce three species with such obviously similar skeletons, but separated by 220 million years. However, one should note that the alternative is also astonishing. If the Triassic species were the antecedents of the three living representatives, it would mean that during the time of "calcite seas," they survived in such a state that very few fossil data are available for an evaluation of phylogenetic continuity of the two faunas.

There is a two-fold explanation of this: there was probably reduced mineralization during the Jurassic and Cretaceous time, but there was also considerable ecologic change. What has been suggested concerning the stromatoporoids is still more likely for these three demosponges. They are now located in submarine caves or rocky slopes of reef fronts

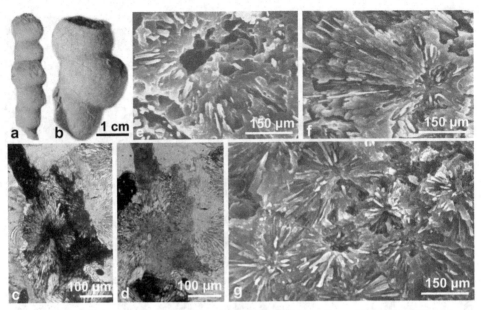

Fig. 7.40. Permian sphinctozoan sponges (*Sollasia* sp.) from the Permian reef at Djebel Tebaga. (a–b) Specimen morphology; (c–d) Thin-section photomicrographs in polarized light; the large calcite crystals include significant remains of the original fibro-spherulitic skeletal units; (e–g) Radiating aragonite fibers visible after careful etching. These confirm the typical spherulitic microstructure.

(where *Vaceletia* was discovered in New Caledonia). With the development of Middle Jurassic reefs that already show rather modern ecologic zonation, it is possible that the prosperous sponge faunas that formed the reefs of the Upper Triassic were displaced to deeper, less-favorable zones in the younger reefs mostly constructed by corals and algae. If true, there would have been no basic biological changes in forming present-day populations.

Reitner and Wörheide (2002) emphasized the extremely conservative character of the biomineralization mechanism in these sponges. Nowhere else is the contrast more striking than in these "biologically simple" organisms, as they are sometimes characterized, and the perfect precision shown in their method of skeletal formation through extremely long time intervals.

Permian aragonitic sponges of the Djebel Tebaga (South Tunisia) and their diagenesis

The survival of Triassic sponges through the Mesozoic calcite-sea episode may seem a surprising phenomenon; some similarity can be found in considering the Permo-Triassic transition through the most pronounced of the big five biological crises. Calcareous Permian outcrops at Djebel Tebaga (South Tunisia) contain a rich fauna of calcifying demosponges that have varied architectures. Their microstructure is of particular interest, here illustrated by fossils of the genus *Sollasia*, a typical sphinctozoan (Figs. 7.40a–b). Their wall is formed of recrystallized spherulites (Figs. 7.40c–d) included within large calcite crystals as a result of diagenetic modification of carbonate cements. Etching of them allows remains of initial fibrous spherulites to become visible, permitting an analysis of their original organization (Fig. 7.40e–g).

This is an excellent example of aragonite biomineralization in the Permian (Wendt 1977), just prior to the great extinction, the most complete such event seen in the fossil record.

With respect to the Sandberg and Hardie oscillations, the Permian sponges of the Djebel Tebaga occur within the aragonitic-sea interval, although conditions were oscillating according to Hardie's calculations. We can thus perhaps assume that their disappearance as a group at the end Permian extinction was due to a different factor than was their regression at the end Triassic event. However, the similarity in their well-defined micro-structure (and the correlated manner of growth) through two major biological crises makes it highly improbable that this form of control of skeletal growth was reproduced exactly by chance only, without any biological continuity. These examples suggest that the biomineralization mechanism is so well controlled in these organisms that after two major crises involving probably distinct environmental modifications, the process of skeletogenesis was maintained, even retaining all of its minor details. At the same time, the sensitivity of calcareous skeletogenesis to environmental variations is a property that makes these sponges ideal indicators of the latter, justifying a better understanding of their biomineralization mechanisms.

7.5.2 Evolutionary history of the corals: evidence of a growing complexity in Permian and Triassic time

Since Haeckel's theory (1896), it has been common for coral researchers to assume that all of the Rugosa (Paleozoic corals) became extinct in Late Permian time, and that the Scleractinia (the modern corals) appeared during Middle Triassic time. Since all of these corals are placed in the Zoantharia, they are all characterized by the presence of paired mesenteries, and those zoantharians that form carbonate septa do so within up-pocketed basal flesh between pairs of mesenteries. There are also branches of the Zoantharia that lack skeletons, but contain polyps that illustrate the same characteristics as polyps of skeletal corals, as outlined by Oliver and Coates (in Boardman *et al.* 1987). The Subclass Zoantharia is divided on the basis of the appearance of mesenteries within the gut of polyps during ontogeny; these are added in series of four, one next to another in the Rugosa, and in cycles of six in the Scleractinia, added midway between existing mesenteries and septa. This is apparently a simple and valid criterion for separating the Rugosa and the Scleractinia, as well as other orders of the Zoantharia (Oliver 1980).

Septal ontogeny

The Rugosa thus have been characterized as having serial insertion of septa in four sectors of their skeleton (corallite), following the emplacement of an initial set of six primary septa. These septa appear in the early post-larval ontogeny of the corals, first as one pair of septa dividing the calice in two. These are followed by the insertion of a pair of septa appearing (or "inserted") bilaterally, followed by a second pair of septa inserted bilaterally in the corallite to form the six primary septa. Thereafter, septa are inserted into the corallite as series of septa, next to a preexisting septum within four quadrants formed by five of the six primary septa. On the other hand, the few modern species of the Scleractinia that have been closely studied – most notably *Pocillopora damicornis* by Vandermeulen and Watabe (1973), and *Porites lutea* by Jell (1980) – are characterized by the approximately simultaneous appearance of all six primary septa, followed by addition of more septa in cycles, with each new septum bisecting the space between two preexisting septa, and appearing within six somewhat equal sectors of the corallite.

Some corals from the Triassic Period differ from at least some other early species of the Scleractinia in that they do not insert their six primary septa simultaneously, as do advanced forms such as *Porites* and *Acropora*, but rather, these primary septa appear in the same sequence (Fig. 7.30), as do those in the Rugosa. Among the very diverse Triassic fauna, these corals form a small group, first represented by *Zardinophyllum* (Montanaro-Gallitelli 1973) from the Dolomites (Italy), followed by *Gallitellia* (Cuif 1975a), also from the Dolomites, and a more diversified group discovered in the Lower Norian of Turkey (South Anatolia) gathered in a new Family Pachythecalidae (Cuif 1975b). Figures 7.27, 7.30 and 7.31 provide some evidence of the nontypical scleractinian mode of development of the septal system in these corals from the Lower Mesozoic.

On the other hand, Ezaki (1997) has also clarified details of septal insertion in the scleractinian-like Permian species *Numidiaphyllum*, also from Tunisia (Djebel Tebaga), and here illustrated as clear evidence of the cyclic insertion of septa in six areas of its fasciculate corallites.

Skeletal mineralogy

In the past, it was generally accepted that the Rugosa had calcitic skeletons and the Scleractinia have aragonitic ones. As noted above (Section 7.4), this is no longer strictly true, as Stolarski *et al.* (2007) have presented compelling evidence that the genus *Coelosmilia* contains at least one species that had a calcitic skeleton in the Cretaceous Period. Additionally, a group of Upper Carboniferous and Permian rugosans had a high intermediate content of Mg in their skeletal calcite, making them liable for diagenetic alteration to form the zigzag skeletal structure first studied by Schindewolf (1941), containing up to 7–8 mol% Mg (Webb and Sorauf 2002; Sorauf and Webb 2003). Modification of skeletal composition did occur in late Paleozoic rugose corals, probably as a reaction to the changing seawater chemistry at that time, apparently connected to a rise in the Mg/Ca ratio towards the end of the Paleozoic "calcite sea." In addition, the presence of aragonitic skeletons in the Permian genus *Numidiaphyllum* was suggested by Wendt (1990), based on the mode of its recrystallization, and this has been reinforced by observations of Ezaki (2000) that their skeletal materials are recrystallized into coarsely crystalline calcite.

Further study of numidiaphillid corals from the Permian of China by Ezaki (2000) indicates that the genus *Houchanocyathus* has septal patterns similar to those in Triassic genera. This indicates that change in some skeletal characteristics had already begun in the Permian, and that Permian corals were not exclusively calcitic in their mineralogy. Examined at the nanoscale, some Permian (Polycoelidae) and Triassic corals (Pachythecalidae) exhibit the same coated, granular nanostructure (Fig. 7.30). The genome of modern Cnidaria enables them to produce both calcite and aragonite, as in the calcitic Order Octocorallia, where the genus *Heliopora* produces an aragonitic skeleton. It is clear that simple mineralogy thus is no longer useful as an ordinal character, and utilizing differences in coral symmetry (4-fold versus 6-fold) is complicated by the possible exceptions on both sides, both in late Rugosa and in early Scleractinia.

Thus, in the skeletonized orders of the Subclass Zoantheria, the primary basis for separation between orders remains the different method of appearance ("insertion") of mesenteries in their gut: cyclic or serial in the two major skeletal groups, the Scleractinia and the Rugosa. Modification of this should be made only in conjunction with evaluation of other bases that can also be utilized for separation of orders of nonskeletonized zoantherians.

Skeletal microstructure indicating overall control on the mode of growth

Microstructural information that has emerged to date from research on Triassic Scleractinia emphasizes that these do not represent a single Triassic group ancestral to modern scleractinians. There are multiple microstructural types in the Triassic, as described above in Section 7.4, and it is necessary to take these into account in reaching any conclusions regarding the biology and evolution of the scleractinians. On the Triassic side of the Permo-Triassic gap, the diversity of the Anisian fauna (lower part of the Middle Triassic), and the even more diverse microstructural characteristics in corals from Ladinian and Carnian strata of the Dolomites, the Norian of the Lycian Taurus (Turkey) and the Rhaetian of the northern Tyrol, have been pointed out by several authors (Cuif 1965, 1975c; Qi 1984; Reidel 1991). It is impressive that in both classic fossiliferous areas and in more recently discovered ones, isolated coral species – *Gigantostylis* from the Rhaetian of the Fischerwiese (Salzkammergut, Austria), *Zardinophyllum* from the Dolomites (Italy), and the Pachythecalidae (Norian of Turkey) – contain corals that clearly have distinct skeletal patterns that differ greatly from those usually observed in the Scleractinia. Frech (1890) had interpreted the genus *Gigantostylis* as representing "ancient corals," as Montanaro-Gallitelli did with *Zardinophyllum* a century later. This great diversity of corals in the Triassic must enter into any consideration of Paleozoic ancestors of modern forms.

The molecular approach

A number of recent studies (Chen *et al.* 1995; Romano and Palumbi 1996; Romano and Cairns 2000; Fukami *et al.* 2008) used the molecular approach to infer relationships between scleractinian corals and their possible evolutionary history. Their results, obtained by using different molecular sequences, generally favor a more or less clear monophyly of the modern corals.

This conclusion epitomizes the importance of the contribution of paleontological approaches in unraveling an evolutionary problem. The diversity of the coral fauna during this crucial Permian–Triassic time cannot be understood through studies dealing only with extant corals. The majority of the coral faunas involved in this evolutionary process have been subjected to two extreme mass extinctions, the Permo-Triassic and the Triassic–Jurassic.

The widely shared statement that, when trying to solve a complex question, the simplest solution requiring the least-complex scenario is generally the correct one, probably does not apply in this instance. The simplest scenario would state that all scleractinians evolved from an actinarian anemone that developed the capability for skeletogenesis in Middle Triassic time. This straightforward solution is too simple to account for the complexities of the known data on Triassic corals.

To be honest, we do not know quite how complicated it would have been to evolve and change zoantherian patterns of insertion of mesenteries from serial to cyclic. It is simplest to do this with a single ancestor earlier in the Paleozoic, and that may indeed have occurred in the Upper Carboniferous. However, the multiplicity of microstructures in the Triassic suggests that either there was an astonishing amount of change taking place during the "nonskeletal" times in the early part of the Triassic Period, or there were multiple Permian ancestors with the capability of producing an aragonite skeleton, but which did not do so until the Middle Triassic when conditions became favorable for their basal ectoderm to commence skeletogenesis.

At present, details of what groups were involved, and how they evolved are lacking. The 10 Ma gap between the last skeletonized Paleozoic corals and the earliest Mesozoic corals was undoubtedly a time of major and perhaps accelerated evolution of the Zoantharia. It seems apparent that one or more small groups of these cnidarians survived this interval while either retaining or developing cyclical insertion of mesenteries and the septa between them, thus producing the successful radiation of the Middle Triassic.

7.5.3 Microstructure and mineralogy of the Coleoidea (Mollusca Cephalopoda)

The evolutionary history of the coleoid cephalopods is somewhat enigmatic, due to a decrease in their forming an external calcareous shell, sometimes leading to its complete disappearance, as in the squid and octopus. In present-day seas, the calcareous shell is only preserved in a highly evolved form as the cuttlefish bone in the important Family Sepiidae (see Fig. 1.18), and also in the Spirulida. Fossil coleoids are known since the Lower Carboniferous (Mississipian), but their time of origin may be even earlier. In the Hunsrück Slate (Lower Devonian, Germany), X-ray radiography has revealed the pyritized soft body of a *Loligo*-like animal (close to the squid), but no consensus exists regarding the formal identification of this organism (Stürmer 1985).

However, in the Lower Devonian, many fossil forms exhibit calcareous skeletons in which the original features of the ancestral cephalopod shell are recognizable. They are united in the Subclass Coleoidea. These early coleoid cephalopods are distinct from the long-lived and diversified Nautiloidea (sometimes said to be a sister group of the Belemnitida) by the new solution they brought to the persistent question that cephalopods face, due to their way of life: the control of buoyancy. As actively swimming organisms and primarily predators, cephalopods have to maintain a suitable orientation of their shells while in motion. The new shell archetype is essentially conical and linear, with internal compartments separated by septa, forming a phragmocone. The animal body is located only in the last compartment. Previous compartments are partly filled by liquid, the volume of which is regulated to adjust vertical displacement; this is the role of the siphon that extends through the series of chambers to the apex of the shell. But, with shell growth, disequilibrium increases between the compartment inhabited and the more apical part of the shell. In ancient nautiloid cephalopods (the "orthocerids"), additional deposits of calcareous material inside apical chambers are commonly observed that bring compensatory weight to act against the upward buoyancy of the apical part of the shell. In the course of the lower Paleozoic, a purely geometrical solution occurred when spiral growth was invented. Spiral growth allows the gravity center of the organisms to be permanently maintained in the vertical axis passing through the buoyancy center. Such a solution is illustrated by the extant *Nautilus* (Fig. 1.17).

An innovative solution was created in the coleoid Order Belemnitida in the Upper Carboniferous. By extending the mantle (the mineralizing organ in the Phylum Mollusca) toward the rear of the shell, the mineralizing side of the mantle was in place against the external surface of the phragmocone. Thus, the ancestral orthocerid shell became internal and was covered (at least partly) by newly produced mineral material deposited over the exterior of the phragmocone. This is the origin of the rostrum (or guard) of the belemnites. Greatly developed at its apex, this hyper-developed outer layer of the phragmocone was able to maintain the overall equilibrium of the animal body in the same way as the internal chamber deposits did in the orthocerids. Figure 7.41 shows the appearance of the original shell (the ancestral phragmocone), recognizable beneath the thick fibrous carbonate layers of the rostrum. Thus, the belemnite rostrum is not a new structure, but the greatly thickened outer prismatic layer of the cephalopod phragmocone.

Regarding their evolution, it should be recognized that parallel to the occurrence of this new solution to the buoyancy question, a few, very rare fossils, such as the questionable squid-like Hunsrück fossil, suggest that the trend toward shell reduction may have produced the early representatives of the nonskeletal coleoids during this same time period. This raises the question of the rapidity of evolutionary processes and the value of hypothesizing lineages that are only documented by scattered occurrences of fossils.

Regarding biomineralization, the Class Cephalopoda is remarkable for its continuous use of aragonite in forming skeletal structures. Shells of ancient nautiloids have always been aragonitic, just as they are in modern *Nautilus*. All structures of the specimens of Fig. 7.41 are aragonite, and the entire phylum can perhaps be considered as a counter-example to the usual sensitivity of the biomineralization process to environmental conditions. An exception exists, however: the belemnites that were very abundant during the Jurassic and Cretaceous Periods, among the first fossils illustrated (Gessner 1565). They are generally preserved as calcite rostra (or guards). The phragmocone itself continued to be formed with the fundamental aragonite mineralogy and microstructure of other cephalopod shells (see Figs. 1.17 and 1.18).

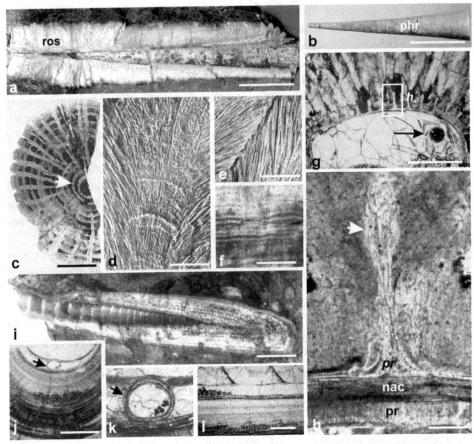

Fig. 7.41. Belemnite rostra as seen in Triassic specimens. (a–f) The conical rostrum (a) is an "additional" feature covering the initial shell (b: phragmocone "phr"). In this species the rostrum is made first of concentric layers (c: arrow) and then of distinct radial sectors. These sectors are formed of aragonite fibers (d–e), and are thickened by distinct growth layers; (g–h) Closer observation of the contact between the shell and the rostrum (h) shows that the wall of the phragmocone is comprised of the normal three-layered sequence of cephalopod shells: a nacreous layer ("nac") between two prismatic layers ("pr"). Note the elongation of the external prismatic layer producing a series of longitudinal crests (arrow), a remnant of ancient ornamentation that is later invisible due to being covered by the "rostrum," the thickened outer prismatic layer of the phragmocone; (i–l) Another species in which the septa of the axial cone are visible due to erosion of the cone wall and half of the rostrum: (j–l) show transverse and longitudinal sections of the siphon. a, c–d: Dauphin and Cuif (1980); b, g: Dauphin (1983).

The Belemnoidea are rostrum-bearing coleoids and include two groups, the belemnites and aulacocerids (Doyle *et al.* 1994). Aulacocerids were first described as primitive belemnite-like organisms (Stolley 1919), having an originally organic rostrum in life. Belemnites (Fig. 7.42) are entirely fossil organisms, and from a phylogenetic point of view, the possibility of a relationship between aulacocerids and the belemnites has long been a matter of dispute, in large part based on the significance attributed to mineralogic criteria. Jeletzky (1966) formalized the concept of a fully distinct origin of these taxa (at the order level), as he considered that the calcite mineralogy of belemnite rostra excludes the Aulacocerida as their possible ancestors. Thus, when aragonitic "belemnitid rostra" were discovered, they were excluded from the Belemnitida, and new taxa were created. Additional distinctive characters were commonly used to separate aulacocerids and belemnites. Donovan (1964), Jeletzky (1966), and Teichert (1967) have discussed the significance of the apical angle of the phragmocone (about 8–10° in aulacocerids, but wider in belemnites), different spacing between septa within phragmocones, or detailed features of the passage of the siphon through the septa. Recent restudy of key specimens (including those from the Paleozoic) led Doyle *et al.* (1994) to reconsider this concept, at present accepting the derivation of belemnites from the aulacocerids. In support of this view, there is a correlation between the passage of aragonite to calcite rostra and Sandberg's/Stanley–Hardie's oscillations at the Triassic–Jurassic boundary. The Belemnitida thus provide a good case history, and collecting data regarding the mineralogy, fine-scale structures, and potential impact of diagenesis becomes particularly important.

Optical and SEM observation of the rhombohedral fracture of crystalline calcite in the radial prisms of the belemnites provides clear signs of diagenesis (Figs. 7.42f–h). Higher-magnification images obtained with the atomic force microscope (Figs. 7.42j–l) indicate that, in contrast to the usual disordered arrangement of granular elements forming growth layers, grain clusters with rather parallel limits are seen developing here. Further development of this process might lead to formation of crystallographic patterns at broader scales in these radial units. Note here that in the Devonian brachiopod fibers illustrated earlier (Fig. 7.17), signs of diagenetic modification by formation of rhombohedral calcite faces are absent, in spite of their much older age (see Fig. 7.17). Among echinoderm skeletons, also exclusively calcitic, recrystallization strictly follows consistent crystal orientations of biomineralization-induced orientations of each skeletal unit, whatever its size (Figs. 4.47–4.51). A question thus arises regarding the origin of individual radial skeletal units within the belemnite guard. No indications have been reported of organic envelopes surrounding prismatic units that could cause development of a recrystallization process. As pointed out by Gustomesov (1976, 1978), information about the formation of belemnite rostra is still limited to descriptions of the mode of growth.

Aragonitic belemnites

Reports by Spaeth (1971, 1975) of rostral aragonite layers, and more recent study by Bandel and Spaeth (1988) describing sequential aragonite/calcite mineralization, suggest that belemnite rostra underwent some additional, diagenetic, processes. Such a hypothesis is additionally supported by comparison of the calcium isotope ratio in various fossilized biogenic calcites. The $^{40}Ca/^{44}Ca$ ratio in the calcium carbonate of the belemnites is clearly distinct from those of typical calcite within calcitic shells of organisms such as the brachiopods (Gussone *et al.* 2005).

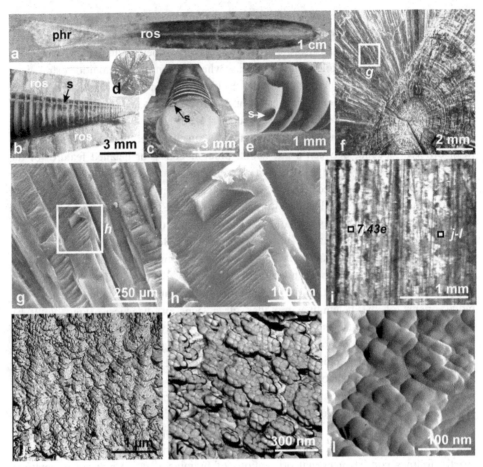

Fig. 7.42. Overall organization and microstructure of radiating prisms of belemnite rostra. (a–e) Rostrum (or guard) ("ros") and phragmocone ("phr") are the two main components of belemnites. The axial section (a) and a lateral view of a fractured longitudinal section (b) indicate the position of the phragmocone (NB: the apex of the guard is at the rear of the animal body). The concavity of the septa is readily visible, as is the marginal siphon ("s"); (f–h) Radiating fibers show calcite cleavage, typical for recrystallized material. Note that the etched portion of the surface (f) also shows concentric growth layering, preserved in spite of the recrystallization (see also Fig. 2.11); (i–l) The intensity of recrystallization largely varies from specimen to specimen. Even within short distances in a moderately recrystallized sample (i) a progression of diagenetic severity can be observed at the growth-layer level. The formation of small clusters of grains following a common orientation may correspond to the early stage of the process that produces the rhombohedral cleavage as the end result. In this process, phase-contrast imaging (k) shows that the organic component is expelled to the exterior of the developing clusters, suggesting fusion of their crystal lattices into broader units.

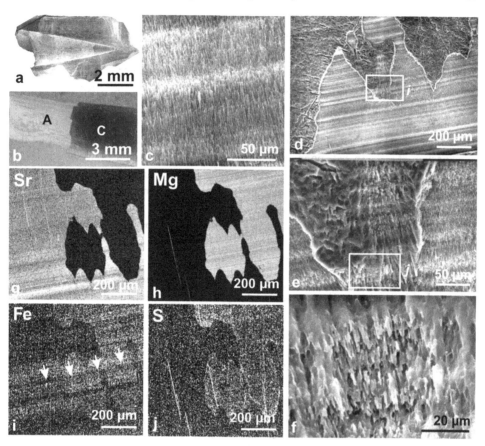

Fig. 7.43. Examples of the mixed-mineral belemnites from the Taymyr Peninsula (Northern Siberia). (a–b) Morphology of two specimens: note the irregular contact line between aragonite ("A") and calcite ("C"); (c–f) Layered aspect of the aragonite fibrous structure (c), comparable to other examples of fibrous aragonite in cephalopods (see also Fig. 7.45). Progression of the diagenetic front (d) showing the inversion from aragonite to calcite leading to formation of a more compact structure. As aragonite is the denser polymorph, its transformation to calcite occupies the space between the aragonite fibers (e–f); (g–j) Distribution of four minor elements in the area of contact between calcite and aragonite. Note the continuity of direction between distribution of Fe (common to the two polymorphs), whereas concentrations of Mg and Sr are much more exclusive. a, b, g–j: Dauphin *et al.* (2007c).

Recently Dauphin, Williams and Barskov (2007c) also studied a small series of belemnites from the Turonian of the Taymyr Peninsula (Northern Siberia) that have a mixed calcite/ aragonite composition (Figs. 7.43a–b), with growth layering clearly visible within fibrous aragonite parts, both structural (Figs. 7.43c–d) and chemical (Figs. 7.43g–j). Chemical mapping of minor elements such as Fe additionally shows that the aragonite–calcite inversion seen here has occurred without perturbation of their distribution patterns, and that mineral layering passes from aragonite to calcite without changes in crystal direction (Fig. 7.43i).

These multiple reports of aragonitic or mixed aragonitic and calcitic belemnites are of special interest because they are concentrated in the Cretaceous, the same period of time during which Stolarski *et al.* (2007) reported scleractinian corals with calcitic skeletons. These two "abnormal" examples of biomineralization – aragonite produced by belemnites and calcite produced by corals – form a significant contrast. They clearly demonstrate that, although water chemistry has an influence on the chosen polymorph in biologically produced carbonates, this selection of a given polymorph does not directly reflect only water chemistry itself. Change in Sandberg or Hardie conditions during the Upper Cretaceous caused this modification of the carbonate polymorph utilized in cephalopods and the opposite change in scleractinian corals, perhaps because mineralization is primarily under the control of organic matrix. We are as yet unable to explain how a single environmental change can result in reverse results in the cephalopods and corals; however, this does in fact disprove any purely chemical application of Sandberg's or Stanley–Hardie's oscillations to living organisms.

The Tertiary coleoids

The end of the Mesozoic was another time of enigmatic evolutionary changes in the history of the coleoids. During the Eocene, several distinct species developed calcified structures of aragonite, all of which contain a conical cavity. Their most surprising aspect is the micro-structural organization of the calcified structures.

Beloptera is the most frequently occurring of these in Eocene strata of the Paris Basin (Figs. 7.44a–c). The elongate, cylindrical calcareous structure developed behind the hollow cone (as a consequence, equivalent to the rostrum in its position) is not formed of concentric growth layers, but rather by thick, parallel laminae (Fig. 7.44d). These laminae are com-posed of fibrous aragonite organized as elongated spherulites (Figs. 7.44e–f).

These elongate spherulites are also seen in the rostrum of *Belopterina* (Figs. 7.44g–j), and can be seen in cross section to have a radial arrangement with a fibrous organization that, in some respects, is reminiscent of the aulacocerid type of organization. In *Vasseuria* (Fig. 7.44k–l), construction of the conical-tube wall also has a comparable microstructural organization. The microstructures of these fossils have been analyzed in a series of com-parative papers (Dauphin 1984 to 1986a), with the goal of interpreting their possible role in the origin of modern-day coleoid faunas, and more specifically their relationship to the cuttlefish bone. Different opinions have been published regarding the position of these enigmatic and short-lived organisms. Teichert (1967) considered their possible origin in the Belemnoidea, an audacious proposal at a time when aragonitic species of belemnites were still unknown or ignored. More recently suggested phylogenetic classifications of living coleoids, now partly based on molecular data, also found any consensus lacking.

In spite of the evolutionary uncertainties, there is a very strong suggestion that the sudden proliferation of multiple calcifying forms of coleoid cephalopods in the Early Tertiary is related to changing overall environmental conditions. Stable favorable conditions for fossilization in sediments of the Paleocene and Eocene of northwestern Europe, also an area where fossil faunas have been intensively studied for two centuries, most likely explain the apparent concentration of these varied and innovative coleoids within this region. A

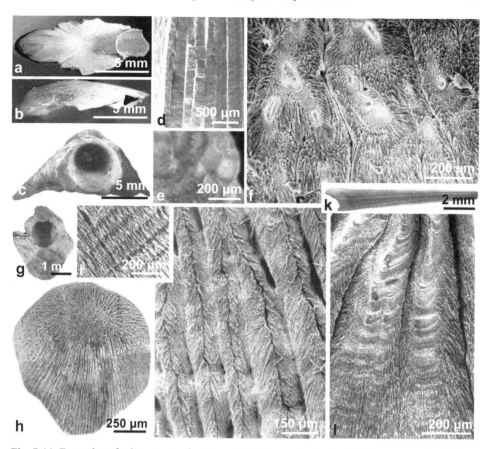

Fig. 7.44. Examples of microstructural patterns in Tertiary coleoids. (a–f) *Beloptera*: external views (a–b) and the hollow cone interpreted as a phragmocone-like structure; (g–j) *Belopterina*: the cone (g) and the fibrous structure with a visible radial arrangement. Growth layering and the arrangement of aragonite fibers in the thick slices indicate similarities in microstructural control and biomineralization mechanism to those of the Mesozoic coleoids; (k–l) *Vasseuria*: the conical tube (k) and the structure of the wall also correspond to a comparable growth process of calcareous units in other, more ancient Coleoidea. a, b, f; Dauphin (1985b); g, h, j: Dauphin (1986b).

major paleogeographic change also occurred when these epicontinental marine areas, formerly dependent on seawater from Tethys during the entire Mesozoic, became coextensive with North Atlantic waters. In the yet largely unknown evolution of the coleoids, the calcified structures developed by these organisms during a brief portion of geologic time would indicate that these diversified groups had improved chances for survival. This aspect of permanence was apparently improved through diversification of their morphologies, reflecting that they had the appropriate mechanisms of calcification. This is exemplified by these enigmatic cephalopods of the Early Cenozoic. When environmental conditions again led to their calcifying, they apparently repeated microstructural patterns in such a way to indicate that these patterns are deeply rooted in the group's genome pool.

Fig. 7.45. Microstructural patterns in Paleozoic coleoids. (a) Section perpendicular to the long axis in a silicified guard (*Dictyoconites* sp.; Smithsonian Institute collections); (b–c) Longitudinal section in *Stenoconites idahoensis*. Note the presence of internal deposits (arrows) in addition to the external rostrum ("phr": phragmocone; "rost": rostrum); (d–f) *Hematites*, showing the precise control of fiber growth in radiating sectors.

Comparing the calcareous structures produced by these early Cenozoic cephalopods to those of their ancestors from the Carboniferous provides us with another example of this astonishing ability of particular groups to maintain their specialized skeletal patterns throughout very long periods of time. Figures 7.45a and 7.45(d–f) illustrate structures formed in early coleoids of Lower Carboniferous age. The radiating organization of external deposits was already present at this time (Fig. 7.45a), and was still utilized in early Cenozoic time (Fig. 7.44h–j).

It is to be expected that improving our understanding of particular methods of biomineralization control particular to each major phylum or group will change these fossils, with still enigmatic structures, into a useful evolutionary sequence.

7.6 Generalizations on diagenesis and biomineralization

In interpreting the fossil record and fossilization processes, the application of a biochemically driven model of crystallization to these fossils is very useful in separating the respective influences of diagenesis and biomineralization.

The oscillations outlined by Sandberg (1983), and by Stanley and Hardie (1998) rank among those rare major concepts that have resulted in an in-depth reexamination of long-standing questions in various areas of Earth history. Interpretation of the fossil record obviously is part of such research. It is important to note that introduction of this concept of a world undergoing chemical change provides an additional source of complexity into study of the fossil record. It should be obvious that variations in seawater composition can influence both of the two steps in the formation of the fossil record: biomineralization and diagenesis. Although it was already difficult to resolve these chemical influences within a stable ocean concept, research became even more difficult with the perspective of oscillating environmental parameters. To what extent are organisms able to modify their skeletons before complete suppression of calcification occurs? This would seem to be one of the first issues to be addressed. A reasonable and reciprocal question also arises from the last observation. What conditions are necessary before noncalcifying organisms from modern seas begin skeletal calcification? This question suggests a broad field of experimentation – an area of research in its earliest stages of development. Disappearance of the calcifying demosponges for multiple geological periods presents an opportunity to study long-term natural experiments. One outcome of such study would be to justify a conclusion that switching off of the skeletogenesis process may be a common answer to environmental change, and might be present in other goups. The small number of case histories presented in the last chapter of this book simply aims at publicizing the types of analytical work required before forming conclusions on environmental influences on skeletogenesis. Oversimplification, as always, constitutes a potential cause of error in evaluating the mechanisms of biomineralization in different groups of living organisms.

Qualitative data gathered in our study of calcareous biominerals do not support a concept that anatomically simple organisms exert less control in forming skeletal parts than do more complex organisms. What we have observed in the calcareous sponge skeleton, thus in a "biologically simple" organism, as well as in the corals, reveals not only a remarkable diversity in their mineralized products, but also, no trace of a decreased level of control over their formation. More precisely, what is known to date regarding organic materials involved in the growth and mineralization process suggests that diversity of composition is equal among most groups. This conclusion is important to methodology. When dealing with assemblages of fossils from a given outcrop, the interpretation of analytical results should not be treated separately, based on the data source, i.e., whether from a "biologically simple" or a "more complex" organism. If controls on skeletogenesis were exerted at a variety of levels, a specific and separate treatment for each taxonomic group would be needed.

Figure 7.46 (from Dauphin *et al.* 1996) presents a statistical expression of compositional diversity among sponges, corals, and cephalopods from the Norian strata of the Lycian Taurus (Turkey), compared to equivalent analyses carried out on modern representatives of these three phyla. Although all were initially aragonitic, the groups are now distinct in their composition. These initial differences are visibly tied to specific trends in their fossilization. Respective changes in the surfaces and positions of sectors representing statistical

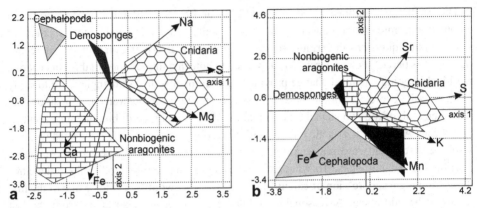

Fig. 7.46. Graphic expression of diagenetic trends among the three main groups of organisms with aragonite skeletons (cephalopods, corals, and demosponges) found in Norian patch reefs of the Lycian Taurus (Alakir river valley). (a) Positions of modern representatives of the three main groups (plus nonbiogenic aragonite) based on concentration of chemical elements (b) Distinct compositional changes in aragonite of the Triassic representatives of these three groups: note the important change affecting the cephalopods, whereas coral skeletons are only slightly modified in their composition. a, b: Dauphin *et al.* (1996).

measurements indicate that each of these materials, submitted to identical diagenetic conditions, have undergone a specific and different change.

As a final example, the interpretation of belemnite rostra epitomizes the type of changes in successful methodologies that have resulted from the exceptional development of analytical technology during the past two decades and illustrates how rewarding the high-resolution approach to fossil materials can be. In most cases, a heterogeneity in diagenetic processes results in varying patterns of fossilization within a given specimen. This is familiar to most researchers. Long a major analytical obstacle, disparity in diagenetic change in distinct areas now becomes a major advantage in determining diagenetic pathways by which the initial structures have been progressively transformed. This is illustrated by the two belemnites of Fig. 7.47. Both have areas in which a granular organization is still visible, with abundant traces of the organic grain cortex (Figs. 7.47b, e). However, places also exist where more advanced recrystallization occurs (Figs. 7.47c, f). Thus, AFM images indicate that here, for the first time, a localized and realistic evaluation of the structural changes can be made. The diagenetic changes undergone by a given portion of a skeleton can be linked to chemical or isotopic measurements made in the same place. In applied research, numerical values acquired in these conditions are considerably improved in their reliability. Reciprocally, while studying fossilization processes, it now becomes possible to accurately assess the geochemical consequences of structural modification.

The microstructural observations on belemnite rostra recall isotopic investigations, starting with the Urey *et al.* paper (1951) that marks the beginnings of this approach to using the fossil record. No other example provides a more significant illustration of the considerable changes that have occurred within the past half-century, both in concepts pertaining to

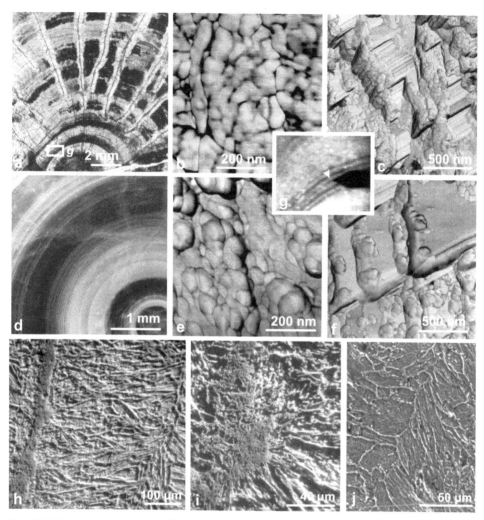

Fig. 7.47. Preservation state in different portions of belemnite rostra. (a–c) Fibrous structures of *Aulacoceras* (specimen also illustrated in Fig. 7.41) here show well-preserved granular structures (b), not seen where the central parts are recrystallized (c); (d–f) Equivalent differences are apparent among calcite rostra of the belemnite (d): rhombohedral units can be observed close to areas containing remains of the original granular material (e–f); (g) This picture of the tip-bearing unit at the end of the AFM cantilever (arrow) during mapping the surface of the *Aulacoceras* sample (a) shows how precise the localization of the imaged areas can be; (h–j) Diagenetic patterns in a coral skeleton from the Santonian of Gosau (Austria); median septal line and limits of the clusters of fibers are clearly defined by an early diagenetic process that has preserved original organic matter. How this process has developed and to what extent this interferes with the original skeletal microstructures can now be investigated at the scale of the biomineralization process itself, the crystallization growth layer. h–j: Sorauf (1999).

biocrystallization and in analytical methods that now allow information to be collected at progressively increasing resolutions. Taken together, these two areas have made unequal progress. However, we have now progressed to the point where further improvement of analytical data requires the precise understanding of calcified structures as biological objects, with an equivalent evaluation of their diagenetic condition as geological samples. Both are now possible and provide unprecedented opportunities for new research.

Innumerable examples are present in the paleontological literature that emphasize the potential of extremely well-preserved fossils to provide information that cannot be found elsewhere. These deserve to be restudied as representatives of organisms living a brief time in the biogeochemical history of the Earth (Figs. 7.47g–i). Many of them are still accessible in museum collections, and therefore are a significant source of information for reconstructions of life history that are accurate, reliable, and significant.

8

Results and perspectives

The result of the initial step that began simply by comparing two types of calcium-carbonate material deposited by biological mechanisms (generally accepted as distinct), this study has moved far afield. Although not yet finished, this research provides truly convergent data regarding the methods of skeletal development that have resulted in calcareous structures produced by organisms. Now it is possible to suggest a generalized model for this bio-mineralization. The model essentially covers structures produced by epithelial mineraliza-tion, but, as this is a broadly shared type of calcification, the model pertains to most invertebrate animals. Coccoliths, unicellular algae with mineralization occurring within Golgi-derived vesicles, are the most notable exception.

In contrast to classical analysis of shell structures that were and are based on descriptions of the sizes and shapes of various mineralized units forming different and distinct shell layers, the methods used here focus rather on the introduction of *time being an essential element of microstructural analysis*. Determining what is synchronous in diverse structures, and still more informative, imaging the mineralized layers that contain actual evidence of the activity of mineralizing organs, now leads to examination of these deposits from a fresh viewpoint, far removed from the traditional approach.

According to this method, the actual growth unit is clearly a mineralized layer and a significant part of this study has been dedicated to the description of the structure and composition of exactly what the biomineralization unit is. This approach is fully applicable to the formation of clearly visible and long-recognized structures, such as prisms or fibers, and also a provides a valid basis for understanding the most complex three-dimensional features, such as cross-lamellar structures (the most widely seen of all microstructural types). Since it summarizes changes in skeletal growth manner and the diversity of organic envelopes, the resulting model also accounts for the laminar elements that occur widely, and form the internal layers of many shells.

The primary result of isolating this elementary growth layer is the recognition of the similarity of its structure in widely separated groups. This is based on the exact observation of several tens of distinct examples distributed over the majority of the major biological groups of invertebrates. Regardless of taxonomy, mineralogy or geometrical complexity, the skeletal growth unit is seen as a mineralized layer having a thickness in the micrometer range, and characterized by the reticulate nature of its crystallized mineral material. Atomic force microscopy (phase-contrast imaging) complemented by high-resolution, synchrotron-based mapping, indicates that organic components are distributed peripherally, and surround granular units that have dimensions in the 10 nm size range. This nanoscale organization of biogenic Ca-carbonate occurs *within every structure* thus far studied.

The rounded surface of reticulate mineral matter never displays any faceted crystal surfaces to reflect crystalline conditions. This is a paradox in that crystalline characteristics plainly do exist. Data on the crystalline state of growth layer components at the ultrastructural or nanometer scale are available using X-ray diffraction methods in high-voltage transmission electron microscopy. These data establish that, in each layered growth unit, the orientation of crystal lattices is continuous between neighboring rounded mineral elements within each growth layer. This indicates to us that *crystallization occurs as the final step in forming the mineralized growth layer*, and thus transforms grains into units having a continuous crystallographic orientation. Therefore, within each microstructural unit, the mineral material is reticulate from a morphological point of view, but crystalline and continuously oriented.

These data have led us to accept a two-step model for growth layer formation. The first step is a secretion phase, during which the mineralizing organ (the molluscan mantle, the basal ectoderm of corallites, etc.) accumulates mineral and organic material between the mineralizing cell layer and the growth surface of previously formed calcareous skeleton. When thickness accumulates to a critical amount, crystallization occurs. Multiple data of microstructures, as well as chemical and biochemical maps of the distribution of skeletal components, demonstrate this continuity of growth layering between adjacent skeletal units. As a result, we conclude that the sequence of events that form one growth layer and its crystallization, is perfectly coordinated over the entire mineralizing area.

Not only is the continuity of these micrometer-thick growth layers over large surfaces striking evidence that such control is exerted at the level of entire organisms, but also, this layered mode of growth explains how organisms successfully develop species-specific sizes and shapes in their skeletal components. The coral skeleton illustrates this especially well. Corallite septa are formed by superposed locally modulated growth layer thicknesses to build specific morphologies. In addition, this precise control on growth layering is exerted throughout the life of the animal. Furthermore, during aging, septal microstructures of septa are commonly progressively modified, thus forming a true ontogenetic sequence at the microstructural level. Newly appearing septa reproduce the microstructural sequence from its original state in the group, then continuously change, finally reaching that level which corresponds to their hierarchical rank in the corallite, and reflect the evolutionary status of the coral species. We can therefore follow the genetic control of biomineralization as it operates in determining the morphology of this structure at the growth layer level. Simultaneously, it exerts control over the development and evolution of growth layers throughout the complete animal. If we understand this level of overall control on coral skeletogenesis, then it should be no surprise that agreement is found between a molecular phylogeny and proposed taxonomic systems based on microstructure.

The layered growth and crystallization process that is seen through very high magnification of skeletal materials also helps explain some long-standing questions with regard to the control of crystallization of microstructural units. Rather than crystals freely growing to specific sizes and shapes within spaces filled by liquid positioned external to the epithelium, organisms effectively limit the dimensions of crystallizing units. More precisely, they limit the thickness of each layer as it is prepared for crystallization. Within the largest microstructural units, such as shell prisms in *Pinna*, a robust organic network is repeatedly produced during each mineralization cycle. It exists within these crystal-like units – the prisms – and can be seen by etching and using fixative dyes. Prisms of *Pinctada* have provided another method of successfully limiting dimensions and volumes that are crystallized. While these prisms are growing in diameter, mineral material is subdivided into three to four distinct sectors, and each crystallizes as a distinct subunit of the prism. It is clear that effective, species-specific arrangements prevent uncontrolled crystallization from occurring. Sometimes however, a failure of this controlling mechanism results in the formation of chemical crystals, such as those rather commonly occurring in fish otoliths. This is reflected by formation of faceted chemically precipitated crystals that have morphologies that are never produced during the normal development of biocrystals in any other group.

Nowhere else is the separation between the formation of mineralized material and shaping of microstructural units more evident than in the structure of the pearl layer, as outlined above. The pearl-sac is an isolated feature with no functional connection to an animal body, so that in the absence of regulation, there frequently results the formation of abnormalities (see above, subsection 5.3.2). Therefore, we see that complete independence does not exist from control by the organic compounds that determine crystallization. Since layered growth results in the superposition of micrometer-thick layers, variations in the polymerization of the specialized molecules that form the limits for microstructural units shape the crystal-like units forming the pearl layer, without any effects of crystallization forces of the mineral. From nanometer to micrometer scales and larger, permanent, biochemical controls at various scales virtually always prevent purely chemical crystallization from happening.

The similarities and differences between mineralizing systems reviewed here indicate the extremely sophisticated nature of the mechanisms that regulate crystallization of biocarbonates. Siliceous structures bring about species-specific mineralized units through deposition of amorphous mineral materials on an organic framework. Precise controls that result in the fascinating and long-recognized level of precision in constructing silicified structures are aided by the poorly reactive nature of silica. The precise control of the size and arrangement of nanometer-sized phosphatic units in bones is less simple, producing complex mineral compositions in bony material. Integration of mineral units into complex bony architectures is controlled at the cellular level. Thus, the concept of microstructure in bone has a very different meaning than when applied to carbonate structures. But it is an interesting fact that the layered mechanism producing enamel provides much better control than those in bone do, with species-specific enamel structures the result.

The specificity of Ca-carbonate biocrystallization mechanisms relies on the ability of the mineralizing organ to form an organic framework to control crystallization over large areas, producing easily identifiable growth layers. From nanoscale granules in the micrometer-thick growth layer to distinct areas mineralized by a particular mineralizing organ extending over meters, as in coral colonies, the mineralizing process is thus controlled. It repeats itself numerous times, and is regulated to be species-specific, thus leading to taxonomy-linked architectures. Since it commonly occurs during formation of the calcareous invertebrate skeleton, such an integration of these levels would appear to be a key concept for interpreting the fossil record.

At the present time, technical capabilities provide ready access to the most basic levels of biomineralization processes. The distinct crystallized and repeated growth layers are actual environment recording units. Tied to this is an expectation that future analytical instruments will be developed to provide data with spatial resolutions of one or several micrometers. This will permit measuring chemical properties in each of these growth layers individually. Additionally, with little added time, the diagenetic condition of these same mineral layers could be checked, thus allowing the selection of optimal spots for more localized measurements. Accurate and reliable measurements will result in much greater understanding of the fine-scale structures and more knowledgeable acquisition of their included data on environmental parameters.

The acceptance of evidence that mineralizing mechanisms are highly sensitive to environmental conditions is fundamental to interpreting fossil faunas over extended time intervals, i.e., geologic time. It is commonly accepted that statistical studies of fossil data provide us with a reliable expression of the diversity of life through geologic time. Now there are an increasing number of experiments that involve the cultivation of calcifying organisms under "nonactualistic conditions" (thus simulating past conditions, or predicted near-future conditions). When examined at the level of individual crystallization layers, the results of these experiments should provide insight into those conditions that determine the presence or absence of particular organisms in the fossil record during any particular geologic time interval. It can be expected that the nanometer-scale interplay between organic and mineral components also will provide data for fine-scale studies of diagenetic processes, and lead to their fuller understanding. In these two critical research areas, the development of a fossil record and its interpretation, the layered crystallization model will contribute to future analysis of fossil carbonates as repositories of biological and/or environmental information.

Understanding the methods of control over the placing of microstructural units into three-dimensional arrangements, and clarifying their role in recording life history now becomes even more important. Research during the past few decades has shown how conservative these controlling mechanisms usually are, and thus they enable phylogenetic lineages to survive major environmental crises. The history of calcifying demosponges is particularly demonstrative of this. Even though skeletal characteristics do not always allow us to precisely place certain microstructural types into commonly accepted taxonomic frameworks, the preservation of their skeletal patterns, including the finest details, does permit our assessing their continuity through time. Thus, a truly fascinating property of biogenic calcareous structures emerges, that their highly sophisticated mechanisms for control of crystallization processes at temporal scales of hours or days can provide us with additional opportunities for understanding their paleobiology. Tied to this is a much improved capacity for analyzing and following evolutionary lineages through geologic time.

References

Abelson P. H. (1954) Amino acids in fossils. *Science* **119**(3096): 576.

Acil Y., Mobasseri A. E., Warnake P. H. *et al.* (2005) Detection of mature collagen in human dental enamel. *Calc. Tissue Int.* **76**: 121–126.

Addadi L., Moradian J., Shaye E. *et al.* (1987) A chemical model for the cooperation of sulfates and carboxylates in calcite crystal nucleation: relevance to biomineralization. *Proc. Nat. Acad. Sci. USA* **84**: 2732–2736.

Addadi L., Raz S., Weiner S. (2003) Taking advantage of disorder: amorphous calcium carbonate and its roles in biomineralization. *Adv. Mater.* **15**(12): 959–970.

Adkins J. F., Boyle E. A., Curry W. B. *et al.* (2003) Stable isotopes in deep-sea corals and a new mechanism for "vital effects". *Geochim. Cosmochim. Acta* **67**(6): 1129–1143.

Aharon P., Chappell J. (1986) Oxygen isotopes, sea level changes and the temperature history of a coral reef environment in New Guinea over the last 10.5 years. *Palaeogeogr. Palaeoclimatol. Palaeoecol.* **56**: 337–379.

Aizenberg J. (2000) Patterned crystallization of calcite in vivo and in vitro. *J. Crystal Growth* **211**: 143–148.

Aizenberg J., Hanson J., Koetzle T. F. *et al.* (1997) Control of macromolecule distribution within synthetic and biogenic single calcite crystals. *J. Am. Chem. Soc.* **119**(5): 881–886.

Aizenberg J., Weaver J. C., Thanawanam M. S. *et al.* (2005) Skeleton of *Euplectella sp.*: structural hierarchy from the nanoscale to the macroscale. *Science* **309**: 275–278.

Albeck S., Aizenberg J., Addadi L. *et al.* (1993) Interactions of various skeletal intracrystalline components with calcite crystals. *J. Am. Chem. Soc.* **115**: 11 691–11 697.

Allemand D., Benazet-Tambutté S. (1996) Dynamics of calcification in the Mediterranean red coral, *Corallium rubrum* (Linnaeus) (Cnidaria, Octocorallia). *J. Exp. Zool.* **276**: 270–278.

Allemand D., Cuif J. P., Watabe N. *et al.* (1994) The organic matrix of skeletal structures of the mediterranean red coral, *Corallium rubrum. Bull. Inst. Océanogr. Monaco*, sp. issue **14**: 129–139.

Alloiteau J. (1952) Madréporaires post-paléozoiques. In *Traité de Paléontologie*, J. Piveteau (ed.). Paris: Masson, Vol. 1, pp. 539–684.

Alloiteau J. (1957) *Contribution à la Systématique des Madréporaires Fossiles*. C.N.R.S. publ., Paris, 462pp.

Al-Sawalmih A. (2007) Crystallographic texture of the arthropod cuticle using synchrotron wide angle X-ray diffraction. MSc dissertation, RWTH Aachen University, 136pp.

Ameye L., Hermann R., Killian C. *et al.* (1999) Ultrastructural localization of proteins involved in sea urchin biomineralization. *J. Histochem. Cytochem.* **47** (9): 1189–1200.

Ameye L., De Becker G., Killian C. *et al.* (2001) Proteins and saccharides of the sea urchin organic matrix of mineralization: characterization and localization in the spine skeleton. *J. Struct. Biol.* **134**: 56–66.

Anderson O. R. (1981) Radiolarian fine structure and silica deposition. In *Silicon and Siliceous Structures in Biological Systems*, T. Simpson & B. E. Volcani (eds.). New York: Springer, pp. 346–379.

Anderson O. R. (2007) Fine structure of silica deposition and the origin of shell components in a testate Amoeba *Netzelia tuberculata*. *J. Eukaryotic Microbiol.* **35**(2): 204–211.

Andrews P. (1990) *Owls, Caves and Fossils*. London: Natural History Museum Publications, 231pp.

Andrews P., Nesbit-Evans E. M. (1983) Small mammal bone accumulations produced by mammalian carnivores. *Paleobiology* **9**(3): 289–307.

Ans R. A., Meyler B. L., Sammelsel'g V. A. (1982) Otolith growth zone structure in the Baltic sprat *Sprattus sprattus balticus* (G. Schneider) (Clupeidae). *J. Ichthyol.* **22**: 156–160.

Arias J. L., Fernandez M. S. (2003) Biomimetic processes through the study of mineralized shells. *Mater. Charact.* **50**: 189–195.

Arias J. L., Carrino D. A., Fernandez M. S. *et al.* (1992) Partial biochemical and immunochemical characterization of avian eggshell extracellular matrices. *Arch. Biochem. Biophys.* **298**(1): 293–302.

Arnaud-Haond S., Goyard E., Vonau V. *et al.* (2007) Pearl formation: persistence of the graft during the entire process of biomineralization. *Mar. Biotechnol.* **9**(1): 113–116.

Azam B., Hemmingsen B., Volcani B. E. (1974) Role of silicon in diatom metabolism. *Arch. Microbiol.* **97**: 103–114.

Bandel K., Spaeth Ch. (1988) Structural differences in the ontogeny of some belemnite rostra. In *Cephalopods – Present and Past*, J. Wiedmann & J. Kullmann (eds.). Stuttgart: Schweizerbart'sche Verlag, pp. 247–271.

Barnes D. J. (1970) Coral skeletons: an explanation of their growth and structure. *Science* **170**: 1305–1308.

Baron R. (1989) Molecular mechanism of bone resorption by the osteoclast. International Association for Dental Research, General Session 66. Canadian Association for Dental Research. Ann. Meet. 12. Satellite Symp. Vol. **224**(2), pp. 317–324.

Baronnet A., Cuif J. P., Dauphin Y. *et al.* (2008) Crystallization of biogenic Ca-carbonate within organo-mineral micro-domains. Structure of the calcite prisms of the pelecypod *Pinctada margaritifera* (Mollusca) at the submicron to nanometer ranges. *Mineral. Mag.* **72**: 617–626.

Barrick R. E., Showers W. J. (1995) Thermophysiology of *Tyrannosaurus rex*: evidence from oxygen isotopes. *Science* **265**: 222–224.

Barskov I. S. (1973) Protoconch structure and shell ontogeny of the belemnites (Coleoidea, Cephalopoda) (in Russian). *Dokl. Akad. Nauk CCCP* **208**(2): 439–442.

Bassett M. G., Kaljo D., Teller L. (1989) The Baltic region. In *A Global Standard for the Silurian System*, C. H. Holland & M. G. Bassett (eds.). Nat. Mus. Wales Geol. Ser. 9, 158–170.

Bates R. L., Jackson J. A. (1980) *Glossary of Geology*. Falls Church: American Geological Institute, 751pp.

Bathurst R. G. C. (1971) Carbonate sediments and their diagenesis. In *Developments in Sedimentology*, Vol. 12. Amsterdam: Elsevier, 620pp.

Bauch D., Darling K., Simstich J. *et al.* (2003) Palaeoceanographic implications of genetic variation in living North Atlantic *Neogloboquadrina pachyderma*. *Nature* **424**: 299–302.

Bé A. W. H., Ericson D. B. (1963) Aspects of calcification in planktonic Foraminifera. In: Comparative biology of calcified tissues, S. L. Leroy & M. L. Moss (eds.). *Ann. New York Acad. Sci.* **109**: 337–350.

Belcher A. M., Wu X. H., Christensen R. J. *et al.* (1996) Control of crystal phase switching and orientation by soluble mollusc-shell protein. *Nature* **381**: 56–58.

Benayahu Y. (1985) Faunistic composition and patterns in the distribution of soft corals (Octocorallia, Alcyonacea) along the coral reefs of Sinai penninsula. In *Proc. 5th Int. Coral Reef Congress*, Tahiti, Vol. **6**, pp. 255–260.

Beniash E., Aizenberg J., Addadi L. *et al.* (1997) Amorphous calcium carbonate transforms into calcite during sea urchin larval spicule growth. *Proc. R. Soc. London B* **264**: 461–465.

Benson S., Benson N. C., Wilt F. H. (1986) The organic matrix of the skeletal spicule of sea urchin *Strongylocentrotus purpuratus* embryos. *J. Cell Biol.* **102**: 1878–1886.

Benson S., Jones E., Crise-Benson N. *et al.* (1983) Morphology of the organic matrix of the spicule of the sea urchin larva. *Exp. Cell Res.* **148**: 249–253.

Bergquist P. R. (1978) *Sponges.* London: Hutchinson & Co, 268pp.

Berman A., Addadi L., Kvick A. *et al.* (1990) Intercalation of sea urchin proteins in calcite: study of a crystalline composite material. *Science* **250**: 664–667.

Berner R. A. (1980) *Early Diagenesis: A Theoretical Approach.* Princeton series in geochemistry. Princeton, NJ: Princeton University Press, 256pp.

Bevelander G., Nakahara H. (1969) An electron microscope study of the formation of the nacreous layer in the shell of certain Bivalve mollusks. *Calc. Tissue Res.* **3**: 84–92.

Bevelander G., Nakahara H. (1980) Compartment and envelope formation in the process of biological mineralization. In: The mechanisms of biomineralization in animals and plants*, Proc. 3rd Biomineralization Symp.*, M. Omori & N. Watabe (eds.). Kanagawa: Tokai University Press, pp. 19–27.

Bezares J., Asaro R. J., Hawley M. (2008) Macromolecular structure of the organic framework of nacre in *Haliotis rufescens*. Implications for growth and mechanical behavior. *J. Struct. Biol.* **163**: 61–75.

Bidder I. G. P. (1898) The skeleton and classification of calcareous sponges. *Proc. R. Soc. London* **64**: 61–76.

Blamart D., Escoubeyrou K., Juillet-Leclerc A. *et al.* (2002) Composition isotopique δ^{18}O-δ^{13}C des otolithes des populations de poissons récifaux de Taiaro (Tuamotu, Polynésie française): implications isotopiques et biologiques. *C. R. Biologie* **325**: 99–106.

Blamart D., Rollion-Bard C., Cuif J. P. *et al.* (2005) C and O isotopes in a deep-sea coral (*Lophelia pertusa*) related to skeletal microstructure. In *Cold-water Corals and Ecosystems*, A. Freiwald & J. M. Roberts (eds.). Berlin: Springer-Verlag, pp. 1005–1020.

Bocquet-Védrine J. (1965) Etude du tégument et de la mue chez le Cirripède operculé *Elminus modestis* Darwin. *Arch. Zool. Exper. Gen.* **105**: 30–76.

Bøggild O. B. (1930) The shell structure of the molluscs. *D. Kgl. Danske Vidensk. Selsk. Skr., Naturvidensk. og Mathem.* **9**(2/2): 231–326.

Borelli G., Mayer-Gostan N., de Pontual H. *et al.* (2001) Biochemical relationships between endolymph and otolith matrix in the trout (*Oncorhynchus mykiss*) and turbot (*Psetta maxima*). *Calc. Tissue Int.* **69**(6): 356–364.

Botha J., Chinsamy A. (2000) Growth patterns deduced from the bone histology of the cynodonts *Diademodon* and *Cynognathus*. *J. Vert. Pal.* **20**(4): 705–711.

Bouligand Y. (1966) Le tégument de quelques Copépodes et ses dépendances musculaires et sensorielles. *Mém. Mus. Natn. Hist. Nat Paris*, sér. A, **40**: 189–206.

Bouligand Y. (1972) Twisted fibrous arrangement in biological materials and cholesteric mesophases. *Tissue & Cell* **4**(2): 189–217.

Bourget E. (1987) Barnacle shells composition, structure and growth. In *Crustacean, Issue 5: Barnacle Biology*, A. J. Southward (ed.). Rotterdam: Balkema, pp. 267–287.

Bourne G. C. (1899) Studies on the structure and formation of the calcareous skeleton of the Anthozoa. *Quart. J. Microsc. Sci.* **41**: 499–547.

Bovee E. C. (1981) Distribution and forms of siliceous structures among Protozoa. In *Silicon and Siliceous Structures in Biological Systems*, T. L. Simpson & B. E. Volcani (eds.). New York: Springer, pp. 233–279.

Bowerbank J. S. (1844) On the structure of the shells of molluscan and conchyferous animals. *Trans. Microsc. Soc. London* **1**: 123–152.

Bradford E. W. (1967) Microanatomy and histochemistry of dentine. In *Structural and Chemical Organization of Teeth*, A. E. W. Miles (ed.). New York: Academic Press, Vol. II, pp. 3–34.

Briggs D. E. G. (2003) The role of decay and mineralization in the preservation of soft-bodied fossils. *Annu. Rev. Earth Planet. Sci.* **31**: 275–301.

Briggs D. E. G., Kear J. (1994) Decay and mineralization of shrimps. *Palaios* **9**(5): 431–456.

Bryan W. H., Hill D. (1941) Spherulitic crystallization as a mechanism of skeletal growth in the hexacorals. *Proc. R. Soc. Queensland* **52**(9): 78–91.

Bubel A. (1975) An ultrastructural study of the mantle of the barnacle *Elminus modestis* Darwin in relation to shell formation. *J. Exp. Mar. Biol. Ecol.* **20**: 287–334.

Buddemeier R. W., Kinzie R. A. (1975) The chronometric reliability of contemporary corals. In *Growth Rhythms and the History of the Earth*, G. D. Rosenberg & S. K. Runcorn (eds.). New York: Wiley, pp. 135–147.

Burger C., Zhou H., Wang H. *et al.* (1966) Le tégument de quelques Copépodes et ses dépendances musculaires et sensorielles. *Mém. Mus. Natn. Hist. Nat. Paris*, Sér. A, **40**: 189–206.

Bütschli O. (1901) Einige beobachtungen über Kiesel-und Kalknadelm von Spongien. *Z. Wiss. Zool.* **69**: 235–286.

Butterfield N. J. (1990) Organic preservation of non-mineralizing organisms and the taphonomy of the Burgess Shale. *Paleobiology* **16**: 272–286.

Cachon J., Cachon M. (1972) Les modalités du dépôt de la silice chez les radiolaires. *Archiv fur Protistenkunde* **114**: 1–13.

Cameron J. N. (1989) Post-molt calcification in the blue crab *Callinectes sapidus*: timing and mechanism. *J. Exp. Biol.* **143**: 285–304.

Carpenter K. (2007) How to make a fossil: Part 2. Dinosaur mummies and other soft tissues. *J. Paleont. Sci.* C.07.002, 23pp.

Carpenter S. J., Lohmann K. C. (1992) Sr/Mg ratios of modern marine calcite: empirical indicators of ocean chemistry and precipitation rate. *Geochim. Cosmochim. Acta* **56**: 1837–1849.

Carpenter S. J., Lohmann K. C. (1995) $\delta^{18}O$ and $\delta^{13}C$ values of modern brachiopod shells. *Geochim. Cosmochim. Acta* **59**: 3749–3764.

Carpenter W. (1845) On the microscopic structure of shells: Part I. *Br. Assoc. Adv. Sci.*, rep. **14**: 1–24.

Carpenter W. (1847) On the microscopic structure of shells: Part II. *Br. Assoc. Adv. Sci.*, rep. **17**: 93–134.

Carr A., Kemp A., Tibbetts I. *et al.* (2006) Microstructure of pharyngeal tooth enameloid in the parrotfish *Scarus rivulatus* (Pisces: Scaridae). *J. Microsc.* **221**(1): 8–16.

Carter H. J. (1879) On the structure of Stromatopora. *Ann. Mag. Nat. Hist.*, ser. 5(4): 253–265.

Cary L. R. (1918) The Gorgonacea as a factor in the formation of a coral reef. *Carnegie Inst. Wash., Pap. Mar. Biol.* 9: 341–362.

Castaing R., Slodzian G. (1962) Microanalyse par émission ionique secondaire. *J. Microscopie* 1: 395–410.

Cayeux L. (1916) Introduction à l'étude pétrographique des roches sédimentaires. *Mém. Carte Géol. France, vol. 1 texte, vol. 2 atlas.*

Chave K. E. (1954) Aspects of the biogeochemistry of magnesium: 1. Calcareous marine organisms. *J. Geol.* 62: 266–283.

Chave K. E. (1984) Physics and chemistry of biomineralization. *Annu. Rev. Earth Planet. Sci.* 12: 203–205.

Checa A. G., Esteban-Delgado F. J., Ramirez-Rico J. *et al.* (2009a) Crystallographic reorganization of the calcitic prismatic layer of oysters. *J. Struct. Biol.* doi: 10.1016/j.jsb.2009.06.009.

Checa A. G., Ramirez-Rico J., Gonzalez-Segura A. *et al.* (2009b) Nacre and false nacre (foliated aragonite) in extant monoplacophorans (=Tryblidiia: Mollusca). *Naturwissenschaften* 96: 111–122.

Chen C. A., Odorico D., Ten Lohuis M. *et al.* (1995) Systematic relationships within the Anthozoa (Cnidaria, Anthozoa) using the 5″ end of the 28S rRNA. *Mol. Phylogen. Evol.* 4(2): 175–183.

Child A. M. (1995) Towards an understanding of the decomposition of bone in the archaeological environment. *J. Archaeol. Sci.* 22: 165–174.

Chombard C., Tillier A., Boury-Esnault N. *et al.* (1997) Polyphyly of "sclerosponges" (Porifera, Demospongiae) supported by 28S ribosomal sequences. *Biol. Bull.* (Woods Hole) 193: 359–367.

Clarke F. W., Wheeler W. C. (1915) The composition of brachiopod shells. *Proc. Nat. Acad. Sci. USA* 1(5): 262–266.

Clarke F. W., Wheeler W. C. (1922) The inorganic constituents of marine invertebrates. *U.S. Geol. Surv. Prof. Pap.* 124, p. 62.

Cobabe E. A., Pratt L. M. (1995) Molecular and isotopic compositions of lipids in bivalve shells: a new prospect for molecular palaeontology. *Geochim. Cosmochim. Acta.* 59(1): 87–95.

Cohen A. L., McConnaughey T. A. (2003) Geochemical perspectives on coral mineralization. In *Mineralogy & Geochemistry: Biomineralization*, P. M. Dove, J. J. De Yoreo & S. Weiner (eds.). Washington DC: Mineralogical Society of America, Vol. 54, pp. 151–187.

Cohen A. L., Gaetani G. A., Lundäl V. T. *et al.* (2006) Compositional variability in a cold-water scleractinian, *Lophelia pertusa*: new insights into "vital effects". *Geochem. Geophys. Geosyst.* 7: doi: 10.1029/2006GC001354.

Cölfen H. (2007) Non classical crystallization. In *Biomineralization from Paleontology to Material Sciences. Proc. 9th Int. Symp. on Biomineralization*, J. L. Arias & M. S. Fernandez (eds.). Santiago: Editorial Universitaria Santiago, pp. 515–526.

Cölfen H., Antonietti M. (2008) *Mesocrystals and Nonclassical Crystallization.* New York: Wiley, 276pp.

Collins M. J., Curry G. B., Muyzer G. *et al.* (1988) Serotaxonomy of skeletal macromolecules in living terebratulid brachiopods. In *Immunological Approaches in Geological Research*, G. Muyzer (ed.). Meppel, The Netherlands: Krips repro., pp. 41–59.

Collins M. J., Muyzer G., Westbroek P. *et al.* (1991) Preservation of fossil biopolymeric structures: conclusive immunological evidence. *Geochim. Cosmochim. Acta* **55**: 2253–2257.

Collins M. J., Riley M. S., Child A. M. *et al.* (1995a) A basic mathematical simulation of the chemical degradation of ancient collagen. *J. Archaeol. Sci.* **22**: 175–183.

Collins M. J., Child A. M., Riley M. S. *et al.* (1995b) Comparing the survival of the bone proteins osteocalcin and collagen. In *Organic Geochemistry: Developments and Applications to Energy, Climate, Environment and Human History*, 17th Int. Meet. Organic Geochem., San Sebastian, 4–8 Sept. 1995, J. O. Grimalt & C. Dorronsoro (eds.), AIGOA, pp. 719–722.

Collins M. J., Nielsen-Marsh C. M., Hiller J. *et al.* (2002) The survival of organic matter in bone: a review. *Archaeometry* **44**(3): 383–394.

Compère P., Jaspar-Versali M. F., Goffinet G. (2002) Glycoproteins from the cuticle of the atlantic shore crab *Carcinus maenas*: I. electrophoresis and western-blot analysis by use of lectins. *Biol. Bull.* **202**: 61–73.

Constantz B. R. (1986) Coral skeleton construction: a physiochemically dominated process. *Palaios* **1**: 152–157.

Constantz B., Weiner S. (1988) Acidic macromolecules associated with the mineral phase of Scleractinian coral skeletons. *J. Exp. Zool.* **248**: 253–258.

Corrigan J. J. (1969) D-amino acids in animals. *Science* **164**: 142–149.

Costlow J. D. Jr. (1956) Shell development in *Balanus improvisus* Darwin. *J. Morph.* **99**: 359–415.

Crawford S. A., Higgins M. J., Mulvaney P. *et al.* (2001) Nanostructure of the diatom frustule as revealed by atomic force and scanning electron microscopy. *J. Phycol.* **37**: 543–557.

Crenshaw M. A. (1972a) The inorganic composition of molluscan extrapallial fluid. *Biol. Bull.* **143**: 506–512.

Crenshaw M. A. (1972b) The soluble matrix from *Mercenaria mercenaria* shell. *Biomineralization* **6**: 6–11.

Crenshaw M. A. (1980) Mechanisms of shell formation and dissolution. In *Skeletal Growth of Aquatic Organisms*, D. C. Rhoads & R. A. Lutz (eds.). New York: Plenum Press, pp. 115–132.

Crenshaw M. A., Ristedt H. (1975) The histochemical localization of reactive groups in septal nacre from *Nautilus pompilius* L. In *The Mechanisms of Mineralization in the Invertebrates and Plants*, N. Watabe & K. M. Wilbur (eds.). Belle Baruch Library of Marine Science, Number 5. Columbia, SC: University of South Carolina Press, pp. 355–367.

Croce G., Frache A., Milanesio M. *et al.* (2003) Fibre diffraction study of spicules from marine sponges. *Microsc. Res. Tech.* **62**: 378–381.

Croce G., Frache A., Milanesio M. *et al.* (2004) Structural characterization of siliceous spicules from marine sponges. *Biophysical J.* **86**: 526–534.

Cuif J. P. (1965) Microstructure du genre *Gigantostylis* Frech. *C. R. Acad. Sc. Paris* **261**: 1046–1049.

Cuif J. P. (1972) Recherches sur les Madréporaires du Trias: I. Famille des Stylophyllidae. *Bull. Mus. Natn. Hist. Nat. Paris*, Sér. 3, **97**: 211–291.

Cuif J. P. (1973) Mise en évidence des premières sclérosponges fossiles dans le Trias des Dolomites. *C. R. Acad. Sci. Paris*, Sér. D, **277**: 2333–2336.

Cuif J. P. (1974) Recherches sur les Madréporaires du Trias: II. Astraeoida. Révision des genres *Montlivaltia* et *Thecosmilia*. Etude de quelques types structuraux du Trias de Turquie. *Bull. Mus. Natn. Hist. Nat. Paris*, Sér. 3, **275**: 293–400.

Cuif J. P. (1975a) Caractères et affinités de *Gallitellia*, nouveau genre de madréporaires du Ladino-carnien des Dolomites. In: 2ème Congr. Int. Cnidaires Fossiles, Paris. *Mém. B.R.G.M.* **89**: 256–263.

Cuif J. P. (1975b) Recherches sur les Madréporaires du Trias: III. Etude des structures pennulaires chez les madréporaires triasiques. *Bull. Mus. Natn. Hist. Nat. Paris*, 3è Sér., **44**: 45–127.

Cuif J. P. (1975c) Caractères morphologiques, microstructuraux et systématiques des Pachythecalidae, nouvelle famille de Madréporaires triasiques. *Geobios* **8**(3): 157–180.

Cuif J. P. (1976) Recherches sur les madréporaires du Trias: IV. Formes cério-méandroides et thamnastéroides du Trias des Alpes et du Taurus sud-anatolien. *Bull. Mus. Natn. Hist. Nat. Paris*, Sér. 3, **381**: 65–196.

Cuif J. P. (1977) Arguments pour une relation phylétique entre les madréporaires paléozoiques et ceux du Trias. Implications de l'analyse microstructurale des Madréporaires triasiques. *Mém. Soc. Geol. Fr.* N. S. **56**(129): 1–54.

Cuif J. P. (1980) Microstructure versus morphology in the skeleton of triassic scleractinian corals. *Acta Palaeont. Pol.* **25**(3–4): 361–374.

Cuif J. P. (1983) Chaetetida à microstructure sphérolitique dans le Trias supérieur de Turquie. *C. R. Acad. Sc. Paris*, Sér. II, **296**: 1469–1472.

Cuif J. P., Dauphin Y. (1994) La perliculture polynésienne: recherche sur les facteurs de qualité des produits. *Revue de Gemmologie* A.F.G. **120**: 2–5.

Cuif J. P., Dauphin Y. (1996) Occurrence of mineralization disturbances in nacreous layers of cultivated pearls produced by *Pinctada margaritifera* var. *Cumingi* from French Polynesia. Comparison with reported shell alterations. *Aquatic Liv. Res.* **9**: 187–193.

Cuif J. P., Dauphin Y. (1998) Microstructural and physico-chemical characterizations of the "centers of calcification" in the septa of some recent Scleractinian corals. *Paläont. Zeit.* **72** (3/4): 257–270.

Cuif J. P., Dauphin Y. (2005a) The environmental recording unit in coral skeletons: a synthesis of structural and chemical evidences for a biochemically driven, stepping-growth process in fibres. *Biogeosciences* **2**: 61–73.

Cuif J. P., Dauphin Y. (2005b) The two-step mode of growth in the scleractinian coral skeletons from the micrometre to the overall scale. *J. Struct. Biol.* **150**: 319–331.

Cuif J. P., Gautret P. (1991) Taxonomic value of microstructural features in calcified skeletons of fossil sponges. In *Fossil and Recent Sponges*, J. Reitner & H. Keupp (eds.) Berlin: Springer Verlag, pp. 159–169.

Cuif J. P., Raguideau A. (1982) Observations sur l'origine de l'individualité cristallographique des prismes de *Pinna nobilis* L. *C. R. Acad. Sc. Paris*, Sér. II, **295**: 415–418.

Cuif J. P., Dauphin Y., Lefèvre R. (1977) Rapport de la localisation du strontium, magnesium et sodium avec la minéralisation et la microstructure de trois rostres d'Aulacocerida triasiques. *C. R. Acad. Sc. Paris*, Sér. D **285**: 1033–1036.

Cuif J. P., Dauphin Y., Denis A. *et al.* (1980) Continuité et périodicité du réseau organique intra-prismatique dans le test de *Pinna muricata* L. (Lamellibranche). *C. R. Acad. Sc. Paris D* **290**: 759–763.

Cuif J. P., Denis A., Gaspard D. (1981) Recherche d'une méthode d'analyse ultrastructurale des tests carbonatés d'invertebrés. *Bull. Soc. Géol. Fr.* 9, 28, **5**: 525–534.

Cuif J. P., Dauphin Y., Denis A. *et al.* (1983) Etude des caractéristiques de la phase minérale dans les structures prismatiques du test de quelques mollusques. *Bull. Mus. Natn. Hist. Nat. Paris*, 4è Sér. A 3: 679–717.

Cuif J. P., Dauphin Y., Denis A. *et al.* (1987) Résultats récents concernant l'analyse des biocristaux carbonatés: implications biologiques et sédimentologiques. *Bull. Soc. Géol. Fr., Sér.* **8** III(2): 269–288.

Cuif J. P., Gautret P., Laghi G. F. *et al.* (1990) Recherche sur la fluorescence UV du squelette aspiculaire chez les Démosponges calcitiques triasiques. *Geobios* **23**(1): 21–31.

Cuif J. P., Gautret P., Marin F. (1991) Correlation between the size of crystals and the molecular weight of organic fractions in the soluble matrices of mollusc, coral and sponge carbonate skeletons. In *Mechanisms and Phylogeny of Mineralization in Biological Systems*, S. Suga & H. Nakahara (eds.). New York: Springer Verlag, pp. 391–395.

Cuif J. P., Denis A., Gautret P. *et al.* (1992) Recherches sur l'altération diagénetique des biominéralisations carbonatées: évolution de la phase organique intrasquelettique dans des polypiers aragonitiques de Madréporaires du Cénozoique (Bassin de Paris) et du Trias supérieur (Dolomites et Turquie). *C. R. Acad. Sc. Paris*, Sér. II, **314**: 1097–1102.

Cuif J. P., Dauphin Y., Denis A. *et al.* (1996) The organo-mineral structure of coral skeletons: a potential source of new criteria for Scleractinian taxonomy. *Bull. Inst. Océanogr. Monaco*, special issue **14**(4): 359–367.

Cuif J. P., Dauphin Y., Gautret P. (1997a) Biomineralization features in scleractinian coral skeletons: source of new taxonomic criteria. *Bol. R. Soc. Esp. Hist. Nat.* (Sec. Geol.) **92**(1–4): 129–141.

Cuif J. P., Dauphin Y., Denis A. *et al.* (1997b) Facteurs de la diagenèse précoce des biominéraux: exemple d'un polypier de *Porites* de Nouvelle Calédonie. *Geobios* **20**: 171–179.

Cuif J. P., Dauphin Y., Freiwald A. *et al.* (1999a) Biochemical markers of zooxanthellae symbiosis in soluble matrices of skeleton of 24 Scleractinia species. *Comp. Biochem. Physiol.* A **123**(3): 269–278.

Cuif J. P., Dauphin Y., Gautret P. (1999) Compositional diversity of soluble mineralizing matrices in some recent coral skeletons compared to fine-scale growth structures of fibres: discussion of consequences for biomineralization and diagenesis. *Int. J. Earth Sciences* **88**: 582–592.

Cuif J. P., Dauphin Y., Denis A. (2003) Biomineralization patterns in fibres and centres of calcification in coral skeletons. In *Biomineralization (BIOM2001): Formation, Diversity, Evolution and Application*, I. Kobayashi & H. Ozawa (eds.). Kanagawa: Tokai University Press, pp. 45–49.

Cuif J. P., Dauphin Y., Doucet J. *et al.* (2003a) XANES mapping of organic sulfate in three scleractinian coral skeletons. *Geochim. Cosmochim. Acta* **67**(1): 75–83.

Cuif J. P., Lecointre G., Perrin C. *et al.* (2003b) Patterns of septal biomineralization in Scleractinia compared with their 28S rRNA phylogeny: a dual approach for a new taxonomic framework. *Zool. Scripta* **32**(5): 459–473.

Cuif J. P., Dauphin Y., Berthet P. *et al.* (2004) Associated water and organic compounds in coral skeletons: quantitative thermogravimetry coupled to infrared absorption spectrometry. *Geochem. Geophys. Geosyst.* **5**(11): doi: 10.1029/2004/GC000783.

Cuif J. P., Ball A. D., Dauphin Y. *et al.* (2008a) Structural, mineralogical, and biochemical diversity in the lower part of the pearl layer of cultivated seawater pearls from Polynesia. *Microsc. Microanal.* **14**: 405–417.

Cuif J. P., Dauphin Y., Farre B. *et al.* (2008b) Distribution of sulphated polysaccharides within calcareous biominerals indicates a widely shared layered growth-mode for the invertebrate skeletons and suggests a two-step crystallization process for the mineral growth units. *Mineral. Mag.* **72**(1): 233–237.

Cuif J. P., Dauphin Y., Meibom A. *et al.* (2008c) Fine-scale growth patterns in coral skeletons: biochemical control over crystallization of aragonite fibres and assessment of early diagenesis. In *Biogeochemical Controls on Palaeoceanographic Environmental Proxies*, W. E. N. Austin & R. H. James (eds.), Geol. Soc. Lond. Spec. Pub. 303: 87–96.

Currey J. D., Nichols D. (1967) Absence of organic phase in echinoderm calcite. *Nature* **214**: 81–83.

Curry G. B., Quinn R., Collins M. J. *et al.* (1991) Immunological responses from brachiopod skeletal macromolecules: a new technique for assessing taxonomic relationships using shells. *Lethaia* **24**: 399–407.

Cusack M., Williams A. (2001) Chemico-structural differentiation of the organo-calcitic shells of Rhynchonellate brachiopods. In *Brachiopods Past and Present*, C. H. C. Brunton, L. R. M. Cocks & S. L. Long (eds.), The Systematics Association, Sp. Vol. Ser. 63., Chapter 3. London: Taylor & Francis, pp. 17–27.

Cusack M., Williams A., Buckman J. O. (1999) Chemico-structural evolution of linguloid brachiopod shells. *Palaeontology* **42**(5):799–840.

Cusack M., Fraser A. C., Stachel T. (2003) Magnesium and phosphorus distribution in the avian eggshell. *Comp. Biochem. Physiol. B* **134**: 63–69.

Cusack M., Dalbeck P., Lee M. R. *et al.* (2007) Carbonate EBSD: a new tool for understanding shell growth and ancient visual systems. *Scanning* **29**(2): 84–85.

Cusack M., Dauphin Y., Cuif J. P. *et al.* (2008a) Micro-XANES mapping of sulphur and its association with magnesium and phosphorus in the shell of the brachiopod, *Terebratulina retusa*. *Chem. Geol.* **253**: 172–179.

Cusack M., Parkinson D., Perez-Huerta A. *et al.* (2008b) Relationship between $\delta^{18}O$ and minor element composition of *Terebratalia transversa*. *Trans. R. Soc. Edinburgh* **98**: 443–449.

Cusack M., Pérez-Huerta A., Chung P. *et al.* (2008c) Oxygen isotope equilibrium in brachiopod shell fibres in the context of biological control. *Miner. Mag.* **72**(1): 239–242.

Cusack M., Dauphin Y., Chung P. *et al.* (2008d) Multiscale structure of calcite fibres of the shell of the brachiopod *Terebratulina retusa*. *J. Struct. Biol.* **164**: 96–100.

Cushman J. A. (1925) An introduction to the morphology and classification of the Foraminifera. *Smithsonian Misc. Coll.* **77**(4): 1–75.

Dalbeck P., Cusack M. (2006) Crystallography (Electron Backscatter Diffraction) and chemistry (Electron Probe Microanalysis) of the avian eggshell. *Cryst. Growth Design* **6**(11): 2558–2562.

Dauphin Y. (1979) Organisation microstructurale de l'os de seiche (Cephalopoda – Dibranchiata). *C. R. Acad. Sc. Paris, Sér.* D **288**: 619–622.

Dauphin Y. (1981) Microstructures des coquilles de Céphalopodes: II. La seiche (Dibranchiata, Decapoda). *Palaeontographica A* **176**: 35–51.

Dauphin Y. (1983) Microstructure du phragmocône du genre triasique *Aulacoceras* (Cephalopoda – Coleoidea): remarques sur les homologies des tissus coquilliers chez les Céphalopodes. *N. Jb. Geol. Paläont. Abh.* **165**(3): 418–437.

Dauphin Y. (1984) Microstructures des coquilles de Céphalopodes: IV. Le "rostre" de *Belosepia* (Dibranchiata). *Paläont. Zeit.* **58**(1–2): 99–117.

Dauphin Y. (1985a) Implications of a microstructural comparison in some fossil and recent coleoid cephalopod shells. *Palaeontographica A* **191**(1–3): 69–83.

Dauphin Y. (1985b) Microstructural studies on cephalopod shells: V. The apical part of *Beloptera* (Dibranchiata, Tertiary). *N. Jb. Geol. Paläont. Abh.* **170**(3): 323–341.

Dauphin Y. (1986a) Microstructural studies on Cephalopod shells: VII. The rostrum of *Vasseuria* (Dibranchiata). *Revue de Paléobiologie* **5**(1): 47–56.

Dauphin Y. (1986b) Microstructures des coquilles de Céphalopodes: VI. La partie apicale de *Belopterina* (Coleoidea). *Bull. Mus. Natn. Hist. Nat. Paris*, 4è Sér. 8, sect. C, **1**: 53–75.

Dauphin Y. (1989a) Implications de l'analyse chimique élémentaire de dents de reptiles actuels et fossiles. *C. R. Acad. Sci. Paris*, Sér. II **309**: 927–932.

Dauphin Y. (1989b) Microstructures et morphologie fonctionnelle: exemple de l'émail dentaire. *Revue de Paleobiologie* **8**(2): 357–363.

Dauphin Y. (1990) Microstructures et composition chimique des coquilles d'oeufs d'oiseaux et de reptiles: 1. Oiseaux actuels. *Palaeontographica A* **214**(1/2): 1–12.

Dauphin Y. (1991) Chemical composition of reptilian teeth: 2. Implications for paleodiets. *Palaeontographica A* **219**(4/6): 97–105.

Dauphin Y. (1996) The organic matrix of coleoid cephalopod shells: molecular weights and isoelectric properties of the soluble matrix in relation to biomineralization processes. *Mar. Biol.* **125**(3): 525–529.

Dauphin Y. (1998) Comparaison de l'état de conservation des phases minérales et organiques d'os fossiles. Implications pour les reconstitutions paléoenvironnementales et phylétiques. *Ann. Paléontol.* **84**(2): 215–239.

Dauphin Y. (1999a) Infrared spectra and elemental composition in recent biogenic calcites: relationships between the v4 band wavelength and Sr and Mg concentrations. *Appl. Spectrosc.* **53**(2): 184–190.

Dauphin Y. (1999b) Evolution des teneurs en Mg de la dentine des dents à croissance continue de mammifères au cours des différentes phases de la formation d'un site fossilifère. *N. Jb. Geol.Paläont. Mh.* **2**: 101–121.

Dauphin Y. (2001a) Comparative studies of skeletal soluble matrices from some scleractinian corals and molluscs. *Int. J. Biol. Macromol.* **28**: 293–304.

Dauphin Y. (2001b) Caractéristiques de la phase organique soluble des tests aragonitiques des trois genres de céphalopodes actuels. *N. Jb. Geol. Paläont. Mh.* **2**: 103–123.

Dauphin Y. (2001c) Nanostructures de la nacre des tests de céphalopodes actuels. *Paläont. Zeit.* **75**(1): 113–122.

Dauphin Y. (2002a) Structures, organo-mineral compositions and diagenetic changes in biominerals. *Curr. Opin. Colloid Interface Sci.* **7**: 133–138.

Dauphin Y. (2002b) Implications de la diversité de composition des phases organiques solubles extraites des squelettes carbonatés. *Bull. Soc. Géol. Fr.* **173**(4): 307–315.

Dauphin Y. (2002c) Fossil organic matrices of the Callovian aragonitic ammonites from Lukow (Poland): location and composition. *Int. J. Earth Sci.* **91**: 1071–1080.

Dauphin Y. (2003) Soluble organic matrices of the calcitic prismatic shell layers of two pteriomorphid bivalves: *Pinna nobilis* and *Pinctada margaritifera. J. Biol. Chem.* **278**(17): 15 168–15 177.

Dauphin Y. (2005) Biomineralization. In *Encyclopedia of Inorganic Chemistry*, R. B. King (ed.). New York: Wiley & Sons, Vol. I, pp. 391–404.

Dauphin Y. (2006a) Mineralizing matrices in the skeletal axes of two *Corallium* species (Alcyonacea). *Comp. Biochem. Physiol. A* **145**: 54–64.

Dauphin Y. (2006b) Structure and composition of the septal nacreous layer of *Nautilus macromphalus* L. (Mollusca, Cephalopoda). *Zoology* **109**: 85–95.

Dauphin Y. (2008) The nanostructural unity of mollusc shells. *Mineral. Mag.* **72**(1): 243–246.

Dauphin Y., Cuif J. P. (1980) Implications systématiques de l'analyse microstructurale des rostres de trois genres d'Aulacocerida triasiques (Cephalopoda-Coleoidea). *Palaeontographica* **A169**: 28–50.

Dauphin Y., Denis A. (2000) Structure and composition of the aragonitic crossed lamellar layers in six species of Bivalvia and Gastropoda. *Comp. Biochem. Physiol.* A **126**: 367–377.

Dauphin Y., Denys C. (1994) Teneurs en Mg de la dentine et mode de croissance des dents: le cas des Rongeurs et des Lagomorphes. *C. R. Acad. Sci. Paris*, Sér. II **318**: 705–711.

Dauphin Y., Dufour E. (2003) Composition and properties of the soluble organic matrix of the otolith of a marine fish: *Gadus morhua* Linne, 1758 (Teleostei, Gadidae). *Comp. Biochem. Physiol.* A **134**: 551–561.

Dauphin Y., Keller J. P. (1982) Mise en évidence d'un type microstructural coquillier spécifique des Céphalopodes dibranchiaux. *C. R. Acad. Sci. Paris*, Sér. II **294**: 409–412.

Dauphin Y., Kervadec G. (1994) Comparaison des diagenèses subies par les phases minérale et protéique soluble des tests de Mollusques Céphalopodes Coleoides. *Palaeontographica* A**232**(4/6): 85–98.

Dauphin Y., Marin F. (1995) The compositional analysis of recent cephalopod shell carbohydrates by Fourier transform infrared spectrometry and high performance anion exchange-pulsed amperometric detection. *Experientia* **51**: 278–283.

Dauphin Y., Perrin C. (1992) Mise en évidence de la présence de matière organique dans un ciment d'aragonite botryoidale par spectrométrie infrarouge à transformée de Fourier (FTIR). *N. Jb. Geol. Paläont. Abh.* **186**(3): 301–319.

Dauphin Y., Williams C. T. (2007) The chemical compositions of dentine and enamel from recent reptile and mammal teeth: variability in the diagenetic changes of fossil teeth. *Cryst. Eng. Comm.* **9**: 1252–1261.

Dauphin Y., Cuif J. P., Mutvei H. *et al.* (1989a) Mineralogy, chemistry and ultrastructure of the external shell layer in ten species of *Haliotis* with reference to *Haliotis tuberculata* (Mollusca: Archaeogastropoda). *Bull. Geol. Inst. Univ. Uppsala* **N.S 15**: 7–38.

Dauphin Y., Denys C., Denis A. (1989b) Les mécanismes de formation des gisements de microvertébrés: 2. Composition chimique élémentaire des os et dents de rongeurs provenant de pelotes de régurgitation. *Bull. Mus. Natn. Hist. Nat. Paris*, Sect. A, Zoologie, 4è Sér. **11**(1): 253–269.

Dauphin Y., Kowalski C., Denys C. (1994) Assemblage data and bone and teeth modifications as an aid to paleoenvironmental interpretations of the open-air pleistocene site of Tighenif (Algeria). *Quaternary Research* **42**: 340–342.

Dauphin Y., Gautret P., Cuif J. P. (1996) Evolution diagénetique de la composition chimique des aragonites biogéniques chez les spongiaires, coraux et céphalopodes triasiques du Taurus lycien (Turquie). *Bull. Soc. Géol. Fr.* **167**(2): 247–256.

Dauphin Y., Denys C., Kowalski K. (1997) Analysis of accumulations of rodent remains: role of the chemical composition of skeletal elements. *N. Jb. Geol. Paläont. Abh.* **203**(3): 295–315.

Dauphin Y., Denis A., Denys C. (1999) Diagenèse des micromammifères de trois niveaux du Plio-Pleistocène de Tighenif (Algérie): comparaison avec des pelotes actuelles de régurgitation de rapaces. *Kaupia* **9**: 35–51.

Dauphin Y., Cuif J. P., Doucet J. *et al.* (2003a) *In situ* chemical speciation of sulfur in calcitic biominerals and the simple prism concept. *J. Struct. Biol.* **142**: 272–280.

Dauphin Y., Cuif J. P., Doucet J. *et al.* (2003b) In situ mapping of growth lines in the calcitic prismatic layers of mollusc shells using X-ray absorption near-edge structure (XANES) spectroscopy at the sulphur edge. *Mar. Biol.* **142**: 299–304.

Dauphin Y., Guzman N., Denis A. *et al*. (2003c) Microstructure, nanostructure and composition of the shell of *Concholepas concholepas* (Gastropoda, Muricidae). *Aquat. Liv. Res.* **16**: 95–103.

Dauphin Y., Andrews P., Denys C. *et al*. (2003d) Structural and chemical bone modifications in a modern owl pellet assemblage from Olduvai Gorge (Tanzania). *J. Taphonomy* **1**(4): 209–231.

Dauphin Y., Cuif J. P., Salomé M. *et al*. (2005) Speciation and distribution of sulfur in a mollusk shell as revealed by in situ maps using X-ray absorption near-edge structure (XANES) spectroscopy at the S K-edge. *Am. Mineral.* **90**: 1748–1758.

Dauphin Y., Cuif J. P., Salomé M. *et al*. (2006) Microstructures and chemical composition of giant avian eggshells. *Anal. Bioanal. Chem.* **386**: 761–771.

Dauphin Y., Williams C. T., Salomé M. *et al*. (2007a) Microstructures and compositions of multilayered shells of *Haliotis* (Mollusca, Gastropoda). In *Biomineralization: From Paleontology to Materials Science. Proc. 9th Inter. Symp. Biomin.*, J. L. Arias & M. S. Fernandez (eds.). Santiago: Editorial Universitaria Santiago, pp. 265–272.

Dauphin Y., Montuelle S., Quantin C. *et al*. (2007b) Estimating the preservation of tooth structures: towards a new scale of observation. *J. Taphonomy* **5**(1): 43–56.

Dauphin Y., Williams C. T., Barskov I. S. (2007c) Aragonitic rostra of the Turonian belemnitid *Goniocamax*: arguments from diagenesis. *Acta Palaeont. Pol.* **52**(1): 85–97.

Dauphin Y., Ball A. D., Cotte M. *et al*. (2008a) Structure and composition of the nacre–prism transition in the shell of *Pinctada margaritifera* (Mollusca, Bivalvia). *Anal. Bioanal. Chem.* **390**: 1659–1169.

Dauphin Y., Cuif J. P., Williams C. T. (2008b) Soluble organic matrices of aragonitic skeletons of Merulinidae (Cnidaria, Anthozoa). *Comp. Biochem. Physiol.* B **150**: 10–22.

Dauphin Y., Massard P., Quantin C. (2008c) *In vitro* diagenesis of teeth of *Sus scrofa* (Mammalia, Suidae): micro- and nanostructural alterations and experimental dissolution. *Palaeogeog. Palaeocl. Palaeoecol.* **266**: 134–141.

Davis A. K., Hildenbrand M. (2008) Molecular processes of biosilicification in diatoms. In *Metal Ions in Life Science: 4. Biomineralization. From Nature to Application*, A. Sigel, H. Sigel & R. K. O. Sigel (eds.). Chichester, UK: Wiley, pp. 255–294.

Davis K. J., Dove P. M., De Yoreo J. J. (2000) The role of Mg^{2+} as an impurity in calcite growth. *Science* **290**(5494): 1134–1137.

Davis K. J., Dove P. M., Wasylenski L. E. *et al*. (2004) Morphological consequences of differential Mg^{2+} incorporation at structurally distinct steps on calcite. *Am. Mineral.* **89**: 714–720.

De Jong E. W., Westbroek P., Westbroek J. F. (1974) Preservation of antigenic properties in macromolecules over 70 myr old. *Nature* **252**: 63–64.

De Jong E. W., Bosch L., Westbroek P. (1976) Isolation and characterization of a Ca^{2+} binding polysaccharide associated with coccoliths of *Emiliania huxleyi* (Lohmann) Kamptner. *Eur. J. Biochem.* **70**: 611–621.

De Niro M. J., Weiner S. (1988) A chemical, enzymatic and spectroscopic characterization of "collagen" and other organic fractions. *Geochim. Cosmochim. Acta* **52**: 2415–2424.

Debrenne F. (1992) Diversification of Archaeocyatha. In *Origin and Early Evolution of the Metazoa*, J. H. Lipps & P. W. Signor (eds.). New York: Plenum Press, pp. 425–443.

Deflandres G. (1936). Microfossiles des silex crétacés. Première partie. Généralités – Flagellés. *Ann. Paléont.* **25**: 151–191.

Degens E. T. (1979) Why do organisms calcify? *Chem. Geol.* **25**: 257–269.

Degens E. T., Johannesson B. W., Meyer R. W. (1967) Mineralization processes in Molluscs and their paleontological significance. *Naturwissenschaften* **54**: 638–640.

Degens E. T., Deuser W. G., Haedrich R. L. (1969) Molecular structure and composition of fish otoliths. *Mar. Biol.* **2**: 105–113.

Dendinger J. E., Alterman A. (1983) Mechanical properties in relation to chemical constituents of postmolt cuticle of the blue crab, *Callinectes sapidus. Comp. Biochem. Physiol.* A **75**(3): 421–424.

Denys C., Kowalski K., Dauphin Y. (1992) Mechanical and chemical alterations of skeletal tissues in a recent Saharian accumulation of faeces from *Vulpes rueppelli* (Carnivora, Mammalia). *Acta Zool. Cracov.* **35**(2): 265–283.

Denys C., Williams T., Dauphin Y. *et al.* (1996) Diagenetical changes in Pleistocene small mammal bones from Olduvai Bed I. *Palaeogeogr. Palaeoclimatol. Palaeoecol.* **126**: 121–134.

Diekwisch T. G. H., Berman B. J., Anderton X. *et al.* (2002) Membranes, minerals, and proteins of developing vertebrate enamel. *Microsc. Res. Tech.* **59**: 373–395.

Dodd J. R. (1967) Magnesium and strontium in calcareous skeletons: a review. *J. Paleontol.* **41**(6): 1313–1329.

Dodson P., Wexlar D. (1979) Taphonomic investigations of owl pellets. *Paleobiology* **5**: 275–284.

Donovan D. T. (1964) Cephalopod phylogeny and classification. *Biol. Rev.* **39**: 259–287.

Doroudi M. S., Southgate P. C. (2003) Embryonic and larval development of *Pinctada margaritifera* (Linnaeus, 1758). *Moll. Res.* **23**: 101–107.

Dos Santos P. R., Added N., Rizzutto M. A. *et al.* (2006) Measurement of Sr/Ca ratio in bones as a temperature indicator. *Braz. J. Phys.* **36**(4), doi: 10.1590/S0103–97332006000800012.

Doyle P., Donovan D. T., Nixon M. (1994) Phylogeny and systematics of the Coleoidea. *The University of Kansas Paleontological Contributions* n.s **5**: 1–15.

Drach P. (1939) Mue et cycle d'intermue chez les Crustacés Décapodes. *Ann. Inst. Océanog. Monaco.* **19**: 103.

Drever J. I. (1988) *Geochemistry of Natural Waters*, 2nd edn. Englewood Cliffs, NJ: Prentice-Hall, 388pp.

Dufour E., Capetta H., Denis A. *et al.* (2000) La diagenèse des otolithes par la comparaison des données microstructurales, minéralogiques et géochimiques: application aux fossiles du Pliocène du Sud-Est de la France. *Bull. Soc. Géol. France* **175**(5): 521–532.

Dunham R. J. (1970) Stratigraphic reefs versus ecologic reefs. *Am. Assoc. Petrol. Geol. Bull.* **54**: 1931–1932.

Eckert C., Schröder H. C., Brandt D. *et al.* (2006) A histochemical and electron microscopic analysis of the spiculogenesis in the demosponge *Suberites domuncula. J. Histochem. Cytochem.* **54**: 1031–1040.

Ehrenberg C. G. (1836) Beobachtungen ueber die organisation der armpolypen. *Mitt. Naturf. Ges. Berlin* **2**: 27–29.

Ehrenberg C. G. (1838) *Die Infusionsthierchen als vollkommene Organismen. Leipzig: Woss,* 547pp., 64pl.

Ehrlich H., Eresrkovsky A. V., Vyalikh D. V. *et al.* (2005) Collagen in natural fibres of deep-sea glass sponge. In *Biomineralization: From Paleontology to Materials Science,* J. L. Arias & M. S. Fernandez (eds.). Santiago: Editorial Universitaria Santiago, pp. 439–448.

Elderfield H., Ganssen G. (2000) Past temperature and $\delta^{18}O$ of surface ocean waters inferred from foraminiferal Mg/Ca ratios. *Nature* **405**: 442–445.

Elliot M., de Menocal P. B., Linsley B. K. *et al.* (2003) Environmental controls on the stable isotopic composition of *Mercenaria mercenaria*: potential application to paleoenvironmental studies. *Geochem. Geophys. Geosyst.* **4**(7): 1056, doi: 10.1029/2002GC000425.

Endo K., Curry G. B., Quinn R. *et al.* (1994) Re-interpretation of Terebratulide phylogeny based on immunological data. *Palaeontology* **37**(2): 349–373.

England J., Cusack M., Lee M. R. (2006) Magnesium and sulphur in the calcite shells of two Brachiopods, *Terebratulina retusa* and *Novocrania anomala*. *Lethaia* **40**: 2–10.

England F., Cusack M., Dalbeck P. *et al.* (2007) Comparison of the crystallographic structure of semi nacre and nacre by electron backscatter diffraction. *Cryst. Growth Des.* **7**(2): 307–310.

Enlow D. H. (1963) *Principles of Bone Remodeling*. Springfield, IL: C. C. Thomas Pub., 123pp.

Epstein S., Buchsbaum R., Lowenstam H. A. *et al.* (1951) Carbonate-water isotopic temperature scale. *Geol. Soc. Am. Bull.* **62**: 417–426.

Epstein S., Buchsbaum R., Lowenstam H. A. *et al.* (1953) Revised carbonate-water isotopic temperature scale. *Geol. Soc. Am. Bull.* **64**: 1315–1326.

Erben H. K. (1972) Uber die Bildung und das Wachstum von Perlmutt. *Biomineralization* **4**: 16–36.

Erben H. K., Watabe N. (1974) Crystal formation and growth in bivalve nacre. *Nature* **248**: 128–130.

Ezaki Y. (1997) The Permian coral *Numidiaphyllum*: new insight into anthozoan phylogeny and Triassic scleractinian origin. *Paleontology* **40**: 1–14.

Ezaki Y. (2000) Paleoecological and phylogenetic implications of a new scleractiniomorph genus from Permian sponge reefs, south China. *Paleontology* **43**: 199–217.

Falini G., Albeck S., Weiner S. *et al.* (1996) Control of aragonite or calcite polymorphism by mollusk shell macromolecules. *Science* **271**: 67–69.

Farre B., Dauphin Y. (2009) Lipids from the nacreous and prismatic layers of two Pteriomorpha Mollusc shells. *Comp. Biochem. Physiol.* B **152**: 103–109.

Fearnhead R. W. (1979) Matrix-mineral relationships in enamel tissues. *J. Dent. Res.* **58**(B): 909–916.

Feigl F. (1937) *Qualitative Analysis by Spot Test*. New York: Nordmann Pub. Co., pp. 161–163.

Fernandez M. S., Araya M., Arias J. L. (1997) Eggshells are shaped by a precise spatio-temporal arrangement of sequentially deposited macromolecules. *Matrix Biol.* **16**: 13–20.

Fernández-Jalvo Y., Andrews P. (1992) Small mammal taphonomy of Gran Dolina, Atapuerca (Burgos), Spain. *J. Archaeol. Sci.* **19**: 407–428.

Fischer J. C. (1970) Révision et essai de classification des Chaetetida (Cnidaria) post-paléozoiques. *Ann. Paléontol. Inv.* **61**(2): 151–220.

Fratzl P., Schreiber S., Boyde A. (1996) Characterization of bone mineral crystals in horse radius by small angle X-ray diffraction. *Calcif. Tissue Int.* **58**: 341–346.

Frech F. (1890) Die Korallen fauna der Trias: I. Die Korallen der Juvavischen Triasprovinz. *Palaeontographica* **7**(1): 116.

Freeman J. A., Wilbur K. M. (1948) Carbonic anhydrase in molluscs. *Biol. Bull.* **94**: 55–59.

Freiwald A., Wilson J. (1998) Taphonomy of modern deep cold-water coral reefs. *Historical Biol.* **13**: 37–52.

Frémy E. (1855) Recherches chimiques sur les os. *Ann. Chim. Paris*, Sér.13 **43**: 47–107.

Frérotte B., Raguideau A., Cuif J. P. (1983) Dégradation *in vitro* d'un test carbonaté d'invertébrés, *Crassostrea gigas* (Thunberg), par action de cultures bactériennes. Intérêt pour l'analyse ultrastructurale. *C. R. Acad. Sci. Paris*, Sér. II **297**: 383–388.

Fricke M., Volkmer D. (2007) Crystallization of calcium carbonate beneath insoluble monolayers: suitable models of mineral–matrix interactions in biomineralization? *Top. Curr. Chem.* **270**: 1–41.

Fripiat F., Corvaisier R., Navez J. *et al.* (2009) Measuring production–dissolution rates of marine biogenic silica by ^{30}Si-isotope dilution using a high-resolution sector field inductively coupled plasma mass spectrometer. *Limnol. Oceanogr.: Methods* **7**: 470–478.

Fu G., Valiiaveetil S., Wopenka B. *et al.* (2005) $CaCO_3$ biomineralization: acidic 8-kDa proteins isolated from aragonitic abalone shell nacre can specifically modify calcite crystal morphology. *Biomacromolecules* **6**: 1289–1298.

Füchtbauer H., Hardie L. A. (1976) Experimentally determined homogeneous distribution coefficients for precipitated magnesian calcites: application to marine carbonate cements. *Geol. Soc. Am. Abst. Prog.* **8**: 877.

Füchtbauer H., Hardie L. A. (1980) Comparison of experimental and natural magnesian calcites. International Society of Sedimentologists Meeting Abstracts (Bochum), pp. 167–169.

Fukami H., Chen C. A., Budd N. *et al.* (2008) Mitochondrial and nuclear genes suggest that stony corals are monophyletic but most families of stony corals are not (Order Scleractinia, Class Anthozoa, Phylum Cnidaria). *Plus One* **3**(9): 3222, doi: 0.1371/journal.pone.0003222.

Furla P., Galgani I., Durand I. *et al.* (2000) Sources and mechanisms of inorganic carbon transport for coral calcification and photosynthesis. *J. Exp. Biol.* **203**: 3445–3457.

Furuhashi T., Schwartzinger C., Miksik I. *et al.* (2009) Molluscan shell evolution with review of shell calcification hypothesis. *Comp. Biochem. Physiol.* B **154**: 351–371.

Gaetani G. A., Cohen A. L. (2006) Element partitioning during precipitation of aragonite from seawater: a framework for understanding paleoproxies. *Geochim. Cosmochim. Acta* **70**: 4617–4634.

Gaffey S. (1988) Water in skeletal carbonates. *Sedim. Petrol.* **58**(3): 397–414.

Garrone R. (1969) Collagen, spongin and mineral skeleton in sponge *Haliclona rosea* (Demospongiae, Haplosclerina). *J. Microscopy* **8**: 581–598.

Gautret P. (1985) Recherche sur la valeur taxonomique des caractéristiques du squelette carbonaté aspiculaire des Spongiaires. Thèse Doctorat Faculté des Sciences, Université Paris-Sud Orsay, 230pp.

Gautret P., Cuif J. P., Stolarski J. (2000) Organic component of the skeleton of scleractinian corals: evidence from in situ Acridine Orange staining. *Acta Palaeont. Pol.* **45**(2): 107–118.

Gautron J., Hincke M. T., Mann K. *et al.* (2001) Ovocalyxin-32, a novel chicken eggshell matrix protein. *J. Biol. Chem.* **276**(42): 39 243–39 252.

Gayathri S., Lakshminarayanan R., Weaver J. C. *et al.* (2007) In vitro study of magnesium-calcite biomineralization in the skeletal materials of the seastar *Pisaster giganteus*. *Chem. Eur. J.* **13**: 3262–3268.

Gessner C. (1565) De Omni Rerum Fossilivm, Lapidvm et Gemmarvm maximè, figuris & similitudinibus Liber: non solùm Medicis, sed omnibus rerum Naturae ac Philologiae studiosis, vtilis & iucundus futurus. Zurich, 368pp.

Gilkeson C. F. (1997) Tubules in Australian marsupials. In *Tooth Enamel Microstructure*, W. v. Koenigswald & P. M. Sander (eds.). Rotterdam: Balkema, pp. 113–122.

Gillis J. A., Donoghue P. C. J. (2007) The homology and phylogeny of Chondrichthyan tooth enameloid. *J. Morph.* **268**: 33–49.

Giraud M. M. (1977) Rôle du complexe chitino-protéique et de l'anhydrase carbonique dans la calcification tégumentaire de *Carcinus maenas* L. Thèse Doctorat Faculté des Sciences, Biologie animale: mention cytologie, Université de Paris VI, 77pp.

Giraud-Guille M. M., Belamie E., Mosser M. (2004) Organic and mineral networks in carapaces, bones and biomimetic materials. *C. R. Palevol.* **3**: 503–513.

Gladfelter E. H. (1982) Skeletal development in *Acropora cervicornis*: I. Patterns of calcium carbonate accretion in the axial corallite. *Coral Reefs* **1**: 45–51.

Glazer A. N., Apell G. S., Hixson C. S. *et al.* (1976) Biliproteins of cyanobacteria and Rhodophyta: homologous family of photosynthetic accessory pigments. *Proc. Nat. Acad. Sci. USA* **73**: 428–431.

Glimcher M. J. (2006) Bone: nature of calcium phosphate crystals and cellular, structural, and physical chemical mechanisms in their formation. In *Reviews in Mineralogy and Geochemistry: Medical Mineralogy and Geochemistry*, N. Sahai & M. A. A. Schoonen (eds.). Washington DC: Mineralogical Society of America, Vol. 64, pp. 223–282.

Goodwin D. H., Flessal K. W., Schöne B. R. *et al.* (2001) Cross-calibration of daily growth increments, stable isotope variation, and temperature in the Gulf of California bivalve mollusk *Chione cortezi*: implications for paleoenvironmental analysis. *Palaios* **16**: 387–398.

Goreau T. (1956) Histochemistry of mucopolysaccharide-like substances and alkaline phosphatase in Madreporaria. *Nature* **4518**: 1029–1030.

Gotliv B. A., Aaddadi L., Weiner S. (2003) Mollusk shell acidic proteins: in search of individual functions. *ChemBioChem* **4**: 522–529.

Goulletquer P., Wolowicz M. (1989) The shell of *Cardium edule, Cardium glaucum* and *Ruditapes philippinarium*: organic content, composition and energy value, as determined by different methods. *J. Mar. Biol. Ass. U.K.* **69**: 563–572.

Grachev M. A., Denikina N. N., Belikov S. I. *et al.* (2002) Elements of the active center of silicon transporters in diatoms. *Mol. Biol. (Mosk)* **36**: 534–536.

Grassmann O., Neder R. B., Putnis A. *et al.* (2003) Biomimetic control of crystal assembly by growth in an organic hydrogel network. *Am. Mineral.* **88**(4): 647–652.

Greenfield E., Wilson D. C., Crenshaw M. A. (1984) Ionotropic nucleation of calcium carbonate by molluscan matrix. *Am. Zool.* **24**: 925–932.

Grefsrud E. S., Dauphin Y., Cuif J. P. *et al.* (2008) Modifications in microstructure of cultured and wild scallop shells (*Pecten maximus*). *J. Shellfish Res.* **27**(4): 633–642.

Grégoire C. (1960) Further studies on structure of the organic components in mother-of-pearl, especially in Pelecypods. *Bull. Inst. R. Sci. Nat. Belg.* **36**(23): 1–22.

Grégoire C. (1961) Sur la structure de la nacre septale des Spirulidae, étudiée au microscope électronique. *Arch. Internat. Physiol. Bioch.* **49**(3): 374–377.

Grégoire C. (1967) Sur la structure des matrices organiques des coquilles de mollusques. *Biol. Rev.* **42**: 653–687.

Grégoire C. (1972a) Experimental alteration of the *Nautilus* shell by factors involved in diagenesis and in metamorphism: Part III. Thermal and hydrothermal changes in the organic and mineral components of the mural mother-of-pearl. *Bull. Inst. R. Sci. Nat. Belg.* **48**(6): 1–85.

Grégoire C. (1972b) Structure of the molluscan shell. In *Chemical Zoology, Mollusca*, M. Florkin & B. T. Scheer (eds.). New York: Academic Press, Vol. 7, pp. 45–145.

Grégoire C., Duchateau G., Florkin M. (1955) La trame protidique des nacres et des perles. *Ann. Inst. Océanogr.* **31**: 1–36.

Griesshaber E., Schmahl W. W., Neuser R. D. *et al.* (2007) Crystallographic texture and microstructure of terebratulide brachiopod shell calcite: an optimized materials design with hierarchical architecture. *Am. Mineral.* **92**: 722–734.

Griesshaber E., Kelm K., Sehrbrock A. *et al.* (2009) Amorphous calcium carbonate in the shell material of the brachiopod *Megerlia truncata*. *Eur. J. Mineral.* **21**: 715–723.

Grobben K. (1908) Die systematsiche Einteilung des Tierreiches. *Verh. Bot. Ges. Osterreich* **58**: 491–511.

Gross W. (1934) Die typen des mikroskopischen Knochenbaues bei fossilen Stegocephalen und Reptilien. *Zeit. Anatomie* **103**: 731–764.

Gunatilaka A. (1975) The chemical composition of some carbonate secreting marine organisms from Connemara. *Proc. R. I. A. Sect. B* **75**: 543–556.

Gussone N., Böhm F., Eisenhauer A. *et al.* (2005) Calcium isotope fractionation in calcite and aragonite. *Geochim. Cosmochim. Acta* **69**(18): 4485–4494.

Gustomesov V. A. (1976) Basic aspects of belemnoid phylogeny and systematic. *Paleont. J.* **2**: 170–179.

Gustomesov V. A. (1978) The pre-Jurassic ancestry of the Belemnitida and the evolutionary changes in the Belemnoidea at the boundary between the Triassic and the Jurassic (in Russian). *Paleont. J.* **3**: 3–13.

Guzmann N., Ball A. D., Cuif J. P. *et al.* (2007) Subdaily growth patterns and organo-mineral nanostructure of the growth layers in the calcitic prisms of the shell of *Concholepas concholepas* Bruguière, 1789 (Gastropoda, Muricidae). *Microsc. Microanal.* **13**(5): 397–403.

Haeckel E. (1862) Die Radiolarian (rhizopoda Radiara). *Eine Monographie*. Berlin: Reimer, 2 vol.

Haeckel E. (1896) Systematische Phylogenie. *Wirbellose Tiere*. Berlin: G. Reimer, pp. 1–720.

Hämmerling J. (1931) Entwicklung und Formbildungsvermögen von *Acetabularia mediterranea*: I. Die normale Entwicklung. *Biol. ZbL.* **51**: 633–647.

Hämmerling J. (1963) Nucleo-cytoplasmic interactions in *Acetabularia* and other cells. *Ann. Rev. Pl. Physiol.* **14**: 65–92.

Hansen H. J. (1999) Shell construction in Foraminifera. In *Modern Foraminifera*, B. K. Sen Gupta (ed.). Dordrecht, The Netherlands: Kluwer Academic Publishers, pp. 57–70.

Hardie L. A. (1996) Secular variation in seawater chemistry: an explanation for the coupled secular variation in the mineralogies of marine limestones and potash evaporites over the past 600 m.y. *Geology* **24**(3): 279–283.

Hare P. E. (1963) Amino acids in the proteins from aragonite and calcite in the shells of *Mytilus californianus*. *Science* **139**: 216–217.

Harris R. C. (1965) Trace element distribution in molluscan skeletal material: I. Magnesium, iron, manganese, and strontium. *Bull. Mar. Sci.* **15**(2): 265–273.

Hartmann W. D. (1982) Porifera. In *Synopsis and Classification of Living Organisms*, S. P. Parker (ed.). New York: McGraw-Hill, Vol. 1, pp. 641–666.

Hartman W. D., Goreau T. F. (1970) Jamaican coralline sponges: their morphology, ecology, and fossil relatives. *Symp. Zool. Soc. London* **25**: 205–243.

Hartman W. D., Goreau T. F. (1975) A Pacific tabulate sponge, living representative of a new order of sclerosponges. *Postilla* **167**: 1–21.

Hay W. W., Towe K. M., Wright R. C. (1963) Ultramicrostructure of some selected foraminiferal tests. *Micropaleontology* **9**(2): 171–195.

Hazelaar S., van der Strate H. J., Gieskes W. W. C. *et al.* (2005) Monitoring rapid valve formation in the pinnate diatom *Navicula salinarum*. *J. Phycology* **4**: 354–358.

Heckel P. H. (1974) Carbonate buildings in the geological record: a review. *Soc. Econ. Pal. Min. Spec. Publ.* **18**: 90–154.

Hedges R. E. (2002) Bone diagenesis: an overview of processes. *Archaeometry* **44**(3): 319–328.

Heider A. R. von (1886) Korallenstudien. *Zeitsch. der. Wissensch. Zoolog.* **46**: 507–535.

Henisch H. K. (1988) *Crystals in Gels and Liesengang Rings.* Cambridge: Cambridge University Press, 197pp.

Hickson S. J. (1911) On *Ceratopora*, the type of a new family of Alcyonaria. *Proc. R. Soc. London B* **84**: 195–200.

Hildebrand M., Volcani B. E., Gassmann W. *et al.* (1997) A gene family of silicon transporters. *Nature* **385**: 688–689.

Hildebrand M., Dahlin M. K., Volcani B. E. (1998) Characterization of a silicon transporter gene family in *Cylindrotheca fusiformis*: sequences, expression analysis, and identification of homologs in other diatoms. *Mol. Gen. Genet.* **260**: 480–486.

Hill D. (1956) Rugosa. In: Moore 1956, Hill, D. 1956. In *Treatise on Invertebrate Paleontology, Part F, Coelenterata*, R. C. Moore (ed.). Lawrence, KS: Geological Society of America and University of Kansas Press, pp. F234–F324 (revised 1981).

Hincke M. T. (1995) Ovalbumin is a component of the chicken eggshell matrix. *Connect. Tissue Tes.* **31**: 227–233.

Hincke M. T., Tsang C. P., Courtney M. *et al.* (1995) Purification and immunochemistry of a soluble matrix protein of the chicken eggshell (ovocleidin 17). *Calc. Tissue Int.* **56**: 578–583.

Hincke M. T., Gautron J., Tsang C. P. W. *et al.* (1999) Molecular cloning and ultrastructural localization of the core protein of an eggshell matrix proteoglycan, ovocleidin-116. *J. Biol. Chem.* **274**(16): 32 915–32 923.

Hincke M. T., Saint Maurice M., Nys Y. *et al.* (2000) Eggshell proteins and shell strength: molecular biology of eggshell matrix proteins and industry applications. In *CAB Intern. 2000, Egg Nutrition and Biotechnology*, J. S. Sim, S. Nakai & W. Guenter (eds.). Wallingford, UK: CAB International, pp. 447–461.

Hodge A. J. (1967) Structure at the electron microscopic level. In *Treatise on Collagen*, G. N. Ramachandran (ed.). Orlando, FL: Academic Press, Vol. 1, pp. 185–205.

Hodge A. J., Petruska J. A. (1963) Recent studies with the electron microscope on ordered aggregates of the tropocollagen macromolecule. In *Aspects of Protein Structure*, G. N. Ramachandran (ed.). New York: Academic Press, Vol. 1, pp. 289–300.

Hoek C. van den, Mann D. G., Jahns H. M. (1995) *Algae: An Introduction to Phycology.* Cambridge: Cambridge University Press, 623pp.

Hollande A., Enjumet M. (1960) Cytologie, évolution et systématique des Sphaeroidés (Radiolaires). *Archiv. Mus. Natn. Hist. Nat., Sér.* **7**(7): 1–134.

Hooke R. (1665) Micrographia: or some physiological descriptions of minute bodies made by magnifying glasses with observations and inquiries thereupon. *London*: J. Martin & J. Allestry, 488pp.

Hooker J. D. (1847) Algae. In *Flora Antarctica*. London: Reeve Broths, pp. 454–502.

Houlbrèque F., Meibom A., Cuif J. P. *et al.* (2009) Strontium-86 labelling experiments show spatially heterogeneous skeletal formation in the scleractinian coral *Porites porites*. *Geophys. Res. Let.* **36**, L04604, doi: 10.1029/2008GL036782.

Huxley T. H. (1879) On the classification and the distribution of the crayfishes. *Proc. Zool. Soc. London* **46**(1): 751–788.

Iler R. K. (1979) *The Chemistry of Silica: Solubility, Polymerization, Colloidal and Surface Properties and Biochemistry.* New York: J. Wiley and Sons, 866pp.

Inage T., Shimokawa H., Teranishi Y. *et al.* (1989) Immuno-cytochemical demonstration of amelogenins and enamelins secreted by ameloblasts during the secretory and maturation stages. *Arch. Histol. Cytol.* **52**(3): 213–229.

Iwata K. (1981) Ultrastructure and calcification of the shell of *Lingula unguis* Linné (Inarticulate Brachiopod). *J. Fac. Sci. Hokkaido Univ., Ser. IV* **20**: 33–65.

Jackson D. J., McDougall C., Green K. *et al.* (2006) A rapidly evolving secretome builds and patterns a sea shell. *BMC Biol.*, doi: 10.1186/1741-7007-4-40.

James N. P. (1983) Reef environments. *Am. Assoc. Petrol. Geol. Mem.* **33**: 345–462.

Jefferies R. P. S. (1986) *The Ancestry of the Vertebrates*. London: British Museum (Natural History), 376pp.

Jeletzky J. A. (1966) Comparative morphology, phylogeny and classification of fossil Coleoidea. *University of Kansas Paleontological Contributions* **7**: 1–162.

Jell J. S. (1980) Skeletogenesis of newly settled planulae of the hermatypic coral *Porites lutea. Acta Palaeont. Polonica* **25**(3–4): 311–320.

Jeuniaux C. (1965) Chitine et phylogénie: application d'une méthode enzymatique de dosage de la chitine. *Bull. Soc. Chim. Biol.* **47**(12): 2267–2278.

Johnston I. S. (1977) Aspects of the structure of a skeletal organic matrix, and the process of skeletogenesis in the reef-coral *Pocillopora damicornis*. In *Proceedings of 3rd Int. Coral Reef Symp.* Miami, FL: Univiversity of Miami, pp. 447–453.

Johnston I. S. (1980) The ultrastructure of skeletogenesis in hermatypic corals. *Int. Review Cytology* **67**: 171–214.

Jones C. W. (1979) The microstructure and genesis of sponge biominerals. In *Biologie des Spongiaires,* Colloques Internationaux du C.N.R.S., S. Lévi & N. Boury-Esnault (eds.). Paris: C.N.R.S., Vol. 291, pp. 425–447.

Jones D. L., Knauth L. (1979) Oxygen isotopic and petrographic evidence relevant to the origin of the Arkansas Novaculite. *J. Sedim. Pet.* **49**: 581–597.

Jonsson M., Fredriksson S., Jontell M. *et al.* (1978) Isoelectric focusing of the phosphoproteins of rat incisor dentin in ampholine and acid pH gradients. Evidence for carrier ampholyte-protein complexes. *J. Chromatography* **157**: 235–242.

Jope M. (1971) Constituents of brachiopod shells. In *Comprehensive Biochemistry*, M. Florkin & E. W. Stotz (eds.). Amsterdam: Elsevier, Vol. 26, pp. 749–784.

Jope M. (1973) The protein of brachiopod shell: V. N-Terminal end groups. *Comp. Biochem. Physiol.* **45**B: 17–24.

Juillet-Leclerc A., Reynaud S., Rollion-Bard C. *et al.* (2009) Oxygen isotopic signature of the skeletal microstructures in cultured corals: identification of vital effects. *Geochim. Cosmochim. Acta* **73**: 5320–5332.

Kaufman P. B., Dayanandan P., Takeoka Y. *et al.* (1981) Silica in shoots of higher plants. In *Silicon and Siliceous Structures in Biological Systems*, T. L. Simpson & B. E. Volcani (eds.). New York: Springer Verlag, pp. 409–449.

Kawakami I. K. (1952) Mantle regeneration in pearl oyster (*Pinctada martensii*) *J. Fuji Pearl Inst.* **2**(2): 1–4.

Keith M. L., Weber J. N. (1965) Systematic relationships between carbon and oxygen isotopes in carbonates deposited by modern corals and algae. *Science* **150**: 498–501.

Kidder D. L., Erwin D. H. (2001) Secular distribution of biogenic silica through the Phanerozoic: comparison of silica replaced fossils and bedded cherts at the series Level. *J. Geology* **109**(4): 509–522.

Killian C. E., Wilt F. H. (1996) Characterization of the proteins comprising the integral matrix of *Strongylocentrotus purpuratus* embryonic spicules. *J. Biol. Chem.* **271**(15): 9150–9159.

Kim H. M., Rey C., Glimcher M. J. (1996) X-ray diffraction, electron microscopy, and Fourier transform infrared spectroscopy of apatite crystals isolated from chicken and bovine calcified cartilage. *Calcif. Tissue Int.* **59**: 58–63.

Kingsley R. J., Tsuzaki M., Watabe N. *et al.* (1990) Collagen in the spicule organic matrix of the gorgonian *Leptogorgia virgulata*. *Biol. Bull.* **179**: 207–213.

Kinsmann D. J. J., Holland H. D. (1969) The co-precipitation of cations with $CaCO_3$: IV. The co-precipitation of Sr^{2+} with aragonite between 16° and 96°C. *Geochim. Cosmochim. Acta* **33**: 1–17.

Kirkpatrick R. (1910) On a remarkable Pharetronid sponge from Christmas Island. *Proc. R. Soc.* **B83**: 124–133.

Kitano Y., Hood D. W. (1965) The influence of organic material on the polymorphic crystallization of calcium carbonate. *Geochim. Cosmochim. Acta* **29**: 29–41.

Kitano Y., Kanamori N., Tokuyama A. (1969) Effects of organic matter on solubilities and crystal form of carbonates. *Am. Zool.* **9**: 681–688.

Klein R. T., Lohmann K. C., Thayer C. W. (1996a) Sr/Ca and $^{13}C/^{12}C$ ratios in skeletal calcite of *Mytilus trosulus*: proxies of metabolic rate, salinity and carbon isotopic composition of seawater. *Geochim. Cosmochim. Acta* **60**: 4207–4221.

Klein R. T., Lohmann K. C., Thayer C. W. (1996b) Bivalve skeletons record sea-surface temperature and salinity via Mg/Ca and $^{18}O/^{16}O$ ratios. *Geology* **24**: 415–418.

Klepal W., Barnes H. (1975a) A histological and scanning electron microscope study of the formation of the wall plates in *Chthamalus depressus* (Poli). *J. Exp. Mar. Biol. Ecol.* **20**(2):183–198.

Klepal W., Barnes H. (1975b) The structure of the wall plate in *Chthamalus depressus*. *J. Exp. Mar. Biol. Ecol.* **20**: 265–285.

Klug C., Schulz H., De Baets K. (2009) Red Devonian trilobites with green eyes from Morocco and the silicification of the trilobite exoskeleton. *Acta Palaeont. Polonica* **54** (1): 117–123.

Kobayashi I. (1980) Various patterns of biomineralization and its phylogenetic significances in bivalve molluscs. In *The Mechanisms of Biomineralization in Animals and Plants*, M. Omori & N. Watabe (eds.). Kanagawa: Tokai University Press, pp. 145–155.

Koch G. von (1886) Ueber das Verhältniss von Skelet und Weichtheilen bei den Madreporaren. *Morph. Jahrb.* **12**: 154–162.

Koch P. L., Tuross N., Fogel M. L. (1997) The effects of sample treatment and diagenesis on the isotopic integrity of carbonate in biogenic hydroxylapatite. *J. Archaeol. Sc.* **24**: 417–429.

Koenigswald W. v., Sander P. M. (1997) Glossary of terms used for enamel microstructures. In *Tooth Enamel Structure*, W. Koenigswald & P. M. Sander (eds.). Rotterdam: Balkema, pp. 267–280.

Kragh M., Molbak L., Andersen S. O. (1997) Cuticular proteins from the lobster, *Homarus americanus*. *Comp. Biochem. Physiol. B* **118**: 147–154.

Krampitz G., Witt W. (1979) Biochemical aspects of biomineralization. *Topics Curr. Chem.* **78**: 57–144.

Krampitz G., Engels J., Cazaux C. (1976) Biochemical studies on water-soluble proteins and related components of gastropod shells. In *The Mechanisms of Mineralization in the Invertebrates and Plants*, N. Watabe & K. M. Wilbur (eds.). The Belle Baruch Library in Marine Science, Number 5. Columbia, SC: University of South Carolina Press, pp. 155–173.

Krampitz G., Weise K., Potz A. *et al.* (1977) Calcium-binding peptide in dinosaur eggshells. *Naturwissenschaften* **64**: 583.

Kröger N., Deutzmann R., Sumper M. (1999) Polycationic peptides from diatom biosilica that direct silica nanosphere formation. *Science* **286**: 1129–1132.

Kröger N., Deutzmann R., Bergsdorf R. *et al.* (2000) Species specific polyamines from diatoms control silica morphology. *Proc. Nat. Acad. Sci. USA* **97**: 14 133–14 138.

Krumbein W. E. (1974) On the precipitation of aragonite on the surface of marine bacteria. *Naturwissenschaften* **61**(4): 167.

Kunioka D., Shirai K., Takahata N. *et al.* (2006) Microdistribution of Mg/Ca, Sr/Ca, and Ba/Ca ratios in *Pulleniatina obliquiloculata* test by using a NanoSIMS: implication for the vital effect mechanism. *Geochem. Geophys. Geosyst.* **7**: Q12P20, doi: 10.1029/2006GC001280.

Lacaze-Duthiers H. de (1864) Histoire naturelle du corail. *J. Exp. Zool.* 371pp.

Lamnie D., Bain M. M., Wess T. J. (2005) Microfocus X-ray scattering investigations of eggshell nanostructure. *J. Synchrotron Rad.* **12**: 721–726.

Landis W. J., Song M. J., Leith A. *et al.* (1993) Mineral and organic matrix interaction in normally calcifying tendon visualized in three dimensions by high voltage electron microscopic tomography and graphic image reconstruction. *J. Struct. Biol.* **110**: 39–54.

Landis W. J., Hodgens K. J., Min J. A. *et al.* (1996) Mineralization of collagen may occur on fibril surfaces: evidence from conventional and high-voltage electron microscopy and three-dimensional imaging. *J. Struct. Biol.* **117**: 24–35.

Le Geros R. Z., Pan C. M., Suga S. *et al.* (1985) Crystallo-chemical properties of apatite in Atremate brachiopod shells. *Calc. Tissue Intern.* **37**: 98–100.

Le Goff R., Gauvrit G., Pinczon du Sel G. *et al.* (1998) Age group determination by analysis of the cuttlebone of the cuttlefish *Sepia officinalis* L. In: Reproduction in the Bay of Biscay. *J. Moll. Studies* **64**: 183–193.

Lecompte M. (1951) Les stromatoporoïdes du Dévonien moyen et supérieur du bassin de Dinant. 1è partie. *Inst. R. Sci. Nat. Belg., Mém.* **116**: 1–215.

Lecompte M. (1952) Les stromatoporïdes du Dévonien moyen et supérieur du bassin de Dinant. 2è partie. *Inst. R. Sci. Nat. Belg., Mém.* **117**: 219–358.

Lécuyer C., O'Neil J. R. (1994) Composition isotopique (H, O) de l'eau en inclusion dans les carbonates biogéniques. *Bull. Soc. Géol. Fr.* **165**(6): 573–581.

Lee-Thorp J., Sponheimer M. (2003) Three case studies used to reassess the reliability of fossil bone and enamel isotope signals for paleodietary studies. *J. Anthrop. Archaeol.* **22**: 208–216.

Lemberg J. (1892) Zur microchemischen Untersuchung einiger Minerale. *Zeit. Deutschen Geologischen Gesellschaft* **40**: 357–359.

Leo R. F. (1975) *Silicification of wood*. Ph.D. dissertation. Harvard University, Cambridge, MA.

Lévi C. (1973) Systématique de la classe des Demospongiaria (Démosponges). In *Traité de Zoologie, Spongiaires*, P. P. Grassé (ed.). Paris: Masson & Cie, pp. 577–632.

Lévi C., Barton J. L., Guillemet C. *et al.* (1989) A remarkably strong natural glassy rod: the anchoring spicule of the *Monorhaphis* sponge. *J. Mat. Sci. Letters* **8**: 337–339.

Levi-Kalisman Y., Falini G., Addadi L. *et al.* (2001) Structure of the nacreous organic matrix of a bivalve mollusk shell examined in the hydrated state using cryo-TEM. *J. Struct. Biol.* **135**: 8–17.

Liao S. S., Cui F. Z., Zhang W. (2005) Hierarchically biomimetic bone scaffold materials: nano-HA/collagen/PLA composite. *J. Applied Biomaterials B* **69**(2): 158–165.

Linde A. (1889) Dentin matrix proteins: composition and possible functions in calcification. *Anat. Rec.* **224**: 154–166.

Lippmann F. (1960) Versuche zur Aufklarung der Bildungsbedingungen von Calcit and Aragonit. *Fortschr. Mineral.* **38**: 156–161.

Lisitzin A. P. (1972) *Sedimentation in the World Ocean*. Tulsa: Society of Economic Paleontologists and Mineralogists, Special Publication No. 17, 218pp.

Livingston B. T., Kilian C. E., Wilt F. *et al.* (2006) A genome-wide analysis of biomineralization-related proteins in the sea urchin *Strongylocentrotus purpuratus*. *Developmental Biology* **300**: 335–348.

Livingstone D. A. (1963) Chemical composition of rivers and lakes. In: Data on geochemistry, 6th edition. *U.S. Geol. Surv. Prof. Paper* **440**, Chapter G, 64pp.

Loeblich A. R. Jr., Tappan H. (1964) Foraminifera classification and evolution. *J. Geol. Soc. India* **5**: 5–39.

Loeblich A. R., Tappan H. (1974) Recent advances in the classification of the Foraminiferida. In *Foraminifera*, R. H. Hedley & C. G. Adams (eds.). London: Academic Press, Vol. 1, 276pp.

Longinelli A. (1984) Oxygen isotopes in mammal bone phosphate: a new tool for paleohydrological and paleoclimatological research? *Geochim. Cosmochim. Acta* **48**: 385–390.

Lough J. M. (2004) A strategy to improve the contribution of coral data to high-resolution paleoclimatology. *Palaeogeogr. Palaeoclimatol. Palaeoecol.* **204**: 115–143.

Lowenstam H. A. (1954) Factors affecting the aragonite: calcite ratios in carbonate-secreting marine organisms. *J. Geol.* **62**: 284–322.

Lowenstam H. A. (1961) Mineralogy, O^{18}/O^{16} ratios, and strontium and magnesium contents of recent and fossil brachiopods and their bearing on the history of oceans. *J. Geol.* **69**: 241–260.

Lowenstam H. A. (1963) Biologic problems relating to the composition and diagenesis of sediments. In *The Earth Sciences: Problems and Progress in Current Research*, T. W. Donnelly (ed.). Chicago, IL: University of Chicago Press, pp. 137–195.

Lowenstam H. A. (1981) Minerals formed by organisms. *Science* **211**: 1126–1131.

Lowenstam H. A., Weiner S. (1989) *On Biomineralization*. Oxford: Oxford University Press, 324pp.

Luquet G., Testenière O., Graf F. (1996) Characterization and n-terminal sequencing of a calcium-binding protein from the calcareous concretion organic matrix of the terrestrial crustacean *Orchestia cavimana*. *Biochim. Biophys. Acta* **1293**(2): 272–276.

Lyell C. (1830–1833) *Principles of Geology*. London: J. Murray, 3 vol.

Machii A. (1968) Histological studies on the pearl-sac formation. *Bull. Natl. Pearl Res. Lab.* **13**: 1489–1539.

Macintyre I. G., Bayer F. M., Logan M. A. *et al.* (2000) Possible vestige of early phosphatic biomineralization in gorgonian octocorals. *Geology* **28**(5): 455–458.

Mackenzie F. T., Pigott J. D. (1981) Tectonic controls of Phanerozoic sedimentary rock cycling. *J. Geol. Soc. Lond.* **38**: 183–196.

Magdans U., Gies H. (2004) Single crystal structure analysis of sea urchin spine calcites: systematic investigations of the Ca/Mg distribution as a function of habitat of the sea urchin and the sample location in the spine. *Eur. J. Mineral.* **16**: 261–268.

Maliva R. G., Siever R. (1988a) Diagenetic replacement controlled by force of crystallization. *Geology* **16**(8): 688–691.

Maliva R. G., Siever R. (1988b) Mechanism and controls of silicification of fossils in limestones. *J. Geology* **96**(4): 387–398.

Maliva R. G., Knoll A. H., Siever R. (1989) Secular change in chert distribution: a reflection of evolving biological precipitation in the silica cycle. *Palaios* **4**: 519–532.

Manigault P. (1939) Recherche sur le calcaire chez les Mollusques: Phosphatases et précipitations calciques. Thèse, Faculté des Sciences, Université de Paris, 331pp.

Mann K., Macek B., Olsen J. V. (2006) Proteomic analysis of the acid-soluble organic matrix of the chicken calcified eggshell layer. *Proteomics* **6**: 3801–3810.

Mann S., Heywood B. R., Rajam S. *et al.* (1989) Interfacial control of nucleation of calcium carbonate under organized stearic acid monolayers. *Proc. R. Soc. London A* **423**(1865): 457–471.

Mao-Che L., Golubic S., Le Campion-Alsumard T. *et al.* (2001) Developmental aspects of biomineralization in the Polynesian pearl oyster *Pinctada margaritifera* var. *cumingi*. *Oceanol. Acta* **24**: 35–49.

Marin F., Dauphin Y. (1992) Malformations de la couche nacrée de l'huitre perlière *Pinctada margaritifera* (L.) de la Polynésie française: rapports entre altérations microstructurales et composition en acides aminés. *Annales Sciences Naturelles. Zoologie* **13**(4): 157–168.

Marin F., Muyzer G., Dauphin Y. (1994) Caractérisations électrophorétique et immunologique des matrices organiques solubles des tests de deux Bivalves Ptériomorphes actuels, *Pinna nobilis* L. et *Pinctada margaritifera* (L.). *C. R. Acad. Sci. Paris*, Sér. II **318**: 1653–1659.

Marschal C., Garrabou J., Harmelin J. G. *et al.* (2004) A new method for measuring growth and age in the precious red coral *Corallium rubrum* (L.). *Coral Reefs* **23**: 423–432.

Martill D. M., Wilby P. R. (1994) Lithified prokaryotes associated with fossil soft-tissues from the Santana Formation (Cretaceous) of Brazil. *Kaupia* **2**: 71–77.

Martin J. W., Davis G. E. (2001) *An Updated Classification of the Recent Crustacea*. Los Angeles, CA: Natural History Museum of Los Angeles County, Science Series, Vol. 39, 134pp.

Masters-Helfman P., Bada J. L. (1976) Aspartic acid racemisation in dentine as a measure of ageing. *Nature* **262**: 279–281.

Masuda F., Hirano M. (1980) Chemical composition of some modern marine pelecypod shells. *Sci. Rept. Inst. Geosc., Univ. Tsukuba, sect. B,* **1**: 163–177.

Matter P. III, Davidson F. D., Wyckoff R. W. G. (1969) The composition of fossil oyster shell proteins. *Proc. Nat. Acad. Sci. USA* **64**(3): 970–972.

McClintock Turbeville J., Schulz J. R., Raff R. A. (1994) Deuterostome phylogeny and the sister group of the chordates: evidence from molecules and morphology. *Mol. Biol. Evol.* **11**: 648–655.

McConnaughey T. (1989a) ^{13}C and [18]O isotope disequilibrium in biological carbonates: I. Patterns. *Geochim. Cosmochim. Acta* **53**: 151–162.

McConnaughey T. (1989b) ^{13}C and [18]O isotope disequilibrium in biological carbonates: II. In vitro simulation of kinetic isotope effects. *Geochim. Cosmochim. Acta* **53**: 163–171.

McConnell D. (1963) Inorganic constituents of the shell of the living brachiopod *Lingula*. *Geol. Soc. Am. Bull.* **74**: 363–364.

McCrea J. M. (1950) On the isotopic chemistry of carbonates and a paleotemperature scale. *J. Chem. Phys.* **18**: 849–857.

McGregor H. V., Gagan M. K. (2003) Diagenesis and geochemistry of *Porites* corals from Papua New Guinea: implications for paleoclimate reconstruction. *Geochim. Cosmochim. Acta* **37**: 2147–2156.

McLaren D. J. (1970) Time, life and boundaries. *Presid. Address at the Pal. Soc. Am. J. Paleontology* **44**: 801–815.

McMillan J., Miller D. J. (1990) Highly repeated DNA sequences in the scleractinian coral genus *Acropora*: evaluation of cloned repeats as taxonomic probes. *Mar. Biol.* **104**: 483–487.

Medina M., Weil E., Szmant A. M. (1999) Examination of the *Montastraea annularis* species complex (Cnidaria: Scleractinia) using ITS1 and CO1 sequences. *Mar. Biotechnol.* **1**: 89–97.

Meenakshi V. R., Hare P. E., Wilbur K. M. (1971) Amino acids of the organic matrix of neogastropod shells. *Comp. Biochem. Physiol. B* **40**: 1037–1043.

Meibom A., Cuif J. P., Hillion F. *et al.* (2004) Distribution of magnesium in coral skeleton. *Geophys. Res. Lett.* **31**: L23306, doi: 10.1029/2044GL021313.

Meibom A., Yurimoto H., Cuif J. P. *et al.* (2006) Vital effects in coral skeletal composition display strict three-dimensional control. *Geophys. Res. Lett.* **33**: L11608, doi: 10.1029/2006GL025968.

Meibom A., Mostefaoui S., Cuif J. P. *et al.* (2007) Biological forcing controls the chemistry of reef-building coral skeleton. *Geophys. Res. Lett.* **34**: L02601, doi: 10.1029/2006GL028657.

Meldrum F. C., Cölfen H. (2008) Controlling mineral morphologies and structures in biological and synthetic systems. *Chem. Rev.* **108**(11): 4332–4432.

Meldrum N. U., Roughton F. J. (1933) The state of carbon dioxide in blood. *J. Physiol.* **80**(2): 143–170.

Melnikova G. K. (2001) Coelenterata. In *Atlas of the Triassic Invertebrates from Pamir*, A. U. Rozanov & R. V. Severev (eds.). Moscow: Nauka, pp. 1–80 (in Russian).

Meyers W. J. (1977) Chertification in the Mississippian Lake Valley Formation, Sacramento Mountains, New Mexico. *Sedimentology* **24**: 75–105.

Milliman J. D. (1974) Marine carbonates. In *Recent Sedimentary Carbonates*. Berlin: Springer-Verlag, Vol. 1, 375pp.

Milne-Edwards H., Haime J. (1857) *Histoire Naturelle des Coralliaires ou Polypes Proprement Dits. Tome 2: Classification et Description des Zoanthaires Sclerodermés de la Section des Madréporaires Apores*. Paris: Librairie Encyclopédique Roret, 633pp.

Minchin E. A. (1909) Sponges spicules: a summary of the present knowledge. *Ergebnisse und Fortschritte der Zoologie* **2**: 171–274.

Mitterer R. M. (1978) Amino acid composition and metal binding capability of the skeletal protein of corals. *Bull. Marine Sci.* **28**(1): 173–180.

Montanaro-Gallitelli E. (1973) Microstructure and septal arrangement in a primitive Triassic coral. *Boll. Soc. Paleont. Ital.* **12**: 8–22.

Mori K. (1976) A new sclerosponge from Ngargol, Palau Island, and its fossil relatives. *Tohoku Univ. Sci. Report 2, Ser. Geol.* **46**: 1–9.

Morse J. W., Mackenzie F. T. (1990) Geochemistry of sedimentary carbonates. In *Developments in Sedimentology*. Amsterdam: Elsevier, Vol. 48, 696pp.

Moss M. L. (1977) Skeletal tissues in sharks. *Am. Zool.* **17**: 335–342.

Moss-Salentijn L., Moss M. L., Yuan M. S. (1997) The ontogeny of mammalian enamel. In *Tooth Enamel Structure*, W. Koenigswald & P. M. Sander (eds.). Rotterdam: Balkema, pp. 5–30.

Moynier de Villepoix R. (1892) Note sur le mode de production des formations calcaires du test des Mollusques. *Mém. Soc. Biol.* **4**: 35–42.

Mucci A., Morse J. W. (1983) The incorporation of Mg^{2+} and Sr^{2+} into calcite overgrowths: influence of growth rate and solution composition. *Geochim. Cosmochim. Acta* **47**: 217–233.

Mugiya Y. (1965) Calcification in fish and shell-fish: IV. The differences in nitrogen content between the translucent and opaque zones of otolith in some fish. *Bull. Jpn. Soc. Sci. Fish.* **31**: 896–901.

Müller W. E. G. (2003) The origin of Metazoan complexity: Porifera as integrated animals. *Integr. Comp. Biol.* **43**: 3–10.

Müller W. E. G., Boreiko A., Wang X. *et al.* (2007a) Silicateins, the major biosilica forming enzymes present in demosponges: protein analysis and phylogenetic relationship. *Gene* **395**(12): 62–71.

Müller W. E. G., Eckert C., Kropf K. *et al.* (2007b) Formation of giant spicules in the deep-sea hexactinellid *Monorhaphis chuni* Schulze 1904: electron microscopic and biochemical studies. *Cell Tissue Res.* **329**: 363–378.

Müller W. E. G., Wang X., Burghard Z. *et al.* (2009) Bio-sintering processes in hexactinellid sponges: fusion of biosilica in giant basal spicules from *Monorhaphis chuni*. *J. Struct. Biol.* **168**: 548–561.

Münster G. von (1841) *Beïträge zur Geognosie und Petrefacteden kunde des südöstlichen Tirols*. Berlin: Planzenthiere, Bd. 1, pp. 25–39.

Murshed M., Harmey D., Millan J. L. *et al.* (2005) Unique coexpression in osteoblasts of broadly expressed genes accounts for the spatial restriction of ECM mineralization to bone. *Genes Dev.* **19**: 1093–1104.

Muscatine L., Porter J. W. (1977) Reef corals: mutualistic symbioses adapted to nutrient-poor environments. *Bioscience* **27**: 454–460.

Muscatine L., Goiran C., Land L. *et al.* (2005) Stable isotopes (δ^{13}C and δ^{15}N) of organic matrix from coral skeletons. *Proc. Nat. Acad. Sci. USA* **102**(5): 1525–1530.

Mutvei H. (1964) On the shells of *Nautilus* and *Spirula* with notes on the shell secretion in non cephalopod mollusks. *Arkiv Zool.* **16**(14): 221–278.

Mutvei H. (1970) Ultrastructure of the mineral and organic components of molluscan nacreous layers. *Biomineralization* **2**: 48–72.

Mutvei H. (1977) The nacreous layer in *Mytilus, Nucula*, and *Unio* (Bivalvia). *Calcif. Tissue Res.* **24**: 11–18.

Mutvei H. (1978) Ultrastructural characteristics of the nacre of some Gastropods. *Zoologica Scripta* **7**: 287–296.

Mutvei H. (1979) On the internal structures of the nacreous tablets in molluscan shells. *Scanning Electron Microscopy* **II**: 451–462.

Mutvei H., Dauphin Y., Cuif J. P. (1985) Observations sur l'organisation de la couche externe du test des *Haliotis* (Gastropoda): un cas exceptionnel de variabilité minéralogique et microstructurale. *Bull. Mus. Natn. Hist. Nat. Paris, Sér.* 4, sect. A **1**: 73–91.

Nakahara H. (1979) An electron microscope study of the growing surface of nacre in two gastropod species, *Turbo cornutus* and *Tegula pfeifferi*. *Venus* **38**(3): 205–211.

Nakahara H. (1983) Calcification of gastropod nacre. In *Biomineralization and Biological Metal Accumulation*, P. Westbroek & E. W. De Jong (eds.). Dordrecht, The Netherlands: Reidl D. Publishers, pp. 225–230.

Nakahara H., Bevelander G., Kakai M. (1982) Electron microscopic and amino acid studies on the outer and inner shell layers of *Haliotis rufescens*. *Venus* **41**(1): 33–46.

Necker de Saussure L. A. (1839) Note sur la nature minéralogique des coquilles terrestres, fluviatiles et marines. *Ann. Sci. Nat. Sér.* 2 (Zoologie) **9**: 52–55.

Nelson D. M., Treguer P., Brzezinski M. A. *et al.* (1995) Production and dissolution of biogenic silica in the ocean: revised global estimates, comparisons with regional data and relationship to biogenic sedimentation. *Glob. Biogeochem. Cycle* **9**: 359–372.

Neville A. C. (1975) *Biology of the Arthropod Cuticle*. Berlin: Springer-Verlag, 448pp.

Newell, N. D. (1967) Revolution in the history of life. *Geol. Soc. Am.* spec. pap. **89**: 63–91.

Nicholson H. A. (1886) On some new or imperfectly known species of Stromatoporoids. *II.*
	Ann. Mag. Nat. Hist. Ser. **5**(18): 8–22.
Nissen H. U. (1963) Röntgengefügeanalyse am Kalzit von Echinodermenskeletten. *N. Jb.*
	Geol. Abh. **117**: 230–234.
Noll W. (1934) Geochemie des Strontiums. *Chem. d. Erde* **8**: 507–600.
Nothdurft L. D., Webb G. (2007) Microstructure of common reef-building coral genera
	Acropora, Pocillopora, Goniastrea and *Porites*: constraints on spatial resolution in
	geochemical sampling. *Facies* **53**: 1–26.
Nudelman F., Gotliv B. A., Addadi L. *et al.* (2006) Mollusk shell formation: mapping the
	distribution of organic matrix components underlying a single aragonitic tablet in
	nacre. *J. Struct. Biol.* **153**: 176–187.
Nudelman N., Chen H. H., Goldberg H. A. *et al.* (2007) Lessons from biomineralization:
	comparing the growth strategies of mollusc shell prismatic and nacreous layers in
	Atrina rigida. Faraday Discuss. **136**(9), doi: 10.1039/b704418f.
Odum H. T. (1957) Biochemical deposition of strontium. *Inst. Marine Sci.* **4**: 38–114.
Ogilvie M. M. (1895) Microscopic and systematic study of madreporarian types of corals.
	Proc. R. Soc. London **59**: 9–18.
Ogilvie M. M. (1896) Microscopic and systematic study of madreporarian types of corals.
	Phil. Trans. R. Soc. London B **187**: 83–345.
Ohde S., Kitano Y. (1984) Co-precipitation of strontium with marine Ca-Mg carbonates.
	J. Geochem. **18**: 143–146.
Oliver W. A. Jr. (1980) On the relationship between Rugosa and Scleractinia. *Acta*
	Palaeontol. Pol. **25**(3–4): 395–402.
Oliver W. A. Jr., Coates A. (1987) Phylum Cnidaria. In *Fossil Invertebrates*, R. S. Boardman,
	A. H. Cheetham & A. J. Rowell (eds.). Oxford: Blackwell Scient. Pub., pp. 140–193.
Osborn J. W. (1965) The nature of the Hunter-Schreger bands in enamel. *Arch. Oral Biol.*
	10: 929–993.
Ouizat S., Barroug A., Legrouri A. *et al.* (1999) Adsorption of bovine serum albumin
	on poorly crystalline apatite: influence of maturation. *Mater. Res. Bull.* **34**: 2279–2289.
Owen R., Kennedy H., Richardson C. A. (2002) Isotopic partitioning between scallop shell
	calcite and seawater: effect of shell growth rate. *Geochim. Cosmochim. Acta* **66**(10):
	1727–1737.
Oxman D. S., Barnett-Johnson R., Smith M. E. *et al.* (2007) The effect of vaterite deposition
	on sound reception, otolith morphology, and inner ear sensory epithelia in hatchery-
	reared Chinook salmon. *Can. J. Fish. Aquat. Sci.* **64**(11): 1469–1478.
Pappenhöfer G. A., Harris R. P. (1979) Laboratory cultures of marine holozooplankton
	and its contribution to studies of marine planktonic food webs. *Adv. Marine Biol.*
	16: 211–299.
Parkinson D., Curry G. B., Cusack M. *et al.* (2005) Shell structure, patterns and trends of
	oxygen and carbon stable isotopes in modern brachiopod shells. *Chemical Geol.*
	219: 193–235.
Pasteris J. D., Wopenka B., Freeman J. J. *et al.* (2004) Lack of OH in nanocrystalline apatite
	as a function of degree of atomic order: implications for bone and biomaterials.
	Biomaterials **25**: 229–238.
Pasteris J. D., Yoder C. H., Rogers K. D. *et al.* (2007) Bone apatite: the secret is in the
	carbonate. *Geol. Soc. Am., Ann. Meet. Denver* **39**(6): 295.
Paul C. R. C. (1979) Early echinoderm radiation. In *The Origin of Major Invertebrate*
	Groups, M. R. House (ed.), Systematics Association, Special Publication 21. New
	York: Academic Press, pp. 72–88.

Paul C. R. C. (1988) The phylogeny of the cystoids. In *Echinoderm Phylogeny and Evolutionary Biology*, C. R. C. Paul & A. B. Smith (eds.). Oxford: Clarendon Press, pp. 199–213.

Pérez-Huerta A., Cusack M., Jeffries T. *et al.* (2008) High resolution distribution of magnesium and strontium and the evaluation of Mg/Ca thermometry in recent brachiopod shells. *Chem. Geol.* **247**: 229–241.

Peters W. (1972) Occurrence of chitin in Mollusca. *Comp. Biochem. Physiol.* B **41**: 541–550.

Petit H. (1978) Recherches sur des séquences d'évènements périostracaux lors de l'élaboration de la coquille d'*Amblema plicata Conrad*, 1834. Thèse, Laboratoire de Zoologie, Université de Bretagne occidentale, 76pp.

Pingitore N. E., Meitzner G., Love K. M. (1995) Identification of sulfate in natural carbonates by X-ray absorption spectroscopy. *Geochim. Cosmochim. Acta* **59**(12): 2477–2483.

Pisa M., Jammet C., Laurent D. (2002) First steps of otolith formation of the zebrafish: role of glycogen? *Cell Tissue Res.* **310**: 163–168.

Poole D. F. G. (1967) Phylogeny of tooth tissues: enameloid and enamel in recent vertebrates, with a note on the history of cementum. In *Structural and Chemical Organization of Teeth*, A. E. W. Miles (ed.). New York: Academic Press, Vol. I, pp. 111–149.

Popp B. N., Podosek F. A., Brannon J. C. *et al.* (1986) ^{87}Sr/^{86}Sr ratios in Permo-Carboniferous sea water from the analyses of well-preserved brachiopod shells. *Geoch. Cosmochim. Acta* **50**(7): 1321–1328.

Pratoomchat B., Sawangwong P., Guedes R. *et al.* (2002) Cuticle ultrastructure changes in the crab *Scylla serrata* over the molt cycle. *J. Exp. Biol.* **293**: 414–426.

Pratz E. (1882) Über die verwandschaftlichen Beziehungen einiger Korallengattungen mit hauptsächlicher Berücksichtigung ihrer Septalstructur. *Palaeontographica* **29**: 81–122.

Prenant M. (1925) Contributions à l'étude cytologique du calcaire. *Bull. Biol. Fr. Belg.* **58**: 403–434.

Putnis A., Prieto M., Fernández-Díaz L. (1995) Fluid supersaturation and crystallisation in porous media. *Geol. Mag.* **132**: 1–13.

Puura I., Nemliher J. (2001) Apatite varieties in recent and fossil linguloid brachiopod shells. In *Brachiopods, Past and Present*, C. H. C. Brunton, L. R. M. Cocks, S. M. Long & S. L. Long (eds.). London: Taylor & Francis, pp. 7–16.

Puverel S. (2004) La biominéralisation chez les coraux Scléractiniaires. Etude de la matrice organique et des transports ioniques. Thèse, Université de Nice – Sophia Antipolis, mémoire du centre scientifique de Monaco, 178pp.

Qi W. (1984) An Anisian coral fauna in Guizhou, South China. In: Fourth International Symposium on Fossil Cnidaria, Washington D.C., August 1983, *Palaeontographica Americana* **54**: 187–190.

Raabe D., Romano P., Sachs C. *et al.* (2006) Microstructure and crystallographic texture of the chitin-protein network in the biological composite material of the exoskeleton of *Homarus americanus. Mater. Sci. Eng.* A **421**: 143–153.

Raczynski J., Ruprecht A. L. (1974) The effect of digestion on the osteological composition of owl pellets. *Acta Ornithologica* **14**: 25–38.

Ranner H., Ladriére O., Navez J. *et al.* (2005) Do echinoderms store temperature changes in their skeleton? *Geophys. Res. Abstr.* **7**: 01098, 1607–7962/gra/EGU05-A-01098.

Raup D. M. (1962) Crystallographic data in echinoderm classification. *Systematic Zoology* **11**(3): 97–108.

Raup D. M., Sepkowski J. J. Jr. (1982) Mass extinctions in the marine fossil record. *Science* **215**: 1501–1503.

Rauzer-Chernousova D. M., Fursenko A. V. (1959) The sub-class Foraminifera. *Osnovy Paleontologii, Moscow Izd. Akad SSSR*, 12–211.

Read B. A., Wahlund T. M. (2007) Molecular approach to *Emiliana huxleuyi* coccolith formation. In *Handbook of Biomineralization: Biological Aspects and Structure Formation*, P. Behrens & E. Beaeuerlein (eds.). Weinheim, Germany: Wiley-VCH, pp. 227–240.

Reidel P. (1991) Triassic corals of the Tethys: stratigraphical range, diversity patterns, evolutionary trends and their significance as reef building organisms. *Mitt. Ges. Geol. Bergbaustud. Österr.* **37**: 97–118.

Reiswig H. M. (1971) The axial symmetry of sponge spicules and its phylogenetic significance. *Cahiers Biol. Mar.* **12**: 505–514.

Reitner J. (1991) Phylogenetic aspects and new descriptions of spicule-bearing Hadromerid sponges with a secondary calcareous skeleon (Tetractinomorpha, Demospongiae). In *Fossil and Recent Sponges*, J. Reitner & F. Keupp (eds.). Berlin: Springer, pp. 179–211.

Reitner J., Engeser T. (1987) Skeletal structures and habitats of recent and fossil *Acanthochaetetes* (subclass Tetractinomorpha, Demospongiae, Porifera). *Coral Reefs* **6**: 151–157.

Reitner J., Wörheide G. (2002) Non-lithistid fossil Demospongiae: origin of their paleobiodiversity and highlights in history of preservation. In *Systema Porifera: A Guide to the Classification of Sponges*, J. N. A. Hooper & R. W. M. Van Soest (eds.). New York: Kluwer Akademic/Plenum Pub., pp. 52–68.

Rensberger J. M. (1997) Mechanical adaptation in enamel. In *Tooth Enamel Microstructure*, W. Koenigswald & P. M. Sander (eds.). Rotterdam: Balkema, pp. 237–257.

Richards A. G. (1951) *The Integuments of Arthropods*. Minneapolis, MN: University of Minnesota Press, 324pp.

Ricqlès de A. (1975) Recherches paléohistologiques sur les os longs des tétrapodes: VII. Sur la classification, la signification fonctionelle et l'histoire des tissus osseux des tétrapodes. 1è partie: structures. *Annales de Paléontol.* **61**: 49–149.

Robach J. S., Stock S. R., Veis A. (2005) Transmission electron microscopy characterization of macromolecular domain cavities and microstructure of single-crystal calcite tooth plates of the sea urchin *Lytechinus variegatus*. *J. Struct. Biol.* **151**: 18–29.

Roche J., Ranson G., Eysseric-Lafon M. (1951) Sur la composition des scléroprotéines des coquilles des mollusques (conchiolines). *C. R. Séances Soc. Biol. Fr.* **145**(19–20): 1474–1477.

Rollion-Bard C. (2001) Variability of oxygen isotopes in *Porites* corals: development and implications of stable isotopes (B, C and O) microanalysis by ion microprobe. Thèse, Institute National Polytechnique de Lorraine, Nancy, 165pp.

Rollion-Bard C., Blamart D., Cuif J. P. *et al.* (2003) Microanalysis of C and O isotopes of azooxanthellate and zooxanthellate corals by ion microprobe. *Coral Reefs* **22**: 405–415.

Rollion-Bard C., Blamart D., Cuif J. P. *et al.* (2010) *In situ* measurements of oxygen isotopic composition in deep-sea coral, *Lophelia pertusa*: re-examination of the current geochemical models of biomineralization. *Geochim. Cosmochim. Acta* **74**: 1338–1349.

Romano S. L., Cairns S. D. (2000) Molecular phylogenetic hypotheses for the evolution of scleractinian corals. *Bull. Mar. Sci.* **67**: 1043–1068.

Romano S. L., Palumbi S. R. (1996) Evolution of scleractinian corals inferred from molecular systematics. *Science* **271**: 640–642.

Romano S. L., Palumbi S. R. (1997) Molecular evolution of a portion of the mitochondrial 16S ribosomal gene region in scleractinian corals. *J. Mol. Evol.* **45**: 397–411.

Roniewicz E., Stolarski J. (1999) Evolutionary trends in the epithecate scleractinian corals. *Acta Paleont. Pol.* **44**: 131–166.

Rose G. (1858) Über die heteromorphen Zustände des kohlensauren Kalkerde: II. Vorkommer des Aragonits und Kalkspaths in der orgnaischen Natur. *Abhandl. König. Akad. Wiss. Berlin, Abt. Physik* **81**: 63–111.

Rosenberg G. D. (1980) An ontogenic approach to the environmental significance of bivalve shell chemistry. In *Skeletal Growth of Aquatic Organisms. Biological Records of Environmental Change*, D. C. Rhoads & R. A. Lutz (eds.). New York: Plenum Press, pp. 133–168.

Rosenberg G. D., Hughes W. W., Tkachuk R. D. (1989) Shell form and metabolic gradients in the mantle of *Mytilus edulis. Lethaia* **22**(3): 229–344.

Rousseau M., Bedouet L., Lati E. *et al.* (2006) Restoration of stratum corneum with nacre lipids. *Comp. Bichem. Physiol. B* **145**: 1–9.

Rudall K. M. (1963) The chitin/protein complexes of insect cuticles. *Adv. Insect Physiol.* **1**: 257–313.

Runcorn S. K. (1966) Change in the moment of inertia of the Earth as a result of a growing core. In *The Earth-Moon System*, B. G. Marsden & A. G. W. Cameron (eds.). New York: Plenum Press, pp. 82–92.

Sakae T., Suzuki K., Kozawa Y. (1997) A short review of studies on chemical and physical properties of enamel. In *Tooth Enamel Structure*, W. Koenigswald & P. M. Sander (eds.). Rotterdam: Balkema, pp. 31–39.

Samata T. (1990) Ca-binding glycoproteins in molluscan shells with different types of ultrastructure. *The Veliger* **33**(2): 190–201.

Samata T., Sanguansri P., Cazaux C. *et al.* (1980) Biochemical studies on components of mollusc shells. In *The Mechanisms of Biomineralization in Animals and Plants, Proc. 3rd Intern. Biomin. Symp.*, M. Omori & N. Watabe (eds.). Kanagawa: Tokai University Press, pp. 37–47.

Sandberg P. A. (1983) An oscillating trend in Phanerozoic non-skeletal carbonate mineralogy. *Nature* **305**: 19–22.

Sandford F. (2003) Physical and chemical analysis of the siliceous skeletons in six sponges of two groups (Demospongiae and Hexactinellida). *Microsc. Res. Tech.* **62**: 336–355.

Sasagawa I. (2002) Mineralization patterns in Elasmobranch fish. *Microsc. Res. Techn.* **59**: 396–402.

Sasagawa I., Ishiyama M. (1999) The features of enameloid formation during odontogenesis in teleosts. In *Dental Morphology*, J. T. Mayhall & T. Heikkinen (eds.). Oulu, Finland: Oulu University Press, pp. 285–292.

Sasagawa I., Ishiyama M. (2005) Fine structural and cytochemical mapping of enamel organ during the enameloid formation stages in gars, *Lepisosteus oculatus*, Actinopterigii. *Arch. Oral Biol.* **50**: 373–391.

Sasaki T. (1990) Cell biology of tooth enamel formation. In *Monographs in Oral Science*, H. M. Myers (ed.). Basel: Karger, pp. 1–204.

Satchell P. G., Aanderton X., Ryu O. H. *et al.* (2002) Conservation and variation in enamel protein distribution during vertebrate tooth development. *J. Exp. Zool. (Mol. Dev. Evol.)* **294**: 91–106.

Satchell P. G., Shuler C. F., Diekwisch T. G. H. (2000) True enamel covering in teeth of the Australian lungfish *Neoceratodus forsteri*. *Cell Tissue Res.* **299**: 27–37.

Schindewolf O. H. (1942) *Zur Kentniss des Polycoelien und Plerophyllen.* Berlin: Reichsamt f. Bodenforschung, 324pp.

Schleiden M. (1838) *Beiträge zur Phytogenesis*. Leipzig: Archiv für Anatomie, Physiologie und wissenschaftliche Medicin, "Muller's Archiv", pp. 137–176.

Schlossenberger J. E. (1856) *Erster Versuch einer Allgemeiner und Vergleichenden Thier-Chemie.* Leipzig & Heidelberg: Winter, 364pp.

Schmahl W. W., Griesshaber E., Neuser R. *et al.* (2004) The microstructure of the fibrous layer of Terebratulide brachiopod shell calcite. *Eur. J. Mineral.* **16**: 693–697.

Schmahl W. W., Griesshaber E., Merkel C. *et al.* (2008) Hierarchical fibre composite structure and micromechanical properties of phosphatic and calcitic brachiopod shell biomaterials: an overview. *Min. Mag.* **72**(2): 541–562.

Schmid A-M., Schultz D. (1979) Wall morphogenesis in diatoms: deposition of silica by cytoplasmic vesicles. *Protoplasma* **100**: 267–288.

Schmidt W. J. (1924) *Die Bausteine des Tierkorpers in Polarisiertem Lichte.* Bonn: Cohen Verlag, 528pp.

Schmidt W. J., Keil A. (1958) *Die gesunden und die erkankten Zahngewebe des Ménschen und der Wirbeltiere im Polarisationsmikroskop.* Munich: C. Hanser Verlag, 386pp.

Schoeninger M. J., Deniro M. J. (1982) Carbon isotope ratios of apatite from fossil bone cannot be used to reconstruct diets of animals. *Nature* **197**: 577–578.

Schröder H. C., Natalio F., Shukoor I. *et al.* (2007) Apposition of silica lamellae during growth of spicules in the demosponge *Suberites dommuncula*: biological/biochemical studies and chemical/biomimetical confirmation. *J. Struct. Biol.* **159**: 324–334.

Schroder J. H., Purser B. H. (1986) *Reef Diagenesis*. Berlin: Springer-Verlag, 455pp.

Schroeder J. H., Dworki K. E. J., Papike J. J. (1969) Primary protodolomite in echinoid skeletons. *Geol. Soc. Am. Bull.* **80**: 1613–1616.

Schubert J. K., Kidder D., Erwin D. H. (1997) Silica replaced fossils through the Phanerozoic. *Geology* **25**: 1031–1034.

Schwab D. W., Shore R. E. (1971) Fine structure and composition of a siliceous sponge spicule. *Biol. Bull.* **140**: 125–136.

Schwann T. (1839) Mikroskopische Untersuchungen über die Übereinstimmung in der Struktur und dem Wachstum der Thiere und Pflanzen. Berlin: Reimer Buchhandlung, 268pp.

Schweitzer M. H., Wittmeyer J. L., Homer J. R. (2007) Soft tissue and cellular preservation in vertebrate skeletal elements from the Cretaceous to present. *Proc. R. Soc. London B* **274**: 183–197.

Segar D. A., Collins J. D., Riley J. P. (1971) The distribution of the major and some minor elements in marine animals: Part II. Molluscs. *J. Mar. Biol. Ass. U.K.* **51**: 131–136.

Sharp Z. D., Atudorei V., Furrer H. (2000) The effect of diagenesis on oxygen isotope ratios of biogenic phosphates. *Am. J. Sci.* **300**: 222–237.

Shimizu K., Cha J., Stucky G. D. *et al.* (1998) Silicatein alpha: Cathepsin L-like protein in sponge biosilica. *Proc. Nat. Acad. Sci. USA* **95**: 6234–6238.

Siks I., Hsiao B. S., Chi B. *et al.* (2008) Lateral packing of mineral crystals in bone collagen fibrils. *Biophys. J.* **95**: 1985–1992.

Silyn-Roberts H., Sharp R. M. (1985) Preferred orientation of calcite and aragonite in the reptilian eggshells. *Proc. R. Soc. London B* **225**: 445–455.

Simkiss K. (1965) The organic matrix of the oyster shell. *Comp. Biochem. Physiol.* **16**: 427–435.

Simkiss K. (1994) Amorphous minerals in biology. *Mém. Inst. Océanogr. Monaco* **14**(1): 49–54.

Sinclair D. J., McCulloch M. T. (2004) Corals record low mobile barium concentrations in the Burdekin River during the 1974 flood: evidence for limited Ba supply to rivers? *Palaeogeogr. Palaeoclimatol. Palaeoecol.* **214**: 155–174.

Smout A. H. (1955) Reclassification of the Rotaliidae (foraminifera). *J. Wash. Acad. Sci.* **45**: 201–210.

Sorauf J. E. (1972) Skeletal microstructure and microarchitecture in Scleractinia (Coelenterata). *Paleontology* **15**(1): 88–107.

Sorauf J. E. (1999) Skeletal microstructure, geochemistry and organic remnants in Cretaceous scleractinian corals: Santonian Gosau Beds of Gosau, Austria. *J. Paleont.* **74**: 1029–1041.

Sorauf J. E., Jell J. S. (1977) Structure and incremental growth in the ahermatypic coral *Desmophyllum cristagalli* from the north Atlantic. *Palaeontology* **20**(1): 1–19.

Sorauf J. E, Podoff N. (1977) Skeletal structure in deep water ahermatypic corals. 2nd Int. Symp. Corals and Fossil Corals Reefs, Paris, 1975, mém. B.R.G.M., Vol. 89, pp. 2–11.

Sorauf J. E., Webb G. E. (2003) The origin and significance of zigzag microstructure in Late Paleozoic *Lophophyllidium* (Anthozoa, Rugosa). *J. Paleontology* **77**(1):16–30.

Sorby H. C. (1879) The structure and origin of limestones. *Geol. Soc. London Proc.* **35**: 56–95.

Spaeth C. (1971) Aragonitische und calcitische Primärstrukturen im Schalenbau eines belemniten aus der englischen Unterkreide. *Paläont. Zeit.* **45**: 33–40.

Spaeth C. (1975) Zur Frage der Schwimmverhaltnisse bei Belemniten in Abhangigkeit vom Primärgefuge der Hartteile. *Paläont. Zeit.* **49**: 321–331.

Sparks N. H. C., Motta P. J., Shellis R. P. *et al.* (1990) An analytical electron microscopy study of iron-rich teeth from the butterflyfish (*Chaetodon ornatissimus*). *J. Exp. Biol.* **151**: 371–385.

Stanley G. D., Swart P. K. (1995) Evolution of the coral-zooxanthellae symbiosis during the Triassic: a geochemical approach. *Paleobiology* **21**(2): 179–199.

Stanley S. M., Hardie L. A. (1998) Secular oscillations in the carbonate mineralogy of reef-building and sediment-producing organisms driven by tectonically forced shifts in seawater chemistry. *Palaeog. Palaeoclim. Palaeoecol.* **144**: 3–19.

Stearn C. W., Pickett J. W. (1994) The stromatoporoid animal revisited: building the skeleton. *Lethaia* **27**: 1–10.

Stein C. L. (1982) Silica recrystallization in petrified wood. *J. Sediment. Petrol.* **52**: 1277–1282.

Steinmann G. (1882) Pharetronen Studien. *N. Jb. Mineral.* **2**: 141–191.

Steyger P. S., Wiederhold M. L. (1995) Visualization of aragonitic otoconial matrices in the newt using transmission electron microscopy. *Hear. Res.* **92**: 184–191.

Steyger P. S., Wiederhold M. L., Batten J. (1995) The morphogenic features of otoconia during larval development of *Cynops pyrrhogaster*, the Japanese red-bellied newt. *Hear. Res.* **84**(1–2): 61–71.

Stock C. W. (2001) Stromatoporoidea, 1926–2000. *J. Paleont.* **75**: 1079–1089.

Stolarski J., Roniewicz E. (2001) Towards a new synthesis of evolutionary relationships and classification of Scleractinia. *J. Paleontology* **75**(6): 1090–1108.

Stolarski J., Russo A. (2001) Evolution of the post-Triassic pachythecaline corals. *Bull. Biol. Soc. Washington* **10**: 242–256.

Stolarski J., Meibom A., Radoslaw P. *et al.* (2007) A Cretaceous scleractinian coral with a calcitic skeleton. *Science* **318**: 92–94.

Stolkowki J. (1951a) Essai sur le déterminisme des formes minéralogiques du calcaire chez les êtres vivants (calcaires coquilliers). *Ann. Biol.* **27**(11/12): 781–784.

Stolkowski J. (1951b) Essai sur le déterminisme des formes minéralogiques du calcaire chez les êtres vivants (calcaires coquilliers). *Ann. Inst. Océanogr.* **XXVI**: 1–113.

Stolley E. (1919) Die systematik der belemniten. *Jahresb. d. Niedersächs. Geol. Ver.* **11**: 1–59.

Stürmer W. (1985) A small coleoid cephalopod with soft parts from the Lower Devonian discovered using radiography. *Nature* **318**: 53–55.

Su X., Kamat S., Heuer A. H. (2000) The structure of sea urchin spines, large biogenic single crystals of calcite. *J. Mater. Sci.* **35**: 5545–5551.

Suga S., Taki Y., Ogawa M. (1992) Iron in the enameloid of perciform fish. *J. Dent. Res.* **71**(6): 1316–1325.

Sullivan C. H., Krueger H. W. (1981) Carbon isotope analysis of separate chemical phases in modern fossil bone. *Nature* **292**: 333–335.

Sumper M., Brunner E. (2008) Silica biomineralisation. In: Diatoms – The model organism *Thalassiosira pseudonana*. *ChemBioChem* **9**: 1187–1194.

Sumper M., Brunner E., Lehmann G. (2005) Biomineralization in diatoms: characterization of novel polyamines associated with silica. *FEBS Letters* **579**: 3765–3769.

Sundar V. C., Yablon A. D., Grazul J. L. *et al.* (2003) Fibre-optical features of a glass sponge. *Nature* **424**: 899–900.

Swanson R., Hoegh-Guldberg O. (1998) Amino acid synthesis in the symbiotic sea anemone *Aiptasia pulchella*. *Mar. Biol.* **13**: 83–93.

Swart P. K. (1983) Carbon and oxygen isotope fractionation in scleractinian corals: a review. *Earth Sci. Rev.* **19**: 51–80.

Tadashi S., Mugiya Y. (1996) Biochemical properties of water-soluble otolith proteins and the immune-biochemical detection of the proteins in serum and various tissues of the tilapia *Oreochromis niloticus*. *Fish Sci.* **62**: 970–976.

Tambutté E. (1996) Processus de calcification d'un scléractiniaire hermatypique, *Stylophora pistillata* (Esper, 1797). Croissance in situ à Mururoa. Thèse Doctorat, Universite de Nice – Sophia Antipolis Fac. Sc. 292pp.

Tanaka S., Hatano H., Itasaka O. (1960a) Biochemical studies on pearl: IX. Amino acid composition of conchiolin in pearl and shell. *Bull. Chem. Soc. Jap.* **33**(4): 543–545.

Tanaka S., Hatano H., Suzue G. (1960b) Biochemical studies on pearl: VII. Fractionation and terminal amino acids of conchiolin. *J. Biochem.* **47**(1): 117–123.

Taylor J. D., Krennedy W. J., Hall A. (1969) The shell structure and mineralogy of the Bivalvia: I. Introduction. Nuculacae – Trigonacae. *Bull. Br. Mus. Nat. Hist. Zool.* **3**: 1–125.

Taylor J. D., Kennedy W. J., Hall A. (1973) The shell structure and mineralogy of the Bivalvia: II. Lucinacea – Clavagellacea. *Conclusions. Bull. Br. Mus. Nat. Hist. Zool.* **22**: 253–294.

Teichert C. (1967) Major features of cephalopod evolution. In: Essays in Paleontology and Stratigraphy, C. Teichert & E. L. Yochelson (eds.). R. C. Moore comm., University of Kansas Special Paper, Vol. 2, pp. 162–210.

Teng H. H., Dove P. M., De Yoreo J. J. (2000) Kinetics of calcite growth: surface processes and relationships to macroscopic rate laws. *Geochim. Cosmochim. Acta* **64**: 2255–2266.

Termier H., Termier G. (1973) Stromatopores, Sclérosponges et Pharétrones: les Ischyrospongia. *Livre Jub. M. Soulignac, Ann. Mines Géol. Tunisie* **26**: 285–297.

Thamatrakoln K., Hildebrand M. (2005) Approaches for functional characterization of diatom silicic acid transporters. *J. Nanosci. Nanotechnol.* **5**: 158–166.

Thiele H., Awad A. (1969) Nucleation and oriented crystallization in ionotropic gels. *Biomed. Mat. Res.* **3**: 431–441.

Tomàs J., Geffen A. J. (2003) Morphometry and composition of aragonite and vaterite otoliths of deformed laboratory reared juvenile herring. *J. Fish Biology* **63**(6): 1383–1401.

Tomas J., Geffen A. J., Allena I. S. *et al.* (2004) Analysis of the soluble matrix of vaterite otoliths of juvenile herring (*Clupea harengus*): do crystalline otoliths have less protein? *Comp. Biochem. Physiol.* A **139**: 301–308.

Toots H., Voorhies M. R. (1965) Strontium in fossil bones and the reconstruction of food chains. *Science* **149**: 854–885.

Towe K. M. (1967) Echinoderm calcite: single crystal or polycrystalline aggregate. *Science* **57**: 1048–1050.

Towe K. M. (1972) Invertebrate shell structure and the organic matrix concept. *Biomineralization* **4**: 1–7.

Towe K. M., Cifelli R. (1967) Wall ultrastructure in the calcareous foraminifera: crystallographic aspects and model for calcification. *J. Paleontology* **41**(3): 742–762.

Towe K. M., Lowenstam H. A. (1967) Ultrastructure and development of iron mineralization in the radular teeth of *Cryptochiton stelleri* (Mollusca). *J. Ultrastr. Res.* **17**: 1–13.

Traub W., Arad T., Weiner S. (1989) Three-dimensional ordered distribution of crystals in turkey tendon collagen fibers. *Proc. Natl. Acad. Sci. USA* **86**: 9822–9826.

Travis D. F. (1963) Structural features of mineralization from tissue to macromolecular levels of organization in the decapod Crustacea. *Ann. New York Acad. Sci.* **109**: 177–245.

Travis D. F. (1965) The deposition of skeletal structures in the crustacean: V. The histomorphological and histochemical changes associated with the development and calcification of the branchial exoskeleton in the crayfish *Orconectes virilis*. *Hagen Acta Histochem.* **20**: 193–233.

Travis D. F. (1968) The structure and organization of, and the relationships between the inorganic crystals and the organic matrix of the prismatic region of *Mytilus edulis*. *J. Ultrastr. Res.* **23**: 183–215.

Travis D. F., Gonsalves M. (1969) Comparative ultrastructure and organization of the prismatic region of two bivalves and its possible relation to the chemical mechanism of boring. *Am. Zool.* **9**: 635–661.

Travis D. F., François C. J., Bonar L. *et al.* (1967) Comparative studies of the organic matrices of invertebrate mineralized tissues. *J. Ultrastr. Res.* **18**: 518–550.

Trequer P., Nelson M., Van Bennekom A. J. *et al.* (1995) The silica balance in the world ocean: a re-estimate. *Science* **268**: 375–379.

Tsipursky S. J., Buseck P. B. (1993) Structure of magnesian calcite from sea urchins. *Am. Mineral.* **78**: 775–781

Tsuji T., Sharp D. G., Wilbur K. M. (1958) Studies on shell formation: VII. The submicroscopic structure of the shell of the oyster *Crassostrea virginica*. *J. Biophys. Biochem. Cytol.* **4**(3): 275–279.

Turekian K. K., Armstrong R. L. (1960) Magnesium, strontium and barium concentrations and calcite-aragonite ratios of some recent molluscan shells. *J. Mar. Res.* **18**: 133–151.

Tzeng W. N., Severin K. P., Wickstrom H. *et al.* (1999) Strontium bands in relation to age marks in otoliths of european eel *Anguilla anguilla*. *Zool. Studies* **38**(4): 452–457.

Ubaghs G. (1967) Eocrinoidea. In *Treatise on Invertebrate Paleontology, part S Echinodermata 1*, R. C. Moore & C. Teichert (eds.). Lawrence, KS: Geological Society of America and The University of Kansas, pp. 445–495.

Urey H. C., Lowenstam H. A., Epstein S. *et al.* (1951) Measurement of paleotemperatures and temperatures of the Upper Cretaceous of England, Denmark and the Southeastern United States. *Geol. Soc. Am. Bull.* **62**: 399–416.

Vacelet J. (1964) Etude monographique de l'éponge calcaire Pharétronide de Méditerranée, *Petrobiona massiliana* Vacelet & Lévi. Les Pharétronides actuelles et fossiles. Thèse Fac. Sci. Univ. Aix-Marseille, 125pp.

Vacelet J. (1979) Description et affinités d'une éponge sphinctozoaire actuelle. In *Biologie des Spongiaires*. Colloque Internat. C.N.R.S. 291, C. Lévi & N. Boury-Esnault (eds.), Paris, pp. 483–493.

Vacelet J. (1983) Les éponges hypercalcifiées, reliques des organismes constructeurs de récifs du Paléozoique et du Mésozoique. *Bull. Soc. Géol. Fr.* **108**(4): 547–557.

Vacelet J., Lévi C. (1958) Un cas de survivance, en Méditerranée, du groupe d'éponges fossiles des Pharétronides. *C. R. Acad. Sci. Paris* **246**: 318–320.

Vandermeulen J. H., Watabe N. (1973) Studies on reef corals: I. Skeleton formation by newly settled planula larva of *Pocillopora damicornis*. *Mar. Biol.* **23**: 47–57.

Vaughan T., Wells J. (1943) Revision of the suborders, families, and genera of the Scleractinia. *Geol. Soc. Am., Spec. Pap.* **44**: 1–363.

Veis A. (2005) A window on biomineralization. *Science* **307**: 1419–1420.

Veron J. E. N. (1986) *Corals of Australia and the Indo-Pacific*. North Ryde, NSW, Australia: Australian Institute of Marine Science, 644pp.

Vielzeuf D., Garrabou J., Baronnet A. *et al.* (2008) Nano to macroscale biomineral architecture of red coral (*Corallium rubrum*). *Am. Mineral.* **93**(11–12): 1799–1815.

Vinogradov A. P. (1953) *The Elementary Chemical Composition of Marine Organisms*. New Haven, CT: Sears Foundation for Marine Research, Memoir 2, 647pp.

Volkmer D. (2007) Biologically inspired crystallization of calcium carbonate beneath monolayers: a critical overview. In *Handbook of Biomineralization. Biomimetic and Bioinspired Chemistry*, P. Behrens & E. Baeuerlein (eds.). Weinheim, Germany: Wiley VCH, pp. 65–87.

Volz W. (1896) Die Korallen fauna der Trias: II. Die Korallen der Schichten von St. Casian, in Süd-Tirol. *Palaeontographica* **32**: 1–124.

Von Koenigswald W., Sander P. M. (1989) *Tooth Enamel Microstructure*. Rotterdam: Balkema, 280pp.

Voronkov M. G., Zelchan G. I., Lukevitz E. (1975) Biochemie, toxikologie und farmakologie der verbindungen des silicium. In *Silizium und Leben*, K. R. Kuhlmann (ed.). Berlin: Akademie Verlag, 375pp.

Voss-Foucart M. F. (1968) Essais de solubilisation et de fractionnement d'une conchioline (nacre murale de *Nautilus pompilius*, mollusque céphalopode). *Comp. Biochem. Physiol.* **26**: 877–886.

Voss-Foucart M. F., Grégoire C. (1971) Biochemical composition and submicroscopic structure of matrices of nacreous conchiolin in fossil cephalopods (nautiloids and ammonoids) *Bull. Inst. R. Sci. Nat. Belg.* **47**(41): 1–42.

Voss Foucart M. F., Jeuniaux C., Grégoire C. (1974) Résistance de la chitine de la nacre du nautile (mollusque céphalopode) à l'action de certains facteurs intervenant au cours de la fossilisation. *Comp. Biochem. Physiol.* B **48**: 447–451.

Vrieling E. G., Gieskes W. W. C., Beelen T. P. M. *et al.* (2000) Nanoscale uniformity of pore architectures in diatomaceous silica: a combined small angle and wide angle X-ray scattering study. *J. Phycology* **36**(1):146–159.

Waagen W., Wetzel J. (1886) Salt-range fossils: Part 6. Productus limestone fossils – Coelenterata. *Palaeont. Indica* **13**(1): 835–924.

Wada K. (1961) Crystal growth of molluscan shells. *Bull. Natl. Pearl Res. Laboratory* **36**(7):703–828.

Wada K. (1966a) Spiral growth of nacre. *Nature* **211**(505): 1427.

Wada K. (1966b) Studies on the mineralization of the calcified tissue in molluscs: XII. Specific patterns of non-mineralized layer conchiolin in amino acid composition. *Bull. Jap. Soc. Scientific Fish.* **32**(4): 304–311.

Wada K. (1972) Nucleation and growth of aragonite crystals in the nacre of some bivalve molluscs. *Biomineralization* **6**: 141–159.

Wada K., Fujinuki T. (1976) Biomineralization in bivalve molluscs with emphasis on the chemical composition of the extrapallial fluid. In *The Mechanisms of Mineralization in the Invertebrates and Plants*, N. Watabe & K. Wilbur (eds.). Columbia, SC: University of South Carolina Press, pp. 175–190.

Wainwright S. A. (1964) Studies of the mineral phase of coral skeleton. *Exp. Cell Res.* **34**: 213–230.

Walsh A. (1955) The application of atomic absorption spectra to chemical analysis. *Spectrochim. Acta* **7**: 108–117.

Walther J. (1888) Die Korallenriffe der Sinaihalbinsel. *Abhand. d. Math.-Phys. Cl. der König. Sächs. Ges. d. Wiss.* **14**: 438–505.

Walton D., Cusack M., Curry G. B. (1993) Implications of the amino acid composition of recent New Zealand Brachiopods. *Palaeontology* **36**(4): 883–896.

Wang H. C. (1950) A revision of the Zoantharia Rugosa in the light of their minute skeletal structures. *Phil. Trans. R. Soc.* **234**(B 611): 175–246.

Waskowiak R. (1962) Geochemische Untersuchungen an rezenter Molluskenschalen mariner Herkunft. *Freiberger Forschungshefte* **C136**: 1–155.

Watabe N. (1965) Studies on shell formation: XI. Crystal-matrix relationships in the inner layers of mollusk shells. *J. Ultrastr. Res.* **12**: 351–370.

Watabe N. (1990) Calcium phosphate structures in invertebrates and protozoans. In *Skeletal Biomineralization: Patterns, Processes and Evolutionary Trends*, J. G. Carter (ed.). New York: Van Nostrand Reinhold, pp. 35–44.

Watabe N., Wilbur K. M. (1960) Influence of the organic matrix on crystal type in molluscs. *Nature* **184**: 334.

Watanabe T., Juillet-Leclerc A., Cuif J. P. *et al.* (2007) Recent advances in coral biomineralization with implications for paleo-climatology: a brief overview. In *Elsevier Oceanography Series, Global Climate Change and Response of Carbon Cycle in the Equatorial Pacific and Indian Oceans and Adjacent Landmasses*, H. Kawahata & Y. Awaya (eds.) **73**: 239–254.

Weaver J. C., Morse D. E. (2003) Molecular biology of demosponge axial filaments and their role in biosilicification. *Microsc. Res. Techn.* **62**: 356–367.

Weaver J. C., Pietrasanta L. I., Hedin N. *et al.* (2003) Nanostructural features of demosponge biosilica. *J. Struct. Biol.* **144**: 271–281.

Weaver J. C., Aizenberg J., Fantner G. E. *et al.* (2007) Hierarchical assembly of the siliceous skeletal lattice of the hexactinellid sponge *Euplectella aspergillum. J. Struct. Biol.* **158**: 93–106.

Webb G. E., Sorauf J. E. (2002) Zigzag microstructure in rugose corals: a possible indicator of relative seawater Mg/Ca ratios. *Geology* **30**(5): 415–418.

Weber J. N., Woodhead P. M. (1972) Temperature dependence of oxygen-18 concentration in reef coral carbonates. *J. Geophys. Res.* **77**(3): 463–473.

Wefer G., Berger W. H. (1991) Isotope paleontology: growth and composition of extant calcareous species. *Mar. Geol.* **100**: 207–248.

Wehmiller J. F., Hare P. E. (1971) Racemization of amino acids in marine sediments. *Science* **173**: 907–911.

Weiner S. (1979) Aspartic acid-rich proteins: major components of the soluble organic matrix of mollusk shells. *Calc. Tissue Int.* **29**: 163–167.

Weiner S. (1983) Mollusk shell formation: isolation of two organic matrix proteins associated with calcite deposition in the bivalve *Mytilus californianus*. *Biochemistry* **22**: 4139–4145.

Weiner S. (1985) Organic matrix-like macromolecules associated with the mineral phase of sea urchin skeletal plates and teeth. *J. Exp. Zool.* **234**: 7–15.

Weiner S., Hood L. (1975) Soluble protein of the organic matrix of mollusk shells: a potential template for shell formation. *Science* **190**: 987–989.

Weiner S., Traub W. (1984) Macromolecules in mollusc shells and their functions in biomineralization. *Phil. Trans. R. Soc. Lond. B* **304**: 425–434.

Weiner S., Traub W. (1986) Organization of hydroxyapatite crystals within collagen fibrils. *FEBS Lett.* **206**(2): 262–266.

Weiner S., Traub W. (1991) Organization of crystals in bone. In *Mechanisms and Phylogeny of Mineralization in Biological Systems*, S. Suga & H. Nakahara (eds.). Tokyo: Springer Verlag, pp. 247–253.

Weiner S., Lowenstam H. A., Hood L. (1976) Characterization of 80 million year old mollusk shell proteins. *Proc. Nat. Acad. Sci. USA* **73** (8): 25 341–25 345.

Weiner S., Lowenstam H. A., Hood L. (1977) Discrete molecular weight components of the organic matrices of mollusc shells. *J. Exp. Mar. Biol. Ecol.* **30**: 45–51.

Weiner S., Gotliv B. A., Levi-Kalisman Y. *et al.* (2003) Mollusk shell nacre: an overview of the structure and functions of the organic matrix in shell formation. In *Biomineralization (BIOM2001): Formation, Diversity, Evolution and Application, Proc. 8th Int. Symp. on Biomineralization*, I. Kobayashi & H. Ozawa (eds.). Kanagawa: Tokai University Press, pp. 8–13.

Weiss I. M., Renner C., Strigl M. G. *et al.* (2002) A simple and reliable method for the determination and localization of chitin in abalone nacre. *Chem. Mater.* **14**: 3252–3259.

Weissenfels N., Landschoff H. W. (1977) Bau und Funktion des Süss-wasserschwamm *Ephydratia fluviatilis L.* (Porifera): IV. Die Entwicklung der monaxialen SiO_2-Nadeln in Sandwich-Kulturen. *Zool. Jahrb. Anat.* **98**: 355–371.

Wells J. W. (1956) Scleractinia. In *Treatise on Invertebrate Paleontology. Part F (Coelenterata)*, R. C. Moore (ed.). Lawrence, KS: The University of Kansas Press, pp. F328–F344.

Wendt J. (1977) Aragonite in Permian reefs. *Nature* **267**: 335–337.

Wendt J. (1990) The first aragonitic rugose coral. *J. Paleontology* **64**(3): 335–340.

Wenzl S., Hett R., Richthammer P. *et al.* (2008) Silacidins: highly acidic phosphopeptides from diatom shells assist in silica precipitation in vitro. *Angew. Chem. Int. Ed.* **47**: 1729–1732.

West C. D. (1937) Note on the crystallography of the echinoderm skeleton. *J. Paleont.* **11**: 458–459.

Wheeler A. P., Sikes C. S. (1984) Regulation of carbonate calcification by organic matrix. *Am. Zool.* **24**: 933–944.

Wheeler A. P., George J. W., Evans C. A. (1981) Control of calcium carbonate nucleation and crystal growth by soluble matrix of oyster shell. *Science* **212**: 1397–1398.

Wheeler A. P., Rusenko K. W., Sikes C. S. (1988) Organic matrix from carbonate biomineral as a regulator of mineralization. In *Chemical Aspects of Mineralization*, C. S. Sikes & A. P. Wheeler (eds.). Mobile, AL: University of South Alabama, pp. 9–13.

White H. H. (1842) On fossil Xanthidia. *Microsc. J.* **2**: 35–40.

Whitehouse F. W. (1941) The Cambrian faunas of north-eastern Australia: Part 4. Early Cambrian echinoderms similar to the larval stages of recent forms. *Mem. Queensl. Mus.* **12**: 1–28.

Whittington H. B., Evitt W. R. (1954) Silicified middle Ordovician trilobites. *Geol. Soc. America Mem.* **59**: 137pp., 33pls.

Wiens M., Mangoni A., D'Esposito M. *et al.* (2003) The molecular basis for the evolution of the metazoan bodyplan: extracellular matrix-mediated morphogenesis in marine demosponges. *J. Mol. Evol.* **57**: 1–16.

Williams A., Wright A. D. (1970) Valve structure of the Craniacea and other calcareous inarticulate brachiopods. *The Paleontological Association Press, Special papers in Paleontology* **7**: 1–51.

Williams A., Mackay S., Cusack M. (1992) Structure of the organo-phosphatic shell of the brachiopod *Discina*. *Phil. Trans. R. Soc. Lond. B* **337**: 83–104.

Williams A., Cusack M., Mackay S. (1994) Collagenous chitinophosphatic shells of the brachiopod *Lingula*. *Phil. Trans. Biol. Sci.* **346**(1316): 223–266.

Williams A., Carlson S. J., Brunton C. H. C. *et al.* (1996) A supra-ordinal classification of the brachiopods. *Phil. Trans. R. Soc. Lond. B* **351**: 1171–1193.

Williams A., Cusack M., Brown K. (1999) Growth of protein-doped rhombohedra in the calcitic shell of craniid brachiopods. *Proc. R. Soc. Lond. B* **266**: 1601–1607.

Williams A., Lüter C., Cusack M. (2001) The nature of siliceous mosaics forming the first shell of the brachiopod *Discinisca*. *J. Struct. Biol.* **134**: 25–34.

Williamson W. C. (1860) On some histological features in the shells of the crustacea. *Quart. J. Microsc. Sci.* **8**: 35–57.

Wilson J. L. (1975) *Carbonate Facies in Geologic History*. Berlin: Springer, 471pp.

Wilt F. H., Ettensohn C. A. (2007) The morphogenesis and biomineralization of the sea urchin larval skeleton. In *Handbook of Biomineralization: Biological Aspects and Structure Formation*, E. Bauerlein (ed.). Weinheim, Germany: Wiley VCH, pp. 183–210.

Wilt F. H., Killian C. E., Livinston B. T. (2003) Development of calcareous skeletal elements in invertebrates. *Differentiation* **71**: 237–250.

Wise S. W. Jr (1970) Microarchitecture and mode of formation of nacre (mother-of-pearl) in Pelecypods, Gastropods and Cephalopods. *Eclogae Geol. Helv.* **63**(3): 775–797.

Woelkerling W. J. (1990) An introduction. In *Biology of the Red Algae*, K. M. Cole & R. G. Sheath (eds.). Cambridge: Cambridge University Press, pp. 1–6.

Wopenka B., Pasteris J. D. (2005) A mineralogical perspective on the apatite in bone. *Mater. Sci. Engin.C* **25**: 131–143.

Wörheide G. (1997) The reef cave dwelling ultraconservative coralline demosponge *Astrosclera willeyana* LISTER 1900 from the Indo-Pacific: Micromorphology, ultrastructure, biocalcification, isotope record, taxonomy, biogeography, phylogeny. Doctoral thesis, Fakultät für Geowissenschaften, Georg-August Universität Göttingen, p. 91.

Worms D., Weiner S. (1986) Mollusk shell organic matrix: Fourier transform infrared study of the acidic macromolecules. *J. Exp. Zool.* **237**: 11–20.

Wright L., Schwarcz H. (1996) Infrared and isotopic evidence for diagenesis of bone apatite at Dos Pilas, Guatemala: palaeodietary implications. *J. Archaeol. Sci.* **23**(6): 933–944.

Wright P. J., Woodroffe D. A., Gibb F. M. *et al.* (2002) Verification of first annulus formation in the illicia and otoliths of white anglerfish, *Lophius piscatorius* using otolith microstructure. *J. Mar. Sci.* **59**(3): 587–593.

Wyckoff R. W. G. (1972) *The Biochemistry of Animal Fossils.* Bristol, UK: Wright Scientechnica Publications, 145pp.

Xu M., Gratson G. M., Duoss E. B. *et al.* (2006) Biomimetic silicification of 3D polyamine-rich scaffolds assembled by direct ink writing. *Soft Mater.* **2**: 205–209.

Yatsu N. (1902) On the development of *Lingula anatina. J. Coll. Sci. Tokyo* **17**: 1–112.

Yonge C. M. (1931) Study on the physiology of corals: Great Barrier Reef Expedition 1928–29. *Scient. Rep.* **11**: Part 1, pp. 13–65; Part 2, pp. 83–91.

Young J. R., Henriksen K. (2003) Biomineralization within vesicles: the calcite of coccoliths. In *Mineralogy & Geochemistry: Biomineralization*, P. M. Dove, J. J. De Yoreo & S. Weiner (eds.). Washington DC: Mineralogical Society of America, Vol. 54, pp. 189–215.

Young S. D. (1971) Organic material from scleractinian coral skeletons: 1. Variation in composition between several species. *Comp. Biochem. Physiol.* B **40**: 113–120.

Young S. D., O'Connor J. D., Muscatine L. (1971) Organic material from scleractinian coral skeletons: II. Incorporation of ^{14}C into protein, chitin and lipid. *Comp. Biochem. Physiol.* B **40**: 945–958.

Zelenitsky D. K., Modesto S. P. (2003) New information on the eggshell of ratites (Aves) and its phylogenetic implications. *Can. J. Zool.* **81**(6): 962–970.

Zhuravlev A. Y. (1989) Porifean aspects of archaeocyathan skeletal function. *Mem. Ass. Austral. Palaeontol.* **8**: 387–399.

Ziegler A., Miller B. (1997) Ultrastructure of $CaCO_3$ deposits of terrestrial isopods (Crustacea, Oniscidea). *Zoomorphology* **117**:181–187.

Zittel K. A. (1878) Studien über fossile Spongien: III. Monactinellidae, Tetractinellidae und Calcispongiae. *Abhandl. d. Math-Phys. Cl. d. Kön.-Bayer. Akad. der Wissensch.* **XIII**(2): 1–48.

Name index

Subject index

Subject index

Printed in the United States
By Bookmasters